STOCHASTIC
IN PHYSICS AND CHEMISTRY

STOCHASTIC PROCESSES IN PHYSICS AND CHEMISTRY

Third edition

N.G. VAN KAMPEN

*Institute for Theoretical Physics
of the University at Utrecht*

ELSEVIER

Amsterdam • Boston • Heidelberg • London • New York • Oxford • Paris
San Diego • San Francisco • Singapore • Sydney • Tokyo

Elsevier
Radarweg 29, PO Box 211, 1000 AE Amsterdam, The Netherlands
Linacre House, Jordan Hill, Oxford, OX2 8DP, UK

First edition 1981
Second edition 1992
Revised and enlarged edition 1997
Third edition 2007
Reprinted 2008 (twice)

Copyright © 2007 Elsevier BV. All rights reserved

No part of this publication may be reproduced, stored in a retrieval system
or transmitted in any form or by any means electronic, mechanical, photocopying,
recording or otherwise without the prior written permission of the publisher

Permissions may be sought directly from Elsevier's Science & Technology Rights
Department in Oxford, UK: phone (+44) (0) 1865 843830; fax (+44) (0) 1865 853333;
email: permissions@elsevier.com. Alternatively you can submit your request online by
visiting the Elsevier web site at http://elsevier.com/locate/permissions, and selecting
Obtaining permission to use Elsevier material

Notice
No responsibility is assumed by the publisher for any injury and/or damage to persons
or property as a matter of products liability, negligence or otherwise, or from any use
or operation of any methods, products, instructions or ideas contained in the material
herein. Because of rapid advances in the medical sciences, in particular, independent
verification of diagnoses and drug dosages should be made

British Library Cataloguing in Publication Data
A catalogue record for this book is available from the British Library

Library of Congress Cataloging-in-Publication Data
A catalog record for this book is available from the Library of Congress

ISBN: 978-0-444-52965-7

For information on all Elsevier publications
visit our website at www.elsevierdirect.com

Printed and bound in *Hungary*

08 09 10 10 9 8 7 6 5 4 3

Working together to grow
libraries in developing countries

www.elsevier.com | www.bookaid.org | www.sabre.org

ELSEVIER   BOOK AID International   Sabre Foundation

To the memory of
F. ZERNIKE
*whose influence on this work
runs deeper than I can know*

PREFACE TO THE FIRST EDITION

Que nous sert-il d'avoir la panse pleine de viande, si elle ne se digère? si elle ne se transforme en nous? si elle ne nous augmente et fortifie?

Montaigne

The interest in fluctuations and in the stochastic methods for describing them has grown enormously in the last few decades. The number of articles scattered in the literature of various disciplines must run to thousands, and special journals are devoted to the subject. Yet the physicist or chemist who wants to become acquainted with the field cannot easily find a suitable introduction. He reads the seminal articles of Wang and Uhlenbeck and of Chandrasekhar, which are almost forty years old, and he culls some useful information from the books of Feller, Bharucha-Reid, Stratonovich, and a few others. Apart from that he is confronted with a forbidding mass of mathematical literature, much of which is of little relevance to his needs. This book is an attempt to fill this gap in the literature.

The first part covers the main points of the classical material. Its aim is to provide physicists and chemists with a coherent and sufficiently complete framework, in a language that is familiar to them. A thorough intuitive understanding of the material is held to be a more important tool for research than mathematical rigor and generality. A physical system at best only approximately fulfills the mathematical conditions on which rigorous proofs are built, and a physicist should be constantly aware of the approximate nature of his calculations. (For instance, Kolmogorov's derivation of the Fokker–Planck equation does not tell him for which actual systems this equation may be used.) Nor is he interested in the most general formulations, but a thorough insight in special cases will enable him to extend the theory to other cases when the need arises. Accordingly the theory is here developed in close connection with numerous applications and examples.

The second part, starting with chapter IX [now chapter X], is concerned with fluctuations in nonlinear systems. This subject involves a number of conceptual difficulties, first pointed out by D.K.C. MacDonald. They are of a physical rather than a mathematical nature. Much confusion is caused by the still prevailing view that nonlinear fluctuations can be approached from the same physical starting point as linear ones and merely require more elaborate mathematics. In actual fact, what is needed is a firmer physical basis and a more detailed knowledge of the physical system than required for the study of linear noise. This is the subject of the second part, which has more the character of a monograph and inevitably contains much of my own work.

The bulk of the book is written on the level of a graduate course. The Exercises range from almost trivial to rather difficult. Many of them contain applications and others provide additions to the text, some of which are used later on. My hope is that they will not frustrate the reader but stimulate an active participation in the material.

The references to the literature constituted a separate problem. Anything even approaching completeness was out of the question. My selection is based on the desire to be helpful to the reader. To stress this aspect references are given at the bottom of the page, where the reader can find them without having to search for them. My aim will be achieved if they are sufficient as a guide to further relevant literature. Unavoidably a number of important contributions are not explicitly but only indirectly credited. I apologize to their authors and beg them to consider that this is a textbook rather than a historical account.

....

I am indebted to B.R.A. Nijboer, H. Falk, and J. Groeneveld for critical remarks, to the students who reported a number of misprints, and to Leonie J.M. Silkens for indefatigable typing and retyping.

N.G. van Kampen

PREFACE TO THE SECOND EDITION

This edition differs from the first one in the following respects. A number of additions are concerned with new developments that occurred in the intervening years. Some parts have been rewritten for the sake of clarity and a few derivations have been simplified. More important are three major changes.

First, the Langevin equation receives in a separate chapter the attention merited by its popularity. In this chapter also non-Gaussian and colored noise are studied. Secondly, a chapter has been added to provide a more complete treatment of first-passage times and related topics. Finally, a new chapter was written about stochasticity in quantum systems, in which the origin of damping and fluctuations in quantum mechanics is discussed. Inevitably all this led to an increase in the volume of the book, but I hope that this is justified by the contents.

The dearth of relevant literature mentioned in the previous preface has since been alleviated by the appearance of several textbooks. They are quoted in the text at appropriate places. Some of the references appear in abbreviated form; the key to the abbreviations is given over page.

N.G. van Kampen

ABBREVIATED REFERENCES

FELLER I: W. Feller, *An Introduction to Probability Theory and its Applications*, Vol. I (2nd edition, Wiley, New York 1957).
FELLER II: idem Vol. II (Wiley, New York 1966).
BHARUCHA-REID: A.T. Bharucha-Reid, *Elements of the Theory of Markov Processes and their Applications* (McGraw-Hill, New York 1960).
COX AND MILLER: D.R. Cox and H.D. Miller, *The Theory of Stochastic Processes* (Chapman and Hall, London 1972).
WAX: *Selected Papers on Noise and Stochastic Processes* (N. Wax ed., Dover Publications, New York 1954).
STRATONOVICH: R.L. Stratonovich, *Topics in the Theory of Random Noise*, Vol. I (R.A. Silverman transl., Gordon and Breach, New York 1963).
DE GROOT AND MAZUR: S.R. de Groot and P. Mazur, *Non-equilibrium Thermodynamics* (North-Holland, Amsterdam 1962).
GARDINER: C.W. Gardiner, *Handbook of Stochastic Methods* (Springer, Berlin 1983).
RISKEN: H. Risken, *The Fokker–Planck Equation* (Springer, Berlin 1984).

PREFACE TO THE THIRD EDITION

The main difference with the second edition is that the contrived application of the quantum master equation in section 6 of chapter XVII has been replaced with a satisfactory treatment of quantum fluctuations. Apart from that, throughout the text corrections have been made and a number of references to later developments have been included. Of the more recent textbooks, the following are the most relevant.

GARDINER: C.W. Gardiner, *Quantum Optics* (Springer, Berlin 1991).
GILLESPIE: D.T. Gillespie, *Markov Processes* (Academic Press, San Diego 1992).
COFFEY, KALMYKOV AND WALDRON: W.T. Coffey, Yu.P. Kalmykov, and J.T. Waldron, *The Langevin Equation* (2nd edition, World Scientific, 2004).

TABLE OF CONTENTS

PREFACE TO THE FIRST EDITION vii
PREFACE TO THE SECOND EDITION ix
ABBREVIATED REFERENCES x
PREFACE TO THE THIRD EDITION xi

I. STOCHASTIC VARIABLES

1. Definition 1
2. Averages 5
3. Multivariate distributions 10
4. Addition of stochastic variables 14
5. Transformation of variables 17
6. The Gaussian distribution 23
7. The central limit theorem 26

II. RANDOM EVENTS

1. Definition 30
2. The Poisson distribution 33
3. Alternative description of random events 35
4. The inverse formula 40
5. The correlation functions 41
6. Waiting times 44
7. Factorial correlation functions 47

III. STOCHASTIC PROCESSES

1. Definition 52
2. Stochastic processes in physics 55
3. Fourier transformation of stationary processes 58
4. The hierarchy of distribution functions 61
5. The vibrating string and random fields 64
6. Branching processes 69

IV. MARKOV PROCESSES

1. The Markov property 73
2. The Chapman–Kolmogorov equation 78
3. Stationary Markov processes 81
4. The extraction of a subensemble 86
5. Markov chains 89
6. The decay process 93

V. THE MASTER EQUATION

1. Derivation 96
2. The class of W-matrices 100
3. The long-time limit 104
4. Closed, isolated, physical systems 108
5. The increase of entropy 111
6. Proof of detailed balance 114
7. Expansion in eigenfunctions 117
8. The macroscopic equation 122
9. The adjoint equation 127
10. Other equations related to the master equation 129

VI. ONE-STEP PROCESSES

1. Definition; the Poisson process 134
2. Random walk with continuous time 136
3. General properties of one-step processes 139
4. Examples of linear one-step processes 143
5. Natural boundaries 147
6. Solution of linear one-step processes with natural boundaries 149
7. Artificial boundaries 153
8. Artificial boundaries and normal modes 157
9. Nonlinear one-step processes 161

VII. CHEMICAL REACTIONS

1. Kinematics of chemical reactions 166
2. Dynamics of chemical reactions 171
3. The stationary solution 173
4. Open systems 176
5. Unimolecular reactions 178
6. Collective systems 182
7. Composite Markov processes 186

VIII. THE FOKKER–PLANCK EQUATION

1. Introduction 193
2. Derivation of the Fokker–Planck equation 197
3. Brownian motion 200
4. The Rayleigh particle 204
5. Application to one-step processes 207
6. The multivariate Fokker–Planck equation 210
7. Kramers' equation 215

IX. THE LANGEVIN APPROACH

1. Langevin treatment of Brownian motion 219
2. Applications 221

3. Relation to Fokker–Planck equation 224
4. The Langevin approach 227
5. Discussion of the Itô–Stratonovich dilemma 232
6. Non-Gaussian white noise 237
7. Colored noise 240

X. THE EXPANSION OF THE MASTER EQUATION

1. Introduction to the expansion 244
2. General formulation of the expansion method 248
3. The emergence of the macroscopic law 254
4. The linear noise approximation 258
5. Expansion of a multivariate master equation 263
6. Higher orders 267

XI. THE DIFFUSION TYPE

1. Master equations of diffusion type 273
2. Diffusion in an external field 276
3. Diffusion in an inhomogeneous medium 279
4. Multivariate diffusion equation 282
5. The limit of zero fluctuations 287

XII. FIRST-PASSAGE PROBLEMS

1. The absorbing boundary approach 292
2. The approach through the adjoint equation – Discrete case 298
3. The approach through the adjoint equation – Continuous case 303
4. The renewal approach 307
5. Boundaries of the Smoluchowski equation 312
6. First passage of non-Markov processes 319
7. Markov processes with large jumps 322

XIII. UNSTABLE SYSTEMS

1. The bistable system 326
2. The escape time 333
3. Splitting probability 337
4. Diffusion in more dimensions 341
5. Critical fluctuations 344
6. Kramers' escape problem 347
7. Limit cycles and fluctuations 355

XIV. FLUCTUATIONS IN CONTINUOUS SYSTEMS

1. Introduction 363
2. Diffusion noise 365
3. The method of compounding moments 367

4. Fluctuations in phase space density	371
5. Fluctuations and the Boltzmann equation	374

XV. THE STATISTICS OF JUMP EVENTS

1. Basic formulae and a simple example	383
2. Jump events in nonlinear systems	386
3. Effect of incident photon statistics	388
4. Effect of incident photon statistics – continued	392

XVI. STOCHASTIC DIFFERENTIAL EQUATIONS

1. Definitions	396
2. Heuristic treatment of multiplicative equations	399
3. The cumulant expansion introduced	405
4. The general cumulant expansion	407
5. Nonlinear stochastic differential equations	410
6. Long correlation times	416

XVII. STOCHASTIC BEHAVIOR OF QUANTUM SYSTEMS

1. Quantum probability	422
2. The damped harmonic oscillator	428
3. The elimination of the bath	436
4. The elimination of the bath – continued	440
5. The Schrödinger–Langevin equation and the quantum master equation	444
6. A new approach to noise	449
7. Internal noise	451
SUBJECT INDEX	457

Chapter I

STOCHASTIC VARIABLES

This chapter is intended as a survey of probability theory, or rather a catalogue of facts and concepts that will be needed later. Many readers will find it time-saving to skip this chapter and only consult it occasionally when a reference to it is made in the subsequent text.

1. Definition

A "random number" or "stochastic variable" is an object X defined by
  a. a set of possible values (called "range", "set of states", "sample space" or "phase space");
  b. a probability distribution over this set.
  Ad **a.** The set may be *discrete*, e.g.: heads or tails; the number of electrons in the conduction band of a semiconductor; the number of molecules of a certain component in a reacting mixture. Or the set may be *continuous* in a given interval: one velocity component of a Brownian particle (interval $-\infty, +\infty$); the kinetic energy of that particle $(0, \infty)$; the potential difference between the end points of an electrical resistance $(-\infty, +\infty)$. Finally the set may be partly discrete, partly continuous, e.g., the energy of an electron in the presence of binding centers. Moreover the set of states may be *multidimensional*; in this case X is often conveniently written as a vector **X**. Examples: **X** may stand for the three velocity components of a Brownian particle; or for the collection of all numbers of molecules of the various components in a reacting mixture; or the numbers of electrons trapped in the various species of impurities in a semiconductor.
  For simplicity we shall often use the notation for discrete states or for a continuous one-dimensional range and leave it to the reader to adapt the notation to other cases.
  Ad **b.** The probability distribution, in the case of a continuous one-dimensional range, is given by a function $P(x)$ that is nonnegative,

$$P(x) \geqslant 0, \qquad (1.1)$$

and normalized in the sense

$$\int P(x)\,dx = 1, \qquad (1.2)$$

where the integral extends over the whole range. The probability that X has a value between x and $x + dx$ is

$$P(x)\,dx.$$

Remark. Physicists like to visualize a probability distribution by an "ensemble". Rather than thinking of a single quantity with a probability distribution they introduce a fictitious set of an arbitrarily large number $\mathcal{N}$ of quantities, all having different values in the given range, in such a way that the number of them having a value between x and $x + dx$ is $\mathcal{N}P(x)\,dx$. Thus the probability distribution is replaced with a density distribution of a large number of "samples". This does not affect any of its results, but is merely a convenience in talking about probabilities, and occasionally we shall also use this language. It may be added that it can happen that a physical system does consist of a large number of identical replicas, which to a certain extent constitute a physical realization of an ensemble. For instance, the molecules of an ideal gas may serve as an ensemble representing the Maxwell probability distribution of the velocity. Another example is a beam of electrons scattering on a target and representing the probability distribution for the angle of deflection. But the use of an ensemble is not limited to such cases, nor based on them, but merely serves as a more concrete visualization of a probability distribution. To introduce or even envisage a physical interaction between the samples of an ensemble is a dire misconception[*].

In a continuous range it is possible for $P(x)$ to involve delta functions,

$$P(x) = \sum_n p_n\,\delta(x - x_n) + \tilde{P}(x), \tag{1.3}$$

where $\tilde{P}$ is finite or at least integrable and nonnegative, $p_n > 0$, and

$$\sum_n p_n + \int \tilde{P}(x)\,dx = 1.$$

Physically this may be visualized as a set of discrete states x_n with probability p_n embedded in a continuous range. If $P(x)$ consists of delta functions alone, i.e., if $\tilde{P}(x) = 0$, it can also be considered as a probability distribution p_n on the discrete set of states x_n. A mathematical theorem asserts that any distribution on $-\infty < x < \infty$ can be written in the form (1.3), apart from a third term, which, however, is of rather pathological form and does not appear to occur in physical problems.[**]

Exercise. Let X be the number of points obtained by casting a die. Give its range and probability distribution. Same question for casting two dice.

---

[*] E. Schrödinger, *Statistical Thermodynamics* (Cambridge University Press, Cambridge 1946).
[**] FELLER II, p. 139. He calls the first term in (1.3) an "atomic distribution".

1. DEFINITION

Exercise. Flip a coin N times. Prove that the probability that heads turn up exactly n times is

$$p_n = 2^{-N} \binom{N}{n} \quad (n = 0, 1, 2, \ldots, N) \tag{1.4}$$

("binomial distribution"). If heads gains one penny and tails loses one, find the probability distribution of the total gain.

Exercise. Let X stand for the three components of the velocity of a molecule in a gas. Give its range and probability distribution.

Exercise. An electron moves freely through a crystal of volume Ω or may be trapped in one of a number of point centers. What is the probability distribution of its coordinate r?

Exercise. Two volumes, V_1 and V_2, communicate through a hole and contain N molecules without interaction. Show that the probability of finding n molecules in V_1 is

$$p_n = (1+\gamma)^{-N} \binom{N}{n} \gamma^n, \tag{1.5}$$

where $\gamma = V_1/V_2$ (general binomial distribution or "Bernoulli distribution").

Exercise. An urn contains a mixture of N_1 white balls and N_2 black ones. I extract at random M balls, without putting them back. Show that the probability for having n white balls among them is

$$p_n = \binom{N_1}{n}\binom{N_2}{M-n} \bigg/ \binom{N_1+N_2}{M} \quad \text{("hypergeometric distribution"). \tag{1.6}}$$

It reduces to (1.5) in the limit $N_1 \to \infty$, $N_2 \to \infty$ with $N_1/N_2 = \gamma$.

Note. Many more exercises can be found in texts on elementary probability theory, e.g., J.R. Gray, *Probability* (Oliver and Boyd, Edinburgh 1967); T. Cacoulos, *Exercises in Probability* (Springer, New York 1989).

Excursus. As an alternative description of a probability distribution (in one dimension) one often uses instead of $P(x)$ a function $\mathbb{P}(x)$, defined as the total probability that X has any value $\leq x$. Thus

$$\mathbb{P}(x) = \int_{-\infty}^{x+0} P(x')\,dx',$$

where the upper limit of integration indicates that if P has a delta peak at x it is to be included in the integral.[*] Mathematicians call $\mathbb{P}$ the *probability distribution function* and prefer it to the *probability density* P, because it has no delta peaks, because its behavior under transformation of x is simpler, and because they are accustomed to it. Physicists call $\mathbb{P}$ the *cumulative distribution function*, and prefer P,

---

[*] This is, of course, an arbitrary convention; one might also define $\mathbb{P}(x)$ as the probability that X takes a value $<x$.

because its value at x is determined by the probability at x itself, because in many applications it turns out to be a simpler function, because it more closely parallels the familiar way of describing probabilities on discrete sets of states, and because they are accustomed to it. In particular in multidimensional distributions, such as the Maxwell velocity distribution, $\mathbb{P}$ is rather awkward. We shall therefore use throughout the probability density $P(x)$ and not be afraid to refer to it as "the probability distribution", or simply "the probability".

A more general and abstract treatment is provided by axiomatic probability theory.[*] The x-axis is replaced by a set S, the intervals dx by subsets $A \subset S$, belonging to a suitably defined family of subsets. The probability distribution assigns a non-negative number $\mathscr{P}(A)$ to each A of the family in such a way that $\mathscr{P}(S) = 1$, and that when A and B are disjoint

$$\mathscr{P}(A + B) = \mathscr{P}(A) + \mathscr{P}(B).$$

This is called a probability measure. Any other set of numbers $f(A)$ assigned to the subsets is a stochastic variable. In agreement with our program we shall not use this approach, but a more concrete language.

Exercise. Show that $\mathbb{P}(x)$ must be a monotone non-decreasing function with $\mathbb{P}(-\infty) = 0$ and $\mathbb{P}(+\infty) = 1$. What is its relation to $\mathscr{P}$?

Exercise. An opinion poll is conducted in a country with many political parties. How large a sample is needed to be reasonably sure that a party of 5 percent will show up in it with a percentage between 4.5 and 5.5?

Exercise. Thomas Young remarked that if two different languages have the same word for one concept one cannot yet conclude that they are related since it may be a coincidence.[**] In this connection he solved the following "Rencontre Problem" or "Matching Problem": What is the probability that an arbitrary permutation of n objects leaves no object in its place? Naturally it is assumed that each permutation has a probability $n!^{-1}$ to occur. Show that the desired probability p as a function of n obeys the recurrence relation

$$np(n) - (n-1)p(n-1) = p(n-2).$$

Find $p(n)$ and show that $p(n) \to e^{-1}$ as $n \to \infty$.

---

[*] A. Kolmogoroff, *Grundbegriffe der Wahrscheinlichkeitsrechnung*. Ergebn. Mathem. Grenzgebiete **2**, No. 3 (Springer, Berlin 1933) = A.N. Kolmogorov, *Foundations of the Theory of Probability* (Chelsea Publishing, New York 1950). Or any other modern mathematical textbook such as FELLER II, p. 110; or M. Loève, *Probability Theory* I and II (Springer, New York 1977/1978).

[**] Philos. Trans. Roy. Soc. (London 1819) p. 70; M.G. Kendall, Biometrica **55**, 249 (1968). But the problem goes back to N. Bernoulli (1714) and P.-R. de Montmort (1708), see F.N. David, *Games, Gods, and Gambling* (Griffin, London 1962).

2. Averages

The set of states and the probability distribution together fully define the stochastic variable, but a number of additional concepts are often used. The *average* or *expectation value* of any function $f(X)$ defined on the same state space is

$$\langle f(X) \rangle = \int f(x) P(x) \, dx.$$

In particular $\langle X^m \rangle \equiv \mu_m$ is called the *m-th moment* of X, and μ_1 the average or *mean*. Also

$$\sigma^2 = \langle (X - \langle X \rangle)^2 \rangle = \mu_2 - \mu_1^2 \tag{2.1}$$

is called the *variance* or *dispersion*, which is the square of the *standard deviation* σ.

Not all probability distributions have a finite variance: a counterexample is the *Lorentz* or *Cauchy distribution*

$$P(x) = \frac{1}{\pi} \frac{\gamma}{(x-a)^2 + \gamma^2} \quad (-\infty < x < \infty). \tag{2.2}$$

Actually in this case not even the integral defining μ_1 converges, but it is clear from symmetry that one will not be led to wrong results by setting $\mu_1 = a$.

Exercise. Find the moments of the "square distribution" defined by

$$P(x) = 0 \quad \text{for } |x| > a, \qquad P(x) = (2a)^{-1} \quad \text{for } |x| < a. \tag{2.3}$$

Exercise. The *Gauss distribution* is defined by (compare section 6)

$$P(x) = (2\pi)^{-\frac{1}{2}} e^{-\frac{1}{2}x^2} \quad (-\infty < x < \infty). \tag{2.4}$$

Show that $\mu_{2n+1} = 0$ and

$$\mu_{2n} = (2n-1)!! = (2n-1)(2n-3)(2n-5) \cdots 1.$$

Exercise. Construct a distribution whose moments μ_n exist up to but not beyond a prescribed value of n.

Exercise. From (2.1) one reads off the inequality $\mu_2 \geq \mu_1^2$. Similarly, from the obvious fact $\langle |\lambda_0 + \lambda_1 X + \lambda_2 X^2|^2 \rangle \geq 0$ for all $\lambda_0, \lambda_1, \lambda_2$ prove the inequality

$$\begin{vmatrix} 1 & \mu_1 & \mu_2 \\ \mu_1 & \mu_2 & \mu_3 \\ \mu_2 & \mu_3 & \mu_4 \end{vmatrix} \geq 0.$$

Find the analogous inequalities for higher moments.[*]

---

[*] J.A. Shohat and J.D. Tamarkin, *The Problem of Moments* (American Mathematical Society, New York 1943).

Exercise. Convince yourself that the requirement (1.1) can be replaced with the condition:
  $\int f(x) P(x)\,dx \geq 0$ for any nonnegative continuous function f that vanishes outside a finite interval (i.e., has finite support).
This condition also covers the case (1.3), and excludes the occurrence of derivatives of delta functions in P.

Exercise. Show that for each $n = 1, 2, 3, \ldots$ the function

$$P(x) = n \frac{e^{-x}}{x} I_n(x)$$

is a probability density on $0 < x < \infty$ having no average. (I_n denotes the modified Bessel function.)

The *characteristic function* of a stochastic variable X whose range I is the set of real numbers or a subset thereof is defined by

$$G(k) = \langle e^{ikX} \rangle = \int_I e^{ikx} P(x)\,dx. \tag{2.5}$$

It exists for all real k and has the properties

$$G(0) = 1, \quad |G(k)| \leq 1. \tag{2.6}$$

It is also the *moment generating function* in the sense that the coefficients of its Taylor expansion in k are the moments:

$$G(k) = \sum_{m=0}^{\infty} \frac{(ik)^m}{m!} \mu_m. \tag{2.7}$$

This implies that the derivatives of $G(k)$ at $k = 0$ exist up to the same m as the moments. The same function also serves to generate the *cumulants* κ_m, which are defined by

$$\log G(k) = \sum_{m=1}^{\infty} \frac{(ik)^m}{m!} \kappa_m. \tag{2.8}$$

They are combinations of the moments, e.g.,[*]

$$\kappa_1 = \mu_1,$$
$$\kappa_2 = \mu_2 - \mu_1^2 = \sigma^2,$$
$$\kappa_3 = \mu_3 - 3\mu_2 \mu_1 + 2\mu_1^3,$$
$$\kappa_4 = \mu_4 - 4\mu_3 \mu_1 - 3\mu_2^2 + 12\mu_2 \mu_1^2 - 6\mu_1^4. \tag{2.9}$$

---

[*] The general formula is given by Yu.V. Prohorov and Yu.A. Rozanov, *Probability Theory* (Springer, Berlin 1969) p. 165. Also RISKEN p. 18.

2. AVERAGES

Exercise. Compute the characteristic function of the square distribution (2.3) and find its moments in this way.

Exercise. Show that for the Gauss distribution (2.4) all cumulants beyond the second are zero. Find the most general distribution with this property.

Exercise. The *Poisson distribution* is defined on the discrete range $n = 0, 1, 2, \ldots$ by

$$p_n = \frac{a^n}{n!} e^{-a}. \tag{2.10}$$

Find its cumulants.*⁾

Exercise. Take in (1.5) the limit $V_2 \to \infty$, $N \to \infty$, $N/V_2 = \rho = $ constant. The result is (2.10) with $a = \rho V_1$. Thus the number of molecules in a small volume communicating with an infinite reservoir is distributed according to Poisson.

Exercise. Calculate the characteristic function of the Lorentz distribution (2.2). How does one see from it that the moments do not exist?

Exercise. Find the distribution and its moments corresponding to the characteristic function $G(k) = \cos ak$.

Exercise. Prove that the characteristic function of any probability distribution is uniformly continuous on the real k-axis.

Exercise. There is no reason why the characteristic function should be positive for all k. Why does that not restrict the validity of the definition (2.8) of the cumulants?

Equation (2.5) states that $G(k)$ is the Fourier transform of a function $\bar{P}(x)$ that coincides with $P(x)$ inside I and vanishes outside it. Hence

$$\bar{P}(x) = \frac{1}{2\pi} \int_{-\infty}^{\infty} G(k) e^{-ikx} dk.$$

In normal usage this somewhat pedantic distinction between P and $\bar{P}$ would be cumbersome, but it is needed to clarify the following remark.

Suppose x only takes integral values $n = \ldots, -2, -1, 0, 1, 2, \ldots$ with probabilities p_n. In order to construct the characteristic function one first has to write this as a distribution $\bar{P}(x)$ over all real values,

$$\bar{P}(x) = \sum_n p_n \delta(x - n). \tag{2.11}$$

Then the general definition (2.5) states

$$G(k) = \sum_n p_n e^{ikn}.$$

This is a periodic function whose Fourier transform reproduces, of course, (2.11), when k is treated as a variable with range $(-\infty, +\infty)$. In addition, however, one notes that the p_n themselves are obtained by taking the Fourier

---

*⁾ See http://en.wikipedia.org/wiki/Cumulant.

transform over a single period,

$$p_n = \frac{1}{2\pi} \int_0^{2\pi} G(k) \, e^{-ikn} \, dk. \tag{2.12}$$

Any distribution whose range consists of the points

$$na \quad (a > 0; n = \ldots, -2, -1, 0, 1, 2, \ldots) \tag{2.13}$$

is called a *lattice distribution*. For such distributions $|G(k)|$ is periodic with period $2\pi/a$ and therefore assumes its maximum value unity not at $k = 0$ alone. This fact characterizes lattice distributions: for all other distributions the inequality (2.6) can be sharpened to[*]

$$|G(k)| < 1 \quad \text{for } k \neq 0. \tag{2.14}$$

More generally the following question may be asked. When the values of x are confined to a certain subset I of the real axis, how does this show up in the properties of G? If I is the interval $-a < x < a$ it is known that $G(k)$ is analytic in the whole complex k-plane and "of exponential type".[**] If I is the semi-axis $x \geq 0$ the function $G(k)$ is analytic and bounded in the upper half-plane. But no complete answer to the general question is available, although it is important for several problems.

Remark. In practical calculations the factor i in (2.7) and (2.8) is awkward. It may be avoided by setting $ik = s$ and using the characteristic function $\langle e^{sX} \rangle$, provided one bears in mind that its existence is guaranteed only for purely imaginary s. When X only takes positive values it has some advantage to use $\langle e^{-sX} \rangle$, which exists in the right half of the complex s-plane. When X only takes integral values it is convenient to use the *probability generating function* $F(z) = \langle z^X \rangle$, which is uniquely defined for all z on the unit circle $|z| = 1$, and will be employed in chapter VI. When X only takes nonnegative integer values, $F(z)$ is also defined and analytic inside that circle.

Exercise. Actually the most general lattice distribution is not defined by the range (2.13), but by the range $na + b$. With this definition prove that (2.14) holds if and only if $P(x)$ is not a lattice distribution.

Exercise. Take any r real numbers $k_1, k_2, \ldots, k_r$ and consider the $r \times r$ matrix whose i, j element is $G(k_i - k_j)$. Prove that this matrix is positive definite; or semi-definite for some special distributions. Functions G having this property for all sets $\{k\}$ are called "positive definite" or "of positive type".

---

[*] For this and other properties of characteristic functions, see E. Lukacs, *Characteristic Functions* (Griffin, London 1960); P.A.P. Moran, *An Introduction to Probability Theory* (Clarendon, Oxford 1968).

[**] R.E.A.C. Paley and N. Wiener, *Fourier Transforms in the Complex Domain* (American Mathematical Society, New York 1934).

2. AVERAGES

Exercise. When X only takes the values $0, 1, 2, \ldots$ one defines the *factorial moments* ϕ_m by $\phi_0 = 1$ and

$$\phi_m = \langle X(X-1)(X-2)\cdots(X-m+1)\rangle \quad (m \geqslant 1). \tag{2.15}$$

Show that they are also generated by F, viz.,

$$F(1-x) = \sum_{m=0}^{\infty} \frac{(-x)^m}{m!} \phi_m. \tag{2.16}$$

Exercise. The *factorial cumulants* θ_m are defined by

$$\log F(1-x) = \sum_{m=1}^{\infty} \frac{(-x)^m}{m!} \theta_m. \tag{2.17}$$

Express the first few in terms of the moments. Show that the Poisson distribution (2.10) is characterized by the vanishing of all factorial cumulants beyond θ_1.

Exercise. Find the factorial moments and cumulants of (1.5).

Exercise. A harmonic oscillator with levels $nh\nu$ ($n = 0, 1, 2, \ldots$) has in thermal equilibrium the probability

$$p_n = (1-\gamma)\gamma^n \tag{2.18}$$

to be in level n, where $\gamma = \exp[-h\nu/kT]$. This is called the "geometrical distribution" or "Pascal distribution". Find its factorial moments and cumulants and show that its variance is larger than that of the Poisson distribution with the same average.

Exercise. A Hohlraum is a collection of many such oscillators with different frequencies. Suppose there are Z oscillators in a frequency interval $\Delta\nu$ much smaller than kT/h. The probability of finding n quanta in this group of oscillators is[*]

$$p_n = (1-\gamma)^Z \frac{(Z+n-1)!}{(Z-1)!n!} \gamma^n \tag{2.19}$$

("negative binomial distribution"; for $Z = 1$ it reduces, of course, to (2.18)). Derive from (2.19) the familiar formula for the equilibrium fluctuations in a Bose gas.

Exercise. Ordinary cumulants are adapted to the Gaussian distribution and factorial cumulants to the Poisson distribution. Other cumulants can be defined that are adapted to other distributions. For instance, define the π_m by

$$\frac{1}{F(1-x)} = 1 - \sum_{m=1}^{\infty} \frac{(-x)^m}{m!} \pi_m \tag{2.20}$$

and show that all π_m for $m > 1$ vanish if and only if the distribution is (2.18). Find generalized cumulants that characterize in the same way the distributions (2.19) and (1.5).

---

[*] D. ter Haar, *Elements of Statistical Mechanics* (Holt, Rinehart and Winston, New York 1954) p. 74; third ed. (Butterworth-Heinemann, Oxford 1995) p. 95. For other applications, see G. Ekspong, in: *Multiparticle Dynamics* (A. Giovannini and W. Kittel eds., World Scientific, Singapore 1990) p. 467.

3. Multivariate distributions

Let X be a random variable having r components $X_1, X_2, ..., X_r$. Its probability density $P_r(x_1, x_2, ..., x_r)$ is also called the *joint probability distribution* of the r variables $X_1, X_2, ..., X_r$. Take a subset of $s < r$ variables $X_1, X_2, ..., X_s$. The probability that they have certain values $x_1, x_2, ..., x_s$, regardless of the values of the remaining $X_{s+1}, ..., X_r$, is

$$P_s(x_1, ..., x_s) = \int P_r(x_1, ..., x_s, x_{s+1}, ..., x_r) \, dx_{s+1} \cdots dx_r. \qquad (3.1)$$

It is called the *marginal distribution* for the subset.

On the other hand, one may attribute fixed values to $X_{s+1}, ..., X_r$ and consider the joint probability distribution of the remaining variables $X_1, ..., X_s$. This is called the *conditional probability of* $X_1, ..., X_s$, conditional on $X_{s+1}, ..., X_r$ having the prescribed values $x_{s+1}, ..., x_r$. It will be denoted by[*]

$$P_{s|r-s}(x_1, ..., x_s | x_{s+1}, ..., x_r). \qquad (3.2)$$

In physical parlance: from the ensemble representing the distribution in r-dimensional space, one extracts the subensemble of those samples in which $X_{s+1} = x_{s+1}, ..., X_r = x_r$; the probability distribution in this subensemble is (3.2).

The total joint probability P_r is equal to the marginal probability for $X_{s+1}, ..., X_r$ to have the values $x_{s+1}, ..., x_r$, times the conditional probability that, this being so, the remaining variables have the values $x_1, ..., x_s$:

$$P_r(x_1, ..., x_r) = P_{r-s}(x_{s+1}, ..., x_r) P_{s|r-s}(x_1, ..., x_s | x_{s+1}, ..., x_r).$$

This is *Bayes' rule*, usually expressed by

$$P_{s|r-s}(x_1, ..., x_s | x_{s+1}, ..., x_r) = \frac{P_r(x_1, ..., x_r)}{P_{r-s}(x_{s+1}, ..., x_r)}. \qquad (3.3)$$

Suppose that the r variables can be subdivided in two sets $(X_1, ..., X_s)$ and $(X_{s+1}, ..., X_r)$ such that P_r factorizes:

$$P_r(x_1, ..., x_r) = P_s(x_1, ..., x_s) P_{r-s}(x_{s+1}, ..., x_r).$$

Then the two sets are called *statistically independent* of each other. The factor P_s is then also the marginal probability density of the variables $X_1, X_2, ..., X_s$. At the same time it is the conditional probability density

$$P_{s|r-s}(x_1, ..., x_s | x_{s+1}, ..., x_r) = P_s(x_1, ..., x_s).$$

---

[*] The reverse notation, with the prescribed values in front of the bar, is obsolescent.

3. MULTIVARIATE DISTRIBUTIONS

Hence the distribution of $X_1, \ldots, X_s$ is not affected by prescribing values for $X_{s+1}, \ldots, X_r$, and vice versa.

Note. When the denominator in (3.3) vanishes the numerator vanishes as well, as can easily be shown. For such values of $x_{s+1}, \ldots, x_r$ the left-hand side is not defined. The conditional probability is not defined when the condition cannot be met.

Exercise. Prove and interpret the normalization of the conditional probability

$$\int P_{s|r-s}(x_1, \ldots, x_s | x_{s+1}, \ldots, x_r) \, dx_1 \cdots dx_s = 1. \tag{3.4}$$

Exercise. What is the form of the joint probability density if all variables are mutually independent?

Exercise. Maxwell's derivation of the velocity distribution in a gas was based on the assumptions that it could only depend on the speed $|v|$, and that the cartesian components are statistically independent. Show that this leads to Maxwell's law.

Exercise. Compute the marginal and conditional probabilities for the following ring-shaped bivariate distribution:

$$P_2(x_1, x_2) = \pi^{-1} \delta(x_1^2 + x_2^2 - a^2).$$

Exercise. Generalize this ring distribution to r variables evenly distributed on a hypersphere in r dimensions, i.e., the microcanonical distribution of an ideal gas. Find the marginal distribution for x_1. Show that it becomes Gaussian in the limit $r \to \infty$, provided that the radius of the sphere also grows, proportionally to $\sqrt{r}$.

Exercise. Two dice are thrown and the outcome is 9. What is the probability distribution of the points on the first die conditional on this given total? Why is this result not incompatible with the obvious fact that the two dice are independent?

Exercise. The probability distribution of lifetimes in a population is $P(t)$. Show that the conditional probability for individuals of age τ is

$$P(t|\tau) = P(t) \bigg/ \int_\tau^\infty P(t') \, dt' \quad (t > \tau). \tag{3.5}$$

Note that in the case $P(t) = \gamma e^{-\gamma t}$ one has $P(t|\tau) = P(t - \tau)$: the survival chance is independent of age. Show that this is the only P for which that is true.

The moments of a multivariate distribution are

$$\langle X_1^{m_1} X_2^{m_2} \cdots X_r^{m_r} \rangle = \int x_1^{m_1} x_2^{m_2} \cdots x_r^{m_r} P(x_1, x_2, \ldots, x_r) \, dx_1 \, dx_2 \cdots dx_r.$$

(They could be denoted by $\mu_{m_1, m_2, \ldots, m_r}$ but that notation is no longer convenient when more variables occur.) The characteristic function is a function of r auxiliary variables

$$G(k_1, k_2, \ldots, k_r) = \langle e^{i(k_1 X_1 + k_2 X_2 + \cdots + k_r X_r)} \rangle.$$

Its Taylor expansion in the variable k generates the moments

$$G(k_1, k_2, \ldots, k_r) = \sum_0^\infty \frac{(ik_1)^{m_1}(ik_2)^{m_2} \cdots (ik_r)^{m_r}}{m_1! m_2! \cdots m_r!} \langle X_1^{m_1} X_2^{m_2} \cdots X_r^{m_r} \rangle. \qquad (3.6)$$

The cumulants will now be indicated by double brackets; they are defined by

$$\log G(k_1, k_2, \ldots, k_r) = \sum_0^\infty{}' \frac{(ik_1)^{m_1}(ik_2)^{m_2} \cdots (ik_r)^{m_r}}{m_1! m_2! \cdots m_r!} \langle\langle X_1^{m_1} X_2^{m_2} \cdots X_r^{m_r} \rangle\rangle, \qquad (3.7)$$

where the prime indicates the absence of the term with all m's simultaneously vanishing. (The double-bracket notation is not standard, but convenient in the case of more than one variable.)

The second moments may be combined into an $r \times r$ matrix $\langle X_i X_j \rangle$. More important is the *covariance matrix*

$$\langle\langle X_i X_j \rangle\rangle = \langle (X_i - \langle X_i \rangle)(X_j - \langle X_j \rangle) \rangle = \langle X_i X_j \rangle - \langle X_i \rangle \langle X_j \rangle. \qquad (3.8)$$

Its diagonal elements are the variances, its off-diagonal elements are called *covariances*. When normalized the latter are called *correlation coefficients*:

$$\rho_{ij} = \frac{\langle\langle X_i X_j \rangle\rangle}{\sqrt{\langle\langle X_i^2 \rangle\rangle \langle\langle X_j^2 \rangle\rangle}} = \frac{\langle X_i X_j \rangle - \langle X_i \rangle \langle X_j \rangle}{\sqrt{(\langle X_i^2 \rangle - \langle X_i \rangle^2)(\langle X_j^2 \rangle - \langle X_j \rangle^2)}}. \qquad (3.9)$$

Take $r = 2$; the statistical independence of X_1, X_2 is expressed by any one of the following three criteria.

(i) All moments factorize: $\langle X_1^{m_1} X_2^{m_2} \rangle = \langle X_1^{m_1} \rangle \langle X_2^{m_2} \rangle$.
(ii) The characteristic function factorizes:

$$G(k_1, k_2) = G_1(k_1) G_2(k_2). \qquad (3.10)$$

(iii) The cumulants $\langle\langle X_1^{m_1} X_2^{m_2} \rangle\rangle$ vanish when both m_1 and m_2 differ from zero.

The variables X_1, X_2 are called *uncorrelated* when it is merely known that their covariance is zero, which is weaker than statistical independence. The reason why this property has a special name is that in many applications the first and second moments alone provide an adequate description.

Exercise. Consider the marginal distribution of a subset of all variables. Express its moments in terms of the moments of the total distribution, and its characteristic function in terms of the total one.

Exercise. Prove the three criteria for independence mentioned above and generalize them to r variables.

Exercise. Prove $-1 \leq \rho_{ij} \leq 1$. Prove that if ρ_{ij} is either 1 or -1 the variables X_i, X_j are connected by a linear relation.

3. MULTIVARIATE DISTRIBUTIONS

Exercise. Show that for any set $X_1, ..., X_r$ it is possible to find r linear combinations

$$Y_i = \sum_{j=1}^{r} a_{ij} X_j \quad (i = 1, ..., r)$$

such that the new variables Y are mutually uncorrelated (orthogonalization procedure of E. Schmidt).

Exercise. Show that each cumulant $\langle\langle X_1^{m_1} X_2^{m_2} \cdots X_r^{m_r} \rangle\rangle$ is an "isobaric" combination of moments, i.e., a linear combination of products of moments, such that the sum of exponents in each product is the same, viz. $m_1 + m_2 + \cdots m_r$.

Exercise. Prove that "independent" implies "uncorrelated" and construct an example to show that the converse is not true.

Exercise. Find moments and cumulants of the bivariate Gaussian distribution

$$P(x, y) = \text{const.} \; e^{-\frac{1}{2}(ax^2 + 2bxy + cy^2)} \quad (ac - b^2 > 0, a > 0).$$

Show that for this distribution "uncorrelated" and "independent" are equivalent

Exercise. A molecule can occupy different levels $n_1, n_2, ...$ with probabilities $p_1, p_2, ...$. Suppose there are N such molecules. The probability for finding the successive levels occupied by $N_1, N_2, ...$ molecules is given by the multinomial distribution

$$P(N_1, N_2, ...) = \frac{N!}{N_1! N_2! \cdots} p_1^{N_1} p_2^{N_2} \cdots . \tag{3.11}$$

Exercise. The correlation coefficients for three variables obey

$$(1 + \rho_{12})(1 + \rho_{13})(1 + \rho_{23}) \geq \tfrac{1}{2}(1 + \rho_{12} + \rho_{13} + \rho_{23})^2.$$

Exercise. If a distribution is obtained from a set of observations it often consists of a single hump. The first and second cumulant are rough indications of its position and its width. Further information about its shape is contained in its "skewness", defined by $\gamma_3 = \kappa_3/\kappa_2^{3/2}$, and its "kurtosis" $\gamma_4 = \kappa_4/\kappa_2^2$. Prove[*]

$$\gamma_3^2 \leq \gamma_4 + 2.$$

Exercise. *Multivariate factorial moments*, indicated by curly brackets, are defined by an obvious generalization of **(2.16)**:

$$\left\langle \prod_j (1 - z_j)^{X_j} \right\rangle = \sum_{\{m\}} \frac{(-z_1)^{m_1}(-z_2)^{m_2} \cdots}{m_1! \, m_2! \cdots} \{X_1^{m_1} X_2^{m_2} \cdots\}. \tag{3.12}$$

Multivariate factorial cumulants, indicated by square brackets, are

$$\log \left\langle \prod_j (1 - z_j)^{X_j} \right\rangle = \sum_{\{m\}}{}' \frac{(-z_1)^{m_1}(-z_2)^{m_2} \cdots}{m_1! \, m_2! \cdots} [X_1^{m_1} X_2^{m_2} \cdots]. \tag{3.13}$$

---

[*] More inequalities of this type are given by A.A. Dubkov and A.N. Malakhov, Radiophys. Quantum Electron. (USA) **19**, 833 (1977).

Write the first few in terms of the moments, in particular

$$[X_i X_j] = \langle X_i X_j \rangle - \langle X_i \rangle \langle X_j \rangle - \delta_{ij} \langle X_i \rangle. \tag{3.14}$$

Exercise. The factorial cumulant of the sum of two statistically independent variables is the sum of their factorial cumulants. A factorial cumulant involving two mutually independent sets of variables vanishes.

4. Addition of stochastic variables

Let X_1, X_2 be two variables with joint probability density $P_X(x_1, x_2)$. The probability that $Y = X_1 + X_2$ has a value between y and $y + \Delta y$ is

$$P_Y(y) \Delta y = \iint_{y < x_1 + x_2 < y + \Delta y} P_X(x_1, x_2) \, dx_1 \, dx_2.$$

From this follows the formula

$$P_Y(y) = \iint \delta(x_1 + x_2 - y) P_X(x_1, x_2) \, dx_1 \, dx_2$$

$$= \int P_X(x_1, y - x_1) \, dx_1. \tag{4.1}$$

If X_1, X_2 are independent this equation becomes

$$P_Y(y) = \int P_{X_1}(x_1) P_{X_2}(y - x_1) \, dx_1. \tag{4.2}$$

Thus the probability density of the sum of two independent variables is the convolution of their individual probability densities.

One easily deduces the following three rules concerning the moments. First the universal identity

$$\langle Y \rangle = \langle X_1 \rangle + \langle X_2 \rangle$$

states: *the average of the sum is the sum of the averages*, regardless of whether X_1, X_2 are independent or not. The second rule is that, if X_1 and X_2 are uncorrelated,

$$\sigma_Y^2 = \sigma_{X_1}^2 + \sigma_{X_2}^2, \tag{4.3a}$$

or, in our double-bracket notation,

$$\langle\langle (X_1 + X_2)^2 \rangle\rangle = \langle\langle X_1^2 \rangle\rangle + \langle\langle X_2^2 \rangle\rangle. \tag{4.3b}$$

The characteristic function of Y is

$$G_Y(k) = G_{X_1, X_2}(k, k).$$

4. ADDITION OF STOCHASTIC VARIABLES

If X_1, X_2 are independent the right-hand side factorizes according to (3.10), so that

$$G_Y(k) = G_{X_1}(k) G_{X_2}(k). \tag{4.4}$$

This is the third rule: *for independent variables the characteristic function of the sum is the product of their individual characteristic functions.*

Remark. A logician might raise the following objection. In section 1 stochastic variables were defined as objects consisting of a range and a probability distribution. Algebraic operations with such objects are therefore also matters of definition rather than to be derived. He is welcome to regard the addition in this section and the transformations in the next one as definitions, provided that he then shows that the properties of these operations that were obvious to us are actually consequences of these definitions.

Averaging is a different kind of operation since it associates with a stochastic variable a non-stochastic or "sure" number. Alternatively it may be viewed as a projection in the following way. The set of all stochastic variables contains a subset of variables whose probability density is a delta peak. This subset is isomorphic with the sure numbers of the range and may therefore be identified with them. The operation of taking the average is then a projection of the total space of stochastic variables onto this subset.

Exercise. Prove (4.3) and show by an example that the condition that X_1 and X_2 are uncorrelated is indispensable.
Exercise. Generalize these statements to the addition of more than two variables.
Exercise. Formulate the rules for the sum of two or more vector variables, the variance being replaced with the covariance matrix.
Exercise. For *independent* variables the cumulants of the sum are equal to the sum of the cumulants. Equation (**4**.3) is a special case of this rule.
Exercise. All three rules are used as a matter of course in the kinetic theory of gases. Give examples.
Exercise. In the space of stochastic variables a scalar product may be defined by $\langle XY \rangle$. Prove that with this definition the projection onto the average is a Hermitian operator.
Exercise. In the space of $N \times N$ real matrices X define the function

$$X \to \langle X \rangle \equiv (\operatorname{Tr} MX)/(\operatorname{Tr} M),$$

where M is a fixed matrix. It is not an average in our sense, but it is a linear projection of X into the real numbers and it maps the unit matrix onto 1. These properties suffice for establishing the identity

$$\langle e^X \rangle = \exp\left[\langle X \rangle + \tfrac{1}{2}\langle\langle X^2 \rangle\rangle + \frac{1}{3!}\langle\langle X^3 \rangle\rangle + \cdots \right].$$

Exercise. If X, Y are two joint stochastic variables and α, β two parameters,
$$\langle X\, e^{\alpha X + \beta Y} \rangle = \langle\langle X\, e^{\alpha X + \beta Y} \rangle\rangle \langle e^{\alpha X + \beta Y} \rangle.$$
The cumulant is taken after expanding the exponential.

An ancient but still instructive example is the *discrete-time random walk*. A drunkard moves along a line by making each second a step to the right or to the left with equal probability. Thus his possible positions are the integers $-\infty < n < \infty$, and one asks for the probability $p_n(r)$ for him to be at n after r steps, starting from $n = 0$. While we shall treat this example in IV.5 as a stochastic process, we shall here regard it as a problem of adding variables.

To each step corresponds a stochastic variable X_j ($j = 1, 2, \ldots, r$) taking the values 1 and -1 with probability $\tfrac{1}{2}$ each. The position after r steps is
$$Y = X_1 + X_2 + \cdots + X_r.$$
One finds immediately $\langle Y \rangle = 0$, and as the steps are mutually independent
$$\langle Y^2 \rangle = r \langle X^2 \rangle = r. \tag{4.5}$$
The fact that the mean square displacement is proportional to the number of steps is typical for diffusion-like processes. It implies for the displacement per unit time
$$\left\langle \left(\frac{Y}{r}\right)^2 \right\rangle = \frac{1}{r} \to 0 \quad \text{as } r \to \infty.$$
That is, the variance of the mean velocity over a long period tends to zero. This distinguishes diffusive spreading from propagation through particles in free flight or through waves.

In order to find the detailed probability distribution of Y we employ the characteristic function
$$G_Y(k, r) = [G_X(k)]^r = [\tfrac{1}{2} e^{ik} + \tfrac{1}{2} e^{-ik}]^r. \tag{4.6}$$
The probability that Y has the value n is the coefficient of e^{ink}:
$$p_n(r) = \frac{1}{2^r} \binom{r}{\tfrac{1}{2}(r-n)}. \tag{4.7}$$
It is understood that the binomial coefficient equals zero unless $\tfrac{1}{2}(r - n)$ is an integer between 0 and r inclusive.

Exercise. Give a purely combinatorial derivation of (4.7) by counting the number of sequences of r steps that end up in n.

Exercise. In the *asymmetric random walk* there is at each step a probability q to step to the left and $1-q$ to the right. Find $p_n(r)$ for this case.

Exercise. Suppose at each step there is a probability q_v for a step of v units ($v = \pm 1, \pm 2, \ldots$), and a probability q_0 to stay put. Find the mean and the variance of the distance after r steps.

Exercise. Let X_j be an infinite set of independent stochastic variables with identical distributions $P(x)$ and characteristic function $G(k)$. Let r be a random positive integer with distribution p_r and probability generating function $f(z)$. Then the sum $Y = X_1 + X_2 + \cdots + X_r$ is a random variable: show that its characteristic function is $f(G(k))$. [This distribution of Y is called a "compound distribution" in FELLER I, ch. XII.]

Exercise. Consider a set of independent particles, each having an energy E with probability density $p(E) = \beta e^{-\beta E}$. Suppose in a certain volume there are n such particles with probability (2.10). Find for the probability density of the total energy in that volume

$$P(E) = e^{-a}\delta(E) + a e^{-a}\beta e^{-\beta E} \sum_{k=0}^{\infty} \frac{(a\beta E)^k}{k!(k+1)!}$$

$$= e^{-a}\delta(E) + \sqrt{a\beta/E}\, e^{-a-\beta E} I_1(2\sqrt{a\beta E}).$$

(I_1 is a modified Bessel function.) Consider also the case $p(E) = \text{const.}\, E^\gamma e^{-\beta E}$ (as in 5.5).

5. Transformation of variables

Let the continuous, one-component variable X be mapped into a new variable Y by

$$Y = f(X). \tag{5.1}$$

Familiar examples are: plotting on a logarithmic scale ($Y = \log X$) and the transformation from frequencies to wavelengths ($Y = 1/X$). In general the range of Y differs from that of X. The probability that Y has a value between y and $y + \Delta y$ is

$$P_Y(y)\,\Delta y = \int_{y < f(x) < y+\Delta y} P_X(x)\,dx.$$

The integral extends over *all* intervals of the range of X in which the inequality is obeyed. An equivalent formula is

$$P_Y(y) = \int \delta[f(x) - y] P_X(x)\,dx. \tag{5.2}$$

From this one derives for the characteristic function of Y,

$$G_Y(k) = \langle e^{ikf(X)} \rangle. \tag{5.3}$$

These equations remain valid when X stands for a quantity with r components and Y for one with s components, where s may or may not be equal to r.[*] A familiar example of a case with $r = 3$, $s = 1$ is the transformation of the Maxwell distribution for $X = (v_x, v_y, v_z)$ into a transformation of the energy $E = \frac{1}{2}m(v_x^2 + v_y^2 + v_z^2)$:

$$P(E) = \int \delta(\tfrac{1}{2}mv^2 - E)\left(\frac{m}{2\pi kT}\right)^{3/2} e^{-mv^2/2kT}\, dv_x\, dv_y\, dv_z$$

$$= 2\pi^{-1/2}(kT)^{-3/2} E^{1/2} e^{-E/kT}$$

("gamma distribution" or "χ^2 distribution").

In the special case that only one X corresponds to each Y (and hence necessarily $r = s$), one may invert (5.1) to give $X = g(Y)$. In that case the transformation of the probability density reduces to

$$P_Y(y) = P_X(x) J,$$

where J is the *absolute value* of the Jacobi determinant $d(x)/d(y)$. This equation may be memorized as

$$P_Y(y)\, d^r y = P_X(x)\, d^r x,$$

provided that one bears in mind the uniqueness condition, and the fact that the sign may have to be changed.

Remark. Consider in particular the group of linear transformations

$$Y = aX + b, \quad a \neq 0. \tag{5.4}$$

They change P_X into P_Y, but the difference is so minor that they are often considered the same distribution, and are denoted by the same name. The transformation can be used to transform the distribution to a standard form, for instance one with zero average and unit variance. In the case of lattice distributions one employs (5.4) to make the lattice points coincide with integers. In fact, the use of (5.4) to reduce the distribution to a simple form is often done tacitly, or in the guise of choosing the zero and the unit on the scale.

Exercise. Derive the equations for addition of variables as a special case of the transformation formulas in this section.

Exercise. The family of gamma distributions is defined by

$$P(x) = \frac{a^\nu}{\Gamma(\nu)} x^{\nu-1} e^{-ax} \quad (a > 0, \nu > 0, 0 < x < \infty). \tag{5.5}$$

Let the variables $X_1, X_2, \ldots, X_r$ be Gaussian with zero average and variance σ^2, and independent. Prove that $Y = X_1^2 + X_2^2 + \cdots + X_r^2$ is gamma distributed.

---

[*] See also D.T. Gillespie, Amer. J. Phys. **51**, 520 (1983).

5. TRANSFORMATION OF VARIABLES

Exercise. Suppose that the diameters of a collection of eggs are distributed according to (5.5). Suppose they all have the same shape, so that their volume is $y = x^3$. Find the distribution of y, its average and variance, and compare them with the corresponding properties for x.

Exercise. A cannon projects a ball with initial velocity v at angle θ with the horizontal. Both v and θ are subject to uncertainties that can be described by a Gaussian distribution for each. The distributions are centered at v_0 and θ_0, respectively, and so sharp that non-physical values, such as negative v or θ, may be ignored. What is the probability distribution of the distance covered by the cannon ball?

Exercise. How are the cumulants affected by (5.4)?

Exercise. Let X have the range $(0, 2\pi)$ and constant probability density in that range. Find the distribution of $Y = \sin X$. Also when $P(x) = A + B \sin x$ ($|B| < A = 1/2\pi$).

Exercise. A point lies on a circle at a random position with uniform distribution along the circle. When viewed by an eccentric observer, what is the probability distribution of its azimuth?

Exercise. A scattering center is bombarded by a homogeneous beam of particles. A particle that comes in with impact parameter b is deflected by an angle $\Theta(b)$. Find the differential cross-section.

Excursus. The theory of probability is nothing but transforming variables. It comprises a collection of techniques for transforming an *a priori* given distribution into another one, called the *a posteriori* distribution. Any application to real phenomena consists of the following three steps.

a. Postulating an *a priori* distribution.
b. Performing the appropriate mathematical transformation.
c. Comparing the *a posteriori* distribution with the observations.

Ad **c.** A probability distribution cannot be observed itself, but can only be compared with the frequency with which each outcome is observed in a series of repeated experiments. Sometimes the distribution is so sharply peaked that it practically reduces to a single value, which can be tested by a single experiment – as in the statistical calculation of thermodynamic quantities.

Ad **b.** This transformation is pure mathematics. It is the main subject of textbooks on probability theory. The subject has been freed from all contamination of reality by the axiomatics of Kolmogorov[*].

Ad **a.** To establish the *a priori* distribution one has to take into account the actual system. It turns out that many systems have a level of description where a simple guess for the probability distribution can be made. In most cases this amounts to identifying units with equal probability. When throwing two dice one computes the *a posteriori* distribution of the total number of points from the assumed *a priori* distribution made up by equal probabilities for the 36 elementary events. There are good reasons for this assumption, but as always in physics it has to be verified by experiment: no amount of mathematics can show that a die is not loaded.

---

[*] Op. cit. in section 1.

The idea of equal probabilities has been elevated by Laplace[*] to the rank of a philosophical principle, called "principle of insufficient reason". Like many philosophical principles it leaves the essential question unanswered: How do I select the elementary events to which equal *a priori* probabilities are to be assigned? In textbook problems about tossing dice or drawing cards it is obvious what the author has in mind. One knows that he is concerned with the mathematics of step **b** and that the dice and cards merely serve as a ritual way of defining an *a priori* distribution. In actual applications, however, step **a** cannot be dismissed so cavalierly.

The idea that a universal principle should be able to provide the *a priori* probability in each case and thereby establish the link between the mathematical probability and the actual world, lost ground during the nineteenth century. An important argument was that in the case of a continuous state space the prescription is not even well-defined since it depends on the choice of variable. A uniform distribution in velocity space is not the same as a uniform distribution in the energy scale. This difficulty has been beautifully demonstrated by Bertrand[**]. Take a fixed circle of

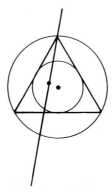

Fig. 1. Bertrand's circle with random chord.

unit radius and draw "at random" a straight line intersecting it. What is the probability that the chord has a length greater than $\sqrt{3}$ (which is length of a side of an equilateral triangle inscribed in the circle)? *First answer*: Take all random lines through an arbitrarily fixed point P of the circle. Apart from the tangent (zero probability) they all intersect the circle. In order that the chord be longer than $\sqrt{3}$, the line has to lie within an angle of 60° out of a total of 180°, hence the probability is $\frac{1}{3}$. *Second answer*: Take all random lines perpendicular to a fixed diameter. The chord is longer than $\sqrt{3}$ when the point of intersection lies on the middle half of

---

[*] Laplace, *Essai philosophique sur les probabilités* (Paris 1812). For general reviews, see: J.R. Lucas, *The Concept of Probability* (Clarendon, Oxford 1970); G. Gigerenzer, Z. Swijtink, T. Porter, L. Daston, J. Beattie, and L. Krüger, *The Empire of Chance* (Cambridge University Press, Cambridge 1989).

[**] J. Bertrand, *Calcul des probabilités* (Gauthier-Villars, Paris 1889); M.G. Kendall and P.A.P. Moran, *Geometrical Probability* (Hafner, New York 1963).

5. TRANSFORMATION OF VARIABLES

the diameter: probability $\frac{1}{2}$. *Third answer*: In order that a chord be longer than $\sqrt{3}$ its midpoint must lie at a distance less than $\frac{1}{2}$ from the center. The area of a circle of radius $\frac{1}{2}$ is a quarter of that of the original circle, hence the probability is $\frac{1}{4}$. – The reader will easily see that each solution is based on a different assumption of equal *a priori* probabilities. The loose phrase "at random" does not sufficiently specify the *a priori* probability to choose between the solutions.

The use of the least squares method for extracting information from imperfect observations assumes a specific *a priori* probability distribution for the errors, viz. the Gauss distribution (2.4). The same assumption, however, cannot be true for all variables that might be used to measure the observed quantity (but at most for one variable, and all those that are linearly connected with it). The method of least squares applied in the frequency scale does not lead to the same result as when applied to the same observations plotted according to wavelengths. The "best value" for the brightness of a star depends on whether one applies the method of least squares to the magnitude or to its intensity in energy measure. The redeeming feature is, as long as the errors are small, that any reasonable transformation is practically linear in the relevant range. But there is no logical foundation for applying it to widely scattered data.

This is only a special aspect of the time-honored problem of induction: how does science manage to deduce general laws from a necessarily finite number of observations? Since classical logic cannot answer this question, many attempts have been made to resort to probability considerations. The aim is to compute the probability for a hypothesized law to be true, given a set of observations. The preceding discussion has shown that this question has no answer unless an *a priori* probability of all possible hypotheses is given or assumed. I draw a ball from an urn and it is black: what is the probability that all balls in that urn are black? The question has no answer, unless it is added that the urn was picked from a specified ensemble of urns containing black and other balls in specified ratios.

When the hypothesis is a scientific theory put forward to explain certain observed facts, no *a priori* probability is given and even "the set of all possible hypotheses" is a hazy concept. The probability of the theory cannot therefore be expressed objectively as a percentage, but is subjective and open to discussion. The reason why nevertheless agreement can often be reached is that, when the number of corroborating observed facts is large, the *a posteriori* probability is also large, even when the chosen *a priori* probability has been small. Yet it should always be borne in mind that scientific induction is beyond the reach of mathematics.[*]

In statistical mechanics of equilibrium one assigns equal probabilities to equal volume elements of the energy shell in phase space.[**] This assignment is determined

---

[*] K.R. Popper, on p. 287 of: *Conjectures and Refutations* (Harper and Row, New York 1968), argues that the least probable theories are the most valuable. He means that the more precise predictions a theory makes, the less one would bet on it *a priori*, but the greater its value when it turns out to be true, i.e., when its *a posteriori* probability after checking against reality is close to unity.

[**] This defines the macrocanonical ensemble; the canonical distribution can be derived from it although it is often postulated as an *a priori* distribution in its own right.

by the underlying physics; the principle of insufficient reason might also mean equal probability for equal surface elements of the energy hypersurface. The physics needed is the Liouville theorem, the fact that energy is constant, and the assumption that no other measurable quantity is constant. This is called ergodicity, but mathematical ergodicity is not enough. One also has to assume that no quantity exists that is practically constant on the time scale of the experiment, such as the ratio of para- to ortho-hydrogen.[*)]

The twentieth century saw a revival of Laplace's idea, in particular in the work of Jaynes.[**)] He gave it the more modern formulation of the "principle of maximum entropy", but it is again an attempt at replacing physics with a philosophical principle, and suffers from the same shortcomings. In quantum mechanics the equal probability assumption is even more alluring, because the discreteness of the set of eigenfunctions makes the prescription unique[†)]; but still physics is needed to tell which eigenstates should be selected, and to justify the assumption.

Exercise. Compute the probability distribution of the length of the chord in the three cases of Bertrand's example.

Exercise. N urns contain b_k black balls and a_k others ($k = 1, 2, \ldots, N$). I pick one urn at random (equal probability for each) and draw one ball at random. When this is black, what is the probability that my urn is the k-th one? What is the answer when I pick my urn with a probability proportional to the total number of balls it contains?

Exercise. Show that a Lorentz distribution in the frequency scale is also a Lorentz distribution in the wavelength.

Exercise. Suppose one expects x to be a function of t of the form $f(t; \alpha)$, where α stands for a set of parameters to be found from the observed values x_i at times t_i. The method of least squares consists in determining them from the condition

$$\sum_i \{x_i - f(t_i; \alpha)\}^2 = \text{minimum}.$$

Now transform the whole problem to a new variable $y = \phi(x)$ and show that the minimalization in this variable leads to different values of the α.

Exercise. One particular distribution of straight lines in the plane is special in that it is invariant for translations and rotations. To which solution of Bertrand's problem does it correspond? [Kendall and Moran, loc. cit.]

[*)] D.A. McQuarrie, *Statistical Thermodynamics* (University Science Books, Mill Valley, CA 1973) p. 104.

[**)] E.T. Jaynes: *Papers on Probability, Statistics and Statistical Physics* (R.D. Rosenkrantz ed., Kluwer, Dordrecht 1989); *Probability Theory* (Cambridge Univ. Press 2003); B. Buck and V.A. Macaulay eds., *Maximum Entropy in Action* (Clarendon, Oxford 1991). A desperate attempt was made by J.M. Keynes, *A Treatise on Probability* (Macmillan, London 1921, 1973).

[†)] R.C. Tolman, *The Principles of Statistical Mechanics* (Clarendon, Oxford 1938) p. 349; K. Huang, *Statistical Mechanics* (Wiley, New York 1963); D.A. McQuarrie, op. cit., p. 36.

6. The Gaussian distribution

The general form of the Gaussian distribution for one variable is

$$P(x) = C\, e^{-\frac{1}{2}Ax^2 - Bx} \quad (-\infty < x < \infty), \tag{6.1}$$

where A is a positive constant that determines the width, B determines the position of the peak, and C is the normalization constant,

$$C = \left(\frac{A}{2\pi}\right)^{1/2} e^{-B^2/2A}. \tag{6.2}$$

It is often convenient to express the parameters A, B in terms of the average $\mu_1 = -B/A$ and variance $\sigma^2 = 1/A$:

$$P(x) = (2\pi\sigma^2)^{-1/2} \exp\left[-\frac{(x-\mu_1)^2}{2\sigma^2}\right]. \tag{6.3}$$

The distribution is named after Gauss[*] and is also called the *normal distribution*[**]. For reference we write its characteristic function

$$G(k) = e^{i\mu_1 k - \frac{1}{2}\sigma^2 k^2}. \tag{6.4}$$

The use of the cumulants is particularly suited for this distribution,

$$\kappa_1 = \mu_1, \quad \kappa_2 = \sigma^2, \quad \kappa_3 = \kappa_4 = \cdots = 0.$$

When $X_1, X_2, \ldots, X_r$ are Gaussian variables and mutually independent, their sum $Y = X_1 + X_2 + \cdots + X_r$ is again Gaussian, as is immediately seen with the aid of (4.4). The average and variance of Y are the sums of the averages and variances of the variables X. This fully determines the distribution of Y.

Exercise. Prove that (6.3) tends to $\delta(x - \mu_1)$ when σ^2 tends to zero. [Multiply with a test function and integrate.]

Exercise. The property that the sum of two independent Gaussian variables is again Gaussian is not unique. Prove that the Lorentz and the Poisson distribution have a similar property. [Compare the Remark in 7.]

Exercise. Prove that the convolution of two gamma distributions (5.5) with the same a is again a gamma distribution.

---

[*] Apart from Gauss (1809) there were others, e.g., de Moivre (1733) and Laplace (1812). See F.M. David, op. cit. in 1; I. Schneider Hrsg., *Die Entwicklung der Wahrscheinlichkeitsrechnung von den Anfängen bis 1933* (Akademie-Verlag, Berlin 1989).

[**] The misleading adjective "normal" was introduced by K. Pearson, who later regretted it: Biometrika **13**, 25 (1920).

The multivariate Gaussian distribution has the form

$$P(x) = C \exp\left[-\frac{1}{2}\sum_{i,j=1}^{r} A_{ij}x_i x_j - \sum_{i=1}^{r} B_i x_i\right], \tag{6.5}$$

with positive-definite symmetric matrix A. In vector notation

$$P(x) = C \exp[-\tfrac{1}{2} x \cdot \mathbb{A} \cdot x - B \cdot x]. \tag{6.6}$$

The normalization constant C is found by transforming to new variables in which $\mathbb{A}$ is diagonal and then using (6.2), with the result

$$C = (2\pi)^{-r/2} (\text{Det } \mathbb{A})^{1/2} \exp[-\tfrac{1}{2} B \cdot \mathbb{A}^{-1} \cdot B]. \tag{6.7}$$

The corresponding characteristic function is

$$G(k) = \exp[-\tfrac{1}{2} k \cdot \mathbb{A}^{-1} \cdot k - i k \cdot \mathbb{A}^{-1} \cdot B]. \tag{6.8}$$

On expanding in powers of k one finds from this

$$\langle X_i \rangle = -\sum_j (\mathbb{A}^{-1})_{ij} B_j, \tag{6.9a}$$

$$\langle\langle X_i X_j \rangle\rangle = (\mathbb{A}^{-1})_{ij}. \tag{6.9b}$$

This shows that for a Gaussian distribution the covariance matrix defined in (3.8) is equal to $\mathbb{A}^{-1}$. It follows that *a Gaussian distribution is fully determined by the averages of the variables and their covariance matrix*. In particular, if the variables are uncorrelated, $\mathbb{A}^{-1}$ is diagonal and hence also $\mathbb{A}$, so that the variables are also independent. Thus, *provided that it is known that the joint distribution is Gaussian, "uncorrelated" implies "independent"* (compare the Exercise in 3). This independence can always be achieved by a linear, and even by an orthogonal, transformation of the variables.

The moments of a multivariate Gauss distribution with zero mean have a remarkable property. Consider the distribution (6.6) with $B = 0$. We write its characteristic function in terms of $s_j = ik_j$ in order to avoid irrelevant factors i:

$$G(-is_1, -is_2, \ldots, -is_r) = \prod_{p,q} \exp[\tfrac{1}{2}\langle X_p X_q \rangle s_p s_q]$$

$$= \prod_{p,q} \{1 + \tfrac{1}{2}\langle X_p X_q \rangle s_p s_q + \cdots\}. \tag{6.10}$$

To find the moment $\langle X_i X_j X_k \cdots \rangle$ one has to collect the terms proportional to $s_i s_j s_k \cdots$. We suppose that the number of factors is even, and that the subscripts are different from each other. Then the dots in (6.10) cannot contribute. Hence the only way in which $s_i s_j s_k \cdots$ occurs in (6.10) is as the result of multiplying suitable pairs $s_p s_q$. On the other hand, each product of pairs $s_p s_q$ that makes up $s_i s_j s_k \cdots$ does occur. The result is

$$\langle X_i X_j X_k \cdots \rangle = \sum \langle X_p X_q \rangle \langle X_u X_v \rangle \cdots. \tag{6.11}$$

6. THE GAUSSIAN DISTRIBUTION

The subscripts $p, q, u, v, \ldots$ are the same as $i, j, k, \ldots$ taken two by two. The summation extends over all different ways in which $i, j, k, \ldots$ can be subdivided in pairs. The factor $\tfrac{1}{2}$ in (6.10) cancels because the product in (6.10) contains each pair twice.

Exercise. The normalizing constant C is found from
$$C^{-1} = \int \exp\left[-\tfrac{1}{2} x \cdot \mathbb{A} \cdot x - B \cdot x\right] d^r x.$$
In order to evaluate the integral it is not necessary to use an *orthogonal* transformation of x: any linear transformation that diagonalizes $\mathbb{A}$ can be used. Derive in this way (6.7) and (6.8).

Exercise. Carry out the following derivation of (6.9b). First shift the origin so that $\langle X_i \rangle = 0$. Then
$$\sum_j \langle X_i X_j \rangle A_{jk} = C \int x_i x_j A_{jk} \exp\left[-\tfrac{1}{2} \sum A_{pq} x_p x_q\right] d^r x$$
$$= C \int x_i \left(-\frac{\partial}{\partial x_k}\right) \exp[\ ]\, d^r x,$$
which by partial integration gives the desired result.*)

Exercise. The cumulants of first and second degree of a multivariate Gaussian distribution are given by (6.9b). Prove that *all higher cumulants vanish*.

Exercise. If X is a multicomponent Gaussian variable with zero mean and $f(X)$ a polynomial,**)
$$\langle X_k f(X) \rangle = \sum_j \langle X_k X_j \rangle \left\langle \frac{\partial f(X)}{\partial X_j} \right\rangle.$$

Exercise. For the bivariate Gaussian distribution with zero mean the standard form analogous to (6.3) is
$$P(x, y) = \frac{1}{2\pi \sigma_x \sigma_y \sqrt{1-\rho^2}} \exp\left[-\frac{1}{2}\left\{\frac{x^2}{\sigma_x^2} - 2\rho \frac{xy}{\sigma_x \sigma_y} + \frac{y^2}{\sigma_y^2}\right\} \frac{1}{1-\rho^2}\right], \quad (6.12)$$
where ρ is the correlation coefficient ρ_{12} defined in (3.9). What is the analogous form for n variables?

Exercise. For the case $\sigma_x = \sigma_y = 1$ expand (6.12):
$$P(x, y) = \frac{1}{2\pi} e^{-\tfrac{1}{2}(x^2 + y^2)} \sum_{n=0}^{\infty} \frac{\rho^n}{n!} \mathrm{He}_n(x) \mathrm{He}_n(y), \quad (6.13)$$
where He_n are Hermite polynomials[†].

---
*) L. Onsager, Phys. Rev. **38**, 2265 (1931).
) K. Furutsu, J. Res. NBS D **67, 303 (1963); E.A. Novikov, Soviet Phys. JETP **20**, 1290 (1965).
[†] M. Abramowitz and I.A. Stegun, *Handbook of Mathematical Functions* (National Bureau of Standards, Washington, DC 1964 etc.) p. 775.

7. The central limit theorem

Let $X_1, X_2, \ldots, X_r$ be a set of r independent stochastic variables, each having the same Gaussian probability density $P_X(x)$ with zero average and variance σ^2. Their sum Y has the probability density

$$P_Y(y) = [2\pi r\sigma^2]^{-1/2} \exp\left[-\frac{y^2}{2r\sigma^2}\right].$$

Thus $\langle Y^2 \rangle = r\sigma^2$ grows linearly with r. On the other hand, the distribution of the arithmetic mean of the variables X becomes narrower with $1/r$. It is therefore useful to define a suitably scaled sum

$$Z = \frac{X_1 + X_2 + \cdots + X_r}{\sqrt{r}}.$$

It has a variance σ^2 and hence

$$P_Z(z) = [2\pi\sigma^2]^{-1/2} \exp\left[-\frac{z^2}{2\sigma^2}\right]. \tag{7.1}$$

The central limit theorem states that, even when $P_X(x)$ is *not* Gaussian, but some other distribution with zero average and finite variance σ^2, equation (7.1) is still true in the limit $r \to \infty$. This remarkable fact is responsible for the dominant role of the Gaussian distribution in all fields of statistics. To obtain this result write the characteristic function of an arbitrary P_X:

$$G_X(k) = \int e^{ikx} P_X(x)\, dx = 1 - \tfrac{1}{2}\sigma^2 k^2 + \cdots. \tag{7.2}$$

Hence one finds for the characteristic function of Z

$$G_Z(k) = \left[G_X\left(\frac{k}{\sqrt{r}}\right)\right]^r = \left[1 - \frac{\sigma^2 k^2}{2r}\right]^r \to e^{-\tfrac{1}{2}\sigma^2 k^2}, \tag{7.3}$$

which, indeed, corresponds to the distribution (7.1). The dots in (7.2) give rise in $G_X(k/\sqrt{r})$ to terms of order $r^{-3/2}$ and do not, therefore, contribute in the limit $r \to \infty$.

Example. It is illuminating to see explicitly how the probability distribution tends to its limit.[*] Let X be a stochastic variable that takes the values 0 and 1 with probability $\tfrac{1}{2}$ each. Let Y be the sum of r such variables. Then Y takes the values

---

[*] M. Kac, Amer. Mathem. Monthly **54**, 369 (1947); C. Domb and E.L. Offenbacher, Amer. J. Phys. **46**, 49 (1978).

7. THE CENTRAL LIMIT THEOREM

$0, 1, 2, \ldots, r$ with probability

$$p_n = 2^{-r}\binom{r}{n}. \tag{7.4}$$

To study the limit one first has to find the proper new variable Z. The average and variance follow from (7.4) or can be found more directly to be

$$\langle Y \rangle = r\langle X \rangle = \tfrac{1}{2}r, \qquad \sigma_y^2 = r\sigma_x^2 = \tfrac{1}{4}r.$$

Accordingly set

$$Y = \tfrac{1}{2}r + \tfrac{1}{2}r^{1/2}Z.$$

The probability that Z lies between z and $z + \Delta z$ is

$$P_Z(z)\,\Delta z = \sum_{\tfrac{1}{2}r + \tfrac{1}{2}r^{1/2}z < n < \tfrac{1}{2}r + \tfrac{1}{2}r^{1/2}(z+\Delta z)} p_n. \tag{7.5}$$

When r is large this defines a smooth probability density P_Z, which is the one to be determined.

For large r the expression (7.4) can be written

$$\log p_n = -r\log 2 + (r + \tfrac{1}{2})\log r - (n + \tfrac{1}{2})\log n - (r - n + \tfrac{1}{2})\log(r-n) - \tfrac{1}{2}\log 2\pi$$

$$= -r\log 2 - \tfrac{1}{2}\log r - (n + \tfrac{1}{2})\log(\tfrac{1}{2} + \tfrac{1}{2}r^{-1/2}z)$$
$$\quad - (r - n + \tfrac{1}{2})\log(\tfrac{1}{2} - \tfrac{1}{2}r^{-1/2}z) - \tfrac{1}{2}\log 2\pi$$

$$= \log 2 - \tfrac{1}{2}\log r - (\tfrac{1}{2}r + \tfrac{1}{2}r^{1/2}z + \tfrac{1}{2})(r^{-1/2}z - \tfrac{1}{2}r^{-1}z^2 + \cdots)$$
$$\quad - (\tfrac{1}{2}r - \tfrac{1}{2}r^{1/2}z + \tfrac{1}{2})(-r^{-1/2}z - \tfrac{1}{2}r^{-1}z^2 + \cdots) - \tfrac{1}{2}\log 2\pi$$

$$= \log 2 - \tfrac{1}{2}\log r - \tfrac{1}{2}z^2 - \tfrac{1}{2}\log 2\pi + \mathcal{O}(r^{-1/2}).$$

Use this result in (7.5):

$$P_Z(z)\,\Delta z = \tfrac{1}{2}r^{1/2}\,\Delta z\,\exp\left[\log 2 - \tfrac{1}{2}\log r - \tfrac{1}{2}z^2 - \tfrac{1}{2}\log 2\pi\right]$$
$$= (2\pi)^{-1/2}\,e^{-\tfrac{1}{2}z^2}\,\Delta z. \tag{7.6}$$

The importance of this limiting expression to a physicist is, of course, the implication that for large r he may replace (7.4) with the approximation

$$P_Y(y) = (\tfrac{1}{2}\pi r)^{-1/2}\exp\left[-\frac{(y - \tfrac{1}{2}r)^2}{\tfrac{1}{2}r}\right]. \tag{7.7}$$

The question how a discrete distribution can be approximated by a continuous probability is answered by (7.5): it is a coarse scale description. More precisely, (7.7) gives the probability of finding Y in an interval $y, y + \Delta y$ when $\Delta y \gg 1$. It obviously incorrectly describes the probability in an interval $\Delta y \lesssim 1$.

Another paradox is that (7.7) extends from $-\infty$ to $+\infty$, although by construction Y cannot take negative values. It is not possible to simply cut off (7.7) at zero without violating the normalization. The answer is that (7.7) is an approximation, which does in fact approximate the correct value zero for negative y extremely well when r is large:

$$P_Y(y<0) < (\tfrac{1}{2}\pi r)^{-1/2} \exp[-\tfrac{1}{2}r].$$

Exercise. Show that the result (7.7) is in fact the same as obtained from the central limit theorem.

Exercise. Apply the central limit theorem to the random walk in **4**, comp. (4.7). Compare the result with an explicit calculation as above.

Exercise. Show that the Poisson distribution (2.10) tends to a Gaussian for $a \to \infty$.

Exercise. A similar verification as above can be done when X takes the values 0, 1 with probabilities α, β, where $\alpha + \beta = 1$ (rather than $\tfrac{1}{2}, \tfrac{1}{2}$) so that p_n is given by (1.5) with $\gamma = \alpha/\beta$. See, for instance, FELLER I, pp. 169 ff. However, we now take the limit $N \to \infty$ and *let at the same time α go to zero*, $\alpha = a/N$, so that $\langle Y \rangle = a$ remains fixed. Show that in this limit the binomial distribution tends to the Poisson distribution.

Various sets of mathematical conditions for the rigorous validity of the central limit theorem can be found in the textbooks.[*] The physicist is better served with a qualitative insight into the scope of its applicability. For this purpose we add a few comments.

Firstly, no smoothness properties of $P(x)$ are required. This is clear from the above example, where

$$P(x) = \tfrac{1}{2}\delta(x) + \tfrac{1}{2}\delta(x-1).$$

On the other hand, a minimal smoothness condition of the characteristic function $G(k)$ is needed, namely that its second derivative at the origin exists. That such a condition cannot be ignored with impunity is demonstrated by the Lorentz distribution. If the variables X_i are independent and have the same Lorentz distribution, then their sum Y is again Lorentzian, according to an Exercise in **6**. Hence it does not approach a Gaussian.

Secondly it is clearly not necessary that all variables X have the same distribution. Suppose one has r_1 variables with one distribution, and r_2 variables with another, and let both r_1 and r_2 tend to infinity with a fixed ratio. Then the sums Y_1 and Y_2 can each be approximated by a Gaussian and the total $Y = Y_1 + Y_2$ is again Gaussian. Its characteristic function is in obvious notation,

$$G_Y(k) = e^{i(\mu_1 r_1 + \mu_2 r_2)k - \tfrac{1}{2}(\sigma_1^2 r_1 + \sigma_2^2 r_2)k^2}.$$

On the other hand, a sequence of variables X_j whose average increases

---

[*] FELLER I, ch. X; B.W. Gnedenko and A.N. Kolmogorov, *Grenzverteilungen von Summen unabhängiger Zufallsgrössen* (Akademie-Verlag, Berlin 1959); M. Loève, *Probability Theory* I (4th Ed., Springer, New York 1977) ch. VI.

7. THE CENTRAL LIMIT THEOREM

indefinitely with j may give rise to a non-Gaussian total Y; an example is easily constructed.

Thirdly it is easy to see that the condition that the X are independent is important. If one takes for all r variables one and the same X the result cannot be true. On the other hand, a sufficiently weak dependence does not harm. This is apparent from the calculation of the Maxwell velocity distribution from the microcanonical ensemble for an ideal gas, see the Exercise in 3. The microcanonical distribution in phase space is a joint distribution that does not factorize, but in the limit $r \to \infty$ the velocity distribution of each molecule is Gaussian. The equivalence of the various ensembles in statistical mechanics is based on this fact.

Exercise. Verify by explicit calculation that for Lorentzian variables not only the proof of the central limit theorem breaks down, but also the result is wrong.

Exercise. In a random walk the steps alternate in length: every other step covers two units (left or right). Find the limiting distribution.

Exercise. Take a sequence of variables X_j ($j = 1, 2, \ldots, r$) with distributions $P_j(x) = f(x - j)$ with fixed f. Show that the central limit property does not apply, but that the variable Z defined by

$$\sum_{j=1}^{r} X_j = \tfrac{1}{2} r(r + 1) + Z$$

does tend to a Gaussian. How can this be seen *a priori*?

Exercise. An example of variables that are not independent is the *random walk with persistence*. Suppose that after a step to the right the probability for the next step is α to the right and β to the left. Similarly a step to the left has a probability α to persist and β to revert. One has $\langle X_j \rangle = 0$, $\langle X_j^2 \rangle = 1$ and $\langle X_j X_{j+1} \rangle = \alpha - \beta \equiv \rho$. One also finds $\langle X_j X_{j+k} \rangle = \rho^k$. Hence $\langle Y \rangle = 0$ and

$$\langle Y^2 \rangle = \sum_{j=1}^{r} \langle X_j^2 \rangle + 2 \sum_{j=2}^{r} \sum_{k=1}^{j-1} \langle X_j X_k \rangle = \frac{1+\rho}{1-\rho} r + \mathcal{O}(r^0). \tag{7.8}$$

It will be found in IV.5 that Y is again Gaussian in the limit.

Remark. The unfortunate name "stable" is used for distributions having the property that the sum of two variables so distributed has again the same distribution (possibly shifted and rescaled as in (5.4)). The general class of stable distributions has been characterized by P. Lévy.[*] The Gauss and the Lorentz distribution are special cases. All these distributions, except, of course, the Gauss distribution itself, violate the central limit theorem. The proof does not apply to them because their variance is infinite. The gamma distributions (5.5) are *not* stable: the sum of two variables with the same gamma distribution has *another* gamma distribution.

---

[*] FELLER II, pp. 165 and 540; E.W. Montroll and J.T. Bendler, J. Statist. Phys. **34**, 129 (1984).

Chapter II

RANDOM EVENTS

This chapter describes a more sophisticated class of random variables, which occur in certain situations in physics and other fields. They can be viewed alternatively as random functions, so that it seemed logical to place this chapter here. On the other hand, for pedagogical reasons it would be better to relegate it to a later stage, since the work is rather advanced, and the results are not needed until chapter XV. Only section **2** should not be skipped.

1. Definition

The counts in a Geiger counter, the arrivals at the anode of a vacuum tube, or of customers at a serving window, are events which may be registered as dots on a time axis. Other examples are the eigenvalues of a random Hermitian matrix on the real line*[)], and the energies of cosmic ray particles marked on the energy scale. The random character of the position of these dots leads to the study of the following kind of stochastic variable, called a "random set of points"**[)], or of events, or a "point process"[†)].

a. The sample space consists of states, each consisting of
(i) a nonnegative integer $s = 0, 1, 2, \ldots$; and
(ii) for each s a set of s real numbers τ_σ obeying

$$-\infty < \tau_1 < t_2 < \cdots < \tau_s < \infty. \tag{1.1}$$

The reader will recognize the analogy with the grand ensemble in statistical mechanics and with the Fock space in field theory.

b. The probability distribution over these states is given by a sequence of nonnegative functions[††)] $Q_s(\tau_1, \tau_2, \ldots, \tau_s)$ defined in the domain (1.1) and

---

*[)] M.L. Mehra, *Random Matrices* (Academic Press, New York 1967 and 1991); M. Carmeli, *Statistical Theory and Random Matrices* (Marcel Dekker, New York 1983); L.E. Reichl, *The Transition to Chaos* (Springer, New York 1992).

**[)] STRATONOVICH, ch. 6. We prefer the word "dots" to distinguish them from "points", which merely mark certain values of *t*. See also A. Ramakrishnan, in: *Encyclopedia in Physics* 3/2 (S. Flügge ed., Springer, Berlin 1959) section 33.

[†)] D.L. Snyder, *Random Point Processes* (Wiley, New York 1975); D.R. Cox and V. Isham, *Point Processes* (Chapman and Hall, London 1980).

[††)] We use the letter Q for this probability density, since P is already overburdened with various meanings. Ramakrishnan, loc. cit., uses the name "Janossy densities"; compare L. Janossy, Proc. Roy. Irish Acad. **53**, 181 (1950).

1. DEFINITION

normalized according to

$$Q_0 + \int_{-\infty}^{\infty} d\tau_1 \, Q_1(\tau_1) + \int_{-\infty}^{\infty} d\tau_1 \int_{\tau_1}^{\infty} d\tau_2 \, Q_2(\tau_1, \tau_2) + \cdots = 1. \quad (1.2)$$

For many purposes it is convenient to eliminate the restriction (1.1) by the following purely algebraic device. One allows each τ_σ to range from $-\infty$ to $+\infty$, but agrees that all $s!$ sets $\{\tau_1, \tau_2, ..., \tau_s\}$ that are the same apart from a permutation correspond to one and the same state. In addition one extends the definition of $Q_s(\tau_1, \tau_2, ..., \tau_s)$ to the whole s-dimensional space by stipulating that it is a symmetric function of its variables. The normalization condition (1.2) may then be written

$$Q_0 + \sum_{s=1}^{\infty} \frac{1}{s!} \int_{-\infty}^{\infty} d\tau_1 \, d\tau_2 \cdots d\tau_s \, Q_s(\tau_1, \tau_2, ..., \tau_s) = 1. \quad (1.3)$$

In this integral two or more τ's might coincide although the value of the Q_s in those points has not been defined. Fortunately, the set of such points is of measure zero in s-dimensional space, so that they do not contribute to the integral, provided it is agreed that Q_s shall not contain delta functions of the type $\delta(\tau_1 - \tau_2)$. Thus we restrict ourselves to situations in which the dots do not have a positive probability to coincide: the serving window must not be the one for marriage licences.

Averages are defined for functions A on the same state space; such a function consists of a sequence

$$\{A_0, A_1(\tau_1), A_2(\tau_1, \tau_2), ..., A_s(\tau_1, \tau_2, ..., \tau_s), ...\}. \quad (1.4)$$

In principle each A_s need only be defined in the region (1.1), but in order to utilize the extension to the whole s-space we again extend the definition of A_s by stipulating its symmetry. Then

$$\langle A \rangle = A_0 Q_0 + \sum_{s=1}^{\infty} \frac{1}{s!} \int A_s(\tau_1, \tau_2, ..., \tau_s) \, Q_s(\tau_1, \tau_2, ..., \tau_s) \, d\tau_1 \, d\tau_2 \cdots d\tau_s.$$
$$(1.5)$$

Thus the averaging brackets $\langle \ \rangle$ imply a sum over s as well as, for each s, an s-fold integral.

As an example of such a state function take the number N of dots in a given interval (t_a, t_b). To write it in the form (1.4) we define the *indicator*[*]
$\chi(t)$ of the interval by setting $\chi(t) = 1$ for $t_a < t < t_b$, and $\chi(t) = 0$ for other t. The physical quantity N is then represented by the sequence

$$N \to \{0, \chi(\tau_1), \chi(\tau_1) + \chi(\tau_2), \chi(\tau_1) + \chi(\tau_2) + \chi(\tau_3), ...\}.$$

---

[*] The older name "characteristic function" would be confusing in the present context.

Its average is

$$\langle N \rangle = \left\langle \sum_{\sigma=1}^{s} \chi(\tau_\sigma) \right\rangle$$

$$= \sum_{s=1}^{\infty} \frac{1}{s!} \int_{-\infty}^{\infty} d\tau_1 \cdots d\tau_s \, Q_s(\tau_1, \ldots, \tau_s) \sum_{\sigma=1}^{s} \chi(\tau_\sigma)$$

$$= \sum_{s=1}^{\infty} \frac{1}{(s-1)!} \int_{-\infty}^{\infty} \chi(\tau_1) \, d\tau_1 \int_{-\infty}^{\infty} d\tau_2 \cdots d\tau_s \, Q_s(\tau_1, \ldots, \tau_s)$$

$$= \sum_{s=1}^{\infty} \frac{1}{(s-1)!} \int_{t_a}^{t_b} d\tau_1 \int_{-\infty}^{\infty} d\tau_2 \cdots d\tau_s \, Q_s(\tau_1, \ldots, \tau_s). \quad (1.6)$$

The fact that the result is not particularly simple and involves a sum over all Q_s is the reason why in section 3 another way of describing random dots will be developed, which is more adapted to computing such averages.

Exercise. The transition from the limited domain (1.1) to the full domain with symmetric Q_s is especially convenient, if not indispensable, for generalizing the description to random dots on a plane or in space. Write explicitly the functions Q_s for the grand-canonical ensemble of an ideal gas in a fixed volume.

Exercise. Show that the mean square of the number N of dots in the interval (t_a, t_b) is given by

$$\langle N^2 \rangle = \langle N \rangle + \sum_{s=2}^{\infty} \frac{1}{(s-2)!} \int_{t_a}^{t_b} d\tau_1 \int_{t_a}^{t_b} d\tau_2 \int_{-\infty}^{\infty} d\tau_3 \cdots d\tau_s \, Q_s. \quad (1.7)$$

Exercise. A random set of dots on $(0, \infty)$ is constructed according to the following recipe. The probability for the *first* dot to lie in $(\tau_1, \tau_1 + d\tau_1)$ is $w(\tau_1) d\tau_1$, where w is a given nonnegative function with

$$\int_0^{\infty} w(\tau) \, d\tau = \frac{1}{\gamma} < 1. \quad (1.8)$$

The probability density for the second dot is $w(\tau_2 - \tau_1)$ and so on. Calculate the Q_s.

Exercise. Generalize the description to random sets of two (or more) different species of dots ("marked dots").

Exercise. Suppose the dots have a non-zero probability to coincide in pairs. This may be described as a case of two species, namely singles and doubles. Show that the corresponding two-species distribution can be re-arranged as a one-species distribution Q_s, which now does involve delta functions.

Exercise. The objects (1.4) form a linear vector space. Let the scalar product be defined with a weight function $1/s!$, so that (1.5) is the scalar product (A, Q). Write (1.3) and (1.7) as scalar products.

Exercise. Let p be a probability distribution on $(0, \infty)$. Let $\tau_1, \tau_2, \ldots, \tau_N$ be N independent random quantities, each with distribution p. Let x denote the smallest of these N numbers.*⁾ Then the probability distribution of x is

$$-\frac{d}{dx}\left[\int_x^\infty p(\tau)\,d\tau\right]^N.$$

Consider the special case $p(\tau) = e^{-\tau}$.

2. The Poisson distribution

The random dots are called *independent* when each Q_s factorizes in the form

$$Q_s(\tau_1, \tau_2, \ldots, \tau_s) = e^{-\nu} q(\tau_1) q(\tau_2) \cdots q(\tau_s), \quad Q_0 = e^{-\nu}. \tag{2.1}$$

q is some nonnegative integrable function and the normalizing constant $e^{-\nu}$ is determined by

$$\nu = \int_{-\infty}^\infty q(\tau)\,d\tau. \tag{2.2}$$

In this case one easily finds from (**1.6**) for the average of the number N in the interval (t_a, t_b)

$$\langle N \rangle = \int_{t_a}^{t_b} q(\tau)\,d\tau. \tag{2.3}$$

Also from (**1.7**) one finds for the mean square

$$\langle N^2 \rangle = \langle N \rangle^2 + \langle N \rangle. \tag{2.4}$$

It is even possible to compute the explicit probability distribution of N. Its characteristic function is

$$\langle e^{ikN} \rangle = \left\langle \exp ik \sum_{\sigma=1}^s \chi(\tau_\sigma) \right\rangle$$

$$= e^{-\nu} \sum_{s=0}^\infty \frac{1}{s!} \left(\int_{-\infty}^\infty e^{ik\chi(\tau)} q(\tau)\,d\tau \right)^s$$

$$= \exp\left[\int_{-\infty}^\infty (e^{ik\chi(\tau)} - 1) q(\tau)\,d\tau \right]$$

$$= \exp\left[(e^{ik} - 1) \int_{t_a}^{t_b} q(\tau)\,d\tau \right]$$

---

*⁾ See F.C. Hoppensteadt, *An Introduction to the Mathematics of Neurons* (Cambridge University Press, Cambridge 1986) p. 112.

$$= \exp\left[(e^{ik} - 1)\langle N \rangle\right]$$

$$= e^{-\langle N \rangle} \sum_{N=0}^{\infty} \frac{\langle N \rangle^N}{N!} e^{ikN}. \tag{2.5}$$

Comparing the coefficients of e^{ikN} on both sides of this equation one finds: *The probability that exactly N of the independent random dots fall in the given interval is*

$$p_N = \frac{\langle N \rangle^N}{N!} e^{-\langle N \rangle}. \tag{2.6}$$

This is how the Poisson distribution, previously mentioned in (I.2.10), enters into physics. It generally describes the probability distribution of the number of independent events found in a limited region, such as the raindrops hitting a given tile or radioactive decays during a given period. It is determined by a single parameter, which is the average and, according to (2.4), also the variance. The Poisson distribution for nonnegative integers enjoys the same popularity as the Gauss distribution for the continuous range $(-\infty, \infty)$. Yet it should never be regarded as universal, since it is derived only for *independent* events. In VII.3 we shall see that in chemical reactions it is not always true. Nor is it universally true that the variance of a random number of particles is equal to the average of that number. Nevertheless as a useful rule of thumb one may often guess that the variance is of the same order as the average.

As a further specialization suppose that $q(\tau)$ is constant in an interval $(-T, T)$ and vanishes outside it. The constant is necessarily equal to $v/2T \equiv \rho$ and represents the average number of events per unit time. In the limit $T \to \infty$, $v \to \infty$, with fixed ρ, one approaches a stationary distribution of dots, called *shot noise*.[*)] The fact that stationary distributions can only be described by a limiting process is another drawback of the present treatment of random dots which will be overcome in the next section.

Exercise. Verify that (2.2) is the correct normalization and show that v is the average of the total number of dots, $v = \langle s \rangle$.

Exercise. Generalize the formulas to independent random dots in three-dimensional space. Show that the number N of dots in an arbitrary region is again distributed according to (2.6).

Exercise. Generalize the formalism to more than one species of dots and show that the multivariate analog of the Poisson distribution is merely a product of single-variable distributions (2.6).

[*)] Occasionally this name is applied to non-stationary sets of independent events as well. The precise definition of "stationary" is given in (3.13).

Exercise. When N_1 and N_2 are two statistically independent variables, each with a Poisson distribution, their sum $N_1 + N_2$ is again Poissonian.

Exercise. Suppose there are N primary individuals, where N is random with a Poisson distribution. Each primary produces M secondaries, where M is again Poissonian. Find an expression for the distribution of the total number of individuals.

Exercise. Consider a superposition of Poisson distributions:

$$p_n = \int_0^\infty \phi(\alpha) \frac{\alpha^n}{n!} e^{-\alpha} \, d\alpha, \tag{2.7}$$

where $\phi(\alpha)$ is itself a probability distribution. Derive

$$\langle N \rangle_p = \langle \alpha \rangle_\phi, \qquad \langle N^2 \rangle_p - \langle N \rangle_p = \langle \alpha^2 \rangle_\phi.$$

Thus the variance is always greater than that of the pure Poisson distribution with the same average. Also express the probability generating function of p_n in the characteristic function of $\phi(\alpha)$ and conclude that the moments of α are equal to the factorial moments of n; compare (I.2.15).

Exercise. Any distribution p_n can be represented by (2.7) if one drops the condition that ϕ should be a probability density and permits other integration paths.[*]

3. Alternative description of random events

Equations (1.6) and (1.7) for the number of dots in a given interval suggest that it is useful to define a sequence of functions $f_n(t_1, t_2, \ldots, t_n)$ for $n = 1, 2, \ldots$ by setting

$$f_1(t_1) = \sum_{s=1}^\infty \frac{1}{(s-1)!} \int_{-\infty}^\infty d\tau_2 \cdots d\tau_s \, Q_s(t_1, \tau_2, \ldots, \tau_s),$$

$$f_2(t_1, t_2) = \sum_{s=2}^\infty \frac{1}{(s-2)!} \int_{-\infty}^\infty d\tau_3 \cdots d\tau_s \, Q_s(t_1, t_2, \tau_3, \ldots, \tau_s),$$

$$\cdots\cdots\cdots\cdots\cdots\cdots\cdots\cdots\cdots\cdots\cdots\cdots\cdots\cdots$$

$$f_n(t_1, t_2, \ldots, t_n) = \sum_{s=n}^\infty \frac{1}{(s-n)!} \int_{-\infty}^\infty d\tau_{n+1} \cdots d\tau_s \, Q_s(t_1, t_2, \ldots, t_n, \tau_{n+1}, \ldots, \tau_s).$$

$$\tag{3.1}$$

One may then write (1.6) and (1.7) in the simpler form

$$\langle N \rangle = \int_{t_a}^{t_b} f_1(t_1) \, dt_1, \tag{3.2}$$

$$\langle N^2 \rangle = \langle N \rangle + \int_{t_a}^{t_b} f_2(t_1, t_2) \, dt_1 \, dt_2. \tag{3.3}$$

---

[*] C.W. Gardiner and S. Chaturvedi, J. Statist. Phys. **17**, 429 (1977) and **18**, 501 (1978).

The intuitive meaning of these functions f_n can be gleaned from the defining equations (3.1):

$f_n(t_1, t_2, \ldots, t_n) \, dt_1 \, dt_2 \cdots dt_n =$ the probability that the intervals $(t_1, t_1 + dt_1)$ and $(t_2, t_2 + dt_2)$, etc., each contain a dot, regardless of how many there may be outside these intervals.

The probability that one of these intervals contains two or more dots is negligibly small, because we agreed not to admit situations in which the Q_s involve delta functions. This agreement also has the effect that no delta functions occur in the f_n. Again there is no need to assign a value to the f_n for coinciding arguments.

The following properties of the f_n are obvious.
(i) $f_n \geq 0$;
(ii) f_n is symmetric in its arguments $t_1, t_2, \ldots, t_n$.

The f_n obey no normalization, however, and are therefore themselves not probability densities. This may seem a bit confusing inasmuch as they were interpreted as probabilities. But their probability interpretation only applies to *infinitesimal* intervals, or at least intervals so small that there is no appreciable chance to find two or more dots in them. For a longer interval (t_a, t_b) one sees from (3.2) that the integrated f_1 is the average number of dots, *not* the total probability of finding a dot in the interval. Thus f_1 represents the average density of dots. Similarly the higher f_n are something like "average joint densities". Only for infinitesimal intervals do they reduce to probabilities. Stratonovich[*] called them *distribution functions*, but to avoid misunderstanding we prefer to refer to them as "the functions f_n".

Exercise. Compute the f_n for independent dots and show that

$$f_n(t_1, t_2, \ldots, t_n) = f_1(t_1) f_1(t_2) \cdots f_1(t_n). \tag{3.4}$$

Exercise. Suppose the arrivals of photons in a counter constitute a random set with known stochastic properties. Each photon has a probability α to be counted. Express the f_n for the counting events in those for the arrivals.

Exercise. The basic reason why the f_n are more useful than the Q_n for describing dots is the following. Most quantities A in whose average one is interested are "sum functions", i.e., they consist of a single-particle function $a(\tau_\sigma)$ summed over all particles, or of a pair function $a(\tau_\sigma, \tau_{\sigma'})$ summed over all pairs; etc.[**] In general,

$$A = \sum_{\sigma_1, \sigma_2, \ldots, \sigma_n} a(\tau_{\sigma_1}, \tau_{\sigma_2}, \ldots, \tau_{\sigma_n}),$$

---

[*] STRATONOVICH, ch. 6.
[**] The importance of this fact for statistical mechanics was stressed by A.J. Khinchin, *Mathematical Foundations of Statistical Mechanics* (G. Gamow, transl., Dover Publications, New York 1949) p. 63. But he called A a sum function only if $n = 1$.

3. ALTERNATIVE DESCRIPTION OF RANDOM EVENTS

where a is a function of only a small number n of particles. Examples are (3.2) and (3.3). Show that the average of A involves only $f_1, f_2, \ldots, f_n$ and no higher distribution functions. Under which condition does it involve f_n alone?

Suppose an apparatus reacts on the arrival of photons or electrons with a sensitivity which, owing to external influences, depends on time. Let the response to an arrival at τ be $u(\tau)$. (It is essential, however, that the response is not affected by the arrival of other particles: no dead time or recovery time.) When s arrivals occur at times $\tau_1, \tau_2, \ldots, \tau_s$ the total output is $U = \Sigma_{\sigma=1}^{s} u(\tau_\sigma)$. When the arrivals are random the average total output is, according to (1.5)

$$\left\langle \sum_{\sigma=1}^{s} u(\tau_\sigma) \right\rangle = \sum_{s=1}^{\infty} \frac{1}{(s-1)!} \int_{-\infty}^{\infty} d\tau_1 \cdots d\tau_s \, Q_s(\tau_1, \ldots, \tau_s) u(\tau_1)$$

$$= \int_{-\infty}^{\infty} u(t_1) f_1(t_1) \, dt_1. \tag{3.5}$$

The mean square of U is

$$\left\langle \sum_\sigma u(\tau_\sigma) \sum_{\sigma'} u(\tau_{\sigma'}) \right\rangle = \left\langle \sum_\sigma u(\tau_\sigma)^2 \right\rangle + \left\langle \sum_{\sigma \neq \sigma'} u(\tau_\sigma) u(\tau_{\sigma'}) \right\rangle$$

$$= \int u(t_1)^2 f_1(t_1) \, dt_1 + \int\int u(t_1) u(t_2) f_2(t_1, t_2) \, dt_1 \, dt_2.$$

It is possible to continue to higher moments, and in fact, to compute in this way the entire characteristic function of U. However, the same result is obtained in an easier way when we first introduce the generating function for the f_n.

In chapter I it was shown how the handling of moments and cumulants was facilitated by the use of a moment generating function. A similar tool will now be introduced with respect to the f_n. Instead of the auxiliary variable k we now need an auxiliary *function*, or "test function" $v(t)$, and instead of a generating function we have therefore a *functional*, i.e., a quantity depending on all the values that v takes for $-\infty < t < \infty$ (indicated by $[v]$). The *generating functional* for the f_n is

$$L([v]) = \left\langle \prod_{\sigma=1}^{s} \{1 + v(\tau_\sigma)\} \right\rangle. \tag{3.6}$$

II. RANDOM EVENTS

Working out the product one obtains

$$L([v]) = 1 + \left\langle \sum_\sigma v(\tau_\sigma) \right\rangle + \left\langle \sum_{\sigma < \sigma'} v(\tau_\sigma) v(\tau_{\sigma'}) \right\rangle + \cdots$$

$$= 1 + \int v(t_1) f_1(t_1) \, dt_1 + \tfrac{1}{2} \int v(t_1) v(t_2) \, f_2(t_1, t_2) \, dt_1 \, dt_2 + \cdots$$

$$= 1 + \sum_{n=1}^\infty \frac{1}{n!} \int v(t_1) v(t_2) \cdots v(t_n) \, f_n(t_1, t_2, \ldots, t_n) \, dt_1 \, dt_2 \cdots dt_n. \tag{3.7}$$

It appears that the f_n are the coefficients in the expansion of L in powers of the test function v, so that the knowledge of $L([v])$ for all functions v uniquely determines all f_n.

To find a corollary to this result we express v in another function u by

$$v(t) = e^{iu(t)} - 1.$$

Then the identity (3.7) states

$$\left\langle \exp\!\left[i \sum_\sigma u(\tau_\sigma) \right] \right\rangle = 1 + \sum_{n=1}^\infty \frac{1}{n!} \int \{e^{iu(t_1)} - 1\} \cdots \{e^{iu(t_n)} - 1\}$$

$$\times f_n(t_1, \ldots, t_n) \, dt_1 \cdots dt_n. \tag{3.8}$$

Exercise. Show that this result leads immediately to an expression for the characteristic function $\langle e^{ikU} \rangle$ in terms of the f_n, where U is the same as above.

Exercise. Verify (3.7) and (3.8) for independent dots.

Exercise. Prove (3.8) by brute force, i.e., expand the left-hand side in a Taylor series and express each term as a sum involving a number of f_n.

Exercise. Let $\{t_\mu\}$ be a set of m time points. Then the $m \times m$ matrix $f_2(t_\mu, t_\nu)$ is positive definite, or at least nonnegative.

Exercise. Derive for the characteristic function of the number N of dots in a given interval (t_a, t_b)

$$\langle e^{ikN} \rangle = 1 + \sum_{n=1}^\infty \frac{(e^{ik} - 1)^n}{n!} \int_{t_a}^{t_b} f_n(t_1, t_2, \ldots, t_n) \, dt_1 \, dt_2 \cdots dt_n. \tag{3.9}$$

Hence the probability for having no dots in (t_a, t_b) is

$$p_0(t_a, t_b) = 1 + \sum_{n=1}^\infty \frac{(-1)^n}{n!} \int_{t_a}^{t_b} f_n(t_1, t_2, \ldots, t_n) \, dt_1 \, dt_2 \cdots dt_n \tag{3.10}$$

$$= L([-\chi]),$$

where χ is the indicator of (t_a, t_b).

Exercise. Take v non-overlapping intervals and express the characteristic function $G(k_1, k_2, \ldots, k_v)$ of the joint distribution function of their occupation numbers $N_1, N_2, \ldots, N_v$ in terms of the f_n.

Exercise. Prove the following identity for the functional derivatives of L,

$$\frac{\delta^r L}{\delta v(\theta_1) \cdots \delta v(\theta_r)} = \sum_{n=0}^{\infty} \frac{1}{n!} \int v(t_1) \cdots v(t_n) f_{r+n}(\theta_1, \ldots, \theta_r, t_1, \ldots, t_n) \, dt_1 \cdots dt_n. \quad (3.11)$$

On substituting $v \equiv 0$ this expression reduces to $f_r(\theta_1, \ldots, \theta_r)$.

Exercise. The relation (3.1) between the Q_s and the f_n is a linear mapping in the vector space of objects of the form (1.4). The matrix Ω of this mapping is given by

$$(n; t_1, t_2, \ldots, t_n | \Omega | s; \tau_1, \tau_2, \ldots, \tau_s) = \frac{1}{(s-n)!} \delta(t_1 - \tau_1) \delta(t_2 - \tau_2) \cdots \delta(t_n - \tau_n), \quad (3.12)$$

with the understanding that $(s-n)!^{-1} = 0$ when $s < n$.

Although all physical processes are, of course, of limited duration there are many situations in which this limitation is not of practical interest. When studying the noise in an electronic device one normally is not interested in effects connected with the switching on and off. A description in which the duration does not enter is then simpler and more appropriate. One is thus led to examining sets of dots whose density does not tend to zero for $t \to \pm \infty$. Such sets cannot be described in terms of the Q_s, because the normalization condition (1.3) requires Q_s to vanish at infinity. It is true that this shortcoming can be overcome by introducing a fictitious long time interval T, but that burdens the equations with an irrelevant quantity. The description in terms of the f_n, however, carries over without additional artifice.

Of special interest are random events whose stochastic properties do not change with time. They are called *stationary*, defined by

$$f_n(t_1 + \tau, t_2 + \tau, \ldots, t_n + \tau) = f_n(t_1, t_2, \ldots, t_n) \quad \text{(all } n, t_j, \tau\text{)}. \quad (3.13)$$

In particular, the density of events f_1 is constant. As mentioned in **2**, when the set is both stationary and independent it is called *shot noise*. In that case one has according to (3.4)

$$f_n(t_1, t_2, \ldots, t_n) = (f_1)^n. \quad (3.14)$$

Hence shot noise is fully determined by a single parameter, viz., its density. The alternative name "Poisson process" indicates that it may be regarded as a stochastic process, as will be seen in IV.2.

Exercise. Compute $L([v])$ for the case of shot noise.
Exercise. Apply the results (3.9) and (3.10) to shot noise.

Exercise. A cathode is heated by a varying current such as to have a probability $\phi(\tau)\,d\tau$ to emit an electron in a time interval $(\tau, \tau + d\tau)$, regardless of other emitted electrons. Find the functions f_n describing these emission events.

4. The inverse formula

Equation (3.1) defined the successive f_n as linear combinations of the Q_s. We shall now derive the inverse relations (for $s \geq 1$)

$$Q_s(\tau_1, \tau_2, \ldots, \tau_s) = \sum_{n=s}^{\infty} \frac{(-1)^{n-s}}{(n-s)!} \int dt_{s+1} \cdots dt_n \, f_n(\tau_1, \tau_2, \ldots, \tau_s, t_{s+1}, \ldots, t_n). \tag{4.1}$$

For that purpose first note that if one inserts into the definition (3.6) of L the test function $v(\tau) = w(\tau) - 1$ one gets

$$L([w-1]) = \left\langle \prod_{\sigma=1}^{s} w(\tau_\sigma) \right\rangle.$$

It is understood that for $s = 0$ the product is taken to be unity. Then according to (1.5)

$$L([w-1]) = Q_0 + \sum_{s=1}^{\infty} \frac{1}{s!} \int w(\tau_1) w(\tau_2) \cdots w(\tau_s)$$
$$\times Q_s(\tau_1, \tau_2, \ldots, \tau_s) \, d\tau_1 \, d\tau_2 \cdots d\tau_s. \tag{4.2}$$

Thus L, as a functional of the test function w, is the generating functional of the Q_s.

On the other hand, one has from (3.7)

$$L([w-1]) = 1 + \sum_{n=1}^{\infty} \frac{1}{n!} \int \{w(t_1) - 1\} \cdots \{w(t_n) - 1\} f_n(t_1, \ldots, t_n) \, dt_1 \cdots dt_n.$$

From this one obtains Q_s by collecting all terms with s factors w. It is readily seen that the result is (4.1).[*]

Exercise. Verify the result explicitly for the case of independent dots.
Exercise. Find the expression for Q_0, which is needed to complete the inverse relations (4.1).
Exercise. The relation (3.1) maps the vectors Q_s linearly onto the vectors f_n. The vector $(Q_0 = 1, 0, 0, \ldots)$ is mapped into the null vector $(f_1 = 0, 0, \ldots)$. Yet the mapping has an inverse. How is this paradox resolved?

---

[*] This replaces the two-page proof in the first edition.

Example. A photon traverses a medium in which it has a probability β per unit time to create a secondary photon by stimulated emission. One wants to know the probability distribution p_s of the number s of secondaries. The probability that n secondaries are created at $t_1, t_2, \ldots, t_n$, regardless of what happens at other times, is

$$f_n(t_1, t_2, \ldots, t_n) = \beta^n.$$

Using (4.1) one obtains

$$Q_s(\tau_1, \tau_2, \ldots, \tau_s) = \beta^s \, e^{-\beta T},$$

where T is the traversal time. Hence

$$p_s = \frac{1}{s!} \int_0^T Q_s(\tau_1, \ldots, \tau_s) \, d\tau_1 \cdots d\tau_s = \frac{(\beta T)^s}{s!} e^{-\beta T}.$$

This result is obvious when one notices that the emission events are statistically independent.

Next suppose that the medium is infinite, but that the primary photon has a probability α per unit time to be absorbed. Then

$$f_n(t_1, t_2, \ldots, t_n) = \beta^n \exp[-\alpha \, \text{Max} \, t_\nu].$$

The same calculation now yields

$$p_s = \frac{1}{s!} \sum_{n=0}^{\infty} \frac{(-1)^n}{n!} (n+s)! \left(\frac{\beta}{\alpha}\right)^{n+s} = \frac{(\beta/\alpha)^s}{(1+\beta/\alpha)^{s+1}}.$$

The corresponding probability generating function is

$$F(z) = \sum_{s=0}^{\infty} z^s p_s = \frac{\alpha}{\alpha + \beta - \beta z}. \tag{4.3}$$

The total number of photons produced in the cascade (secondaries, tertiaries, etc.) will be computed in (III.6.11).

5. The correlation functions

In addition to the distribution functions f_n a second sequence of symmetric functions g_m, called *correlation functions*, will be useful. They are defined in terms of the f_n by the following cluster expansion:

$$f_1(t_1) = g_1(t_1),$$
$$f_2(t_1, t_2) = g_1(t_1)g_1(t_2) + g_2(t_1, t_2),$$
$$f_3(t_1, t_2, t_3) = g_1(t_1)g_1(t_2)g_1(t_3)$$
$$\qquad + g_1(t_1)g_2(t_2, t_3) + g_1(t_2)g_2(t_1, t_3) + g_1(t_3)g_2(t_1, t_2)$$
$$\qquad + g_3(t_1, t_2, t_3),$$
$$\cdots\cdots\cdots\cdots\cdots\cdots\cdots\cdots\cdots\cdots\cdots\cdots\cdots\cdots\cdots\cdots \tag{5.1}$$

The general expansion for f_n is obtained by the following rule.

(i) Subdivide the variables $t_1, t_2, \ldots, t_n$ in all possible different ways in subsets (not counting the empty set, but including the total set as one particular subdivision).

(ii) For each subdivision take the product of the functions g for the separate subsets.

(iii) Add these products over all subdivisions.

For example, f_6 will contain the term (see fig. 2)

$$g_1(t_6)g_2(t_1, t_3)g_3(t_2, t_4, t_5). \tag{5.2}$$

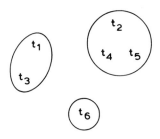

Fig. 2. The term (5.2) in the cluster expansion of f_6.

The following equivalent formulation of the rule will be used presently.

(i)' Partition the integer n, i.e., write n as a sum of positive integers, not necessarily different. To put it differently, choose nonnegative integers k such that

$$k_1 \cdot 1 + k_2 \cdot 2 + k_3 \cdot 3 + \cdots + k_n \cdot n = n.$$

(By this condition all k's beyond k_n are automatically zero.)

(ii)' Form the product of k_1 factors g_1, k_2 factors g_2, etc.,

$$g_1(t_1)g_1(t_2) \cdots g_1(t_{k_1})g_2(t_{k_1+1}, t_{k_1+2}) \cdots$$

(iii)' Construct all different terms that, given the partition, can be obtained from this by permuting the variables t. Terms are not rated as different if they merely differ by the order of variables inside the individual g's, or by the order of factors g. The number of terms obtained in this way is

$$\frac{n!}{k_1! k_2! \cdots k_n! (1!)^{k_1} (2!)^{k_2} \cdots (n!)^{k_n}}. \tag{5.3}$$

(iv)' Sum over all these terms and subsequently over all partitions of N:

$$f_n(t_1, t_2, \ldots, t_n) = \sum_{\text{partitions}} \sum_{\text{permutations}} g_1^{(k_1)} g_2^{(k_2)} \cdots g_n^{(k_n)}. \tag{5.4}$$

5. THE CORRELATION FUNCTIONS

The abbreviated notation is meant to indicate that there are k_1 factors g_1 with different arguments, k_2 factors g_2, and so on.

We shall now prove the fundamental identity

$$1 + \sum_{n=1}^{\infty} \frac{1}{n!} \int v(t_1)v(t_2) \cdots v(t_n) f_n(t_1, t_2, \ldots, t_n) \, dt_1 \, dt_2 \cdots dt_n$$

$$= \exp\left[\sum_{m=1}^{\infty} \frac{1}{m!} \int v(t_1)v(t_2) \cdots v(t_m) g_m(t_1, t_2, \ldots, t_m) \, dt_1 \, dt_2 \cdots dt_m \right]. \quad (5.5)$$

Substitute in the first line the f_n as given by (5.4). The various terms obtained under (iii)', belonging to the same partition, give the same contribution to the integral. Hence the first line becomes

$$1 + \sum_{n} \sum_{\text{partitions}} \frac{1}{k_1!} \left\{ \int v(t) g_1(t) \, dt \right\}^{k_1}$$

$$\times \frac{1}{k_2!} \left\{ \frac{1}{2!} \int v(t')v(t'') g_2(t', t'') \, dt' \, dt'' \right\}^{k_2} \cdots . \quad (5.6)$$

The sum extends over all values of n, and for each individual value over all its partitions. But every set of integers k is a partition of some n. Vice versa, all partitions of all n are obtained by assigning to each k in (5.6) a nonnegative integer value. The only exception is the case that all k's vanish simultaneously. Hence the sum may be written as a multiple sum over all values of the k's; the one missing term is supplied by the 1 in front. Thus we find for the first line in (5.5)

$$\sum_{k_1=0}^{\infty} \frac{1}{k_1!} \left\{ \int v(t) g_1(t) \, dt \right\}^{k_1} \sum_{k_2=0}^{\infty} \frac{1}{k_2!} \left\{ \frac{1}{2!} \int v(t')v(t'') g_2(t', t'') \, dt' \, dt'' \right\}^{k_2} \cdots .$$

This is identical with the second line of (5.5); Q.E.D.

The first line of (5.5) is the generating functional (3.7) of the f_n. Hence (5.5) may also be written

$$\log L[v] = \sum_{m=1}^{\infty} \frac{1}{m!} \int v(t_1)v(t_2) \cdots v(t_m) g_m(t_1, t_2, \ldots, t_m) \, dt_1 \, dt_2 \cdots dt_m. \quad (5.7)$$

This shows that $\log L$ is the generating functional of the g_m, just as the cumulants were generated by the logarithm of the moment generating function. Of course, one may *define* the g_m by (5.7) and then prove that they obey (5.1).

The following property (which can easily be proved) is the main reason for introducing the correlation functions: if the dots are independent all g_m for $m > 1$ vanish. When the physical situation suggests that the dots are almost independent one may expect the g_m to decrease rapidly in magnitude, but this is a physical rule of thumb rather than a mathematical truth.

In many situations one expects the dots to be statistically dependent only over short time intervals. A formal expression of this "cluster property" is

$$\lim_{\tau \to \infty} g_{m+m'}(t_1, t_2, \ldots, t_m, t_{m+1} + \tau, t_{m+2} + \tau, \ldots, t_{m+m'} + \tau) = 0 \quad (5.8)$$

(all $m, m', t_1, \ldots, t_{m+m'}$). The same property expressed in the f_n reads

$$\lim_{\tau \to \infty} [f_{n+n'}(t_1, t_2, \ldots, t_n, t_{n+1} + \tau, t_{n+2} + \tau, \ldots, t_{n+n'} + \tau)$$
$$- f_n(t_1, t_2, \ldots, t_n) f_{n'}(t_{n+1} + \tau, t_{n+2} + \tau, \ldots, t_{n+n'} + \tau)] = 0 \quad (5.9)$$

(all $n, n', t_1, \ldots, t_{n+n'}$) and is sometimes called the "product property".

Exercise. Verify (5.5) for independent dots.
Exercise. Show that (5.1) and (5.4) are equivalent.
Exercise. Using the result (3.9) show that the characteristic function for the number of N of dots in the interval (t_a, t_b) is

$$\langle e^{ikN} \rangle = \exp\left[\sum_{m=1}^{\infty} \frac{(e^{ik} - 1)^m}{m!} \int_{t_a}^{t_b} g_m(t_1, t_2, \ldots, t_m) \, dt_1 \, dt_2 \cdots dt_m \right]. \quad (5.10)$$

In particular, the probability for having no dots in the interval is

$$p_0(t_a, t_b) = \exp\left[\sum_{m=1}^{\infty} \frac{(-1)^m}{m!} \int_{t_a}^{t_b} g_m(t_1, t_2, \ldots, t_m) \, dt_1 \, dt_2 \cdots dt_m \right]. \quad (5.11)$$

6. Waiting times

Suppose a random set of dots representing a sequence of events is given. The following question may be asked. If I start observing at some time t_0, how long do I have to wait for the next event to occur? Of course, the time θ from t_0 to the next event is a random variable with values in $(0, \infty)$ and the quantity of interest is its probability density, $w(\theta; t_0)$ (which depends parametrically on t_0 unless the random set of events is stationary). This question is of particular interest in queuing problems. The function $w(\theta; t_0)$ has also been measured electronically for the arrivals of photons produced by luminescence.

Our aim is to express $w(\theta; t_0)$ in terms of the quantities that specify the random set. Let $p_0(t_0, t_0 + \theta)$ denote the probability that no event occurs between t_0 and $t_0 + \theta$. Then the probability $w(\theta; t_0) \, d\theta$ that the first event after t_0 occurs between $t_0 + \theta$ and $t_0 + \theta + d\theta$ is $p_0(t_0, t_0 + \theta) -$

$p_0(t_0, t_0 + \theta + d\theta)$. Hence

$$w(\theta; t_0) = -\frac{d}{d\theta} p_0(t_0, t_0 + \theta). \tag{6.1}$$

Now $p_0(t_0, t_0 + \theta)$ has been expressed in terms of the distribution functions by (3.10). Substituting that result one finds

$$w(\theta; t_0) = f_1(t_0 + \theta) + \sum_{n=1}^{\infty} \frac{(-1)^n}{n!} \int_{t_0}^{t_0+\theta} f_{n+1}(t_0 + \theta, t_1, \ldots, t_n) \, dt_1 \cdots dt_n. \tag{6.2}$$

This expresses the waiting time distribution in terms of the sequence of distribution functions f_n of the random set.

Exercise. Find $w(\theta; t_0)$ for independent events. In particular, for shot noise

$$w(\theta; t_0) = f_1 \exp[-\theta f_1]. \tag{6.3}$$

Exercise. Derive the following expression for $w(\theta; t_0)$ in terms of the correlation functions

$$w(\theta; t_0) = p_0(t_0, t_0 + \theta)\bigg[g_1(t_0 + \theta)$$
$$+ \sum_{m=1}^{\infty} \frac{(-1)^m}{m!} \int_{t_0}^{t_0+\theta} g_{m+1}(t_0 + \theta, t_1, t_2, \ldots, t_m) \, dt_1 \, dt_2 \cdots dt_m \bigg]. \tag{6.4}$$

Exercise. Show

$$w(\theta; t_0) = \left[\frac{\delta L([v])}{\delta v(t_0 + \theta)} \right]_{v = -\chi}; \tag{6.5}$$

the right-hand side means that first the functional derivative with respect to v at $t_0 + \theta$ is taken, and afterwards for v the negative indicator of $(t_0, t_0 + \theta)$ is substituted.

Another question may be asked: suppose *I have observed an event at* t_a, what is the probability distribution $w(\theta | t_a)$ of the time I have to wait until the next event? The joint probability for having one event between $t_a - dt_a$ and t_a, and one in $(t_b, t_b + dt_b)$ with no events between them is

$$[p_0(t_a, t_b) - p_0(t_a, t_b + dt_b)] - [p_0(t_a - dt_a, t_b) - p_0(t_a - dt_a, t_b + dt_b)]$$
$$= -dt_a \, dt_b \frac{\partial^2 p_0(t_a, t_b)}{\partial t_a \partial t_b}. \tag{6.6}$$

The conditional probability for having an event in $(t_b, t_b + dt_b)$, knowing that there was one in $(t_a - dt_a, t_a)$, is obtained on dividing (6.6) by the probability $f_1(t_a) \, dt_a$ for an event to occur in $(t_a - dt_a, t_a)$ (Bayes' rule). Thus the probability distribution for the waiting time $t_b - t_a$ after the event t_a is

$$w(t_b - t_a | t_a) = -\frac{1}{f_1(t_a)} \frac{\partial^2 p_0(t_a, t_b)}{\partial t_a \partial t_b}. \tag{6.7}$$

With the aid of (3.10) this can be expressed in the distribution functions f_n,

$$w(t_b - t_a | t_a) = \frac{1}{f_1(t_a)} \Bigg[f_2(t_a, t_b)$$

$$+ \sum_{n=1}^{\infty} \frac{(-1)^n}{n!} \int_{t_a}^{t_b} f_{n+2}(t_a, t_b, t_1, \ldots, t_n) \, dt_1 \cdots dt_n \Bigg]. \tag{6.8}$$

Remark. The following alternative derivation provides further insight. The random set of dots on the time axis can be visualized as an ensemble of numerous individual sample sets. From this ensemble extract the subensemble of those sample sets that have one dot between $t_a - dt_a$ and t_a. This subensemble represents again a random set of dots. The quantities belonging to this new random set will be distinguished by a tilde. The required waiting time distribution $w(\theta | t_a)$ is the same as the quantity $\tilde{w}(\theta; t_a)$, i.e., the analog of (6.1) applied to the subensemble and with t_a substituted for t_0. [Note the semicolon in (6.1) versus the bar in (6.7)!]

In order to find $\tilde{w}(\theta; t_a)$ the previous result (6.2) can be utilized, provided that first the $\tilde{f}_n$ are determined. Now the distribution function $\tilde{f}_n$ has the following interpretation.

$\tilde{f}_n(t_1, t_2, \ldots, t_n) \, dt_1 \, dt_2 \cdots dt_n$ = the probability that the intervals $(t_1, t_1 + dt_1)$ and $(t_2, t_2 + dt_2)$, etc., each contain a dot, conditional on the fact that one dot lies in $(t_a - dt_a, t_a)$.

Hence by Bayes' rule

$$\tilde{f}_n(t_1, t_2, \ldots, t_n) = \frac{f_{n+1}(t_a, t_1, t_2, \ldots, t_n)}{f_1(t_a)} \quad (n = 1, 2, \ldots). \tag{6.9}$$

Thus we have expressed the distribution functions of the subensemble in terms of the original ones. According to (3.10) one now has

$$\tilde{p}_a(t_a, t_a + \theta) = 1 + \frac{1}{f_1(t_a)} \sum_{n=1}^{\infty} \frac{(-1)^n}{n!} \int_{t_a}^{t_a + \theta} f_{n+1}(t_a, t_1, \ldots, t_n) \, dt_1 \cdots dt_n, \tag{6.10}$$

and according to (6.1)

$$\tilde{w}(\theta; t_a) = \frac{1}{f_1(t_a)} \Bigg[f_2(t_a, t_a + \theta) + \sum_{n=1}^{\infty} \frac{(-1)^n}{n!} \int_{t_a}^{t_a + \theta} f_{n+2}(t_a, t_a + \theta, t_1, \ldots, t_n) \, dt_1 \cdots dt_n \Bigg].$$

This is the same formula as (6.8).*⁾

---

*⁾ An extensive review of different distributions of time points and their consequences is given by R. Metzler and J. Klafter in Physics Reports **339**, 1 (2000).

Exercise. Find the distribution of time intervals for independent events.

Exercise. For shot noise, $w(\theta; t_0) = w(\theta)$ is independent of t_0. Show that also $w(\theta | t_a) = w(\theta)$. Thus the waiting time when starting from an arbitrary time is the same as when starting from one of the events. On the one hand this is obvious, because selecting an event is merely one way of picking an arbitrary time point t_0, and does not affect the statistics of the remaining events. On the other hand it is paradoxical, because the average time elapsed *since the last event previous to t_0* is also the same. Hence the average time between the events on both sides of t_0 is twice that amount. Explain this paradox. [See the discussion of the "Weglängenparadoxon" by F. Zernike in: Handbuch der Physik 4 (Geiger and Scheel eds., Springer, Berlin 1929) p. 440.]

Exercise. Show that for shot noise the distribution of the time interval between an event and its k-th successor is

$$w_k(\theta) = v^k \theta^{k-1} e^{-v\theta}/(k-1)! \tag{6.11}$$

(χ^2-distribution – in this connection also called "Erlang-k distribution"*⁾).

Exercise. Show (compare (6.5))

$$w(t_b - t_a | t_a) = \frac{1}{f_1(t_a)} \left[\frac{\delta^2 L([v])}{\delta v(t_a) \, \delta v(t_b)} \right]_{v=-\chi}.$$

Exercise. Generalize (6.8) for the case that more than one event is prescribed.

Exercise. Prove (6.9) by first determining the conditional probabilites $\tilde{Q}_s$ for the subensemble and subsequently deriving the corresponding $\tilde{f}_n$ from them.

Exercise. If an event has been observed at t_a the probability density for some other event (not necessarily the next one) to occur at t_b is $f_2(t_a, t_b)/f_1(t_a)$. One defines the *pair distribution function* by

$$g(t_a, t_b) = \frac{f_2(t_a, t_b)}{f_1(t_a) f_1(t_b)} = 1 + \frac{g_2(t_a, t_b)}{f_1(t_a) f_1(t_b)}.$$

Generalize this definition to dots in three dimensions and verify that it is the pair distribution function of statistical mechanics.

Exercise. The conditional probability $q_k(\tau | t_a)$ that, having observed an event at t_a, one will have exactly k events between t_a and $t_a + \tau$ is called a Palm function.**⁾ Express $q_k(\tau | t_a)$ in the f_n.

7. Factorial correlation functions

Suppose one is interested in a quantity of the form of an output $U = \Sigma\, u(\tau_\sigma)$. The average is given in (3.5). We want to find the higher

---

*⁾ L. Kosten, *Stochastic Theory of Service Systems* (Pergamon, Oxford 1973).
**⁾ FELLER I, p. 413, footnote.

cumulants $\langle\langle U^n\rangle\rangle$. For that purpose consider the characteristic function

$$\langle e^{ikU}\rangle = \left\langle \exp\left[ik \sum_{\sigma=1}^{s} u(\tau_\sigma)\right]\right\rangle.$$

Equation (3.8) expresses this quantity as a series of the f_n. According to (5.5) this can be transformed into a series of the g_m. For brevity we absorb the factor ik in the quantities U and u,

$$\log\langle e^U\rangle = \sum_{m=1}^{\infty} \frac{1}{m!} \int (e^{u(t_1)} - 1) \cdots (e^{u(t_m)} - 1) g_m(t_1, \cdots, t_m) \, dt_1 \cdots dt_m. \quad (7.1)$$

To find the cumulants of U, this must be rearranged according to the powers of u:

$$\sum_{m=1}^{\infty} \frac{1}{m!} \sum_{n_i=1}^{\infty} \int \frac{[u(t_1)]^{n_1}}{n_1!} \cdots \frac{[u(t_m)]^{n_m}}{n_m!} g_m(t_1, \cdots, t_m) \, dt_1 \cdots dt_m. \quad (7.2)$$

$\langle\langle U^n\rangle\rangle$ is made up of all terms with $n_1 + n_2 + \cdots = n$.

To find a general expression we define *factorial correlation functions* $h_n(t_1, \ldots, t_n)$ by the following scheme

$$h_1(t_1) = g_1(t_1)$$
$$h_2(t_1, t_2) = \delta(t_1 - t_2) g_1(t_1) + g_2(t_1, t_2)$$
$$h_3(t_1, t_2, t_3) = \delta(t_1 - t_2)\delta(t_1 - t_3) g_1(t_1) + \delta(t_1 - t_2) g_2(t_1, t_2)$$
$$\qquad + \delta(t_1 - t_3) g_2(t_1, t_2) + \delta(t_2 - t_3) g_2(t_1, t_2) + g_3(t_1, t_2, t_3)$$
$$\cdots\cdots\cdots\cdots\cdots\cdots\cdots\cdots\cdots\cdots\cdots\cdots\cdots\cdots\cdots\cdots\cdots\cdots \quad (7.3)$$

The general formula for $h_n(t_1, \ldots, t_n)$ is given by the following prescription.

(i) Subdivide the set of variables $t_1, \ldots, t_m$ in all possible ways into clusters. Each subdivision, say in n clusters, is going to produce a term in h_n.

(ii) For each of the n clusters in the subdivision write a product of delta functions that ankyloses the variables t in that cluster. That leaves us with n independent variables, which we call temporarily $\tau_1, \ldots, \tau_n$.

(iii) Multiply the product of the delta functions with $g_n(\tau_1, \ldots, \tau_n)$. This constitutes the term belonging to the subdivision.

(iv) The sum of the terms belonging to all possible subdivisions is h_n. For example, h_6 contains among others the term (see fig. 2)

$$\delta(t_1 - t_3)\delta(t_2 - t_4)\delta(t_2 - t_5) g_3(t_1, t_2, t_6).$$

7. FACTORIAL CORRELATION FUNCTIONS

The reader can now verify that with this definition one has the identity, see (7.2),

$$\log\langle e^U \rangle = \sum_{n=1}^{\infty} \frac{1}{n!} \int u(t_1) \cdots u(t_n) h_n(t_1, \ldots, t_n) \, dt_1 \cdots dt_n. \tag{7.4}$$

From it follows that the n-th cumulant of U is given by

$$\langle\langle U^n \rangle\rangle = \int u(t_1) \cdots u(t_n) h_n(t_1, \ldots, t_n) \, dt_1 \cdots dt_n. \tag{7.5}$$

Exercise. The number N of dots in the interval (t_a, t_b) is a random variable with cumulants

$$\langle\langle N^n \rangle\rangle = \int_{t_a}^{t_b} h_n(t_1, \ldots, t_n) \, dt_1 \cdots dt_n.$$

Exercise. Find the cumulants of U for the case that the τ_σ are statistically independent. Rederive in this way for the quantity N above the Poisson distribution.

Exercise. For the same case of independence, rescale the density by setting $f_1(t) = \lambda \bar{f}_1(t)$, and rescale the individual contributions by setting $u(\tau) = \lambda^{-1/2} \bar{u}(\tau)$. Show that U becomes Gaussian in the limit $\lambda \to \infty$. It is necessary that u has zero average in the sense

$$\int u(t) f_1(t) \, dt = 0. \tag{7.6}$$

Suppose that there are *two* functions U, V and one is interested in their mixed moments and cumulants. These can be obtained by the following device. Replace the U in the preceding work with $pU + qV$ and compare the various powers of p and q. The mixed cumulants are then obtained from (7.5) as follows:

$$\langle\langle (pU + qV)^n \rangle\rangle = \int [pu(t_1) + qv(t_1)] \cdots [pu(t_n) + qv(t_n)]$$
$$\times h_n(t_1, \ldots, t_n) \, dt_1 \cdots dt_n.$$

The terms with $p^k q^l$ on both sides give

$$\langle\langle U^k V^l \rangle\rangle = \int u(t_1) \cdots u(t_k) v(t_{k+1}) \cdots v(t_{k+l}) h(t_1, \ldots, t_{k+l}) \, dt_1 \cdots dt_{k+l}. \tag{7.7}$$

Exercise. The notation can be abbreviated by writing the variables $t_1, t_2, \ldots$ as subscripts 1, 2, ... Then

$$\langle\langle UV \rangle\rangle = \int u_1 v_2 g_{12} + \int u_1 v_1 g_1, \tag{7.8}$$

$$\langle UV \rangle = \int u_1 v_2 f_{12} + \int u_1 v_1 f_1.$$

Exercise. The same device can be used for more functions. Express $\langle\langle UVW \rangle\rangle$ in the correlation functions.

Exercise. Let U and V have zero average as in (7.6). One then has

$$\langle\langle U^2 V^2 \rangle\rangle = \int u_1 u_2 v_3 v_4 g_{1234}$$

$$+ \int (u_1^2 v_2 v_3 + u_1 u_2 v_3^2 + 4 u_1 v_1 u_2 v_3) g_{123}$$

$$+ \int (2 u_1^2 v_1 v_2 + 2 u_1 u_2 v_2^2 + 2 u_1 v_1 u_2 v_2 + u_1^2 v_2^2) g_{12}$$

$$+ \int u_1^2 v_1^2 g_1. \tag{7.9}$$

The importance of these equations is the following.[*] In optics, U may represent a complex electric field amplitude and V its conjugate. If U results from a large number of coherent sources or scatterers, each endowed with a time lag τ_σ, one has

$$U = \sum_\sigma u(\tau_\sigma) = \sum_\sigma e^{i\omega \tau_\sigma}, \qquad V = U^*.$$

In most cases the distribution of the τ_σ is practically stationary; more precisely, the functions f_n do not vary noticeably when $t_1, t_2, \ldots, t_n$ are simultaneously shifted by an amount $\Delta t \sim \omega^{-1}$. Consequently (7.6) is satisfied. More generally all products of factors U and U^* average out to zero except those that have as many factors U as U^*. According to (7.8) the mean square field is

$$\langle UU^* \rangle = \langle\langle UU^* \rangle\rangle = \int e^{i\omega(t_1 - t_2)} g(t_1, t_2) \, dt_1 \, dt_2 + \int g_1(t_1) \, dt_1.$$

Note that for a strictly stationary distribution both integrals diverge since the number of contributing terms $u(\tau_\sigma)$ is infinite. One therefore has to

---

[*] P.N. Pusey and W. van Megen, Physica A **157**, 705 (1989).

7. FACTORIAL CORRELATION FUNCTIONS

restrict the τ_σ to some large interval T, divide the equation by T, and regard

$$J = \lim_{T \to \infty} \langle UU^* \rangle / T$$

as the relevant physical quantity, namely, the average intensity. Similarly one obtains from (7.9)

$$\langle\langle (UU^*)^2 \rangle\rangle = \int u_1 u_2 u_3^* u_4^* g_{1234} + 2 \operatorname{Re} \int u_1^2 u_2^* u_3^* g_{123}$$

$$+ 4 \int u_2 u_3^* g_{123} + 4 \int u_1 u_2^* g_{12} + \int (u_1 u_2^*)^2 g_{12}$$

$$+ 2 \int g_{12} + \int g_1.$$

All these integrals involve a factor T, assuming that the correlations decrease sufficiently rapidly for large time differences.

Remark. When U is a field strength one may regard UU^* as an instantaneous intensity I (ignoring for the moment that it should be divided by T). Its moments $\langle I^k \rangle$ are the same as those of UU^*. However, the cumulants are not the same because their definition depends on whether UU^* is regarded as a product of two quantities or as a single quantity I. For example,

$$\langle\langle U^2 U^{*2} \rangle\rangle = \langle (UU^*)^2 \rangle - 2\langle UU^* \rangle^2.$$
$$\langle\langle I^2 \rangle\rangle = \langle I^2 \rangle - \langle I \rangle^2$$
$$= \langle\langle U^2 U^{*2} \rangle\rangle + \langle UU^* \rangle^2.$$

Chapter III

STOCHASTIC PROCESSES

In this chapter stochastic processes are defined, together with a number of auxiliary concepts. The reason why they occur in physics is discussed. General properties of them are derived and some examples are treated in detail.

1. Definition

Once a stochastic variable X has been defined, an infinity of other stochastic variables derives from it, namely all quantities Y that are defined as functions of X by some mapping f. These quantities Y may be any kind of mathematical object, in particular also functions of an additional variable t,

$$Y_X(t) = f(X, t).$$

Such a quantity $Y(t)$ is called a *random function*, or, since in most cases t stands for time, a *stochastic process*. Thus a stochastic process is simply a function of two variables, one of which is the time t, and the other a stochastic variable X as defined in chapter I. On inserting for X one of its possible values x, an ordinary function of t obtains:

$$Y_x(t) = f(x, t),$$

called a *sample function* or *realization* of the process. In physical parlance one regards the stochastic process as an "ensemble" of these sample functions.

It is easy to form averages, on the basis of the given probability density $P_X(x)$ of X. For instance,

$$\langle Y(t) \rangle = \int Y_x(t) P_X(x)\, dx.$$

More generally, take n values $t_1, t_2, \ldots, t_n$ for the time variable (not necessarily all different) and form the n-th moment

$$\langle Y(t_1) Y(t_2) \cdots Y(t_n) \rangle = \int Y_x(t_1) Y_x(t_2) \cdots Y_x(t_n) P_X(x)\, dx. \tag{1.1}$$

Of particular interest is the *autocorrelation function*

$$\kappa(t_1, t_2) \equiv \langle\langle Y(t_1) Y(t_2) \rangle\rangle = \langle \{Y(t_1) - \langle Y(t_1)\rangle\}\{Y(t_2) - \langle Y(t_2)\rangle\} \rangle$$
$$= \langle Y(t_1) Y(t_2) \rangle - \langle Y(t_1) \rangle \langle Y(t_2) \rangle. \tag{1.2}$$

1. DEFINITION

For $t_1 = t_2$ it reduces to the time-dependent variance $\langle\langle Y(t)^2 \rangle\rangle = \sigma^2(t)$.

A stochastic process is called *stationary* when the moments are not affected by a shift in time, i.e., when

$$\langle Y(t_1 + \tau)Y(t_2 + \tau) \cdots Y(t_n + \tau) \rangle = \langle Y(t_1)Y(t_2) \cdots Y(t_n) \rangle \tag{1.3}$$

for all n, all τ, and all $t_1, t_2, \ldots, t_n$. In particular, $\langle Y \rangle$ is independent of time. It is often convenient to substract this constant from $Y(t)$ and to deal with the zero-mean process $\tilde{Y}(t) = Y(t) - \langle Y \rangle$. The autocorrelation function $\kappa(t_1, t_2)$ of a stationary process depends on $|t_1 - t_2|$ alone and is not affected by this subtraction. Often there exists a constant τ_c such that $\kappa(t_1, t_2)$ is zero or negligible for $|t_1 - t_2| > \tau_c$; one then calls τ_c the *autocorrelation time*.

As remarked in II.3, strictly stationary processes do not exist in nature, let alone in the laboratory, but they may be approximately realized when a process lasts much longer than the phenomena one is interested in. One condition is that it lasts much longer than the autocorrelation time. Processes without a finite τ_c never forget that they have been switched on in the past and can therefore not be treated as approximately stationary.

The stochastic quantity $Y(t)$ may consist of several components $Y_j(t)$ ($j = 1, 2, \ldots, r$). The autocorrelation function is then replaced with the correlation matrix

$$K_{ij}(t_1, t_2) \equiv \langle\langle Y_i(t_1) Y_j(t_2) \rangle\rangle. \tag{1.4}$$

The diagonal elements represent *autocorrelations*, the off-diagonal elements are *cross-correlations*[*]. In case of a zero-average stationary process this equation reduces to

$$K_{ij}(\tau) = \langle Y_i(t) Y_j(t + \tau) \rangle = \langle Y_i(0) Y_j(\tau) \rangle.$$

Note the obvious property for stationary processes

$$K_{ij}(\tau) = K_{ji}(-\tau). \tag{1.5}$$

When $Y(t)$ is a complex number (e.g., the amplitude of an oscillation), it may be treated as a two-component process, but it is often more convenient to maintain the complex notation. One may then define a complex autocorrelation function

$$\kappa(\tau) \equiv \langle\langle Y(0)^* Y(\tau) \rangle\rangle = \kappa(-\tau)^*, \tag{1.6}$$

which is often useful although it contains less information than the 2×2 correlation matrix.

Exercise. Let $\psi(t)$ be a given function of t and X a given random variable. Then

$$Y(t) = \psi(t) X \tag{1.7}$$

---

[*] These are the usual terms in physics although not quite in agreement with (I.3.8) and (I.3.9).

is a stochastic process. Compute its n-time averages, its variance and autocorrelation function (i.e., express them in the stochastic properties of X).

Exercise. Determine the relation between the 2×2 correlation matrix of a complex process and its complex autocorrelation function.

Exercise. Take for X a random set of dots (chapter II) and define the stochastic process

$$Y(t) = \sum_{\sigma=1}^{s} \delta(t - \tau_\sigma). \tag{1.8}$$

Compute $\langle Y(t) \rangle$ and $\langle Y(t_1)Y(t_2) \rangle$.

Exercise. Same question for the process

$$Y(t) = \sum_{\sigma=1}^{s} u(t - \tau_\sigma), \tag{1.9}$$

where u is a given function with finite width.

Exercise. When the random set τ_s in (1.9) is independent and stationary, with density v, the result is

$$\langle Y \rangle = v \int_{-\infty}^{\infty} u(\tau) \, d\tau, \qquad \kappa(t) = v \int_{-\infty}^{\infty} u(\tau) \, u(t+\tau) \, d\tau. \tag{1.10}$$

These equations are called "Campbell's theorem"[*], and we shall refer to this process as "Campbell's process".

Exercise. When Y is the Campbell process, find the characteristic function of its value $Y(t)$ at time t. Show that its cumulants are

$$\langle\langle Y(t)^m \rangle\rangle = v \int_{-\infty}^{\infty} \{u(\tau)\}^m \, d\tau.$$

Exercise. Let $Y(t) = \sin(\omega t + X)$, where X has a constant probability density in the range $(0, 2\pi)$. Find the autocorrelation function of Y.

Exercise. For the same $Y(t)$ show that the characteristic function of the joint distribution of $Y(t_1)$, $Y(t_2)$ is, putting $\omega = 1$,

$$J_0(\sqrt{k_1^2 + k_2^2 + 2k_1 k_2 \cos(t_1 - t_2)}) = \sum_{-\infty}^{\infty} (-1)^m J_m(k_1) J_m(k_2) \, e^{im(t_1 - t_2)}.$$

Hence $|y_1| < 1$ and $|y_2| < 1$, and

$$P_2(y_1, t_1; y_2, t_2) = (1 - y_1^2)^{-1/2} (1 - y_2^2)^{-1/2}$$

$$\times \left(1 + \sum_{m=1}^{\infty} 2^{m+1} T_m(y_1) T_m(y_2) \cos m(t_1 - t_2) \right),$$

where T_m are the Chebyshev polynomials.

---

[*] N.R. Campbell, Proc. Camb. Philos. Soc. **15**, 117 (1909); S.O. Rice, Bell System Technical Journal **23**, 282 (1944) and **24**, 46 (1945), reprinted in WAX.

Exercise. A stochastic process $Y(t)$ is defined by

$$Y(t) = X_n \quad \text{for } \xi + n < t < \xi + n + 1,$$

where n runs through all integers, $\{X_n\}$ is an infinite set of independent stochastic variables with identical distributions, and ξ is a stochastic variable in $(0, 1)$ with constant probability density in that range. Show that $Y(t)$ is stationary and find its autocorrelation function.

2. Stochastic processes in physics

The role in physics of probability and stochastic methods is the subject of many a profound study. We here merely make a few down-to-earth remarks about the way stochastic processes enter into the physical description of nature.

a. In nature one encounters many phenomena in which some quantity varies with time in a very complicated and irregular way, for instance, the position of a Brownian particle. There is no hope to compute this variation in detail, but it may be true that certain averaged features vary in a regular way, which can be described by simple laws. For instance, the instantaneous value of the force exerted by the molecules of a gas on a piston varies rapidly and unpredictably, but when integrated over a small time interval (which is automatically done by the inertia of the piston) it becomes a smooth function obeying Boyle's law. Similarly, the instantaneous current fluctuations in an electric circuit are very complicated, but on taking the square and integrating over small time intervals one obtains a quantity that is connected by simple laws to other features of the circuit.

b. The averaging over a suitable time interval is a rather awkward procedure. One therefore resorts to a drastic reformulation. The single irregularly varying function of time is replaced by an ensemble of functions, i.e., it is turned into a stochastic process. All averages are redefined as averages over the ensemble rather than over some time interval of the single function. The proper choice of the ensemble depends on the nature of the observed quantity and on the underlying physics. It must be chosen in such a way that the two kinds of averages coincide.

c. The reformulation of an irregular function of time as a stochastic process is often performed implicitly via an intuitive but vague use of such words as "random", "probability", and "average". To arrive at a better justification consider Brownian motion. One may actually observe a large number of Brownian particles and average the result; that means that one really has a physical realization of the ensemble (provided the particles do not interact). One might also observe one and the same particle on successive days; the results will be the same if one assumes that sections of the trajectory that lie 24 hours apart are statistically independent. In practice, one simply observes the trajectory of a single particle during a long time. The idea is that the irregularly varying function may be cut into a collection of long time intervals and that this collection can serve as the ensemble that defines the stochastic process. The condition for this "self-averaging" to work is that the behavior of the function during one interval does not affect the behavior during the next interval. If this is so the time average equals the ensemble average and the process is called *ergodic*.

d. This justification by means of self-averaging applies to stationary cases, because one may then choose the time intervals as long as one wishes. If the state of the system changes with time one must choose the interval long enough to smooth out the rapid fluctuations, but short compared to the overall change. The basic assumption is that such an intermediate interval exists. If that is not so, different methods are needed.

e. Other efforts to justify the averaging often involve an appeal to the authority of Gibbs[*], or to the uncontrolled interaction with the surroundings (which does not solve the problem but hides it). Or one resorts to anthropomorphic explanations such as the observer's ignorance of the precise macroscopic state[**]. This last argument is particularly insidious because it is half true. It is true that in many cases the observer is unable to see small rapid details, such as the motion of the individual molecules. On the other hand, he knows the experimental fact that there exists a macroscopic aspect, for which one does not need to know these details. Knowing that the details are irrelevant one may as well replace them with a suitable average. However, having said this one has not even begun to explain that experimental fact. The fundamental question is: How is it possible that such a macroscopic behavior exists, governed by its own equations of motion, regardless of the details of the microscopic motion?

f. Having accepted that the irregular motion of a system may be reformulated as a stochastic process, one is faced with the task of choosing the appropriate process. For a closed, isolated system that is usually done as follows. The microscopic deterministic motion may be represented by a trajectory in the phase space Γ. Each point $X \in \Gamma$ is, after a time t, mapped by the motion into a point $X^t \in \Gamma$, where $X^t = f(X, t)$ is uniquely determined. If one now chooses at some initial time $t = 0$ not a single initial state X, but a probability density $P(x)$ in Γ, then $f(X, t)$ is a stochastic process as defined in the preceding section. The initial $P(x)$ is to be chosen so as to reflect the way in which the system was prepared. Any other physical quantity pertain-

---

[*] J.W. Gibbs, *Elementary Principles in Statistical Mechanics* (Yale University Press, New Haven 1902).

[**] R.C. Tolman, op. cit. in I.5; E.T. Jaynes, op. cit. in I.5.

ing to the system is a function $Y(X^t)$ of the phase point X^t and has therefore also become a stochastic process $Y(X, t)$.

This is the usual approach to the stochastic description of nonequilibrium behavior and fluctuations. It is the starting point of the derivation of the so-called "generalized Langevin equation" and of the Kubo relations in linear response theory. It was even advocated in the first edition of this book – but it is wrong. The irregular motion of a Brownian particle can *not* be related to a probability distribution of some initial state. Rather it is brought about by the surrounding bath molecules and is a vestige of all the variables of the total system that have been ignored in order to obtain an equation for the Brownian particle alone, see IV.1 and VIII.3. The proper way of establishing the stochastic description of Brownian motion is therefore the careful elimination of the bath variables from the complete set of microscopic equations for the total system*[)].

g. This conclusion applies to all many-body systems. Macroscopic physics is based on the fact that it is possible to select a small set of variables in such a way that they obey approximately an autonomous set of deterministic equations, such as the hydrodynamic equations, Ohm's law, and the damping of the Brownian particle. Their approximate nature appears in the existence of fluctuation terms, by which the eliminated variables make themselves felt. As a consequence the macroscopic variables are stochastic functions of time. The stochastic description in terms of the macroscopic variables will be called *mesoscopic*. It comprises both the deterministic laws and the fluctuations about them.**[)]

Of course, the macroscopic equations cannot actually be derived from the microscopic ones. In practice they are pieced together from general principles and experience. The stochastic mesoscopic description must be obtained in the same way. This semi-phenomenological approach is remarkably successful in the range where the macroscopic equations are linear, see chapter VIII. In the nonlinear case, however, difficulties appear, which can only be resolved by the improved, but still mesoscopic, method of chapter X.

h. The miracle by which the enormous number of microscopic variables can be eliminated is based on the following idea. They vary so much more rapidly than the few macroscopic ones that they are able to reach almost instantaneously their equilibrium distribution, i.e., the one that belongs to the instantaneous values of the macroscopic variables as if these were fixed. This is the inescapable randomness assumption. Since the macroscopic variables are not fixed but vary slowly one is forced to readjust repeatedly the assumed randomness. This *repeated randomness assumption* is drastic, but indispensable whenever one tries to make a connection between the

---

*[)] N.G. van Kampen and I. Oppenheim, Physica A **138**, 231 (1986).

**[)] Nowadays the term 'mesoscopic' is often used to denote the phenomena on the borderline between classical and quantum mechanics.

microscopic world and the macroscopic or mesoscopic levels. It appears under the aliases "Stosszahlansatz", "molecular chaos", or "random phase approximation", and it is responsible for the appearance of irreversibility. Many attempts have been made to eliminate this assumption[*], usually amounting to hiding it under the rug of mathematical formalism. In the present context it is important because it will have the effect that the stochastic processes are Markov processes (chapter IV). The difficulties and limitations of the repeated randomness assumption are not discussed here. Nor do we go into the connections with ergodic theory and with chaos.[**] Quantum mechanics is considered in chapter XVII.

3. Fourier transformation of stationary processes

Consider a single-component real process $Y(t)$ in some fixed interval $0 < t < T$. Each realization $Y_x(t)$ is an ordinary function of t and may be Fourier-transformed in this interval. To avoid complex coefficients we use the sine transform

$$Y_x(t) = \sum_{n=1}^{\infty} A_{n,x} \sin\left(\frac{n\pi}{T} t\right).$$

The Fourier coefficients $A_{n,x}$ are given by

$$A_{n,x} = \frac{2}{T} \int_0^T \sin\left(\frac{n\pi}{T} t\right) Y_x(t) \, dt \tag{3.1}$$

and obey the Parseval identity

$$\frac{1}{2} \sum_{n=1}^{\infty} A_{n,x}^2 = \frac{1}{T} \int_0^T Y_x(t)^2 \, dt. \tag{3.2}$$

If one considers all possible values of x, with their probability density $P_X(x)$, the $A_{n,x}$ become stochastic variables A_n. For example, their average is simply

$$\langle A_n \rangle = \frac{2}{T} \int_0^T \sin\left(\frac{\pi n}{T} t\right) \langle Y(t) \rangle \, dt,$$

---

[*] For literature, see D. ter Haar, Rev. Mod. Phys. **27**, 289 (1955); P.C.W. Davies, *The Physics of Time Asymmetry* (Surrey University Press, London 1974); H.D. Zeh, *The Physical Basis of the Direction of Time* (Springer, Berlin 1989).

[**] A. Lasota and M.C. Mackey, *Probabilistic Properties of Deterministic Systems* (Cambridge University Press, Cambridge 1985); M.C. Mackey, *Time's Arrow: The Origins of Thermodynamic Behavior* (Springer, New York 1992).

3. FOURIER TRANSFORMATION OF STATIONARY PROCESSES

and by averaging (**3.2**)

$$\frac{1}{2}\sum_{n=1}^{\infty} \langle A_n^2 \rangle = \frac{1}{T}\int_0^T \langle Y(t)^2 \rangle \, dt.$$

Suppose $Y(t)$ is a stationary, zero-average process with a finite autocorrelation time τ_c; then $\langle Y(t)^2 \rangle$ is independent of time and one has

$$\sum_{n=1}^{\infty} \tfrac{1}{2}\langle A_n^2 \rangle = \langle Y^2 \rangle.$$

This equation expresses the mean square of the fluctuations $\langle Y^2 \rangle$ as a sum of terms, each referring to a single sine wave with frequency $\pi n/T$. The question is how this total $\langle Y^2 \rangle$ is distributed over the frequencies. That is, we want to know *the spectral density of fluctuations $S(\omega)$*, defined by

$$S(\omega)\,\Delta\omega = \sum_{\omega < \pi n/T < \omega + \Delta\omega} \tfrac{1}{2}\langle A_n^2 \rangle. \tag{3.3}$$

In order to be able to take $\Delta\omega$ small, one has to choose T large, so that many values n fall inside the interval $\Delta\omega$. The answer to our question is provided by the *Wiener–Khinchin theorem*, which states that $S(\omega)$ is the cosine transform of the autocorrelation function:

$$S(\omega) = \frac{2}{\pi}\int_0^{\infty} \cos(\omega\tau)\,\kappa(\tau)\,d\tau. \tag{3.4}$$

To prove this, substitute in (**3.3**) the explicit expression (**3.1**) for the Fourier coefficients:

$$\langle A_n^2 \rangle = \frac{4}{T^2}\int_0^T dt \int_0^T dt' \sin\frac{\pi n t}{T}\sin\frac{\pi n t'}{T}\langle Y(t)Y(t')\rangle$$

$$= \frac{4}{T^2}\int_0^T \sin\frac{\pi n t}{T}\,dt \int_{-t}^{T-t}\sin\frac{\pi n(t+\tau)}{T}\kappa(\tau)\,d\tau.$$

As we have supposed that $\kappa(\tau)$ decreases rapidly for $|\tau| > \tau_c$,

$$\int_{-\infty}^{\infty}|\kappa(\tau)|\,d\tau = 2\int_0^{\infty}|\kappa(\tau)|\,d\tau < \infty.$$

Furthermore T can be taken large compared to the τ-values over which $\kappa(\tau)$ differs appreciably from zero. Then the second integral may be taken from $-\infty$ to $+\infty$:

$$\langle A_n^2 \rangle = \frac{4}{T^2} \int_0^T \sin^2 \frac{\pi n t}{T} \, dt \int_{-\infty}^{\infty} \cos \frac{\pi n \tau}{T} \kappa(\tau) \, d\tau$$

$$+ \frac{4}{T^2} \int_0^T \sin \frac{\pi n t}{T} \cos \frac{\pi n t}{T} \, dt \int_{-\infty}^{\infty} \sin \frac{\pi n \tau}{T} \kappa(\tau) \, d\tau$$

$$= \frac{4}{T^2} \frac{T}{2} \int_{-\infty}^{\infty} \cos \frac{\pi n \tau}{T} \kappa(\tau) \, d\tau.$$

Substitute this result in (3.3) and suppose that $\kappa(\tau)$ is sufficiently smooth, so that one may put in the interval $\Delta\omega$

$$\int_{-\infty}^{\infty} \cos \frac{\pi n \tau}{T} \kappa(\tau) \, d\tau = \int_{-\infty}^{\infty} \cos(\omega\tau) \kappa(\tau) \, d\tau.$$

The result is, since there are $(T/\pi)\Delta\omega$ terms in this interval $\Delta\omega$,

$$S(\omega)\Delta\omega = \frac{T}{\pi} \Delta\omega \frac{1}{2} \frac{4}{T^2} \frac{T}{2} \int_{-\infty}^{\infty} \cos(\omega\tau) \kappa(\tau) \, d\tau,$$

which is the desired result (3.4).

Exercise. Alternative forms of the Wiener–Khinchin theorem (3.4) are

$$S(\omega) = \frac{1}{\pi} \int_{-\infty}^{\infty} e^{i\omega\tau} \kappa(\tau) \, d\tau, \qquad S(f) = 4 \int_0^{\infty} \cos(2\pi f \tau) \kappa(\tau) \, d\tau,$$

where $f = \omega/2\pi$ is the ordinary frequency. For a complex process Y

$$S(\omega) = \frac{1}{\pi} \int_{-\infty}^{\infty} \cos(\omega\tau) \kappa(\tau) \, d\tau,$$

where κ is the complex autocorrelation function (1.6).

Exercise. When Y has r components, which are real, one naturally defines

$$S_{ij}(\omega) = \frac{2}{\pi} \int_0^{\infty} \cos(\omega\tau) K_{ij}(\tau) \, d\tau. \tag{3.5}$$

Then the fluctuation spectrum of any linear combination $Z = \Sigma_j c_j Y_j$ is $S_Z(\omega) = \Sigma_{ij} S_{ij}(\omega) c_i c_j$. Find a similar formula for r complex components.

Exercise. The spectral density of Campbell's process (defined by (1.9) with stationary independent τ_s) is

$$S(\omega) = 2\nu |\hat{u}(\omega)|^2, \tag{3.6}$$

where $\hat{u}$ is the Fourier transform of u.

Exercise. The response of a critically damped ballistic galvanometer to a current pulse at $t = 0$ is $u(t) = ct \, e^{-\gamma t}$. Find the spectral density of the response to a stationary stream of independent random pulses.

Exercise. A cathode emits electrons at independent random times. Derive for the spectral density of the current fluctuations

$$S_I(f) = 2e\langle I \rangle, \tag{3.7}$$

where f is the frequency $\omega/2\pi$. This is Schottky's theorem for shot noise in the saturated diode current.[*]

Exercise. Let $Y(t)$ be the fluctuating part of an electrical current. It is often easier to measure the transported charge $Z(t) = \int_0^t Y(t')\,dt'$. Show that the spectral density of Y is related to the charge fluctuations by "MacDonald's theorem"[**]:

$$S(\omega) = \frac{\omega}{\pi} \int_0^\infty \sin \omega t \, \frac{d}{dt} \langle Z(t)^2 \rangle \, dt. \tag{3.8}$$

[Hint: First show $(d/dt)\langle Z(t)^2 \rangle = 2\int_0^t \kappa(t')\,dt'$.] The upper limit in (3.8) is defined in the Cesàro sense[†], or with the aid of a converging factor $e^{-\varepsilon \tau}$. In particular for the shot noise described by (3.7) one has[††]

$$\langle\langle \{Z(t)\}^2 \rangle\rangle = e\langle I \rangle t.$$

4. The hierarchy of distribution functions

Suppose a stochastic process $Y_X(t)$ is given in the way described in III.1. Then the probability density for $Y_X(t)$ to take the value y at time t is

$$P_1(y, t) = \int \delta\{y - Y_x(t)\} P_X(x)\,dx. \tag{4.1}$$

Similarly the joint probability density that Y has the value y_1 at t_1, and also the value y_2 at t_2, and so on till y_n, t_n, is

$$P_n(y_1, t_1; y_2, t_2; \ldots; y_n, t_n)$$

$$= \int \delta\{y_1 - Y_x(t_1)\} \delta\{y_2 - Y_x(t_2)\} \cdots \delta\{y_n - Y_x(t_n)\} P_X(x)\,dx. \tag{4.2}$$

In this way an infinite hierarchy of probability densities P_n ($n = 1, 2, \ldots$) is defined. They permit one to compute all averages used hitherto, such as

$$\langle Y(t_1) Y(t_2) \cdots Y(t_n) \rangle = \int y_1 y_2 \cdots y_n P_n(y_1, t_1; y_2, t_2; \ldots; y_n, t_n)\,dy_1\,dy_2 \cdots dy_n.$$

---

[*] W. Schottky, Ann. Physik **57**, 541 (1918); A. van der Ziel, *Noise* (Prentice-Hall, Englewood Cliffs, NJ 1954).
[**] D.K.C. MacDonald, Reports on Progress in Physics **12**, 56 (1948).
[†] E.C. Titchmarsh, *The Theory of Functions* (Oxford University Press, Oxford 1932).
[††] J.M.W. Milatz, Nederl. T. Natuurk. **8**, 19 (1941).

Although the right-hand side of (4.2) has a meaning when some of the times are equal, we shall regard the P_n to be defined only when all times are different. The hierarchy of functions P_n then obeys the following four "consistency conditions":

(i) $P_n \geq 0$;
(ii) P_n does not change on interchanging two pairs (y_k, t_k) and (y_l, t_l);
(iii) $\int P_n(y_1, t_1; \ldots; y_{n-1}, t_{n-1}; y_n, t_n) \, dy_n = P_{n-1}(y_1, t_1; \ldots; y_{n-1}, t_{n-1})$;
(iv) $\int P_1(y_1, t_1) \, dy_1 = 1$.

Inasmuch as the P_n enable one to compute all averages they constitute a complete specification of the stochastic process. (In fact, the specification is highly overcomplete, since according to (iii) any finite number of P_n may be omitted without losing information.) It has been proved by Kolmogorov[*] that any set of functions obeying the four consistency conditions determines a stochastic process $Y(t)$ as defined in III.1. Consequently the hierarchy of joint probability densities constitutes an equivalent alternative to the definition of a stochastic process given in section 1. Admittedly the construction of the variable X that corresponds to a given hierarchy may be rather abstract. In physical applications, therefore, the specification of a process by means of the P_n is often the more suitable one.

The *conditional probability* $P_{1|1}(y_2, t_2 | y_1, t_1)$ is the probability density for Y to take the value y_2 at t_2; given that its value at t_1 is y_1. To put it differently: From all sample functions $Y_x(t)$ of the ensemble select those that obey the condition that they pass through the point y_1 at t_1; the fraction of this *subensemble* that goes through the gate $y_2, y_2 + dy_2$ at t_2 is denoted by $P_{1|1}(y_2, t_2 | y_1, t_1) \, dy_2$. Clearly $P_{1|1}$ is nonnegative and normalized:

$$\int P_{1|1}(y_2, t_2 | y_1, t_1) \, dy_2 = 1.$$

More generally one may fix the values of Y at k different times $t_1, \ldots, t_k$ and ask for the joint probability at l other times $t_{k+1}, \ldots, t_{k+l}$. This leads to the general definition of the conditional probability $P_{l|k}$:

$$P_{l|k}(y_{k+1}, t_{k+1}; \ldots; y_{k+l}, t_{k+l} | y_1, t_1; \ldots; y_k, t_k)$$
$$= \frac{P_{k+l}(y_1, t_1; \ldots; y_k, t_k; y_{k+1}, t_{k+1}; \ldots; y_{k+l}, t_{k+l})}{P_k(y_1, t_1; \ldots; y_k, t_k)}. \quad (4.3)$$

By definition $P_{l|k}$ is symmetric in the set of k pairs of variables, and also in

---

[*] A. Kolmogoroff, op. cit. in I.1; A. Friedman, *Stochastic Differential Equations and Applications* I (Academic Press, New York 1975) ch. I. See, however, the footnote on p. 296 of A. Papoulis, *Probability, Random Variables, and Stochastic Processes* (McGraw-Hill, New York 1965).

4. THE HIERARCHY OF DISTRIBUTION FUNCTIONS

the set of l pairs of variables. This conditional probability is the probability in the subensemble of those sample functions that pass through the k prescribed gates at $t_1, t_2, \ldots, t_k$.

The generalization of the concept of a characteristic function to stochastic processes is the characteristic *functional*. (In a different connection this idea was used in section II.3.) Let $Y(t)$ be a given random process. Introduce an arbitrary auxiliary test function $k(t)$. Then the characteristic or moment generating functional is defined as the following functional of $k(t)$,

$$G([k]) = \left\langle \exp\left[i \int_{-\infty}^{\infty} k(t) Y(t) \, dt\right] \right\rangle. \tag{4.4}$$

The notation $G([k])$ emphasizes that G depends on the whole function k. One need not worry about the convergence of the integral, because the functions k may be restricted to those that vanish for sufficiently large $|t|$.

Expand (4.4) in powers of k,

$$G([k]) = \sum_{m=0}^{\infty} \frac{i^m}{m!} \int k(t_1) \cdots k(t_m) \langle Y(t_1) \cdots Y(t_m) \rangle \, dt_1 \cdots dt_m. \tag{4.5}$$

Thus each moment of the joint distribution of $Y(t_1), Y(t_2), \ldots$ can be found as the coefficient of the term with $k(t_1)k(t_2)\cdots$ in this expression. Similarly the cumulants can be found from

$$\log G([k]) = \sum_{m=1}^{\infty} \frac{i^m}{m!} \int k(t_1) \cdots k(t_m) \langle\langle Y(t_1) \cdots Y(t_m) \rangle\rangle \, dt_1 \cdots dt_m. \tag{4.6}$$

A process is *stationary* when all P_n depend on the time differences alone:

$$P_n(y_1, t_1 + \tau; y_2, t_2 + \tau; \ldots; y_n, t_n + \tau) = P_n(y_1, t_1; y_2, t_2; \ldots; y_n, t_n).$$

A necessary, but by no means sufficient, condition is that $P_1(y_1)$ is independent of time.

A process is called a *Gaussian process* if all its P_n are (multivariate) Gaussian distributions. In that case all cumulants beyond $m = 2$ are zero and

$$G([k]) = \exp\left[i \int k(t_1) \langle Y(t_1) \rangle \, dt_1 \right. \\ \left. - \tfrac{1}{2} \int\int k(t_1)k(t_2) \langle\langle Y(t_1)Y(t_2) \rangle\rangle \, dt_1 \, dt_2\right]. \tag{4.7}$$

Thus a Gaussian process is fully specified by its average $\langle Y(t) \rangle$ and its second moment $\langle Y(t_1)Y(t_2) \rangle$. Gaussian processes are particularly easy to handle and have been extensively studied. They are often used as an approximate description for physical processes, which amounts to assuming that the higher cumulants are negligible. In chapter X it will be shown that this ad hoc assumption can be justified in many cases, but chapters XI and XIII will show that it is by no means justifiable for all cases.

Exercise. The definition of a Gaussian process would be moot if it were incompatible with (iii). Show, however, that when some P_n is Gaussian, so are all the lower ones. Also that the conditional probabilities are Gaussian.

Exercise. Compute the hierarchy of P_n for the process (1.7), assuming $\psi(t) > 0$. Verify that the requirements (i)–(iv) are satisfied.

Exercise. Compute the characteristic functional of the process (1.7) in terms of the characteristic function of X.

Exercise. Compute the characteristic functional of the Campbell process and derive for the cumulants

$$\langle\langle Y(t_1)^{m_1} Y(t_2)^{m_2} \cdots \rangle\rangle = \int_{-\infty}^{\infty} \{u(t_1 - \tau)\}^{m_1} \{u(t_2 - \tau)\}^{m_2} \cdots \, d\tau. \qquad (4.8)$$

Exercise. Let $x(t)$, $p(t)$ be the coordinate and momentum of a free particle. The initial values $x(0)$, $p(0)$ are random variables with given distribution $P_1(x, p, 0)$. Thus $\{x(t), p(t)\}$ constitutes a bivariate random process. Compute $P_1(x_1, p_1, t_1)$, $P_2(x_1, p_1, t_1; x_2, p_2, t_2)$, and $P_{1|1}(x_2, p_2, t_2 | x_1, p_1, t_1)$. Also the higher P_n.

Exercise. The variable $x(t)$ in the previous exercise is a stochastic process by itself. Compute the distribution functions for the case

$$P_1(x, p, 0) = (2\pi)^{-1} \exp[-\tfrac{1}{2}(x^2 + p^2)].$$

Exercise. Let $Y^{(1)}(t)$ and $Y^{(2)}(t)$ be two random processes with hierarchies $P_n^{(1)}$ and $P_n^{(2)}$. Let λ_1 and λ_2 be two nonnegative numbers such that $\lambda_1 + \lambda_2 = 1$. Show that $P_n = \lambda_1 P_n^{(1)} + \lambda_2 P_n^{(2)}$ ("convex addition") is again an admissible hierarchy. What is the random process described by it? [See (XVII.1.11).]

Exercise. Show that the factorial cumulants (1.3.13) for a process Y are given by

$$\log G([k]) = \sum_{m=1}^{\infty} \frac{1}{m!} \int (e^{ik(t_1)} - 1) \cdots (e^{ik(t_m)} - 1) [Y(t_1) \cdots Y(t_m)] \, dt_1 \cdots dt_m. \qquad (4.9)$$

Exercise. Show that the present definition of "stationary" in terms of the P_n is equivalent with (III.1.3).

Exercise. Let $(X(t), Y(t))$ be a bivariate stochastic process with hierarchy P_n. Define

$$\bar{P}_n(x_1, t_1; \ldots; x_n, t_n) = \int P_n(x_1, y_1, t_1; \ldots; x_n, y_n, t_n) \, dy_1 \cdots dy_n. \qquad (4.10)$$

Show that these marginal distribution functions define again a process $X(t)$.

5. The vibrating string and random fields

Most of the examples of stochastic processes in physics and chemistry are of a special type called "Markovian" and belong therefore to the next

5. THE VIBRATING STRING AND RANDOM FIELDS

chapter. The following physical situation, however, provides an illustration of a non-Markovian process. A weightless elastic string is clamped at $x = 0$ and at $x = L$. If $y(x)$ is its transverse displacement, the elastic energy of a particular shape $y(x)$ is

$$E = \frac{1}{2} \int_0^L \left(\frac{dy}{dx}\right)^2 dx. \tag{5.1}$$

When the string is subject to thermal fluctuations (due to the surrounding air) $y(x)$ becomes a random function, x playing the role of the variable called so far t. One expects that the probability for any particular $y(x)$ to materialize will be proportional to

$$e^{-E/kT} = \exp\left[-\tfrac{1}{2}\beta \int_0^L \left(\frac{dy}{dx}\right)^2 dx\right], \tag{5.2}$$

where $\beta = 1/kT$ and T the air temperature. This is only a rough statement, however, as long as we do not know how to distinguish and to count the individual "shapes" $y(x)$, i.e., how to integrate in the space of functions y.

To make these heuristic ideas precise we define a stochastic function $Y(x)$ whose sample functions are the $y(x)$. Take n different points x_ν in the interval $(0, L)$ and label them in increasing order

$$0 < x_1 < x_1 < \cdots < x_n < L. \tag{5.3}$$

At each point x_ν erect a gate $(y_\nu, y_\nu + dy_\nu)$. The energy of a string drawn tight through these gates is

$$\frac{1}{2}\left(\frac{y_1^2}{x_1} + \frac{(y_2 - y_1)^2}{x_2 - x_1} + \cdots + \frac{(y_n - y_{n-1})^2}{x_n - x_{n-1}} + \frac{y_n^2}{L - x_n}\right).$$

For convenience in writing, we formally introduce $x_0 = 0$, $y_0 = 0$, $x_{n+1} = L$, $y_{n+1} = 0$; then this energy expression may be written

$$\tfrac{1}{2} \sum_{\nu=0}^{n} \frac{(y_{\nu+1} - y_\nu)^2}{x_{\nu+1} - x_\nu}.$$

After these preliminaries, define a hierarchy P_n by setting

$P_n(y_1, x_1; \ldots; y_n, x_n)$

$$= \left(\frac{2\pi L}{\beta}\right)^{1/2} \prod_{\nu=0}^{n} \left(\frac{\beta}{2\pi(x_{\nu+1} - x_\nu)}\right)^{1/2} \exp\left[-\frac{\beta}{2}\frac{(y_{\nu+1} - y_\nu)^2}{x_{\nu+1} - x_\nu}\right]. \tag{5.4}$$

This defines P_n when the x_ν obey (5.3). For other orderings, P_n is defined by the symmetry condition (ii). Consistency condition (i) is obviously satisfied; (iii) can be verified by explicit calculation; and (iv) is true owing to the normalizing factor we wrote in front of the product. Moreover, for large n

and a dense subdivision of $(0, L)$ it is clear that P_n agrees with the intuitive formula (5.2). Hence equation (5.4) defines a stochastic process which embodies the physical idea of the thermal equilibrium state of a string.

Each P_n is a multivariate Gaussian distribution, so that we are dealing with a Gaussian process. This enables one to use the equations of I.6. It is then readily found that $\langle Y(x_1) \rangle = 0$, and $x_1 \leqslant x_2$,

$$\langle Y(x_1)Y(x_2) \rangle = \frac{1}{\beta} \frac{x_1(L - x_2)}{L}. \tag{5.5}$$

Note that as a consequence of the fact that the process is not stationary[*] this autocorrelation function does not depend on $|x_1 - x_2|$ alone, and even contains the total length L. Hence the Wiener–Khinchin theorem does not apply directly; yet a similar calculation of the Fourier coefficients A_n yields $\langle A_n \rangle = 0$ and

$$\langle A_n A_m \rangle = \delta_{nm} \frac{2L}{n^2 \pi^2} \frac{1}{\beta}. \tag{5.6}$$

This is simply the statement that the fluctuations of the normal modes are uncorrelated and that each normal mode has an average energy $\frac{1}{2}kT$. (The factor $\frac{1}{2}$ arises because our string is overdamped and has no kinetic energy.)

Remark. According to this calculation the average of the total energy turns out to be infinite. For a physical string this is not a paradox, because the energy expression (5.2) is certainly not correct for too small wave lengths. The same calculation, however, applies also to electromagnetic waves between two reflecting mirrors at $x = 0$ and $x = L$. In that case one expects the equations to remain true for all wave lengths, so that the infinite field energy does present a problem. This is the Rayleigh–Jeans "ultraviolet catastrophe", which was resolved by Planck's introduction of the quantum.

It may be added that the difficulty reappeared later, when it appeared that each oscillator has a zero-point energy. This zero-point energy exists also in empty space and is independent of temperature, and may therefore be subtracted from the total energy without affecting the observed facts. However, the difference between the zero-point energy of the field between both mirrors and the vacuum field does not vanish and depends on L. It therefore gives rise to a force between the mirrors, which is a macroscopic version of the Van der Waals force between molecules, nowadays known as the Casimir effect.[**]

---

[*] Since our variable is x rather than t the word "homogeneous" might have been more suitable, but this term is used in another connection, see IV.4.

[**] D. Langbein, *Theory of the Van der Waals Attraction*, (Ergebn. exacten Naturw. **72**; Springer, Berlin 1974); M.J. Sparnaay, in: *Physics in the Making* (A. Sarlemijn and M.J. Sparnaay eds., North-Holland, Amsterdam 1989).

5. THE VIBRATING STRING AND RANDOM FIELDS

Exercise. The physicist's approach would be to expand $y(x)$ in normal modes and apply (5.6), knowing that the average potential energy of each harmonic oscillator is $\frac{1}{2}kT$. Derive in this way (5.5).

Exercise. Replace in (5.6) the factor $1/\beta$ by the Planck distribution and find in this way the quantum mechanical equivalent of (5.5).

Exercise. Consider a string of point masses with harmonic springs between neighbors and fixed at the ends:

$$\ddot{y}_\nu = y_{\nu+1} + y_{\nu-1} - 2y_\nu, \qquad y_0 = 0, \quad y_{n+1} = 0.$$

Compute $\langle y_\nu y_\mu \rangle$ in thermal equilibrium, and show that no divergence difficulties arise.

A related but more sophisticated concept is the *random field*, which occurs in radiation theory[*]. Let $u(\mathbf{r}, t)$ be a field governed by some linear partial differential equation independent of time, e.g.,

$$\nabla^2 u - \frac{\partial^2 u}{\partial t^2} = 0. \tag{5.7}$$

The solutions of the equation are superpositions of the normal modes $u_q(\mathbf{r}, t)$,

$$u(\mathbf{r}, t) = \sum_q A_q u_q(\mathbf{r}, t). \tag{5.8}$$

Associated with the equation is an expression for the energy of the solution, which is the sum of the energies of the normal modes,

$$E = \sum_q \varepsilon_q A_q^2.$$

Thermal equilibrium is described by an ensemble, i.e., the A_q become random variables with a probability distribution (in classical statistics) proportional with

$$e^{-E/kT} = \prod_q e^{-\beta \varepsilon_q A_q^2}. \tag{5.9}$$

Thus the A_q are independent Gaussian random variables with zero mean. Accordingly $u(\mathbf{r}, t)$ has become a *random field*, i.e., a random function of the *four* variables $\mathbf{r}, t$ rather than of t alone. One is interested in its stochastic properties, for instance, the two-point correlation function

$$\langle u(\mathbf{r}_1, t_1) u(\mathbf{r}_2, t_2) \rangle = \sum_{qq'} \langle A_q A_{q'} \rangle u_q(\mathbf{r}_1, t_1) u_{q'}(\mathbf{r}_2, t_2)$$
$$= \sum_q (2\beta \varepsilon_q)^{-1} u_q(\mathbf{r}_1, t_1) u_q(\mathbf{r}_2, t_2). \tag{5.10}$$

---

[*] Other applications can be found in: C. Preston, *Random Fields* (Lecture Notes in Mathematics 534; Springer, Berlin 1976); R. Kinderman and J.L. Snell, *Markov Random Fields and their Applications* (Amer. Mathem. Soc., Providence, RI 1980).

III. STOCHASTIC PROCESSES

In infinite space the normal modes are a continuous set and (5.8) ought to be an integral. That creates some difficulty in applying (5.9) and one therefore often encloses the whole field in a large cube Ω. As boundary conditions one may put $u = 0$ on the walls of Ω, but the normal modes take a simpler form if one requires u to be periodic with period Ω. The results are not materially affected by these tricks, provided that ultimately Ω goes to infinity. We shall now compute (5.10) for a real field obeying the wave equation (5.7).

The general form of the real solutions of (5.7) is

$$u(r, t) = \sum_q \{a_q e^{i(q \cdot r - qt)} + a_q^* e^{-i(q \cdot r - qt)}\}. \tag{5.11}$$

Here q is a vector with discretely spaced components,

$$q = \left(\frac{2\pi}{L} n_x, \frac{2\pi}{L} n_y, \frac{2\pi}{L} n_z\right), \quad q = |q|,$$

with integers n_x, n_y, n_z running from $-\infty$ to $+\infty$, while L is the edge of the cube Ω. For each q there is one complex coefficient $a_q = a_q' + i a_q''$; its real and imaginary parts are the coefficients previously called A_q. The total energy is

$$E = \tfrac{1}{2} \int_\Omega \{(\nabla u)^2 + (\partial_t u)^2\} \, dr = 2\Omega \sum_q q^2 (a_q'^2 + a_q''^2).$$

In thermal equilibrium one has according to (5.9)

$$\langle a_q'^2 \rangle = \langle a_q''^2 \rangle = (4\Omega\beta q^2)^{-1},$$

or

$$\langle a_q a_q^* \rangle = \delta_{qq'} (2\Omega\beta q^2)^{-1}, \quad \langle a_q a_{q'} \rangle = \langle a_q^* a_{q'}^* \rangle = 0.$$

One now easily finds

$$\langle u(r_1, t_1) u(r_2, t) \rangle = \sum_q (2\Omega\beta q^2)^{-1} \{e^{iq \cdot (r_1 - r_2) - iq(t_1 - t_2)} + e^{-iq \cdot (r_1 - r_2) + iq(t_1 - t_2)}\}.$$

For large Ω the sum over q may be replaced by an integral, taking into account that there are $(L/2\pi)^3$ discrete values in a unit volume of q-space,

$$\sum_q \cdots \to \frac{\Omega}{(2\pi)^3} \int \cdots d^3 q.$$

Thus, setting $r_1 - r_2 = \rho$, $t_1 - t_2 = \tau$,

$$\langle u(r_1, t_1)u(r_2, t_2)\rangle = \frac{1}{2\beta(2\pi)^3} \int \frac{d^3q}{q^2} (e^{i\mathbf{q}\cdot\rho - iq\tau} + e^{-i\mathbf{q}\cdot\rho + iq\tau}). \quad (5.12)$$

Exercise. Prove that (5.11) is the general solution by verifying that it is possible to adjust the a_q to any given initial values of u and $\partial_t u$.

Exercise. Compute $\langle u(x_1, t_1)u(x_2, t_2)\rangle$ for the vibrating string.

Exercise. Show that (5.12) equals zero inside the light cone, and $kT/4\pi|r_1 - r_2|$ outside it.

6. Branching processes

This is a class of processes that need not be Markovian and yet can be treated explicitly to a certain extent*[)]. They occur more often in population problems than in physics. The first example occurred in 1874, when Galton posed the question whether the extinction of upper class family names in England was due to statistics rather than to infertility of the rich.

Consider a population of bacteria or other cells that proliferate by division. A cell of age τ has a probability $\gamma(\tau) \, dt$ to divide into two cells during the next dt, each of which then starts a new branch of the family tree. The question is to determine the probability $P(n, t \,|\, m, 0)$ for having n individuals at time t when starting with m at time zero.

The model can, of course, be varied, e.g., by including a death probability. It also applies to cosmic ray cascades and neutrons in a reactor, if one takes into account the possibility of creating more than two particles at each event (fig. 3). In these two cases, however, γ does not depend on the age, which makes the problem Markovian and therefore amenable to the easier treatment presented in the next chapters.

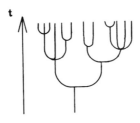

Fig. 3. One realization of a general branching process.

---

*[)] T.E. Harris, *The Theory of Branching Processes* (Springer, Berlin 1963); P. Jagers, *Branching Processes with Biological Applications* (Wiley, London 1975).

The features defining a branching process are: (i) each individual starts a family of descendants; (ii) all these families have the same stochastic properties; (iii) they do not interact with one another. (When the evolution of the families involves intermarriage it is no longer a branching process.) As a consequence of (iii) the conditional probability $P(n, t | m, 0)$ is the convolution of m factors $P(n, t | 1, 0)$. This yields the *first identity* for the probability generating function of a branching process starting with m individuals at $t = 0$,

$$F(z, t | m, 0) = \{F(z, t | 1, 0)\}^m \quad (t \geqslant 0). \tag{6.1}$$

It therefore suffices to study the offspring of a single individual.

The probability for a cell of age τ to divide in the next $d\tau$ is $\gamma(\tau) \, d\tau$. This assumes that the probability depends on the age of the cell alone. Suppose a certain cell was born at $t = 0$. Let $w(\tau)$ be the probability that it reaches the age τ without undergoing a division. Then $w(0) = 1$ and

$$dw(\tau) = -\gamma(\tau) w(\tau) \, d\tau. \tag{6.2}$$

Thus $w(\tau)$ is uniquely determined by $\gamma(\tau)$,

$$w(\tau) = \exp\left[-\int_0^\tau \gamma(\tau') \, d\tau'\right]. \tag{6.3}$$

The probability that there is a single cell present at time t is clearly $w(t)$. The probability that our cell undergoes division between the ages τ and $\tau + d\tau$ is $-dw(\tau)$, as given by (6.2). If that happens the population consists of two newly born cells, each starting its own family branch. Thus the probability for having n cells present at time t, when starting with a single cell at time 0, is

$$P(n, t | 1, 0) = \delta_{n,1} w(t) - \int_0^t dw(\tau) \, P(n, t | 2, \tau). \tag{6.4}$$

This is the *second identity* for branching processes.[*]

Multiply this identity with z^n and sum over $n = 1, 2, 3, \ldots$,

$$F(z, t | 1, 0) = zw(t) - \int_0^t dw(\tau) \, F(z, t | 2, \tau). \tag{6.5}$$

Owing to the homogeneity in time implied by (ii) one has

$$F(z, t | 2, \tau) = F(z, t - \tau | 2, 0) = \{F(z, t - \tau | 1, 0)\}^2.$$

---

[*] It has here been assumed that only division into *two* individuals occurs, see Exercise.

6. BRANCHING PROCESSES

Substitution in (6.5) yields, after rearranging terms,

$$F(z, t \mid 1, 0) - z = -\int_0^t dw(\tau)[\{F(z, t-\tau \mid 1, 0)\}^2 - z]$$

$$= -\int_0^t w'(t-t')[\{F(z, t' \mid 1, 0)\}^2 - z] \, dt'. \qquad (6.6)$$

This is an equation for $F(z, t \mid 1, 0)$ alone, which determines the probability generating function and hence the distribution, once $\gamma(\tau)$ is known. The treatment of the branching process is thereby reduced to solving a nonlinear integral equation. Unfortunately this can only be done explicitly for very few choices of $\gamma(\tau)$.

Exercise. Solve the problem for $\gamma(\tau) = \text{const}$.

Exercise. Suppose an individual of age τ has a probability per unit time $\gamma_\nu(\tau)$ to split up into $\nu = 0, 1, 2, \ldots$ new individuals. Let $\phi(\zeta, \tau) = \sum \zeta^\nu \gamma_\nu(\tau)$ be the corresponding probability generating function. Derive for this case the integral equation analogous to (6.6):

$$F(z, t \mid 1, 0) = zw(t) + \int_0^t w(t-t')\phi\{F(z, t' \mid 1, 0), t-t'\} \, dt'$$

$$= z + \int_0^t w(t-t')[\phi\{F(z, t' \mid 1, 0), t-t'\} - z\phi(1, t-t')] \, dt', \qquad (6.7)$$

with suitable definition of w.

Exercise. Differentiate (6.6) with respect to z and put $z = 1$. The result is an integral equation for $\langle n \rangle_t$, which can be solved.

Exercise. The original Galton–Watson problem was formulated for a discrete time variable $t = 0, 1, 2, \ldots$, indexing successive generations. Show that in this case

$$F_t(z) = F_{t-1}(F_1(z)), \qquad (6.8)$$

so that $F_t(z)$ is the t-th iterate of the function $F_1(z)$. The same equations apply to electron cascades in multiplier tubes[*] and to photons multiplying by stimulated emission[**].

Exercise. In the case of a photon in an infinite medium with absorption and stimulated emission, $F_1(z)$ is given by (II.4.3). The generating function for the m-th

---

[*] P.M. Woodward, Proc. Camb. Philos. Soc. **44**, 404 (1948).
[**] K.M. van Vliet and R.J.J. Zijlstra, Physica A **89**, 353 (1977); K.M. van Vliet, R.J.J. Zijlstra, and N.G. van Kampen, in: *Noise in Physical Systems* (Proc. Fifth Intern. Conf. on Noise; D. Wolf ed., Springer, Berlin 1978).

generation is therefore given by the continued fraction

$$\cfrac{1}{1+\theta - \cfrac{\theta}{1+\theta - \cfrac{\theta}{1+\theta - \cfrac{\ddots}{\cfrac{\theta}{1+\theta-\theta z}}}}}$$

where $\theta = \beta/\alpha$. Find from this

$$F_m(z) = \frac{1 - \theta^m - (\theta - \theta^m)z}{1 - \theta^{m+1} - (\theta - \theta^{m+1})z}. \tag{6.9}$$

Exercise. For the same case of successive generations let $\tilde{F}_t(z)$ be the generating function of the probability distribution of *all* descendants in the generations $1, 2, \ldots, t$. Show

$$\tilde{F}_t(z) = \tilde{F}_{t-1}(z\tilde{F}_1(z)). \tag{6.10}$$

Exercise. Again, for the photon in the infinite medium, $F_1(z)$ is equal to (II.4.3). The generating function of the total number of descendants in the cascade is

$$\tilde{F}_t(z) = \frac{z}{(1+\theta)^t - \{(1+\theta)^t - 1\}z}. \tag{6.11}$$

Exercise. Alternatively (6.11) can be obtained as follows. Let $P(N)$ be the probability that the incident photon has a total number N of descendants in the cascade. Then

$$P(N) = p_0 \delta_{N,0} + \sum_{s=1}^{\infty} p_s \sum_{N_1 + N_2 + \cdots + N_s = N - s} P(N_1)P(N_2)\cdots P(N_s),$$

where p_s is the same as in (II.4.3).

Chapter IV

MARKOV PROCESSES

This chapter defines and describes the subclass of stochastic processes that have the Markov property. Such processes are by far the most important in physics and chemistry. The rest of this book will deal almost exclusively with Markov processes.

1. The Markov property

A Markov process is defined as a stochastic process with the property that for any set of n *successive* times (i.e., $t_1 < t_2 < \cdots < t_n$) one has

$$P_{1|n-1}(y_n, t_n | y_1, t_1; \ldots; y_{n-1}, t_{n-1}) = P_{1|1}(y_n, t_n | y_{n-1}, t_{n-1}). \qquad (1.1)$$

That is, the conditional probability density at t_n, given the value y_{n-1} at t_{n-1}, is uniquely determined and is not affected by any knowledge of the values at earlier times. $P_{1|1}$ is called the *transition probability*.

A Markov process is fully determined by the two functions $P_1(y_1, t_1)$ and $P_{1|1}(y_2, t_2 | y_1, t_1)$; the whole hierarchy can be constructed from them. Indeed, one has for instance, taking $t_1 < t_2 < t_3$,

$$P_3(y_1, t_1; y_2, t_2; y_3, t_3) = P_2(y_1, t_1; y_2, t_2) P_{1|2}(y_3, t_3 | y_1, t_1; y_2, t_2)$$
$$= P_1(y_1, t_1) P_{1|1}(y_2, t_2 | y_1, t_1) P_{1|1}(y_3, t_3 | y_2, t_2).$$
$$\qquad (1.2)$$

Continuing this algorithm one finds successively all P_n. This property makes Markov processes manageable, which is the reason why they are so useful in applications.

Exercise. Argue that the number of neutrons in a molecular reactor is a Markov process.

Exercise. A gambler plays heads and tails. Let Y_t be the amount of his capital after t throws. Show that Y_t is a discrete-time Markov process and find its transition probability.

Exercise. Consider the ordinary differential equation $\dot{x} = f(x)$. Write the solution with initial value x_0 at t_0 in the form $x = \phi(x_0, t - t_0)$. Show that x obeys the definition of the Markov process with

$$P_{1|1}(x, t | x_0, t_0) = \delta[x - \phi(x_0, t - t_0)]. \qquad (1.3)$$

This also holds when x has more than one component. Conclude that any deterministic process is also a Markov process, albeit of a rather singular type.

Exercise. Although the definition of a Markov process appears to favor one time direction, it implies the same property for the reverse time ordering. Prove this with the aid of (1.2).

The oldest and best known example of a Markov process in physics is the Brownian motion.*[)] A heavy particle is immersed in a fluid of light molecules, which collide with it in a random fashion. As a consequence the velocity of the heavy particle varies by a large number of small, and supposedly uncorrelated jumps. To facilitate the discussion we treat the motion as if it were one-dimensional. When the velocity has a certain value V, there will be on the average more collisions in front than from behind. Hence the probability for a certain change ΔV of the velocity in the next Δt depends on V, but not on earlier values of the velocity. Thus the velocity of the heavy particle is a Markov process. When the whole system is in equilibrium the process is stationary and its autocorrelation time is the time in which an initial velocity is damped out. This process is studied in detail in VIII.4.

Yet it turned out that this picture did not lead to agreement with the measurements of Brownian motion. The breakthrough came when Einstein and Smoluchowski realized that it is not this motion which is observed experimentally. Rather, between two successive observations of the position of the Brownian particle the velocity has grown and decayed many times: the interval between two observations is much larger than the auto-correlation time of the velocity. *What is observed is the net displacement resulting after many variations of the velocity.*

Fig. 4. The path of a Brownian particle.

Suppose a series of observations of the same Brownian particle gives a sequence of positions $X_1, X_2, \ldots$. Each displacement $X_{k+1} - X_k$ is subject to chance, but its probability distribution does not depend on the previous history, i.e., it is independent of $X_{k-1}, X_{k-2}, \ldots$. Hence, not only the velocity

---

*[)] For the history of Brownian motion, see G.L. de Haas-Lorentz, Thesis (Leiden 1912) [German transl.: *Die Brownsche Bewegung* (Vieweg, Braunschweig 1913)]; the "Notes" in: A. Einstein, *Investigations on the Theory of Brownian Movement* (A.D. Cowper transl., R. Fürth ed. with notes, Methuen, London 1926; Dover Publications, New York 1956); S.G. Brush, *The Kind of Motion We Call Heat* (North-Holland, Amsterdam 1976) Vol. **2**, ch. 15.

1. THE MARKOV PROPERTY

itself is a Markov process, but on the coarse-grained time scale imposed by the experimental conditions, the position X of the particle is again a Markov process. This picture is the basis of the theory of Brownian motion, which is given in VIII.3.

This example exhibits several features that are of general validity. First it is clear that the Markov property holds only approximately. If the previous displacement $X_k - X_{k-1}$ happened to be a large one, then the chances are slightly in favor of a large velocity at the time when X_k is observed. This velocity will survive for a short time of the order of the autocorrelation time of the velocity, and thereby favor a large value of $X_{k+1} - X_k$. Thus the fact that the autocorrelation time of the *velocity* is not strictly zero gives rise to some correlation between two successive *displacements*. This effect is small, provided that the time between two observations is much longer than the autocorrelation time of the velocity.

Similarly the velocity itself is only approximately a Markov process, because the collisions with the molecules are not instantaneous, but have a certain duration. During the time that a collision is taking place, the change of velocity in the immediate past tells something about the kind of collision, and hence something about the change of velocity in the immediate future. This effect is small provided that each individual collision lasts very short, i.e., if the particles are almost hard spheres. Of course, in addition one has to assume that the motion of the Brownian particle does not create an ordered flow in the surrounding fluid; such a flow would influence the collision probability at later times, and in that way act as a memory storage that violates the Markov assumption. Computer calculations[*] have shown that the velocity of one of the gas molecules themselves is not strictly Markovian for just this reason.[**]

A second feature worth emphasizing is that the same physical system can be described by two different Markov processes, depending on the level of coarseness of the description. Another example of this fact is the dissociation of a gas of binary molecules

$$AB \to A + B.$$

(An inert background gas is supposed to be present to provide the necessary collisions.) On a coarse scale one argues that each molecule AB has a certain probability per unit time to be broken up by some collision. Hence the

---

[*] B.J. Alder and T.E. Wainwright, Phys. Rev. Letters **18**, 988 (1967); W.W. Wood, in: *Fundamental Problems in Statistical Mechanics* III (Proc. Intern. Summer School at Wageningen 1974; North-Holland, Amsterdam 1975).

[**] Even without ordered flow the Markov character is violated by the possibility that the Brownian particle overtakes a molecule with which it has had a previous collision, but this will be rare if the particle is heavy.

change in the concentration between t and $t + \Delta t$ has a certain probability distribution, which depends on the concentration at t, but not on its previous values. Thus on this level the concentration is a Markov process. In a more detailed description of the mechanism of breaking up, one distinguishes the vibrational states of the molecule, and studies the way by which successive collisions kick it back and forth between these states, until it happens to go over the top of the potential barrier. This random walk over the vibrational levels is another Markov process, which will be described in more detail in VII.5.

The concept of a Markov process is not restricted to one-component processes $Y(t)$, but applies to processes $Y(t)$ with r components as well. Examples: the three velocity components of a Brownian particle; the r chemical components of a reacting mixture. The following remark, however, is essential.

For any r-component stochastic process one may ignore a number of components and the remaining s components again constitute a stochastic process. But, if the r-component process is Markovian, the process formed by the $s < r$ components in general is not. One cannot expect that the knowledge of only a few of the components is again sufficient to predict the future probability, not even of these few components themselves. In the first example above each velocity component is itself Markovian; in chemical reactions, however, the future probability distribution of the amount of each chemical component is determined by the present amounts of *all* components.

Vice versa, if a certain physically given process is not Markovian, it is sometimes possible by introducing additional components, to embed it in a Markov process. These additional components serve to describe explicitly information that otherwise would be contained implicitly in the past values of the variables. As an example, consider again the Brownian particle, but let an inhomogeneous external field of force be added. Then the change of velocity during Δt is no longer determined by the collisions alone, but also by the external force and hence by the position of the particle. This position depends on the velocity at all previous times, so that the *velocity* of the Brownian particle is no longer a Markov process. However, the two-component process formed by the velocity *and* position is Markovian, see VIII.7.

In principle any closed isolated physical system can be described as a Markov process by introducing *all* microscopic variables as components of Y. In fact, the microscopic motion in phase space is deterministic and therefore Markovian, compare (1.3). The physicist's question, however, is whether he can find a small set of variables whose behavior in time can be described as a multicomponent Markov process. The well-known, but still miraculous, experimental fact is that this is so for most many-body systems in nature. Of course, such a description is at best approximate and restricted to a macroscopic, coarse-grained level. This reduction to a much smaller number of variables is called "contraction" or "projection", but the justification of this approximation involves the fundamental problems of statistical mechanics and is still the subject of many discussions.[*]

Warning. In the physical literature the epithet "Markov" is often used with regrettable looseness. The term has a magic appeal, which invites its use in an intuitive sense not covered by the definition. The reader should beware of the following pitfalls.

(i) When a physicist talks about a "process" he normally refers to a certain *phenomenon* involving time. Concerning a process defined in this way it is meaningless to ask whether or not it is Markovian, unless one specifies the variables to be used for its description. The art of the physicist is to find those variables that are needed to make the description (approximately) Markovian.

(ii) The criterion (1.1) is a condition on *all* distribution functions P_n of the hierarchy. It is impossible to aver that a process is Markovian if only information about the first few P_n is available. On the other hand, if one knows that the process is Markovian then P_1 and P_2 suffice to specify the entire process.

(iii) Often equations occur of the type

$$\dot{P}(y, t) = \Omega[P(y, t)], \qquad (1.4)$$

where Ω is a linear or nonlinear operator acting on the y-dependence. According to this equation, $P(y, t_0)$ at any one t_0 uniquely determines $P(y, t)$ at all $t > t_0$. From this it is sometimes incorrectly concluded that $Y(t)$ is a Markovian process. First one must ask what $P(y, t)$ stands for. If it is meant to be $P_1(y, t)$ the equation merely tells that the single time probability distribution of $Y(t)$ happens to obey a differential equation, but gives no information about the higher distribution functions that enter into (1.1). For instance, for any stationary process the equation $\dot{P}_1(y, t) = 0$ holds, but surely not all stationary processes are Markovian!

On the other hand, (1.4) may be interpreted as to mean that any solution with initial condition $P(y, t_0) = \delta(y - y_0)$ is identical with the transition probability $P_{1|1}(y, t | y_0, t_0)$. In fact, the master equation derived in the next chapter for Markov processes is of this type. Yet it cannot guarantee Markovian character, inasmuch as it still does not say anything about the higher distribution functions.[**]

(iv) Sometimes equations are produced that express $\dot{P}$ in terms of all earlier values of P, e.g.,

$$\dot{P}(t) = \int_0^t G(t, t') P(t') \, dt'. \qquad (1.5)$$

---

[*] See the literature quoted in III.2.
[**] The operator Ω in the master equation is linear. It can be seen that the transition probability of a Markov process cannot obey a nonlinear equation of the form (1.4). The argument is similar to the one used by D. Polder, Philos. Mag. **45**, 69 (1954).

where G is a linear operator on the y-dependence. The idea is that the solution with initial value $P(y, 0) = \delta(y - y_0)$ represents $P_{1|1}(y, t | y_0, 0)$. One then concludes from this "non-Markovian equation" that $Y(t)$ for $t > 0$ cannot be a Markov process. However, the validity of (1.5) does not mean that the earlier values $P_{1|1}(y, t' | y_0, 0)$ are *indispensable* for knowing its future. A counterexample will be given in (2.9).

Exercise. Should the random walk with persistence (I.7.8) be called a Markov process? [The answer is given in **IV.5**.]

Exercise. As an example of (1.4) solve the equation

$$\frac{\partial P(y, t)}{\partial t} = \int_{-\infty}^{\infty} P(y', t) P(y - y', t) \, dy' - P(y, t) \quad (-\infty < y < \infty),$$

with initial condition $P(y, t_0) = \delta(y - y_0)$. If $Y(t)$ is a stochastic process whose $P_{1|1}(y, t | y_0, t_0)$ is given by this solution it cannot be Markovian, because the solution does not obey the Chapman–Kolmogorov equation (next section).

Exercise. What is wrong with the use of "Markovian" by N. Chomsky, *Syntactic Structures* (Mouton, The Hague 1957)?

2. The Chapman–Kolmogorov equation

Integrating the identity (1.2) over y_2 one obtains for $t_1 < t_2 < t_3$

$$P_2(y_1, t_1; y_3, t_3) = P_1(y_1, t_1) \int P_{1|1}(y_2, t_2 | y_1, t_1) P_{1|1}(y_3, t_3 | y_2, t_2) \, dy_2.$$

Divide both sides by $P_1(y_1, t_1)$,

$$P_{1|1}(y_3, t_3 | y_1, t_1) = \int P_{1|1}(y_3, t_3 | y_2, t_2) P_{1|1}(y_2, t_2 | y_1, t_1) \, dy_2. \quad (2.1)$$

This is called the *Chapman–Kolmogorov equation*[*]. It is an identity, which must be obeyed by the transition probability of any Markov process. The time ordering is essential: t_2 lies between t_1 and t_3. Of course, the equation also holds when y is a vector with r components; or when y only takes discrete values so that the integral is actually a sum.

As noted in **1** a Markov process is fully determined by P_1 and $P_{1|1}$ because the whole hierarchy P_n can be constructed from them. These two functions cannot be chosen arbitrarily, however, but obey two identities:

(i) the Chapman–Kolmogorov equation (2.1);

---

[*] Also "Smoluchowski equation", but we shall use that name in a more special sense, see VIII.1. An earlier reference to Bachelier is given by E.W. Montroll and B.J. West, in: *Studies in Statistical Mechanics* VII (E.W. Montroll and J.L. Lebowitz eds., North-Holland, Amsterdam 1979) p. 76.

2. THE CHAPMAN–KOLMOGOROV EQUATION

(ii) the obviously necessary relation

$$P_1(y_2, t_2) = \int P_{1|1}(y_2, t_2 | y_1, t_1) P_1(y_1, t_1) \, dy_1. \tag{2.2}$$

Vice versa, any two nonnegative functions $P_1, P_{1|1}$ that obey these consistency conditions define uniquely a Markov process.

Exercise. The Chapman–Kolmogorov equation (2.1) expresses the fact that a process starting at t_1 with value y_1 reaches y_3 at t_3 via any one of the possible values y_2 at the intermediate time t_2. Where does the Markov property enter into this argument?

Exercise. Suppose one knows a solution of the Chapman–Kolmogorov equation and wants to use it for constructing a Markov process. How can that be done and how much freedom does one still have?

Exercise. Let Y have the range ± 1. Show that

$$P_{1|1}(y, t | y', t') = \tfrac{1}{2}\{1 + e^{-2\gamma(t-t')}\}\delta_{y, y'} + \tfrac{1}{2}\{1 - e^{-2\gamma(t-t')}\}\delta_{y, -y'} \tag{2.3}$$

obeys the Chapman–Kolmogrov equation. Show that this is consistent with $P_1(y, t) = \tfrac{1}{2}(\delta_{y, 1} + \delta_{y, -1})$. The Markov process so defined is called a *dichotomic* or *two-valued Markov process*, or also *random telegraph process*.

Exercise. Suppose one has a two-valued Y which in each dt has a probability $y \, dt$ to jump. Show that $Y(t)$ is the same process as above.

Exercise. Write (2.3) as a 2×2 matrix and formulate the Chapman–Kolmogorov equation as a property of that matrix.

Exercise. Let $Y(t)$ be a process in which Y takes the value 0, 1 and in which t only takes three values. There are eight sample functions. Out of those eight we attribute a probability $\tfrac{1}{4}$ to each of the following four:

$$1, 0, 0; \quad 0, 1, 0; \quad 0, 0, 1; \quad 1, 1, 1.$$

The other four have zero probability. Show that this process obeys the Chapman–Kolmogorov equation but is not Markovian.[*]

Exercise. The above model can be extended into an infinite sequence of zeroes and ones by stringing such triplets together. Is this sequence a stationary process?

The following two examples of Markov processes are of fundamental importance.

(i) It is easily verified that for $-\infty < y < \infty$ the Chapman–Kolmogorov equation is obeyed by setting for $t_2 > t_1$

$$P_{1|1}(y_2, t_2 | y_1, t_1) = \frac{1}{\sqrt{2\pi(t_2 - t_1)}} \exp\left[-\frac{(y_2 - y_1)^2}{2(t_2 - t_1)}\right]. \tag{2.4}$$

---

[*] E. Parzen, *Stochastic Processes* (Holden-Day, San Francisco 1962) p. 203. Other examples to this effect are given by P. Lévy, Comptes Rendus Seances Acad. Sci. **228**, 2004 (1949) and W. Feller, Annals Mathem. Statist. **30**, 1252 (1959).

If one chooses $P_1(y_1, 0) = \delta(y_1)$ a non-stationary Markov process is defined, called the *Wiener process* or *Wiener–Lévy process*.[*] It is usually considered for $t > 0$ alone and was originally invented for describing the stochastic behavior of the position of a Brownian particle (see VIII.3). The probability density for $t > 0$ is according to (2.2)

$$P_1(y, t) = \frac{1}{\sqrt{2\pi t}} \exp\left[-\frac{y^2}{2t}\right]. \tag{2.5}$$

(ii) $Y(t)$ takes on only the values $n = 0, 1, 2, \ldots$ and $t \geq 0$. A Markov process is defined by $(t_2 \geq t_1 \geq 0)$

$$P_{1|1}(n_2, t_2 \mid n_1, t_1) = \frac{(t_2 - t_1)^{n_2 - n_1}}{(n_2 - n_1)!} e^{-(t_2 - t_1)}, \quad P_1(n, 0) = \delta_{n,0}. \tag{2.6}$$

It is understood that $P_{1|1} = 0$ for $n_2 < n_1$. Thus each sample function $y(t)$ is a succession of steps of unit height and at random moments. It is uniquely determined by the time points at which the steps take place. These time points constitute a random set of dots on the time axis. Their number between any two times t_1, t_2 is distributed according to the Poisson distribution (2.6). Hence $Y(t)$ is called *Poisson process* and describes the same situation as (II.2.6).

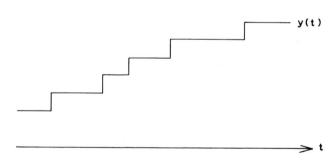

Fig. 5. A sample function of the Poisson process.

Exercise. Write the hierarchy of P_n for the Wiener process.
Exercise. Prove for the Wiener process

$$\langle Y(t_1) Y(t_2) \rangle = \mathrm{Min}(t_1, t_2), \tag{2.7a}$$

$$\langle \{Y(t_1) - Y(t_2)\}\{Y(t_3) - Y(t_4)\} \rangle = (t_1, t_2) \cap (t_3, t_4). \tag{2.7b}$$

The right-hand side of (2.7b) means the length of overlap of both intervals.

---

[*] N. Wiener, J. Math. and Phys. **2**, 131 (1923); COX AND MILLER, p. 205.

Exercise. Also prove for the Wiener process, when $0 < t_1 < t_2$

$$\langle y_2 \rangle_{y_1} = y_1, \qquad \langle\langle y_2^2 \rangle\rangle_{y_1} = t_2 - t_1, \tag{2.7c}$$

$$\langle y_1 \rangle_{y_2} = \frac{t_1}{t_2} y_2, \qquad \langle\langle y_1^2 \rangle\rangle_{y_2} = \frac{t_1}{t_2}(t_2 - t_1). \tag{2.7d}$$

Here y_1, y_2 stand for $Y(t_1), Y(t_2)$; and the symbol $\langle \ \rangle_z$ indicates a conditional average with constant z.

Exercise. Find the moments $\langle Y(t_1)Y(t_2) \cdots Y(t_n) \rangle$ of the Wiener process. [Use (I.6.11).]

Exercise. Show that (2.5) obeys the diffusion equation (or heat conduction equation)

$$\frac{\partial P}{\partial t} = D \frac{\partial^2 P}{\partial y^2} \tag{2.8}$$

for $D = \frac{1}{2}$. What is the solution for arbitrary $D > 0$?

Exercise. Verify that the definition (2.6) is consistent.

Exercise. Find for the Poisson process the same quantities as in (2.7).

Exercise. The transition probability $P_{1|1}(y, t \mid y_0, 0)$ of the Wiener process obeys (1.5) when G is the operator given by the kernel

$$G(y, t \mid y', t') = \frac{\partial}{\partial t} \frac{1}{t} \frac{1}{\sqrt{2\pi(t - t')}} \exp\left[-\frac{(y - y')^2}{2(t - t')} \right]. \tag{2.9}$$

Exercise. It has been remarked in **1** that a Markov process with time reversal is again a Markov process. Construct the hierarchy of distribution functions for the reversed Wiener process and verify that its transition probability obeys the Chapman–Kolmogorov equation.

3. Stationary Markov processes

Stochastic processes that are both Markovian and stationary are of special interest, in particular for describing equilibrium fluctuations. Assume that a closed, isolated physical system has a quantity, or set of quantities, $Y(t)$, which may be treated as a Markov process. When that system is in equilibrium, $Y(t)$ is a stationary Markov process. In particular, P_1 is independent of time and is nothing but the familiar equilibrium distribution of the quantity Y, as determined by equilibrium statistical mechanics. For example, consider the system consisting of a circuit formed by a resistance R and a condenser C, see fig. 6. The fluctuating voltage $Y(t)$ across the condenser is Markovian to a good approximation, provided that the inductance of the circuit is negligible. When R is kept at a constant temperature T, the voltage $Y(t)$ is a stationary Markov process and

$$P_1(y_1) = \left(\frac{C}{2\pi kT} \right)^{1/2} \exp\left[-\frac{C y_1^2}{2kT} \right].$$

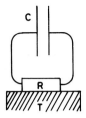

Fig. 6. RC circuit with fixed temperature bath.

An alternative example is formed by the current fluctuations in a circuit consisting of a shunted resistance at constant temperature (fig. 7).

Even when a system is in a steady state other than equilibrium certain physical quantities may be stationary Markov processes. An example are the current fluctuations in the circuit of fig. 7 when a battery is added, which maintains a constant potential difference and therefore a non-zero average current. Another example is a Brownian particle in a homogeneous gravitational field: its vertical velocity is a stationary process, but not its position.

Fig. 7. Shunted resistor at fixed temperature.

For stationary Markov processes the transition probability $P_{1|1}$ does not depend on two times but only on the time interval; for this case we introduce a special notation

$$P_{1|1}(y_2, t_2 | y_1, t_1) = T_\tau(y_2 | y_1) \quad \text{with } \tau = t_2 - t_1. \tag{3.1}$$

The Chapman–Komogorov equation then becomes $(\tau, \tau' > 0)$

$$T_{\tau+\tau'}(y_3 | y_1) = \int T_{\tau'}(y_3 | y_2) T_\tau(y_2 | y_1) \, dy_2. \tag{3.2}$$

If one reads the integral as a product of two matrices, or integral kernels, this equation may be written

$$T_{\tau+\tau'} = T_{\tau'} T_\tau \quad (\tau, \tau' > 0). \tag{3.3}$$

Remark. Although (3.2) and (3.3) only hold for τ and τ' positive, the definition (3.1) is not restricted to positive τ. But the values of T_τ and $T_{-\tau}$ are connected by an identity. Recalling the meaning (III.4.3) of $P_{1|1}$ one has, with $\tau = t_2 - t_1$,

$$P_2(y_1, t_1; y_2, t_2) = T_\tau(y_2 | y_1) P_1(y_1). \tag{3.4}$$

As P_2 is symmetric,
$$T_\tau(y_2|y_1)P_1(y_1) = T_{-\tau}(y_1|y_2)P_1(y_2). \tag{3.5}$$
This is an identity for all stationary Markov processes and may therefore be applied to physical systems in equilibrium without any additional derivation from the equations of motion. It should not, however, be confused with detailed balance, which differs from it by having $+\tau$ in the right-hand member. Detailed balance is a physical property, which does not follow from the mere definition of T_τ but requires a physical derivation, see V.6. In order to avoid erroneous use of the equations (3.2) and (3.3) we now stipulate that in the future the symbol T_τ shall not be used for negative τ.

Exercise. We do not exclude $\tau = 0$, but define
$$T_0(y_2|y_1) = \delta(y_2 - y_1). \tag{3.6}$$
Show that this is in agreement with (3.2) and (3.3).

Exercise. Show that (3.2) would also hold for both τ and τ' negative (if we had not excluded negative values).

Exercise. Verify that the dichotomic Markov process (2.3) is stationary and obeys (3.3). What would $T_{-\tau}$ be in this case? Show that (3.3) does not hold for $\tau > 0$, $\tau' < 0$.

Exercise. The autocorrelation function of a stationary Markov process with zero mean is given by
$$\kappa(\tau) = \iint y_1 y_2 T_\tau(y_2|y_1)P_1(y_1)\,dy_1\,dy_2 \quad (\tau \geq 0). \tag{3.7}$$

Exercise. Prove the identities
$$\int T_\tau(y_2|y_1)\,dy_2 = 1, \tag{3.8}$$
$$\int T_\tau(y_2|y_1)P_1(y_1)\,dy_1 = P_1(y_2). \tag{3.9}$$
Conclude from it that T_τ regarded as an operator has the eigenvalue 1, with left eigenvector $\psi(y) \equiv 1$, and right eigenvector P_1.

The best known example of a stationary Markov process is the *Ornstein–Uhlenbeck* process[*] defined by
$$P_1(y_1) = \frac{1}{\sqrt{2\pi}} e^{-\frac{1}{2}y_1^2}, \tag{3.10}$$
$$T_\tau(y_2|y_1) = \frac{1}{\sqrt{2\pi(1 - e^{-2\tau})}} \exp\left[-\frac{(y_2 - y_1 e^{-\tau})^2}{2(1 - e^{-2\tau})}\right]. \tag{3.11}$$

---

[*] G.E. Uhlenbeck and L.S. Ornstein, Phys. Rev. **36**, 823 (1930); reprinted in WAX. A more explicit description (with pictures) of this and other Markov processes can be found in D.T. Gillespie, *Markov Processes* (Academic Press, San Diego 1992).

The reader will have no difficulty in verifying that the two consistency conditions in **2** are obeyed. This process was originally constructed to describe the stochastic behavior of the *velocity* of a Brownian particle (see VIII.4). Clearly it has zero average and the autocorrelation function is simply

$$\kappa(\tau) = e^{-\tau}. \tag{3.12}$$

The Ornstein–Uhlenbeck process is stationary, Gaussian, and Markovian. *Doob's theorem*[*)] states that it is essentially the only process with these three properties. "Essentially" means that one must allow for linear transformations of y and t, *and* that there is one other, although trivial, process with these properties, see (3.22) below. We sketch the proof.

Let $Y(t)$ be a stationary Gaussian Markov process. By shifting and rescaling we can ensure that $P_1(y)$ is equal to (3.10). The transition probability is Gaussian and has therefore the general form

$$T_\tau(y_2 \mid y_1) = D\, e^{-\frac{1}{2}(A y_2^2 + 2B y_2 y_1 + C y_1^2)},$$

where A, B, C, D are functions of τ. The normalization (3.8) yields

$$D = \sqrt{A/2\pi}, \qquad C = B^2/A,$$

while (3.9) gives in addition $B^2 = A(A-1)$. The one remaining unknown parameter A can be expressed in the equally unknown autocorrelation function using (3.7), which yields $A = (1 - \kappa^2)^{-1}$. Hence

$$T_\tau(y_2 \mid y_1) = \frac{1}{\sqrt{2\pi(1-\kappa^2)}} \exp\left[-\frac{(y_2 - \kappa y_1)^2}{2(1-\kappa^2)}\right]. \tag{3.13}$$

Now take a third time $t_3 = t_2 + \tau'$ ($\tau' > 0$) and use (3.2):

$$\kappa(t_3 - t_1) = \int y_3\, dy_3 \int T_{\tau'}(y_3 \mid y_2)\, dy_2 \int T_\tau(y_2 \mid y_1) y_1 P_1(y_1)\, dy_1$$

$$= \kappa(t_3 - t_2)\kappa(t_2 - t_1). \tag{3.14}$$

This functional relation for $\kappa(\tau)$ shows that

$$\kappa(\tau) = e^{-\gamma\tau}. \tag{3.15}$$

Substitution of this result in (3.13) completes the proof, compare (3.11).

A second theorem is the following. If $Y(t)$ is stationary, Gaussian, and has an exponential autocorrelation function $\kappa(\tau) = \kappa(0) e^{-\gamma\tau}$, then $Y(t)$ is Ornstein–Uhlenbeck and hence Markovian. To prove this we may assume

---

[*)] J.L. Doob, Annals of Math. **43**, 351 (1942); reprinted in WAX. Other theorems about Gaussian processes are given in J.L. Doob, Annals Mathem. Statist. **15**, 229 (1944).

zero mean and unit variance. Then the generating functional (III.4.7) is

$$G([k]) = \exp\left[-\frac{1}{2}\int\int k(t_1)k(t_2)\,e^{-\gamma|t_1-t_2|}\,dt_1\,dt_2\right].$$

This is identical with the generating functional of the Ornstein–Uhlenbeck process.

Exercise. Find the generating functional and all moments of the Ornstein–Uhlenbeck process.

Exercise. Let ϕ be a random angle so that $0 \leqslant \phi < 2\pi$. Let $P_1(\phi) = 1/2\pi$ and $T_\tau(\phi\,|\,\phi_0)$ be the solution of

$$\frac{\partial T_\tau(\phi)}{\partial \tau} = \gamma\,\frac{\partial^2 T_\tau(\phi)}{\partial \phi^2}, \quad T_0(\phi\,|\,\phi_0) = \delta(\phi - \phi_0). \tag{3.16}$$

Find the P_n for this non-Gaussian stationary Markov process.

Exercise. Define the random process $Y(t) = e^{iat + i\phi}$ with fixed a, and ϕ as in the previous Exercise. Show that its spectral density is the Lorentzian (I.2.2).

Exercise. Deduce (3.13) from (I.6.12).

Exercise. Find three examples of processes, each of which has two but not three of the properties stipulated in Doob's theorem.

Exercise. Prove that for any Gaussian process (with zero mean and unit variance) the conditional average at t_2, given the value at t_1, is

$$\langle Y(t_2)\,|\,Y(t_1) = y_1\rangle = \kappa(t_2, t_1) y_1. \tag{3.17}$$

For a Markov process then follows that for $t_1 < t_2 < t_3$

$$\kappa(t_3, t_1) = \kappa(t_3, t_2)\kappa(t_2, t_1). \tag{3.18}$$

Exercise. Let Y be an r-variable Gaussian Markov process. Apply a linear transformation so that $P_1(y, t) \sim \exp[-\tfrac{1}{2} y^2]$. Then derive for the autocorrelation matrix

$$K(t_3, t_1) = K(t_3, t_2)K(t_2, t_1) \quad (t_1 \leqslant t_2 \leqslant t_3). \tag{3.19}$$

If the process is also stationary it follows that $K(\tau) = e^{-\tau G}$ with constant matrix G.*⁾

Exercise. Show that (3.11) obeys the equations

$$\frac{\partial T}{\partial \tau} = \frac{\partial}{\partial y_2} y_2 T + \frac{\partial^2 T}{\partial y_2^2}, \tag{3.20}$$

$$\frac{\partial T}{\partial \tau} = -y_1 \frac{\partial T}{\partial y_1} + \frac{\partial^2 T}{\partial y_1^2}. \tag{3.21}$$

These are the so-called forward and backward Kolmogorov equations for the Ornstein–Uhlenbeck process. Their paramount importance will appear in VIII.4 under the more familiar name of Fokker–Planck equation.

---

*⁾ See Note II of Wang and Uhlenbeck on page 130 of WAX. See also R.F. Fox and G.E. Uhlenbeck, Phys. Fluids **13**, 1893 (1970).

Exercise. $Y(t)$ being the Ornstein–Uhlenbeck process define $Z(t) = \int_0^t Y(t') \, dt'$ for $t \geq 0$. The process $Z(t)$ is Gaussian but neither stationary nor Markovian. Show
$$\langle Z(t_1)Z(t_2)\rangle = e^{-t_1} + e^{-t_2} - 1 - e^{-|t_1 - t_2|} + 2\operatorname{Min}(t_1, t_2).$$

Exercise. For the same $Z(t)$ find the characteristic functional and use it to obtain
$$\langle \cos\{Z(t_1) - Z(t_2)\}\rangle = \exp[-e^{-|t_1-t_2|} + 1 + |t_1 - t_2|].$$

Exercise. In the Ornstein–Uhlenbeck process rescale the variables: $y = \alpha y'$, $t = \beta t'$ and show that in a suitably chosen limit of α and β the $P_{1|1}$ reduces to that of the Wiener process.

Exercise. Prove that the following transition probability obeys the Chapman–Kolmogorov equation
$$T_\tau(y_2 | y_1) = \frac{1}{\pi} \frac{\tau}{(y_2 - y_1)^2 + \tau^2} \quad (-\infty < y < \infty, \tau > 0).$$

The Markov process defined by it is called the *Cauchy process*.[*]

Exercise. A more general version of the second theorem is given in FELLER II, p. 94. Let $Y(t)$ be a non-stationary Gaussian process, $\sigma^2(t)$ its variance, and $\rho(t_1, t_2) = \kappa(t_1, t_2)/\sigma(t_1)\sigma(t_2)$ its correlation coefficient (I.3.9). Then the relation, compare (3.18),
$$\rho(t_3, t_1) = \rho(t_3, t_2)\rho(t_2, t_1) \quad (t_1 \leq t_2 \leq t_3)$$
is necessary and sufficient for $Y(t)$ to be Markovian. Verify that this relation is obeyed by the Wiener process.

Exercise. If $Y(t)$ is the Ornstein–Uhlenbeck process as defined above, and $t > t_1 > t_2 \cdots > t_n$, then
$$\frac{d}{dt}\langle Y(t)Y(t_1)Y(t_2)\cdots Y(t_n)\rangle = -\langle Y(t)Y(t_1)Y(t_2)\cdots Y(t_n)\rangle.$$

Hence if $\Phi(t, [Y])$ is a functional depending on t and all values of Y previous to t one has[**]
$$\frac{d}{dt}\langle Y(t)\Phi(t, [Y])\rangle = \left\langle Y(t)\frac{\partial \Phi}{\partial t}\right\rangle - \langle Y(t)\Phi\rangle.$$

Exercise. The trivial exception to Doob's theorem mentioned above is the "completely random process" defined by
$$P_n(y_1, t_1; y_2, t_2; \cdots; y_n, t_n) = P(y_1, t_1)P(y_2, t_2)\cdots P(y_n, t_n). \tag{3.22}$$
Where did this possibility get lost in our proof?

4. The extraction of a subensemble

It has been remarked in III.4 that by imposing a condition on the sample functions of a stochastic process one defines a subensemble. This concept of

---
[*] A generalization is given by M.-O. Hongler, Phys. Letters A **112**, 297 (1985).
[**] V.E. Shapiro and V.M. Loginov, Physica A **91**, 563 (1978).

4. THE EXTRACTION OF A SUBENSEMBLE

"extracting a subensemble" is particularly useful in the case of stationary Markov processes. We therefore provide a somewhat more formal description for that case.

Let a stationary Markov process $Y(t)$ be given by $P_1(y_1)$ and $T_\tau(y_2|y_1)$. Take a fixed time t_0 and a fixed value y_0. Define a new, non-stationary Markov process $Y^*(t)$ for $y \geq t_0$ by setting

$$P_1^*(y_1, t_1) = T_{t_1-t_0}(y_1|y_0), \tag{4.1a}$$

$$P_{1|1}^*(y_2, t_2|y_1, t_1) = T_{t_2-t_1}(y_2|y_1). \tag{4.1b}$$

Somewhat more generally one may extract a subensemble in which at a given time t_0 the values of $Y(t_0)$ are distributed according to a given probability distribution $p(y_0)$. That amounts to setting

$$P_1^*(y_1, t_1) = \int T_{t_1-t_0}(y_1|y_0) p(y_0) \, dy_0, \tag{4.2a}$$

$$P_{1|1}^*(y_2, t_2|y_1, t_1) = T_{t_2-t_1}(y_2|y_1). \tag{4.2b}$$

This process may be regarded as a weighted mixture of processes of type (4.1).

These processes are non-stationary because the condition singled out a certain time t_0. Yet their transition probability depends on the time interval alone as it is the same as the transition probability of the underlying stationary process. Non-stationary Markov processes whose transition probability depends on the time difference alone are called *homogeneous processes*.[*] They usually occur as subensembles of stationary Markov processes in the way described here. However, the Wiener process defined in 2 is an example of a homogeneous process that cannot be embedded in a stationary Markov process.

Physically the extraction of a subensemble means that one prepares the system in a certain non-equilibrium state. One expects that after a long time the system returns to equilibrium, i.e.,

$$P_1^*(y_1, t_1) \to P_1(y_1) \quad \text{as } t \to \infty.$$

As this should be true for arbitrary preparation $p(y_0)$ one must also have

$$T_{t_1-t_0}(y_1|y_0) \to P_1(y_1). \tag{4.3}$$

One of the principal problems of Markov processes in to show that this is so, see V.3.

---

[*] The name refers to homogeneity in time. Unfortunately it is somewhat confusing because a process may also be homogeneous in space, i.e., invariant for a transition in the space of its states y. We shall therefore often prefer the circumlocution: "Markov process with stationary transition probability".

The extraction of a homogeneous process from a stationary Markov process is a familiar procedure in the theory of linear response. As an example[*] take a sample of a paramagnetic material placed in a constant external magnetic field B. The magnetization Y in the direction of the field is a stationary stochastic process with a macroscopic average value and small fluctuations around it. For the moment we assume that it is a Markov process. The function $P_1(y)$ is given by the canonical distribution

$$P_1(y) = \frac{\text{Tr}[\delta(y - M) e^{-(H - BM)/kT}]}{\text{Tr } e^{-(H - BM)/kT}}. \tag{4.4}$$

Here H is the Hamiltonian operator of the system without external field, M denotes the operator corresponding to the magnetization, and the Trace is the quantum mechanical equivalent of the classical integral over phase space.

Now suppose that for $-\infty < t < t_0$ the field has the value $B + \Delta B$ and at t_0 suddenly jumps to the value B. Then the distribution at t_0 is

$$p(y) = \frac{\text{Tr}[\delta(y - M) e^{-\{H - (B + \Delta B)M\}/kT}]}{\text{Tr } e^{-\{H - (B + \Delta B)M\}/kT}}. \tag{4.5}$$

For $t > t_0$ the magnetization will be the homogeneous process whose initial distribution is given by this p, and whose transition probability is the same as in the equilibrium situation with external B. Hence

$$\langle Y(t) \rangle^* = \int\int y T_{t-t_0}(y \mid y_0) p(y_0) \, dy_0 \, dy. \tag{4.6}$$

Thus the way in which the macroscopic average adjusts itself to the new field B is determined by the transition probability of the equilibrium fluctuations in the new field B.

If one is only interested in the *linear* response a more elegant result may be derived. Expand (4.5) to first order in ΔB,

$$p(y) = P_1(y) + \frac{\Delta B}{kT} \{y - \langle M \rangle_B\} P_1(y) + \mathcal{O}(\Delta B^2),$$

where $P_1(y)$ is the distribution (4.4) and $\langle Y \rangle_B$ is the corresponding average. Substitution in (4.6) yields, after some algebra,

$$\langle Y(t) \rangle^* = \langle Y \rangle_B + \frac{\Delta B}{kT} \kappa(t - t_0).$$

Thus the irreversible relaxation of the average magnetization is (in linear approximation) determined by the autocorrelation function of the equi-

---

[*] R. Kubo and K. Tomita, J. Phys. Soc. Japan **9**, 888 (1954); C.P. Slichter, *Principles of Magnetic Resonance* (Harper and Row, New York 1963); R. Lenk, *Brownian Motion and Spin Relaxation* (Elsevier, Amsterdam 1977).

librium fluctuations. This is an example of the famous fluctuation–dissipation theorem, although that theorem is customarily formulated in frequency language.*⁾

Remark. We assumed that $Y(t)$ is a Markov process. Usually, however, one is interested in materials in which a memory effect is present, because that provides more information about the microscopic magnetic moments and their interaction. In that case the above results are still formally correct, but the following qualification must be borne in mind. It is still true that $p(y_0)$ is the distribution of Y at the time t_0, at which the small field B is switched off. However, it is no longer true that this $p(y_0)$ uniquely specifies a subensemble and thereby the future of $Y(t)$. It is now essential to know that the system has "aged" in the presence of $B + \Delta B$, so that its density in phase space is canonical, not only with respect to Y, but also with respect to all other quantities that determine the future. Hence the formulas cannot be applied to time-dependent fields $B(t)$ unless the variation is so slow that the system is able to maintain at all times the equilibrium distribution corresponding to the instantaneous $B(t)$.

Exercise. From the Wiener process extract the subensemble corresponding to $Y(t_0) = y_0$. Find the evolution of $\langle Y(t) \rangle$ for $t > t_0$. Also find the variance $\langle\langle Y(t)^2 \rangle\rangle$ in this ensemble.

Exercise. Same question for the Ornstein–Uhlenbeck process.

Exercise. A Markov process $Z(t)$ whose transition probability has the form

$$T_\tau(z_2 | z_1) = T_\tau(z_2 - z_1) \tag{4.7}$$

is called a *process with independent increments*.**⁾ It is homogeneous but not stationary, nor a subensemble of a stationary process. In order that (2.1) is obeyed one must have

$$\int e^{ikz} T_\tau(z) \, dz = e^{\tau \Phi(k)}, \tag{4.8}$$

where $\Phi(k)$ is some function with $\Phi(0) = 0$. Show that $P_1(z, t)$ becomes Gaussian for $t \to \infty$.

5. Markov chains

An especially simple class of Markov processes are the Markov chains, which we define by the following properties.[†]

---

*⁾ H.B. Callen and R.F. Greene, Phys. Rev. **86**, 702 (1952) and **88**, 1387 (1952); J.L. Jackson, Phys. Rev. **87**, 471 (1952); DE GROOT AND MAZUR pp. 150 ff.

**⁾ See also IX.6.

[†] This definition is a compromise. Some authors define Markov chains by (i) alone and others postulate only (ii). Most authors do not include (iii) in the definition but treat such cases only. Compare the footnote on page 340 of FELLER I.

(i) The range of Y is a discrete set of states.
(ii) The time variable is discrete and takes only integer values $t = \ldots, -2, -1, 0, 1, 2, \ldots$.
(iii) The process is stationary or at least homogeneous, so that the transition probability depends on the time difference alone.

A *finite Markov chain* is one whose range consists of a finite number N of states. They have been extensively studied, because they are the simplest Markov processes that still exhibit most of the relevant features.[*] The first probability distribution $P_1(y, t)$ is an N-component vector $p_n(t)$ ($n = 1, 2, \ldots, N$). The transition probability $T_\tau(y_2 | y_1)$ is an $N \times N$ matrix. The Markov property (3.3) leads to the matrix equation

$$T_\tau = (T_1)^\tau \quad (\tau = 0, 1, 2, \ldots). \tag{5.1}$$

The probability distribution $p(t)$ originating from the initial distribution $p(0)$ is in matrix notation, omitting superfluous indices,

$$p(t) = T^t p(0). \tag{5.2}$$

Hence the study of finite Markov chains amounts to investigating the powers of an $N \times N$ matrix T of which one knows only that
(i) its elements are nonnegative, and
(ii) each column adds up to unity.

Such matrices are called "stochastic matrices"[**] and have been studied by Perron and Frobenius. It is clear that T has a left eigenvector $(1, 1, \ldots, 1)$ with eigenvalue 1; and therefore a right eigenvector p^s such that $Tp^s = p^s$, which is the $P_1(y)$ of the stationary process. It is not necessarily a physical equilibrium state, but may, e.g., represent a steady state in which a constant flow is maintained. The principal task of the theory is to show that for any initial $p(0)$

$$\lim_{t \to \infty} p(t) = \lim_{t \to \infty} T^t p(0) = p^s. \tag{5.3}$$

From the theorems of Perron and Frobenius it follows that this is true apart from some exceptional cases.[†] However, we shall not pursue this approach here, because in the next chapter all relevant results will be derived for the case of continuous time in a different way.

---

[*] See, e.g., J.G. Kemeny and J.L. Snell, *Finite Markov Chains* (Van Nostrand, Princeton 1960); D.L. Isaacson and R.W. Madsen, *Markov Chains, Theory and Applications* (Wiley, New York 1976).

[**] To be distinguished from "random matrices", i.e., matrices whose elements are stochastic variables.

[†] See, e.g., F.R. Gantmacher, *Applications of the Theory of Matrices* (Interscience, New York 1959), or COX AND MILLER, p. 120.

5. MARKOV CHAINS

Exercise. Find the hierarchy of joint distribution functions $P_n(y_1, t_1; y_2, t_2; ...; y_n, t_n)$ (the y's and t's are integers) for the finite Markov chain defined by a given T and $P_1(y_1, 0)$.

Exercise. For the case $N = 2$ write the most general T and find the corresponding p^s. Then prove that (5.3) holds, with two exceptions.

Exercise. The dichotomic Markov process (2.3) can be reduced to a Markov chain by considering only the probability distribution at a sequence of equidistant time points. Construct the corresponding T and investigate whether the above exceptions can occur.

Exercise. Suppose T decomposes into two blocks as in fig. 8. Show that in this case the eigenvector p^s is not unique, the eigenvalue 1 is degenerate. The system has two sets of states between which no transitions are possible. How should (5.3) be adapted to this case?

Fig. 8. A decomposable transition matrix.

Exercise. Formulate the random walk problem in I.4 as a Markov chain. There is no normalized p^s owing to the infinite range. Remedy this flaw by considering a random walk on a circular array of N points, such that position $N + 1$ is identical with 1. Find p^s for this finite Markov chain. Is it true that every solution tends to p^s?

Exercise. To illustrate the approach to equilibrium, Ehrenfest invented the following model.*) N balls, labelled $1, 2, ..., N$, are distributed over two urns. Every second a numeral is selected at random (equal probabilities) from the set $1, 2, ..., N$ and the ball with that numeral is transferred from its urn to the other. The state of the system is specified by the number n of balls in one of the urns. The process is a Markov chain with

$$T_{nn'} = \frac{n'}{N} \delta_{n+1,n'} + \left(1 - \frac{n'}{N}\right) \delta_{n-1,n'}. \tag{5.4}$$

show that the binomial distribution is a stationary solution.

The random walk model may be generalized by introducing a statistical correlation between two successive steps, in such a way that the probability α for a step in the same direction as the previous step differs from the probability β for a step back ("random walk with persistence"). In this case

---

*) P. and T. Ehrenfest, Mathem.-Naturw. Blätter **3** (1906) = P. Ehrenfest, *Collected Scientific Papers* (M. Klein ed., North-Holland, Amsterdam 1959) p. 128; M. Kac, *Probability and Related Topics in Physical Sciences* (Interscience, London 1959) pp. 73 ff; H. Falk, Physica A **104**, 459 (1980).

the position variable Y is no longer a Markov process, because its probability distribution at time $r+1$ depends not only on its value at r, but also on its value at $r-1$. However, the Markov character can be restored by introducing this previous value explicitly as an additional variable. The two-component process (Y_1, Y_2), in which Y_1 is the position at any time r, and Y_2 the previous position at $r-1$, is again Markovian. The transition matrix is (the value of Y_1 and Y_2 being indicated by n and m, respectively)

$$T(n_2, m_2 | n_1, m_1) = [\delta_{n_2, n_1+1}(\alpha\, \delta_{n_1, m_1+1} + \beta\, \delta_{n_1, m_1-1})$$
$$+ \delta_{n_2, n_1-1}(\beta\, \delta_{n_1, m_1+1} + \alpha\, \delta_{n_1, m_1-1})]\, \delta_{m_2, n_1}. \quad (5.5)$$

Owing to this reduction to a Markov process the model can again be treated in full detail. A stochastic process that can be made Markovian by means of one additional variable is said to be "Markovian of the second degree", and if more variables are needed it is Markovian of some higher degree.[*]

Random walks on square lattices with two or more dimensions are somewhat more complicated than in one dimension, but not essentially more difficult. One easily finds, for instance, that the mean square distance after r steps is again proportional to r. However, in several dimensions it is also possible to formulate the *excluded volume problem*, which is the random walk with the additional stipulation that no lattice point can be occupied more than once. This model is used as a simplified description of a polymer: each carbon atom can have any position in space, given only the fixed length of the links and the fact that no two carbon atoms can overlap. This problem has been the subject of extensive approximate, numerical, and asymptotic studies.[**] They indicate that the mean square distance between the end points of a polymer of r links is proportional to $r^{6/5}$ for large r. A fully satisfactory solution of the problem, however, has not been found. The difficulty is that the model is *essentially* non-Markovian: the probability distribution of the position of the next carbon atom depends not only on the previous one or two, but on *all* previous positions. It can formally be treated as a Markov process by adding an infinity of variables to take the whole history into account, but that does not help in solving the problem.

Exercise. The transition matrix (5.5) for the random walk with persistence acts in too large a space, because only the states with $m = n \pm 1$ have a meaning. Find a simpler reduction of the random walk with persistence to a Markov chain by adding a second variable Y_2 which takes only two values.

---

[*] For these and other generalizations, see M.N. Barber and B.W. Ninham, *Random and Restricted Walks* (Gordon and Breach, New York 1970).
[**] M. Doi and S.F. Edwards, *The Theory of Polymer Dynamics* (Clarendon, Oxford 1986).

Exercise. The mean square distance of the random walk with persistence can easily be found by the following alternative method, compare (I.7.8). It is equal to $\langle (X_1 + X_2 + \cdots + X_r)^2 \rangle$, where each X_k takes the values ± 1, and $\langle X_k X_{k+1} \rangle \neq 0$.

Exercise. Find the mean square distance after r steps of a random walk on a square lattice of d dimensions.

Exercise. Find the mean square distance after r steps of a random walk on a two-dimensional square lattice when U-turns are forbidden. (This is not the excluded volume problem, because each site can be visited many times, provided more than two steps intervene.)

Exercise. Show for the one-dimensional random walk with persistence that the distribution approaches a Gaussian.

Remark. Consider a Markov process that can be visualized as a particle jumping back and forth among a finite number of sites m, with constant probabilities per unit time. Suppose it has a single stationary distribution p_n^s, with the property (5.3). After an initial period it will be true that, if I pick an arbitrary t, the probability to find the particle at n is p_n^s. That implies that p_n^s is the fraction of its life that the particle spends at site n, once equilibrium has been reached. This fact is called *ergodicity*. For a Markov process with finitely many sites ergodicity is tantamount to indecomposability.[*] In (VII.7.13) a more general result for the times spent at the various sites is obtained.

6. The decay process

Consider a piece of radioactive material containing n_0 active nuclei at $t = 0$. The number $N(t)$ of active nuclei surviving at time $t > 0$ is a nonstationary stochastic process. It is clearly Markovian because the probability distribution of $N(t_2)$ at $t_2 > t_1$, conditional on $N(t_1) = n_1$, is independent of the previous history. The same calculation applies to the emission of light by excited atoms, the escape of molecules of a Knudsen gas through a small leak, the killing of enemy troops by random shooting or the destruction of cells by radiation. It has also been used to describe the loss of cosmic ray electrons in an absorbing material, t being the traversed thickness[**].

The process is simply a combination of the mutually independent decay processes of the individual nuclei. Let w be the probability for a single nucleus to survive at time t_1. Even before computing w one may state that

---

[*] This is a simple analogy of Birkhoff's ergodic theorem for dynamical systems, see A.I. Khinchin, *Mathematical Foundation of Statistical Mechanics* (Dover, New York 1949); L.E. Reichl, *A Modern Course in Statistical Physics* (University of Texas Press, Austin, TX 1980) ch. 8.

[**] H.J. Bhabha and W. Heitler, Proc. Roy. Soc. A **159**, 432 (1937); BHARUCHA-REID.

the probability for n_1 nuclei to survive is

$$P_1(n_1, t_1) = \binom{n_0}{n_1} w^{n_1} (1-w)^{n_0 - n_1}. \tag{6.1}$$

Note that this formula is valid for *all* integral values of n_1 provided one takes the binomial coefficient to be zero whenever $n_1 < 0$ or $n_1 > n_0$. From this formula one derives, using a standard computational trick of classical probability theory,

$$\begin{aligned}\langle N(t_1)\rangle &= \left[\sum_{n_1} n_1 \binom{n_0}{n_1} w^{n_1} v^{n_0 - n_1}\right]_{v = 1-w} \\ &= \left[w \frac{\partial}{\partial w}(w+v)^{n_0}\right]_{v=1-w} \\ &= wn_0.\end{aligned} \tag{6.2}$$

Next we have to compute the survival chance $w = w(t_1)$, knowing that the probability per unit time for a still surviving nucleus to decay is a constant γ. This task has already been performed in (III.6.3), which for constant γ yields

$$w(t) = e^{-\gamma t}. \tag{6.3}$$

Substitution in (6.1) now yields the explicit P_1 of our non-stationary process

$$P_1(n_1, t_1) = \binom{n_0}{n_1} e^{-\gamma t_1 n_1} (1 - e^{-\gamma t_1})^{n_0 - n_1}. \tag{6.4}$$

The same formula permits us to write down the transition probability for $t_2 > t_1$,

$$P_{1|1}(n_2, t_2 | n_1, t_1) = \binom{n_1}{n_2} e^{-\gamma(t_2-t_1)n_2} (1 - e^{-\gamma(t_2-t_1)})^{n_1 - n_2}. \tag{6.5}$$

Together they fully determine the Markov process.

Remark. It is seen from (6.4) that in the limit $t_1 \to \infty$

$$P_1(n, \infty) = \delta_{n,0}. \tag{6.6}$$

Thus all probability ends up in the state $N = 0$, which is therefore called an *absorbing state*. All other states ($N \geq 1$) are depleted in the course of time; they are called *transient states*. They can only occur because the decay products disappear into an infinitely large universe. For finite physical systems transient states are excluded, see V.5. If our radioactive sample were enclosed in an impermeable container there would be a non-zero probability for the emitted particles to be reabsorbed. Such a

6. THE DECAY PROCESS

situation is encountered when studying the emission and absorption of phonons by atoms in a laser, or by a single atom in a cavity.[*]

This example may be employed to illustrate a point which will be important in the next chapter. The transition probability (6.5) reduces for $t_2 - t_1 = 0$ to

$$P_{1|1}(n_2, t_1 | n_1, t_1) = \delta_{n_1, n_2}, \tag{6.7}$$

as it should. Now take $t_2 - t_1 = \tau$ and omit second and higher orders of τ. The result is

$$P_{1|1}(n_2, t_1 + \tau | n_1, t_1) = \delta_{n_1, n_2}(1 - \gamma n_1 \tau) + \delta_{n_1 - 1, n_2} n_1 \gamma \tau + \mathcal{O}(\tau^2). \tag{6.8}$$

One recognizes in the last term the probability that one of the n_1 active nuclei decays during τ; the probability for more decays is of higher order in τ. The first represents the probability that no transition took place. The height of the initial Kronecker delta is reduced in agreement with the normalization conditions (3.8)

Exercise. Verify by explicit calculation that (6.5) obeys the Chapman–Kolmogorov equation.

Exercise. Obtain for the probability generating function of $N(t_1)$

$$F(z) = \langle z^{N(t_1)} \rangle = \{1 + w(z - 1)\}^{n_0}, \tag{6.9}$$

and find from it (6.2) and also the variance of $N(t_1)$.

Exercise. Find the factorial cumulants (I.2.17) for the decay process.

Exercise. Formulate the decay process as a branching process and use (III.6.7) for obtaining (6.9).

Exercise. A certain two-state Markov chain has one absorbing and one transient state. What is the form of the transition matrix T?

---

[*] N. Nayak, R.K. Bullough, B.V. Thompson, and G.S. Agarwal, IEEE J. Quantum Electr. **24**, 1331 (1988); H. Dehmelt, Rev. Mod. Phys. **62**, 525 (1990); W. Paul, idem p. 531.

Chapter V

THE MASTER EQUATION

The master equation is an equivalent form of the Chapman–Kolmogorov equation for Markov processes, but it is easier to handle and more directly related to physical concepts. It will be the pivot of most of the work in this book.

1. Derivation

Consider a Markov process, which for convenience we take to be homogeneous, so that we may write T_τ for the transition probability. The Chapman–Kolmogorov equation (IV.3.2) for T_τ is a functional relation, which is not easy to handle in actual applications. The master equation is a more convenient version of the same equation: it is a differential equation obtained by going to the limit of vanishing time difference τ'. For this purpose it is necessary first to ascertain how $T_{\tau'}$ behaves as τ' tends to zero. In the previous section it was found that $T_{\tau'}(y_2|y_1)$ for small τ' has the form[*]

$$T_{\tau'}(y_2|y_1) = (1 - a_0 \tau') \delta(y_2 - y_1) + \tau' W(y_2|y_1) + o(\tau'). \tag{1.1}$$

Here $W(y_2|y_1)$ is the *transition probability per unit time* from y_1 to y_2 and hence

$$W(y_2|y_1) \geq 0. \tag{1.2}$$

The coefficient $1 - a_0 \tau'$ in front of the delta function is the probability that no transition takes place during τ'; hence[**]

$$a_0(y_1) = \int W(y_2|y_1) \, dy_2, \tag{1.3}$$

which also follows immediately from (IV.3.8). We shall here adopt (1.1) as generally valid but promise the reader a further discussion of this essential point in XI.1.

---

[*] The symbol $o(\tau')$ stands for an unspecified term with the property that $o(\tau')/\tau'$ tends to zero as $\tau' \to 0$.

[**] The rationale for the notation $a_0(y)$ will become clear in (V.8.2).

Now in the Chapman–Kolmogorov equation (IV.3.2) insert this expression for $T_{\tau'}$,

$$T_{\tau+\tau'}(y_3|y_1) = [1 - \alpha_0(y_3)\tau']T_\tau(y_3|y_1) + \tau' \int W(y_3|y_2)T_\tau(y_2|y_1)\,dy_2.$$

Divide by τ', go to the limit $\tau' \to 0$, and use (1.3):

$$\frac{\partial}{\partial \tau} T_\tau(y_3|y_1) = \int \left\{ W(y_3|y_2)T_\tau(y_2|y_1) - W(y_2|y_3)T_\tau(y_3|y_1) \right\} dy_2. \quad (1.4)$$

This differential form of the Chapman–Kolmogorov equation is called the master equation.

The equation is usually written in the simplified, more intuitive form

$$\frac{\partial P(y,t)}{\partial t} = \int \left\{ W(y|y')P(y',t) - W(y'|y)P(y,t) \right\} dy'. \quad (1.5)$$

This equation must be interpreted as follows. Take a time t_1 and a value y_1, and consider the solution of (1.5) that is determined for $t \geq t_1$ by the initial condition $P(y, t_1) = \delta(t - t_1)$. This solution is the transition probability $T_{t-t_1}(y|y_1)$ of the Markov process – for any choice of t_1 and y_1. The master equation is *not* meant as an equation for the single-time distribution $P_1(y,t)$![*)]

If the range of Y is a discrete set of states with labels n, the equation reduces to

$$\frac{dp_n(t)}{dt} = \sum_n \left\{ W_{nn'} p_{n'}(t) - W_{n'n} p_n(t) \right\}. \quad (1.6)$$

In this form the meaning becomes particularly clear: *the master equation is a gain–loss equation for the probabilities of the separate states n.* The first term is the gain of state n due to transitions from other states n', and the second term is the loss due to transitions from n into other states. Remember that $W_{nn'} \geq 0$ when $n \neq n'$, and that the term with $n = n'$ does not contribute to the sum.

Terminology. The name "master equation" first appeared in a paper in which it actually had the role of a general equation from which all other results were derived.[**)] It then got stuck onto a special type of equation, namely the above probability balance equation.

[*)] I. Oppenheim and K.E. Shuler, Phys. Rev. B **138**, 1007 (1965); P. Hänggi and H. Thomas, Z. Phys. B **26**, 85 (1977).
[**)] A. Nordsieck, W.E. Lamb, and G.E. Uhlenbeck, Physica **7**, 344 (1940).

One drawback of the name is that it is not specific enough to forbid the use of it for equations of a different type, which creates confusion.*⁾ Sometimes the more specific name "*Pauli master equation*" is used.**⁾ The other drawback is that it cannot be translated. In the Polish literature the name "M-equation" has been introduced (following a suggestion of M. Kac), which does not suffer from these drawbacks. I shall also use it occasionally but it may be too late to make this more international usage prevail.

Not only is the master equation more convenient for mathematical operations than the original Chapman–Kolmogorov equation, it also has a more direct physical interpretation. The quantities $W(y|y')\,\Delta t$ or $W_{nn'}\,\Delta t$ are the probabilities for a transition during a short time Δt. They can therefore be computed, for a given system, by means of any available approximation method that is valid for short times. The best known one is time-dependent perturbation theory, leading to "Fermi's Golden Rule"†⁾

$$W_{nn'} = \frac{2\pi}{\hbar}|H'_{nn'}|^2\rho(E_n).$$

(n, n' label eigenstates of the unperturbed Hamiltonian, $H'_{nn'}$ is the matrix element of the perturbation term in the Hamiltonian, and ρ is the density of unperturbed levels.) The master equation then serves to determine the resulting evolution of the system over long time periods. In this way the two time scales can be treated separately – at the expense of *assuming* the Markov property.

This interpretation of the master equation means that is has an entirely different role than the Chapman–Kolmogorov equation. The latter is a nonlinear equation, which results from the Markov character, but contains no specific information about any particular Markov process. In the master equation, however, one considers the transformation probabilities as given by the specific system, and then has a linear equation for the probabilities which determine the (mesoscopic) state of that system.

As an example we treat the decay process of IV.6 in terms of the master equation. The decay probability γ per unit time is a property of the radioactive nucleus or the excited atom, and can, in principle, be computed by solving the Schrödinger equation for that system. To find the long-time evolution of a collection of emitters write $P(n, t)$ for the probability that there are n surviving emitters at time t. The transition probability for a

---

*⁾ See, e.g., V.M. Kenkre, J. Statist. Phys. **30**, 293 (1983).

**⁾ W. Pauli used an equation of the form (1.6), in: *Probleme der modernen Physik* (Sommerfeld Festschrift, Hirzel, Leipzig 1928).

†⁾ L.I. Schiff, *Quantum Mechanics* (3rd Ed., McGraw-Hill, New York 1968) p. 285. See also XVII.2.

1. DERIVATION

transition from n' to n in a short time Δt is

$$T_{\Delta t}(n|n') = 0 \quad \text{for } n > n',$$
$$= n'\gamma \Delta t \quad \text{for } n = n' - 1,$$
$$= \mathcal{O}(\Delta t)^2 \quad \text{for } n < n' - 1.$$

Hence, according to (1.1)

$$W_{nn'} = \gamma n' \, \delta_{n, n'-1} \quad (n \neq n').$$

Substitution in (1.6) yields the master equation

$$\dot{p}_n(t) = \gamma(n+1) p_{n+1}(t) - \gamma n p_n(t). \tag{1.7}$$

This equation has to be solved with initial condition $p_n(0) = \delta_{n,n_0}$. The explicit solution has been obtained in a different manner in IV.6. We here merely show that some partial results can already be obtained without knowing the complete solution. It should be remarked that the following device is very useful when it works, but is restricted to a specially simple class of master equations, defined in VI.1 as "linear". Multiply (1.7) with n and sum over n:

$$\sum_{n=0}^{\infty} n \dot{p}_n = \gamma \sum_{n=0}^{\infty} n(n+1) p_{n+1} - \gamma \sum_{n=0}^{\infty} n^2 p_n$$
$$= \gamma \sum_{n=0}^{\infty} (n-1) n p_n - \gamma \sum_{n=0}^{\infty} n^2 p_n$$
$$= -\gamma \sum_{n=0}^{\infty} n p_n. \tag{1.8}$$

Thus we have found for the average of the stochastic variable $N(t)$

$$\frac{d}{dt} \langle N(t) \rangle = -\gamma \langle N(t) \rangle. \tag{1.9}$$

Solve this simple equation for $\langle N(t) \rangle$ with initial value $\langle N(0) \rangle = n_0$,

$$\langle N(t) \rangle = n_0 \, e^{-\gamma t}. \tag{1.10}$$

This is the same result as (IV.6.2), compare (IV.6.3).

Exercise. Construct the M-equation for the dichotomic Markov process (IV.2.3).
Exercise. Solve that M-equation and obtain in this way (IV.2.3).
Exercise. It is physically obvious that the solution of (1.7) with $p_n(0) = \delta_{n,n_0}$ must vanish (at all $t \geq 0$) for $n > n_0$. Show that this is in fact a consequence of (1.7). Hence it makes no difference whether the summations in (1.8) are extended to $n = \infty$ or to $n = n_0$.
Exercise. By a computation similar to (1.8) find, for the decay process, $\langle N(t)^2 \rangle$ and $\langle\langle N(t)^2 \rangle\rangle$ and compare the results with (IV.6.9).

Exercise. The random walk with continuous time is defined as follows. The states are all integers n ($-\infty < n < \infty$). A particle can jump between neighboring states. In a short time dt has probability $\tfrac{1}{2}\gamma\, dt$ to jump to the right, and the same probability to jump to the left. Construct the master equation for $p_n(t)$ (compare VI.2).

Exercise. In a population of n bacteria, each individual has a probability α per unit time to die and β to give birth to a new individual. Construct the M-equation ("birth and death process", compare chapter VI).

Exercise. Write the M-equation for processes with independent increments, defined in (IV.4.7), and solve it. Using this solution show again that P_1 becomes asymptotically Gaussian.

Exercise. A Markov process whose transition matrix factorizes, $W(y|y') = u(y)v(y')$ (for $y \neq y'$), is called a "kangaroo process". Show that such M-equations can be solved, i.e., $P(y, t)$ can be expressed in $P(y, 0)$ by means of integrals. [Hint: Derive an integral equation for $\sigma(t) = \int v(y)P(y, t)\, dy$ and solve it by Laplace transformation.]

Exercise. Find the explicit solution of a kangaroo process with $v(y) = $ const. and interpret the result ("Kubo–Anderson process"; the dichotomic Markov process is a special case).

Exercise. Let $(X(t), Y(t))$ be a bivariate Markov process. Suppose its transition probabilities per unit have the property

$$\int W(x, y | x', y')\, dy \quad \text{is independent of } y'.$$

Show that the marginal process $X(t)$ defined in (III.4.10) is again Markovian.

2. The class of W-matrices

It is convenient to use the notation for discrete states; the generalization to the continuous case is straightforward from a formal point of view, while any additional mathematical complications are of no concern to us. The master equation (1.6) can be written in a more compact form when the following matrix $\mathbb{W}$ is defined:

$$\mathbb{W}_{nn'} = W_{nn'} - \delta_{nn'}\left(\sum_{n''} W_{n''n}\right). \tag{2.1}$$

Then (1.6) may be written

$$\dot{p}_n(t) = \sum_{n'} \mathbb{W}_{nn'} p_{n'}(t), \tag{2.2}$$

or, more briefly,

$$\dot{p}(t) = \mathbb{W} p(t), \tag{2.3}$$

where p is the vector with components p_n.

2. THE CLASS OF W-MATRICES

Formally the solution of (2.3) with given initial $p_n(0)$ may be written

$$p(t) = e^{tW} p(0). \tag{2.4}$$

This expression for $p(t)$ is sometimes convenient but does not help to find $p(t)$ explicitly. The familiar method for solving equations of type (2.3) by means of eigenvectors and eigenvalues of W cannot be used as a general method, because W need not be symmetric, so that it is not certain that all solutions can be obtained as superpositions of these eigensolutions (see, however, V.7).

Any general results will have to be based on the following two properties of W,

$$W_{nn'} \geqslant 0 \quad \text{for } n \neq n'; \tag{2.5a}$$

$$\sum_n W_{nn'} = 0 \quad \text{for each } n'. \tag{2.5b}$$

These properties are preserved when a permutation is applied simultaneously to rows and columns, which amounts to relabeling the states. They are not preserved by arbitrary similarity transformations $W \to S^{-1} W S$; such transformations are therefore of no avail in the present context. A matrix with the properties (2.5) will be called a W-matrix. We shall now list a number of consequences of these properties. Our statements are rigorous for finite-dimensional W-matrices. They also often apply to systems whose range is denumerably infinite or continuous. In these cases they constitute a useful, if not universally reliable guide.

Equation (2.5b) states that W has a left eigenvector $\psi = (1, 1, 1, ...)$ with zero eigenvalue. Hence there is also a right eigenvector ϕ with the same eigenvalue,

$$W\phi = 0.$$

Of course, there may be more than one. Each ϕ is a time-independent solution of the master equation. When normalized it represents a stationary probability distribution of the system, provided its components ϕ_n are nonnegative. In the next section we shall show that this provision is satisfied. But first we shall distinguish some special forms of W.

A matrix W is called *completely reducible* or *decomposable* if by a suitable simultaneous permutation of rows and columns it can be cast into the form

$$W = \begin{pmatrix} A & 0 \\ 0 & B \end{pmatrix}, \tag{2.6}$$

where A and B are two square matrices of lower dimensionality. In this case the states decompose into two subsets between which no transition is possible. It is easily seen that A and B are again W-matrices. Hence they have right eigenvectors ϕ^A and ϕ^B with zero eigenvalue. Consequently, (2.6) has at least two linearly independent eigenvectors with eigenvalue zero:

$$\begin{pmatrix} \mathbb{A} & 0 \\ 0 & \mathbb{B} \end{pmatrix} \begin{pmatrix} \phi^A \\ 0 \end{pmatrix} = 0 \quad \text{and} \quad \begin{pmatrix} \mathbb{A} & 0 \\ 0 & \mathbb{B} \end{pmatrix} \begin{pmatrix} 0 \\ \phi^B \end{pmatrix} = 0.$$

The case of a decomposable matrix (2.6) merely means that one has two non-interacting systems, governed by two M-equations with matrices $\mathbb{A}$ and $\mathbb{B}$, respectively. A non-trivial example is a system in which all transitions conserve energy: each energy shell E has its own M-equation and its own stationary distribution ϕ^E. The stationary solutions of the total M-equation are linear superpositions of them with arbitrary coefficients π_E,

$$\phi = \sum_E \pi_E \phi^E.$$

All ϕ constructed in this way with the additional restrictions

$$\pi_E \geqslant 0 \quad \text{and} \quad \sum_E \pi_E = 1$$

are stationary distributions of the total system. They represent states in which the energy is distributed in different ways over the various energy shells.

A matrix $\mathbb{W}$ is called (*incompletely*) *reducible* if it can be cast in the form

$$\mathbb{W} = \begin{pmatrix} \mathbb{A} & \mathbb{D} \\ 0 & \mathbb{B} \end{pmatrix}, \tag{2.7}$$

where $\mathbb{A}$ and $\mathbb{B}$ are again square, but $\mathbb{D}$ in general is not. $\mathbb{A}$ is again a W-matrix but the columns of $\mathbb{B}$ add up to negative amounts. Evidently (2.7) has an eigenvector with zero eigenvalue

$$\phi = \begin{pmatrix} \phi^A \\ 0 \end{pmatrix}.$$

To give this statement more physical content consider the master equation corresponding to (2.7). Let me use the labels a and b for the two sets of components, respectively:

$$\begin{aligned} \dot{p}_a &= \sum_{a'} \mathbb{A}_{aa'} p_{a'} + \sum_{b'} \mathbb{D}_{ab'} p_{b'}, \\ \dot{p}_b &= \sum_{b'} \mathbb{B}_{bb'} p_{b'}. \end{aligned} \tag{2.8}$$

This last line constitutes a closed set of equations for the p_b, i.e., its solutions can be obtained without knowing the p_a. But the p_b loose probability to the

2. THE CLASS OF $\mathbb{W}$-MATRICES

states a:

$$\frac{d}{dt}\sum_b p_b = \sum_{b'}\left(\sum_b \mathbb{B}_{bb'}\right)p_{b'} = -\sum_{b'}\left(\sum_a \mathbb{D}_{ab'}\right)p_{b'}. \quad (2.9)$$

The states b are depleted and a solution of the M-equation can only be stationary if all its components b are zero. Such states b are called *transient*, while the set of states a into which the probability is collected is called *absorbing*.

A $\mathbb{W}$-matrix will be called a *splitting* matrix if it can be cast into the form

$$\mathbb{W} = \begin{bmatrix} \mathbb{A} & 0 & \mathbb{D} \\ 0 & \mathbb{B} & \mathbb{E} \\ 0 & 0 & \mathbb{C} \end{bmatrix}. \quad (2.10)$$

$\mathbb{A}$ and $\mathbb{B}$ are $\mathbb{W}$-matrices, $\mathbb{C}$ is square, and at least some elements of $\mathbb{D}$ and $\mathbb{E}$ are non-zero. A splitting matrix is reducible, with the additional property that when the rows and columns corresponding to the transient states are erased a decomposable matrix is left. There are three sets of states, labelled with a, b, c. The states c are transient and deplete into a and b. There are (at least) two linearly independent eigenvectors with eigenvalue zero, viz.

$$\begin{bmatrix} \phi^A \\ 0 \\ 0 \end{bmatrix} \quad \text{and} \quad \begin{bmatrix} 0 \\ \phi^B \\ 0 \end{bmatrix}.$$

Exercise. If $\mathbb{W}$ is a 2×2 matrix the exponential in (2.4) can be evaluated directly by expanding in powers of $\mathbb{W}$. Use this method for solving the master equation of the dichotomic Markov process.

Exercise. Show that the following matrix is reducible according to the definition in the text

$$\mathbb{W} = \begin{pmatrix} \mathbb{A} & 0 \\ \mathbb{C} & \mathbb{B} \end{pmatrix}.$$

Exercise. As an example of a splitting process consider the π^- decay

$$\pi^- \begin{matrix} \nearrow e^- + \nu \\ \searrow \mu^- + \nu \to e^- + \nu + \nu + \bar{\nu}. \end{matrix}$$

Identify the states a, b, c and write the matrix $\mathbb{W}$ for this process.

Exercise. When $\mathbb{W}$ is of the form (2.6), (2.7), or (2.10) each power $\mathbb{W}^k$ ($k = 0, 1, 2, \ldots$) has the same form.

Exercise. Let $\mathbb{W}$ have an eigenvector with nonnegative components, some of which are zero. Then $\mathbb{W}$ is reducible. Also when $\mathbb{W}$ has a degenerate eigenvalue having two eigenvectors with nonnegative components.

Exercise. Let $\mathbb{W}$ have, at the eigenvalue zero, two linearly independent eigenvectors with positive components. Then $\mathbb{W}$ is decomposable.

Exercise. Let $\mathbb{W}$ have the form (2.7) with no other condition on $\mathbb{D}$ than that its elements are not negative. Then either all b are transient or $\mathbb{W}$ is splitting or even decomposable. There is at least one transient state unless $\mathbb{D} = 0$. The b are certainly all transient when $\mathbb{D}$ has at least one nonzero element in each column.

Exercise. The branching process in III.6 is not Markovian and obeys therefore no master equation unless γ is independent of τ. Write the master equation for this special case. What can be said about transient and absorbing states?

Exercise. Verify that the use of "absorbing" and "transient" in IV.6 is in agreement with the present definitions.

Exercise. A $\mathbb{W}$-matrix can be represented by a graph in which the vertices represent the states and a directed line is drawn from one vertex n to another n' when $W_{n'n} > 0$. How are the transient and absorbing states in such a graph characterized?

Exercise. An example with an infinite set of states to which the analysis of this section does not apply is the infinite random walk with continuous time

$$\dot{p}_n = p_{n+1} + p_{n-1} - 2p_n. \tag{2.11}$$

Show that there is no stationary distribution and that all states are transient (compare VI.2).

3. The long-time limit

A fundamental property of the master equation is: *As $t \to \infty$ all solutions tend to the stationary solution*; or – in the case of decomposable or splitting $\mathbb{W}$ – to one of the stationary solutions. Again this statement is strictly true only for a finite number of discrete states. For an infinite number of states, and *a fortiori* for a continuous state space, there are exceptions, e.g., the random walk (2.11).[*] Yet it is a useful rule of thumb for a physicist who knows that many systems tend to equilibrium. We shall therefore not attempt to give a general proof covering all possible cases, but restrict ourselves to a finite state space. There exist several ways of proving the theorem. Of course, they all rely on the property (2.5), which defines the class of $\mathbb{W}$-matrices.

(i) The mathematicians discretize time by introducing a finite time step Δt and thereby reduce the process to a Markov chain with transition matrix $T_1 = \exp(\mathbb{W} \Delta t)$. The theorems of Perron and Frobenius mentioned in IV.5

---

[*] J.O. Vigfússon, J. Statist. Phys. **27**, 339 (1982).

then provide a complete answer. This approach seems rather artificial to the physicist, and moreover transfers the problem to proving Perron–Frobenius.

(ii) The physicists constructs an entropy function (or H-function or Lyapunov function) having the property that it varies monotonically but is bounded, so that is must tend to a limit. This is, of course, a general principle in physics, but we are only interested in its application to master equations, which will be given in **5**. It should be pointed out that this proof too is strictly correct only for a finite number of states.

(iii) In the special case that $\mathbb{W}$ is a symmetric matrix and therefore diagonalizable it is sufficient to show that all its eigenvalues are negative except the zero eigenvalue belonging to the stationary solution. This symmetry can often be derived from the detailed balance property, see V.**6**, and the corresponding eigenfunction expansion is given by V.**7**. But detailed balance is not universal and there exist many applications of master equations whose $\mathbb{W}$ is not symmetric.

(iv) Other proofs are due to Kirchhoff, who used network theory[*], and to Uhlmann, who used mathematical inequalities involving convexity[**].

(v) The following proof is relatively simple and uses the intuitive idea that the transitions tend to transfer probability from states having more than their equilibrium share into destitute states. A different version of the proof is outlined in the last Exercise of V.**9**.

We first establish two lemmas. Let $\phi(t)$ be any solution of the master equation, not necessarily positive or normalized. *At a given time* t the positive, negative and zero components are distinguished by using an index u, v, w, respectively:

$$\phi_u(t) > 0, \qquad \phi_v(t) < 0, \qquad \phi_w(t) = 0.$$

Let $U(t)$ be the sum of the positive ones,

$$U(t) = \sum_u \phi_u(t). \tag{3.1}$$

Obviously $U(t)$ is positive; or zero if the set of components u is empty. Our FIRST LEMMA asserts that $U(t)$ *is a monotonic non-increasing function of* t.

$U(t)$ varies in time firstly because each of the terms in (3.1) varies, and secondly because at certain instants new components enter into the sum or others leave it. During the time intervals between these instants one has

---

[*] J. Schnakenberg, Rev. Mod. Phys. **48**, 571 (1976).
[**] A. Uhlmann, Wissens. Z. Karl-Marx-Universität Leipzig, Mathem.-Naturw. Reihe **27**, 213 (1978); R. Kubo, in: *Perspectives in Statistical Physics* (H.J. Raveché ed., North-Holland, Amsterdam 1981).

from (2.5b)

$$\dot{U}(t) = \sum_u \dot{\phi}_u = \sum_u \left(\sum_{u'} \mathbb{W}_{uu'} \phi_{u'} + \sum_{v'} \mathbb{W}_{uv'} \phi_{v'} \right)$$
$$= \sum_{u'} \left(-\sum_v W_{vu'} - \sum_w W_{wu'} \right) \phi_{u'} + \sum_{v'} \left(\sum_u W_{uv'} \right) \phi_{v'}. \quad (3.2)$$

Each term is non-positive and therefore

$$\dot{U}(t) \leqslant 0. \quad (3.3)$$

At an instant when a term enters or leaves the sum (3.1) the function $U(t)$ is not differentiable, but it is still continuous since that term is zero at that instant. Thus $U(t)$ is continuous everywhere, and obeys (3.3) except at a discrete set of time points; consequently it cannot increase. This proves the lemma.

Similarly to (3.1) define the negative sum V by

$$V(t) = \sum_v \phi_v(t). \quad (3.4)$$

$V(t)$ is non-decreasing since $U + V = $ const. Hence the COROLLARY: *if initially $\phi_n(0) \geqslant 0$ for all n, then $\phi_n(t) \geqslant 0$ for all $t > 0$.* If the master equation would not have this property of conserving positivity, it could not be the correct equation for the evolution of a probability distribution.

From (3.3) follows that $U(t)$ tends to a limit $U(\infty) \geqslant 0$ and, similarly, $V(t)$ tends to $V(\infty)$. Our SECOND LEMMA asserts that *unless $\mathbb{W}$ is decomposable or of splitting type at least one of these limiting values must vanish*, so that ultimately all $\phi_n(t)$ have the same sign or are zero. The sign is determined by the initial values, because

$$\sum_n \phi_n(t) = \text{const.} = C. \quad (3.5)$$

If $C = 0$ all $\phi_n(t)$ will have to tend to zero. Without loss of generality we may suppose $C \geqslant 0$, and prove $V(\infty) = 0$.

To prove this second lemma we note that $\dot{U}(\infty) = 0$. This is compatible with (3.2) only if each of the three terms vanishes. That can happen in a number of ways.

(i) The set of components u is empty, i.e., $\phi_n(\infty) \leqslant 0$ for all n. As we chose $C \geqslant 0$ this implies $\phi_n(\infty) = 0$ and $U(\infty) = V(\infty) = 0$.

(ii) The set of components u is not empty, but both sets of components v and w are. Then, obviously, $V(\infty) = 0$.

(iii) Neither the set of components u or v are empty, but the set of components w is. Then one must have $W_{vu'} = W_{uv'} = 0$, so that $\mathbb{W}$ has the

form
$$\begin{pmatrix} \mathbb{W}_{uu'} & 0 \\ 0 & \mathbb{W}_{vv'} \end{pmatrix}.$$

Thus $\mathbb{W}$ is decomposable and therefore excluded from the lemma.

(iv) None of the sets u, v, w are empty. Then one must have $\mathbb{W}_{vu'} = \mathbb{W}_{wu'} = \mathbb{W}_{uv'} = 0$, so that $\mathbb{W}$ has the form

$$\begin{bmatrix} \mathbb{W}_{uu'} & 0 & \mathbb{W}_{uw'} \\ 0 & \mathbb{W}_{vv'} & \mathbb{W}_{vw'} \\ 0 & \mathbb{W}_{wv'} & \mathbb{W}_{ww'} \end{bmatrix}.$$

Thus $\mathbb{W}$ is reducible. By writing the analog of eq. (3.2) for $\dot{V}(t)$ one also finds $\mathbb{W}_{wv'} = 0$, so that $\mathbb{W}$ is of splitting type and also excluded from the lemma.

This completes the proof of the second lemma. A COROLLARY is that for a time-independent solution either all components are nonnegative, or all non-positive. For a stationary probability distribution one has, of course, $p_n^s \geq 0$, because $C = 1$.

Now suppose that $p_n^{(1)}(t)$ and $p_n^{(2)}(t)$ are two probability distributions satisfying a master equation that is neither decomposable nor splitting. Then $\phi_n(t) = p_n^{(1)}(t) - p_n^{(2)}(t)$ is a solution for which $C = 0$. Hence

$$\phi_n(t) = p_n^{(1)}(t) - p_n^{(2)}(t) \to 0.$$

From this follows first that there cannot be more than one stationary distribution. Secondly, that any other solution tends to it. This is the desired result.

Exercise. Show for a continuous range that

$$U(t) = \int_{P(y,t)>0} P(y,t)\,dy$$

is again monotonically decreasing and conclude that an initially positive solution remains positive. Why can one – strictly speaking – no longer conclude that the solution must approach a stationary distribution?

Exercise. The average time a system spends in a transient state b is given by

$$\theta_b = \int_0^\infty p_b(t)\,dt.$$

Show that θ_b can be found from the equation

$$\sum_{b'} \mathbb{B}_{bb'} \theta_{b'} = -p_b(0). \tag{3.6}$$

See also VII.7.

Exercise. Suppose $\mathbb{W}$ is of the splitting type (2.10) and the system starts in a transient state c_0. The total probability for ending up in one of the states a is

$$p_A(\infty) = - \sum_{a,c} \mathbb{D}_{ac}(\mathbb{C}^{-1})_{cc_0}.$$

Exercise. Let $n = 0, 1, 2, \ldots$ and

$$\dot{p}_0 = -p_0 + \sum_{v=1}^{\infty} \alpha_v p_v,$$

$$\dot{p}_v = p_{v-1} - (1 + \alpha_v) p_v \quad (v = 1, 2, \ldots).$$

Find the stationary solution for the case that $\Sigma \alpha_v = \infty$. Show that when $\Sigma \alpha_v$ converges no stationary solution exists.

4. Closed, isolated, physical systems

In this title "physical" means that there must be an underlying microscopic description in terms of Hamilton equations or a Schrödinger equation. "Closed" means that no exchange of matter with the external world takes place, so that the set of microscopic variables is fixed. "Isolated" means that no external time-dependent forces act on the system, so that the energy is a constant of the motion and the trajectory in phase space is confined to a single energy shell. Moreover the system has to be supposed finite in the sense that the measure of each individual energy shell is finite. According to equilibrium statistical mechanics for such a system there is, for given value of the energy, an equilibrium distribution – the microcanonical ensemble – which can be found from phase space considerations alone.[*] When there are additional constants of the motion, e.g., in a cylindrical vessel the angular momentum, the energy shell is decomposed into "subshells", each corresponding to fixed values for these constants. No transition between subshells is possible. On the other hand, ergodic theory tells us that once the system is in a certain subshell its motion covers the whole subshell, provided that all constants of the motion were taken into account when defining the subshells.[**]

Now suppose that such a system can be described on a mesoscopic level by a master equation. That means that the subshell to which it is confined

---

[*] Actually an additional stability criterion is needed, see M.E. Fisher, Archives Rat. Mech. Anal. **17**, 377 (1964); D. Ruelle, *Statistical Mechanics, Rigorous Results* (Benjamin, New York 1969). A collection of point particles with mutual gravitational attraction is an example where this criterion is not satisfied, and for which therefore no statistical mechanics exists.

[**] A.I. Khinchin, *Mathematical Foundations of Statistical Mechanics* (G. Gamow transl., Dover Publications, New York 1949); R. Balescu, *Equilibrium and Nonequilibrium Statistical Mechanics* (Wiley–Interscience, New York 1975) Appendix.

4. CLOSED, ISOLATED, PHYSICAL SYSTEMS

can again be subdivided into "phase cells" in such a way that the evolution of the system can be described (approximately) in terms of transition probabilities $W_{nn'}$ between any two cells n, n'. Then these $W_{nn'}$ have certain properties in addition to (2.5), which in general are not true for W-matrices describing open systems, or non-physical systems such as populations. These properties are the subject of this and the next two sections.

First it is clear that the total W decomposes into separate blocks for the separate subshells. Hence we may confine ourselves to a single subshell. According to the ergodic property the remaining block of W is indecomposable. It therefore has a single stationary solution p_n^s.

Next we know that this p_n^s must be identified with the equilibrium distribution p_n^e, as determined by the volume of the phase cell n:

$$\sum_{n'} W_{nn'} p_{n'}^e = \left(\sum_{n'} W_{n'n} \right) p_n^e. \tag{4.1}$$

This is a relation among the transition probabilities, the coefficients p_n^e being known from ordinary statistical mechanics.[*] Moreover no p_n^e vanishes and therefore there are no transient states, so that the W for each subshell is irreducible.

As an example take a gas in a cylindrical vessel. In addition to the energy there is one other constant of the motion: the angular momentum around the cylinder axis. The $6N$-dimensional phase space is thereby reduced to subshells of $6N-2$ dimensions. Consider a small subvolume in the vessel and let $Y(t)$ be the number of molecules in it. According to III.2 $Y(t)$ is a stochastic function, with range $n = 0, 1, 2, \ldots, N$. Each value $Y = n$ delineates a phase cell[**]. One expects that $Y(t)$ is a Markov process if the gas is sufficiently dilute; and that p_n^e is approximately a Poisson distribution if the subvolume is much smaller than the vessel.

Finally we mention *detailed balance*, which will be studied in section **6**. Equation (4.1) merely states the obvious fact that in equilibrium the sum of all transitions per unit time into any state n must be balanced by the sum of all transitions from n into other states n'. Detailed balance is the stronger assertion that *for each pair n, n' separately* the transitions must balance:

$$W_{nn'} p_{n'}^e = W_{n'n} p_n^e. \tag{4.2}$$

This is true for closed, isolated, finite physical systems under certain restric-

---

[*] Throughout we use the superscript e for the thermodynamic equilibrium, and s for any stationary, i.e., time-independent solution of the master equation.

[**] Called "Z-star" by P. and T. Ehrenfest, in: *Encyklopädie der mathematischen Wissenschaften*, Band 4, Nr. 32 (Teubner, Leipzig 1912); translated by M.J. Moravcsik under the title: *Conceptual Foundations of the Statistical Approach in Mechanics* (Cornell University Press, Ithaca 1959).

tions to be specified in **6**. Inasmuch as p_n^e is known from ordinary statistical mechanics it is a property of the transition probabilities.

The detailed balance property can also be applied to a system interacting with a heat bath, by the simple device of treating it as a small subsystem of the larger system that includes the heat bath. Examples are: a gas in contact with a heat bath, a collection of atoms in interaction with the electromagnetic radiation field, and a spin system interacting with the lattice phonons. Let ε_n denote the energies of the various states of the small system in which we are interested. Let E be the energy of the total system, which has, of course, a fixed value. In thermodynamic equilibrium the probability for the small system to be in state n is, apart from normalization, equal to that volume in the phase space of the total system that corresponds to the condition that the small system is in n. That volume is the product of the phase space volume g_n of the small system and the phase space volume of the heat bath with energy $E - \varepsilon_n$. According to statistical mechanics the latter factor is

$$\exp\left[\frac{1}{k}S(E-\varepsilon_n)\right] = \exp\left[\frac{1}{k}S(E) - \frac{1}{k}\frac{\mathrm{d}S}{\mathrm{d}E}\varepsilon_n\right] = \mathrm{const.}\,\exp\left[-\frac{\varepsilon_n}{kT}\right]. \quad (4.3)$$

Here k is Boltzmann's constant, S the entropy of the heat bath, and T its temperature. Hence *detailed balance for a system in contact with a heat bath at temperature T has the form*

$$W_{nn'}g_{n'}\,\mathrm{e}^{-\varepsilon_{n'}/kT} = W_{n'n}g_n\,\mathrm{e}^{-\varepsilon_n/kT}. \quad (4.4)$$

Note that this relation requires no detailed knowledge of the bath, but only its general thermodynamic properties.[*]

Remark. The idea of detailed balance first arose in chemical reaction kinetics. Suppose that in a mixture of chemical compounds a cycle of three possible reactions exists, as in fig. 9. Equation (**4.2**) asserts that, *in equilibrium*, each reaction takes place

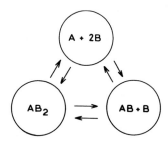

Fig. 9. An example of detailed balance.

---

[*] In particular, it is *not* necessary to invoke quantum mechanics, see J.F. Dobson, Chem. Phys. Letters **61**, 157 (1979).

as often from left to right as from right to left. In this connection the statement may appear intuitively obvious. The reason for this intuitive feeling is that it ought to be possible to block one reaction without affecting the equilibrium and the other reaction rates. And, of course, as soon as the atoms can no longer travel in a cycle, but only back and forth, the detailed balance property is no more than the equilibrium property (4.1). In other applications it is less plausible that one channel between two states can be blocked without affecting the others; in fact, it cannot always be true because detailed balance in the form (4.2) is true only under certain restrictions, specified in **6**.

Exercise. Show that the terms omitted when expanding $S(E - \varepsilon_n)$ in (4.3) tend to zero when the size of the heat bath goes to infinity ("thermodynamic limit").

Exercise. Two volumes are filled with a gas and communicate through a number of apertures of different sizes. What does detailed balance have to say about the gas flows through the apertures?

Exercise. An atom undergoes transitions between its levels E_n through emission and absorption of photons. The transition probabilities $W_{nn'}$ are related by (4.4). Show that this guarantees that in equilibrium the atom has a Boltzmann distribution.

5. The increase of entropy

The following calculation proves the approach to the stationary distribution in an alternative way, which is more familiar to physicists. In addition it provides some information about how this approach takes place, viz., in such a way that a certain functional of the distribution increases monotonically. In fact, it turns out that there are many such functionals. The reasons why one of them has a special role in physics and is honored with the name "entropy" will be discussed.

The present proof is more limited than the one in **3** because we have to assume beforehand that there is a stationary solution that is everywhere positive. For closed, isolated, physical systems one knows that that is so, and we therefore use here the symbol p_n^e for that stationary solution of the master equation. Yet the proof also applies to other cases provided they have no transient states, but the proof does not require detailed balance of any other symmetry relation of the type (4.2).

Consider the M-equation (2.2). Assume that a normalized stationary solution p_n^e exists with $p_n^e > 0$. Take an arbitrary nonnegative *convex* function $f(x)$ defined for positive x:

$$0 \leqslant x < \infty, \quad f(x) \geqslant 0, \quad f''(x) > 0. \tag{5.1}$$

Define a quantity H by

$$H(t) = \sum_n p_n^e f\left(\frac{p_n(t)}{p_n^e}\right) = \sum_n p_n^e f(x_n), \tag{5.2}$$

where x_n is an abbreviation for p_n/p_n^e. Clearly $H(t) \geq 0$ and

$$\frac{dH(t)}{dt} = \sum_{nn'} f'(x_n)(W_{nn'}p_{n'} - W_{n'n}p_n)$$
$$= \sum_{nn'} W_{nn'} p_{n'}^e \{x_{n'} f'(x_n) - x_{n'} f'(x_{n'})\}. \quad (5.3)$$

Now for an arbitrary set of numbers ψ_n one readily verifies the identity

$$\sum_{nn'} W_{nn'} p_{n'}^e (\psi_n - \psi_{n'}) = 0. \quad (5.4)$$

Choose $\psi_n = f(x_n) - x_n f'(x_n)$ and add the resulting identity to (5.3)

$$\frac{dH(t)}{dt} = \sum_{nn'} W_{nn'} p_{n'}^e \{(x_{n'} - x_n) f'(x_n) + f(x_n) - f(x_{n'})\}. \quad (5.5)$$

It is evident from fig. 10 that for any convex f the factor $\{\ \}$ is negative unless $x_n = x_{n'}$. The conclusion is that $H(t)$ decreases monotonically.

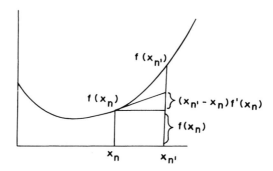

Fig. 10. A property of convex functions.

As $H(t)$ decreases but cannot become negative it must tend to a limit. In this limit (5.5) must vanish.*) That is only possible if $x_n = x_{n'}$ for all pairs of states n, n' for which $W_{nn'} \neq 0$. This implies that each $x_{n'}$ is equal to all those x_n that can be reached from it by a string of transitions with non-zero probabilities. Either n covers all states; or it covers a subset of them which is not connected by any non-zero transition probability to any state outside the subset. In the former case, $p_n(\infty)$ is proportional to p_n^e and must be equal to it because of normalization. In the latter case, $\mathbb{W}$ is decomposable, and one can merely assert that $p_n(\infty)$ is proportional to p_n^e inside the subset, but different proportionality constants may hold in the different subsets.

---

*) This is the customary argument in dealing with H-theorems, but is actually only rigorous if the number of states n is finite.

Exercise. Actually a function f is called "convex" when $f''(x) \geq 0$, and "strictly convex" when $f''(x) > 0$, as was required in (5.1). Show by a counterexample that mere convexity is not enough for our purpose.

Exercise. Prove (5.4).

Exercise. If the detailed balance equation (4.2) holds ony may write for the right-hand side of (5.3)

$$\tfrac{1}{2} \sum_{nn'} W_{nn'} p^e_{n'} (x_{n'} - x_n)\{f'(x_n) - f'(x_{n'})\}.$$

This can immediately be seen to be negative, without the use of (5.4). Compare 7.

For our purpose any convex $f(x)$ that is defined for $x \geq 0$ and bounded below could serve, e.g., $f(x) = x^2$ or $f(x) = x^\alpha$ ($\alpha > 1$) have been used[*]. Yet one customarily chooses

$$f(x) = x \log x, \qquad H = \sum_n p_n \log \frac{p_n}{p^e_n}, \tag{5.6}$$

which is bounded below and has two additional properties, which endow it with more physical content.

The first one is that this particular form of H can also be used to prove the approach to equilibrium in the case of Boltzmann's kinetic equation for dilute gases. The Boltzmann equation is nonlinear and a different technique is needed to prove that all solutions tend to equilibrium. This technique is based on (5.6); other convex functions cannot be used.[**] Incidentally, the Boltzmann equation is *not* a master equation for a probability density, but an evolution equation for the particle density in the six-dimensional one-particle phase space ("μ-space"). The *linearized* Boltzmann equation, however, has the same structure as a master equation (compare XIV.5).

The second additional property of (5.6) is that it makes H an *additive* quantity in the following sense. Take two independent systems with states n and m and probabilities p_n and q_m. They may be viewed as one combined system, whose states are labelled (n, m) with probabilities $p_n q_m$. Then the functional H of the combined system is the sum of those for the separate systems,

$$\sum_{n,m} p_n q_m \log \frac{p_n q_m}{p^e_n q^e_m} = \sum_n p_n \log \frac{p_n}{p^e_n} + \sum_m q_m \log \frac{q_m}{q^e_m}.$$

It can be concluded from this that, if the system consists of a gas or a solid, H is an *extensive* quantity.[†]

---

[*] For a survey, see A. Wehrl, Rev. Mod. Phys. **50**, 221 (1978).
[**] P.A.P. Moran, Proc. Camb. Philos. Soc. **57**, 833 (1961).
[†] This excludes such anomalies as "non-extensive entropy".

These two additional properties of the H in (5.6), together with its monotonic decrease, has led to its identification with the entropy defined by the second law of thermodynamics. It must be realized, however, that H is a functional of a non-equilibrium probability distribution, whereas the thermodynamic entropy is a quantity defined for thermodynamic equilibrium states.[*] The present entropy is therefore a generalization of the thermodynamic entropy; the generalized entropy is

$$S = -kH + S^e, \tag{5.7}$$

where k is a factor determined by the units in which S is measured, and S^e is a constant term independent of the p_n. In equilibrium, $H = 0$, so that S^e is the thermodynamic entropy, which can only be altered by external interference. The quantity H is merely the entropy difference between equilibrium and non-equilibrium states.

Exercise. With the special choice (5.6) one finds (5.5) is negative if it is true that for all positive x, y

$$x \log x - x \log y - x + y \geq 0, \tag{5.8}$$

with equality only for $x = y$. Prove this so-called Klein's inequality[**] directly.

Exercise. Prove the following generalization of the above H-theorem. Let p_n and p'_n be two solutions of the M-equation and set

$$H(t) = \sum p_n (\log p_n - \log p'_n). \tag{5.9}$$

This quantity is positive and tends to zero. Hence $p_n(t) - p'_n(t) \to 0$.

6. Proof of detailed balance

We shall prove detailed balance in closed, isolated, classical systems.[†] The label n denotes phase cells, but it is more convenient to use continuous variables and to write the detailed balance equation (4.2) in the form

$$W(y|y')P^e(y') = W(y'|y)P^e(y). \tag{6.1}$$

Here y stands for the values of a set of macroscopic observables $Y(q, p)$. We shall prove (6.1) under the following two conditions.

(i) The Hamiltonian function of the system, which governs the microscopic motion, is an even function of all momenta p_k. This requirement excludes an external magnetic field and an overall rotation of the entire system.

---

[*] Or at least local equilibrium, see N.G. van Kampen, Physica **25**, 1294 (1959).
[**] O. Klein, Z. Phys. **72**, 767 (1931).
[†] E.P. Wigner, J. Chem. Phys. **22**, 1912 (1954); DE GROOT AND MAZUR, p. 93. For the quantum mechanical proof, see XVII.7.

6. PROOF OF DETAILED BALANCE

(ii) The variables Y are also even functions of the momenta.

If these conditions are not satisfied one cannot generally prove (6.1), but a similar equation still holds, which will be derived afterwards.

Let the system be described by f coordinates q_k and f momenta p_k ($k = 1, 2, \ldots, f$), which form the $2f$-dimensional phase space Γ. The equations of motion are

$$\frac{dq_k}{dt} = \frac{\partial H(q, p)}{\partial p_k}, \quad \frac{dp_k}{dt} = -\frac{\partial H(q, p)}{\partial q_k}. \tag{6.2}$$

As a consequence of (i) one has the following invariance property. If one substitutes in (6.2)

$$t = -\bar{t}, \quad q_k = \bar{q}_k, \quad p_k = -\bar{p}_k, \tag{6.3}$$

the equations in terms of $\bar{t}, \bar{q}_k, \bar{p}_k$ have the same form as in t, q, p. Thus to each possible trajectory in Γ one can find another one by reflecting the values of p_k, which, traversed in the opposite direction, is again a solution of (6.2). Or, to formulate this more precisely, suppose at some time our system is in the point (q, p) of Γ-space. At a time τ later it will be in a point (q', p'), uniquely related to (q, p) by the equations of motion. Then it is also true that if it starts in $(q', -p')$ it will be at a time τ later in $(q, -p)$. This property of the equations of motion is called invariance for time reversal. It is the starting point for the proof of detailed balance. We abbreviate the notation. Let the point (q, p) in Γ-space be denoted by x, and let $x^\tau = (q', p')$ be the point into which the system is transported in the time τ. Furthermore $\bar{x} = (q, -p)$ so that

$$\bar{\bar{x}} = x. \tag{6.4}$$

Time reversal invariance is then expressed by

$$\overline{(x^\tau)^\tau} = \bar{x} \quad \text{or} \quad \overline{x^\tau} = (\bar{x})^{-\tau}. \tag{6.5}$$

It is also clear that the bar operation conserves the volume in Γ

$$d\bar{x} = dx. \tag{6.6}$$

We can now give a formal proof. A consequence of (ii) is

$$Y_{\bar{x}}(0) = Y_x(0), \tag{6.7}$$

from which follows

$$Y_{\bar{x}}(t) = Y_{\bar{x}^t}(0) = Y_{\overline{x^{-t}}}(0) = Y_{x^{-t}}(0) = Y_x(-t). \tag{6.8}$$

Moreover, since the equilibrium distribution is a function of the constants of the motion, which must also be even functions of the velocities, one has

$$P^e_X(\bar{x}) = P^e_X(x). \tag{6.9}$$

After these preliminaries we write

$$P_2(y_1, 0; y_2, \tau) = \int \delta[y_1 - Y_x(0)] \, \delta[y_2 - Y_x(\tau)] P_X^e(x) \, dx.$$

Change the integration variable from x to $\bar{x}$,

$$= \int \delta[y_1 - Y_{\bar{x}}(0)] \, \delta[y_2 - Y_{\bar{x}}(\tau)] P_X^e(\bar{x}) \, d\bar{x}.$$

Use (6.7), (6.8), (6.9), (6.6),

$$= \int \delta[y_1 - Y_x(0)] \, \delta[y_2 - Y_x(-\tau)] P_X^e(x) \, dx$$

$$= P_2(y_1, 0; y_2, -\tau) = P_2(y_2, 0; y_1, \tau).$$

Thus we have found a symmetry relation for P_2, which can immediately be translated into a relation for $P_{1|1}$:

$$P_{1|1}(y_2, \tau \mid y_1, 0) P^e(y_1) = P_{1|1}(y_1, \tau \mid y_2, 0) P^e(y_2). \tag{6.10}$$

If the variables Y are such that $Y(t)$ is a Markov process, this relation may be written

$$T_\tau(y_2 \mid y_1) P^e(y_1) = T_\tau(y_1 \mid y_2) P^e(y_2), \tag{6.11}$$

and with the definition (1.1) it yields (6.1). Q.E.D.

Now suppose the condition (i) is not satisfied. In particular, let there be an external magnetic field B with vector potential A. Then the Hamilton function contains $(p - eA)^2$ instead of p^2, and does not remain the same on replacing p with $-p$. However, this can be remedied by simultaneously changing the sign of the field. The transformation (6.3) then maps the trajectories of (6.2) with field B onto the reversed trajectories of (6.2) with field $-B$. As a consequence one finds instead of (6.1)

$$W(y \mid y'; B) P^e(y') = W(y' \mid y; -B) P^e(y). \tag{6.12}$$

It is here assumed that the variables Y are even functions of the *velocities*, that is of $p - eA$, so that again the phase cells are mapped into themselves. If there is an overall rotation, as in a centrifuge[*], one also has to reverse the angular velocity Ω. The result is then, written in the discrete notation of (4.2),

$$W_{nn'}(B, \Omega) p_{n'}^e = W_{n'n}(-B, -\Omega) p_n^e. \tag{6.13}$$

---

[*] G.J. Hooyman and S.R. de Groot, Physica **21**, 73 (1955); DE GROOT AND MAZUR, pp. 264 ff.

On the other hand, suppose that $H(q, p)$ is even in the p_k, but that some of the $Y(q, p)$ are odd functions of the p_k. The even and odd Y are characterized by

$$Y_+(q, p) = Y_+(q, -p), \qquad Y_-(q, p) = -Y_-(q, -p).$$

In hydrodynamics, for instance, the local particle and energy densities are even, while the three components of the flow velocity are odd. It is clear that equilibrium cannot distinguish between time directions and therefore

$$P^e(y_+, y_-) = P^e(y_+, -y_-) = P^e(y).$$

The same argument as before now leads to

$$W(y_+, y_- | y'_+, y'_-) P^e(y') = W(y'_+, -y'_- | y_+, -y_-) P^e(y). \tag{6.14}$$

Remark. The detailed balance relation (4.2) or (6.1) asserts that the matrix $\mathbb{W}$ is virtually symmetric and will be seen in the next section to guarantee that $\mathbb{W}$ can be diagonalized. The relation (6.14) is also a property of $\mathbb{W}$ but does not by itself guarantee diagonalizability, and wil be referred to as "extended detailed balance". The relations (6.12) and (6.13) are not properties of $\mathbb{W}$ but relate the transition probabilities in one system to those in another system. They will therefore not be honored with the name detailed balance. The extended detailed balance property will be important in XI.4.

Exercise. From (4.2) it follows that the matrix

$$\mathbb{V}_{nn'} = [p_n^e]^{-1/2} \mathbb{W}_{nn'} [p_{n'}^e]^{1/2} \tag{6.15}$$

is symmetric and can therefore be diagonalized. Conclude from this that $\mathbb{W}$ can be diagonalized and express its eigenfunctions and eigenvalues in those of $\mathbb{V}$.

Exercise. Find an operator R such that (6.14) is equivalent to saying that $R\mathbb{W}R^{-1}$ is symmetric. Why can this not be used to diagonalize $\mathbb{W}$?

Exercise. It was understood in (6.3) that the momenta p_k are the actual velocities or linear combinations of them. This need not be true if general canonical transformations are allowed. Show, however, that regardless of the choice of variables there always exists an automorphism $x \to \bar{x}$ having the properties (6.4), (6.5), (6.6), (6.7) and that therefore the proof still holds.

7. Expansion in eigenfunctions

As mentioned before, the properties (V.2.5), which characterize a $\mathbb{W}$-matrix, are not sufficient to guarantee that there exists a matrix S such that $S^{-1}\mathbb{W}S$ is diagonal. The additional detailed balance property (4.2) or (6.1), however, makes $\mathbb{W}$ symmetric in a certain sense and thereby diagonalizable, see (6.15). In this section we develop the consequences.

Without loss of generality we assume that $\mathbb{W}$ is indecomposable. Owing to the symmetry property (**4.2**) or (**6.1**) this implies that it cannot be reducible to the form (**2.7**) either. This would not be a valid conclusion if some of the p_n^e could vanish, but in a closed isolated physical system one knows that all states have a non-zero equilibrium probability.

The equation for eigenfunctions and eigenvectors of $\mathbb{W}$ is

$$\mathbb{W}\Phi_\lambda = -\lambda\Phi_\lambda. \tag{7.1}$$

We have called the eigenvalues $-\lambda$ because it will appear that they are ≤ 0. The same symbol λ is used to label the eigenfunctions, but when degeneracy occurs this notation will have to be refined. There is one eigenvalue $\lambda = 0$ with $\Phi_0 = p^e$, and we know that Φ_0 is positive. It has been proved in V.3 that this eigenvalue is not degenerate. From (7.1) it follows that

$$p(t) = \sum_\lambda c_\lambda \Phi_\lambda e^{-\lambda t}, \tag{7.2}$$

with arbitrary constants c_λ, is a solution of the master equation. In the case of a continuous eigenvalue spectrum the sum has to be replaced with an integral over λ.

Suppose one has found all eigenvalues and eigenvectors obeying (7.1). The question then is whether (7.2) is complete, i.e., whether it is sufficient to represent all solutions of the master equations. In other words, is it possible to find for every initial distribution $p(0)$ suitable constants c_λ such that

$$p(0) = \sum_\lambda c_\lambda \Phi_\lambda. \tag{7.3}$$

If $\mathbb{W}$ is a finite matrix, linear algebra tells us that this can in fact be asserted when $\mathbb{W}$ is symmetric. If $\mathbb{W}$ is an operator in an infinite-dimensional space the mathematical conditions are considerably more complicated, but as a rule of thumb one may also regard any symmetrical operator as diagonalizable – as is customary in quantum mechanics. It will now be shown that the detailed balance property guarantees that the operator $\mathbb{W}$ is symmetrical. We adopt the notation of a continuous range.

In the space of real functions $\Phi(y)$ define a scalar product between any two functions Φ, Ψ by

$$(\Phi, \Psi) = \int \frac{\Phi(y)\Psi(y)}{\Phi_0(y)} \, dy = (\Psi, \Phi). \tag{7.4}$$

The norm of a function Φ is $\sqrt{(\Phi, \Phi)}$. The functions with finite norm constitute our Hilbert space. The detailed balance property (**6.1**) may now be expressed by

$$(\Phi, \mathbb{W}\Psi) = (\Psi, \mathbb{W}\Phi) = (\mathbb{W}\Phi, \Psi), \tag{7.5}$$

which is the relation by which the symmetry of the operator $\mathbb{W}$ is defined.

7. EXPANSION IN EIGENFUNCTIONS

We conclude that the eigenfunctions are complete. Moreover it follows that the eigenvalues are real and that any two eigenvectors are orthogonal in terms of the scalar product (7.4). For discrete eigenvalues they may be normalized so that

$$(\Phi_\lambda, \Phi_{\lambda'}) = \delta_{\lambda\lambda'}. \tag{7.6}$$

In particular for $\lambda = \lambda' = 0$ this normalization reads

$$(\Phi_0, \Phi_0) = \int \Phi_0(y)\,\mathrm{d}y = 1, \tag{7.7}$$

which makes $\Phi_0(y)$ not merely proportional but identical with $P^e(y)$. For a continuous spectrum, or a continuous part of the spectrum, (7.6) is replaced with

$$(\Phi_\lambda, \Phi_{\lambda'}) = \delta(\lambda - \lambda'). \tag{7.8}$$

Thanks to this orthonormality the coefficient c_λ in (7.3) can be found by the usual device:

$$c_\lambda = (\Phi_\lambda(y), P(y, 0)). \tag{7.9}$$

The completeness is expressed by

$$\sum_\lambda \frac{\Phi_\lambda(y)\Phi_\lambda(y')}{\Phi_0(y')} = \delta(y - y'). \tag{7.10}$$

The solution of the M-equation with given initial $P(y, 0)$ is

$$P(y, t) = \sum_\lambda \Phi_\lambda(y)\,\mathrm{e}^{-\lambda t} \int \frac{\Phi_\lambda(y')}{\Phi_0(y')} P(y', 0)\,\mathrm{d}y'. \tag{7.11}$$

Remark. Apart from the question whether the set of all eigenfunctions is complete, one is in practice often faced with the following problem. Suppose for a certain operator W one has been able to determine a set of solutions of (7.1). Are they *all* solutions? For a finite matrix W this question can be answered by counting the number of linearly independent vectors Φ_λ one has found. For some problems with a Hilbert space of infinite dimensions it is possible to show directly that the solutions are a complete set, see, e.g., VI.8. Ordinarily one assumes that any reasonably systematic method for calculating the eigenfunctions will give all of them, but some problems have one or more unsuspected exceptional eigenfunctions.

Exercise. Formulate the above equations for the case that degenerate eigenvalues occur.

Exercise. Show that the $\mathbb{W}$-matrix

$$\mathbb{W} = \begin{bmatrix} -1 & 1 & 3 \\ \tfrac{1}{2} & -2 & 0 \\ \tfrac{1}{2} & 1 & -3 \end{bmatrix} \tag{7.12}$$

does not have a complete set of eigenvectors. Find nevertheless $e^{t\mathbb{W}}$.

Exercise. Show that $c_0 = 1$ so that (7.2) may be written

$$P(y, t) = P^e(y) + \sum_{\lambda > 0} c_\lambda \Phi_\lambda(y) e^{-\lambda t}. \tag{7.13}$$

Exercise. Show that (7.10) implies orthonormality as well.

Exercise. Find for the transition probability

$$T_\tau(y \,|\, y') = \sum_\lambda \frac{\Phi_\lambda(y) \Phi_\lambda(y')}{\Phi_0(y')} e^{-\lambda t}, \tag{7.14}$$

for the autocorrelation function in equilibrium

$$\kappa(\tau) = \sum_{\lambda \neq 0} e^{-\lambda t} \left[\int y \Phi_\lambda(y) \, dy \right]^2, \tag{7.15}$$

and for the fluctuation spectrum

$$S(\omega) = \frac{2}{\pi} \sum_\lambda \frac{\lambda}{\lambda^2 + \omega^2} \left[\int y \Phi_\lambda(y) \, dy \right]^2. \tag{7.16}$$

Exercise. Every $\Phi_\lambda(y)$ other than $\Phi_0(y)$ must become negative for some y. The eigenfunctions are therefore not themselves probability densities. Why does that not violate the corollary in V.3 and why does that do no harm in the general solution (7.2) or (7.14)?

The expansion in eigenfunctions leads to expressions of the various quantities pertaining to the stochastic process – as in equations (7.13) through (7.16). It also simplifies some of the derivations, in particular the proof of the approach to equilibrium. In fact, according to (7.13) it is sufficient to prove that all λ other than $\lambda = 0$ are positive, i.e., that $\mathbb{W}$ is negative semi-definite. In the same notation as in V.5 one has for any vector $p_n = x_n p_n^e$ in the Hilbert space

$$(p, \mathbb{W}p) = \sum_{nn'} \frac{p_n}{p_n^e} (W_{nn'} p_{n'} - W_{n'n} p_n)$$

$$= \sum_{nn'} x_n (W_{nn'} p_{n'}^e x_{n'} - W_{n'n} p_n^e x_n)$$

$$= -\tfrac{1}{2} \sum_{nn'} W_{nn'} p_{n'}^e (x_n - x_{n'})^2.$$

This is, indeed, negative unless all x_n are equal and therefore p_n proportional to p_n^e. The only exception would be that so many of the $W_{nn'}$ vanish that the states n can be subdivided in two groups that are not linked by any transition probability, but then $\mathbb{W}$ would be decomposable. This result constitutes another proof of the approach to equilibrium, but this proof is restricted to those $\mathbb{W}$ that have the symmetry property expressed by detailed balance. It does not exclude the possibility of a continuous spectrum.

For another example of a result obtained by means of the eigenfunction expansion consider the expression (7.16) for the spectral density of the fluctuations. By a slight change of notation it can be cast in the form

$$S(\omega) = \int_0^\infty \frac{g(\tau)\,d\tau}{1+\omega^2\tau^2}, \tag{7.17}$$

which may be interpreted as a superposition of Debye relaxation functions $(1+\omega^2\tau^2)^{-1}$ with relaxation times τ. The weight function $g(\tau)$ of the superposition may be continuous or consist of delta functions, but according to (7.16) it is never negative. *Hence $S(\omega)$ decreases monotonically when ω runs from 0 to ∞.*

It is also possible to obtain an asymptotic expansion of $S(\omega)$ for $\omega \to \infty$. Expand each separate term in (7.16) in powers of $1/\omega^2$,

$$S(\omega) = \frac{2}{\pi}\sum_{\nu=0}^\infty \frac{(-1)^\nu}{\omega^{2\nu+2}} \sum_\lambda \lambda^{2\nu+1} \left[\int y\Phi_\lambda(y)\,dy\right]^2$$

$$= \frac{2}{\pi}\sum_{\nu=0}^\infty \frac{(-1)^\nu}{\omega^{2\nu+2}} \sum_\lambda \int y'\Phi_\lambda(y')\,dy' \int y(-\mathbb{W})^{2\nu+1}\Phi_\lambda(y)\,dy.$$

Define the transpose $\widetilde{\mathbb{W}}$ by $\widetilde{\mathbb{W}}(y|y') = \mathbb{W}(y'|y)$,

$$S(\omega) = \frac{2}{\pi}\sum_{\nu=0}^\infty \frac{(-1)^{\nu+1}}{\omega^{2\nu+2}} \sum_\lambda \int y'\Phi_\lambda(y')\,dy' \int \Phi_\lambda(y)\widetilde{\mathbb{W}}^{2\nu+1}y\,dy.$$

The sum over λ can now be carried out with the aid of (7.10),

$$S(\omega) = \frac{2}{\pi}\sum_{\nu=0}^\infty \left(\frac{-1}{\omega^2}\right)^{\nu+1} \int\int y'\Phi_0(y')\,\delta(y-y')\widetilde{\mathbb{W}}^{2\nu+1}y\,dy\,dy'$$

$$= \frac{2}{\pi}\sum_{\nu=0}^\infty \left(\frac{-1}{\omega^2}\right)^{\nu+1} \langle y\widetilde{\mathbb{W}}^{2\nu+1}y\rangle^e. \tag{7.18}$$

This is the asymptotic expansion. Note that the coefficients of the successive terms can be found without solving the M-equation but merely by applying the operator $\widetilde{\mathbb{W}}$ a finite number of times. The expressions obtained for them are called "sum rules".

Exercise. An alternative definition of $\widetilde{\mathbb{W}}$ is the identity

$$\int \Phi(y)\mathbb{W}\Psi(y)\,dy = \int \Psi(y)\widetilde{\mathbb{W}}\Phi(y)\,dy \quad \text{for all } \Phi, \Psi.$$

Detailed balance is expressed by

$$\mathbb{W} P^e(y)\Psi(y) = P^e(y)\widetilde{\mathbb{W}}\Psi(y) \quad \text{for all } \Psi.$$

Exercise. Show that the coefficients in (7.18) may also be written with $y - \langle y \rangle$ in place of y.

Exercise. Derive the identity

$$S(\omega) = -\frac{2}{\pi}\operatorname{Re}\left\langle y\, \frac{1}{\widetilde{\mathbb{W}} - i\omega}\, y\right\rangle^e,$$

where $(\widetilde{\mathbb{W}} - i\omega)^{-1}$ is the resolvent operator associated with $\widetilde{\mathbb{W}}$.

Exercise. Let the charge Q on a condenser be described by a master equation. The fluctuation spectrum in equilibrium obeys

$$\lim_{\omega \to \infty} S_I(\omega) = -\frac{2}{\pi}\langle Q\widetilde{\mathbb{W}} Q\rangle^e.$$

Exercise. Let Y be the stationary Markov process described by an M-equation with operator $\mathbb{W}$. Let $P_2(y_1, t_1; y_2, t_2)$ be its two-time distribution and $G_2(k_1, t_1; k_2, t_2)$ the characteristic function thereof. Derive

$$G_2(k_1, t_1; k_2, t_2) = \langle e^{ik_2 y}\, e^{(t_2 - t_1)\widetilde{\mathbb{W}}}\, e^{ik_1 y}\rangle^e \quad (t_2 < t_1).$$

The result does not change when k_1 and k_2 are interchanged; what does that mean physically?

Exercise. Derive (7.18) without using the eigenfunction expansion. [Hint: Apply successive partial integrations to (III.3.4) and use (IV.3.7).]

8. The macroscopic equation

Let Y be a physical quantity with Markov character, having a given initial value y_0. As examples one may think of the current in a circuit, or the set of numbers of various molecules participating in a chemical reaction. The master equation determines the probability distribution at all $t > 0$. In ordinary macroscopic physics, however, one ignores fluctuations and treats Y as if it were a non-stochastic single-valued quantity y. The evolution of $y(t)$ is described by a deterministic differential equation for $y(t)$, called the *macroscopic* or *phenomenological equation*. Examples are Ohm's law, the rate equations of chemical kinetics, and the growth equations for populations. What is the logical connection between such macroscopic equations and the master equation? As the M-equation determines the entire probability distribution it must be possible to derive from it the macroscopic equation as an

8. THE MACROSCOPIC EQUATION

approximation for the case that the fluctuations are negligible. The purpose of this section is to give this derivation in a somewhat intuitive way. The systematic treatment is given in X.3.

For definiteness consider a closed, isolated physical system. If at $t=0$ the quantity Y has the precise value y_0 the probability density $P(y, t)$ is initially $\delta(y - y_0)$. It will tend to $P^e(y)$ as t increases. If y_0 is macroscopically different from the equilibrium value of Y it means that y_0 is far outside the width of $P^e(y)$, because macroscopically observed values are large compared to the equilibrium fluctuations. We also know from experience that the fluctuations remain small during the whole process. That means that $P(y, t)$, for each t, is a sharply peaked function of y. The location of this peak is a fairly well-defined number, having an uncertainty of the order of the width of the peak, and is to be identified with the macroscopic value $y(t)$. For definiteness one customarily adopts the more precise definition

$$y(t) = \langle Y \rangle_t = \int y P(y, t) \, dy. \tag{8.1}$$

Yet it should be clear that this is not a logical necessity: any value inside the peak could be used for the macroscopic value $y(t)$.[*]

As t increases the peak slides bodily along the y-axis from its initial location $y(0) = y_0$ to its final location $y(\infty) = \langle Y \rangle^e$. (The width only grows from its initial value zero to its final value, being the width of P^e.) This motion determines the evolution of $y(t)$ and hence the macroscopic equation. Bearing this picture in mind we can now readily derive the macroscopic equation.

First one has the exact identity

$$\frac{d}{dt} \langle Y \rangle_t = \int y \frac{\partial P(y, t)}{\partial t} \, dy$$

$$= \iint y \{ W(y|y') P(y', t) - W(y'|y) P(y, t) \} \, dy \, dy'$$

$$= \iint (y' - y) W(y'|y) P(y, t) \, dy \, dy'.$$

---

[*] The identification of the macroscopic value with the average has sometimes been justified with the argument that measurements are done many times and that the average of their outcomes is taken as final value. However, normally the measurement (of a current or pressure, etc.) is simply too imprecise to observe the fluctuations, and the averaging merely serves to reduce experimental errors.

Define the jump moments[*] $a_\nu(y)$ by

$$a_\nu(y) = \int (y' - y)^\nu W(y'|y)\,dy' \quad (\nu = 0, 1, 2, \ldots). \tag{8.2}$$

Then one has as an exact consequence of the M-equation

$$\frac{d}{dt}\langle Y\rangle_t = \int a_1(y)P(y,t)\,dy = \langle a_1(Y)\rangle_t. \tag{8.3}$$

Now if $a_1(y)$ happens to be a *linear* function of y this equation is also the same as

$$\frac{d}{dt}\langle Y\rangle_t = a_1(\langle Y\rangle_t). \tag{8.4}$$

Hence in this case we find an exact equation for the evolution of the **y**(t) as defined in (8.1).

If, however, $a_1(y)$ is *not* a linear function of y it is no longer true that (8.3) is the same as (8.4). Rather, on expanding $a_1(y)$ around $\langle Y\rangle$ one has

$$\langle a_1(Y)\rangle = a_1(\langle Y\rangle) + \tfrac{1}{2}\langle(Y-\langle Y\rangle)^2\rangle a_1''(\langle Y\rangle) + \cdots. \tag{8.5}$$

This is no longer a closed equation for $\langle Y\rangle$, but higher moments enter as well. The evolution of $\langle Y\rangle$ in the course of time is therefore not determined by $\langle Y\rangle$ itself, but is influenced by the fluctuations around this average. The macroscopic approximation consists in ignoring these fluctuations, and keeping only the first term in the expansion (8.5). With this approximation therefore (8.4) is valid even when $a_1(y)$ is nonlinear. Thus one obtains as macroscopic equation

$$\dot{\mathbf{y}} = a_1(\mathbf{y}). \tag{8.6}$$

Remark. From the linear integro-differential equation for $P(y, t)$ we have derived a nonlinear equation for **y**(t). Thus the *essentially linear* master equation may well correspond to a physical process that in the laboratory would be regarded as a *nonlinear* phenomenon inasmuch as its macroscopic equation is nonlinear. This is not paradoxical provided one bears in mind that the distinction between linear and nonlinear is defined for *equations*. It is wrong to apply it to a physical phenomenon, unless one has agreed upon a specific mathematical description of it. Newton's equations for the motion of the planets are nonlinear, but the Liouville equation of the solar system is linear. This connection between linear and nonlinear equations is not a matter of approximation: the linear Liouville equation is rigorously equivalent with the nonlinear equations of motion of the particles. Generally: any linear partial

---

[*] Originally called "derivate moments" by J.E. Moyal, J. Roy. Statist. Soc. B **11**, 150 (1949). Note that a_0 in (1.3) is simply the lowest jump moment.

8. THE MACROSCOPIC EQUATION

differential equation is mathematically equivalent with the equations for its characteristics, which may well be nonlinear.

It is also true that near the equilibrium value y^e one may approximate the nonlinear macroscopic equation (8.6) by a linear one:

$$\frac{d}{dt}[y(t) - y^e] = a'_1(y^e)[y(t) - y^e]. \tag{8.7}$$

This linearization, however, is an additional approximation, not inherent in the linearity of the master equation. It may be added that approach to equilibrium requires $a'_1(y^e)$ to be negative, which entails the positivity of the various transport coefficients such as the Ohmic resistance of a circuit.

It is also possible to deduce an approximate equation for the width of the distribution. First one has the exact identity

$$\frac{d}{dt} \langle Y^2 \rangle_t = \iint (y'^2 - y^2)W(y'|y)P(y)\,dy\,dy'$$

$$= \iint \left\{(y'-y)^2 + 2y(y'-y)\right\}W(y'|y)P(y)\,dy\,dy'$$

$$= \langle a_2(Y) \rangle_t + 2\langle Y a_1(Y) \rangle_t.$$

Hence the variance $\sigma^2(t) = \langle Y^2 \rangle_t - \langle Y \rangle_t^2$ obeys

$$\frac{d\sigma^2}{dt} = \langle a_2(Y) \rangle_t + 2\langle (Y - \langle Y \rangle_t) a_1(Y) \rangle_t. \tag{8.8}$$

When a_1 and a_2 are linear in y this is identical with

$$\frac{d\sigma^2}{dt} = a_2(y(t)) + 2\sigma^2 a'_1(y(t)), \tag{8.9}$$

where the prime denotes differentiation. When a_1, a_2 are not linear the latter equation may be regarded as an approximation, valid if the fluctuations are small.

It is clear that for the very existence of a macroscopic approximation it is necessary that the fluctuations are small. So far we have appealed to experience to argue that this is the case, but it may now be linked to the properties of the master equation by using (8.9). In this equation $a_2 > 0$ by definition, and $a'_1(y) < 0$ at $y = y^e$ and hence in some neighborhood. (The case of $a'_1(y^e) = 0$ is treated in ch. XI.) It follows that σ^2 tends to increase at a rate a_2, but this tendency is kept in check by the second term. Hence σ^2 will try to become equal to

$$a_2/2|a'_1|. \tag{8.10}$$

The condition for the approximate validity for (**8.6**) is that the second term of (**8.5**) is small,

$$\left|\frac{a_2 a_1''}{4a_1'}\right| \ll |a_1|. \tag{8.11}$$

This says that the second derivatives, which are responsible for the departure from linearity, must be small.

Having obtained the equation (**8.9**) for the variance, one may now include the second term in (**8.5**) to obtain a correction to the macroscopic equation

$$\dot{y} = a_1(y) + \tfrac{1}{2}\sigma^2 a_1''(y). \tag{8.12}$$

It will be found in X that this equation together with (**8.9**) does indeed constitute the first approximation beyond the macroscopic equation (**8.6**).[*]

Comment. The macroscopic equation (**8.6**) is a differential equation for y, which determines $y(t)$ uniquely when the initial value $y(0)$ is given. In the next approximation (**8.12**) the evolution of y depends on the variance of the fluctuations as well. The reason is that y fluctuates around y and thereby feels the value of a_1 not merely at y but also in the neighborhood. This effect is proportional to the *curvature* of a_1; the *slope* of a_1 is ineffective as the fluctuations are symmetric (in this approximation). The magnitude of the fluctuations, however, is determined by the second equation (**8.9**), which does contain the slope of a_1.

The fact that there are now two equations, viz. (**8.12**) and (**8.9**), implies that no longer is $y(t)$ determined by $y(0)$ alone, but by the initial vaue of σ^2 as well. One might hope that, after a short initial transient time, σ^2 adjusts itself by rapidly approaching an asymptotic value depending on the instantaneous $y(t)$ alone, so that a "renormalized" equation for y holds after the initial transient. However, this is not the case: the time scale on which σ^2 approaches (**8.10**) is determined by the coefficient a_1' in (**8.9**) and is therefore comparable to the rate at which y itself varies, see (**8.7**). There is no separation of time scales and therefore no single equation for y by itself.

The situation is similar to the one encountered in the kinetic theory of dilute plasmas.[**] To lowest order in the density[†] the one-particle distribution function of the electrons obeys the Vlasov equation. The next order approximation consists of two coupled equations for the one-particle and two-particle distribution functions. On the other hand, in the kinetic theory of gases Bogolyubov[††] has proposed an

---

[*] N.G. van Kampen, in: *Fundamental Problems in Statistical Mechanics* (Proc. NUFFIC Summer Course; E.G.D. Cohen ed., North-Holland, Amsterdam 1962); R. Kubo, K. Matsuo, and K. Kitahara, J. Statist. Phys. **9**, 51 (1973).

[**] N. Rostoker and M.N. Rosenbluth, Phys. Fluids **3**, 1 (1960).

[†] Or rather the plasma parameter $4\pi e^2/kTd$, where e is the electron charge and d the average distance between the electrons.

[††] N.N. Bogolyubov, J. Phys. USSR **10**, 265 (1946); *Problems of a Dynamical Theory in Statistical Mechanics* (E. Gora transl., Providence College, Providence 1959).

approximation scheme in which all higher orders merely added corrections to the equations for the one-particle distribution function, but his scheme failed[*].

Exercise. Find the jump moments and the macroscopic equation for the decay process and for the Poisson process.

Exercise. Define jump moments for the case that Y has more components. Show that the matrix $a'_1(y^e)$ must be negative definite, or at least semi-definite.

Exercise. Prove the following theorem[**]. The macroscopic equation is linear if and only if the function $Q(y) = y - \langle y \rangle^e$ is a left eigenvector of $\mathbb{W}$.

Exercise. In analogy with (8.11) show that the conditions for (8.12) to be a good approximation are

$$|a''_2| \ll |a'_1| \quad \text{and} \quad (a''_1)^2 \ll |a'_1|^3/a_2. \tag{8.13}$$

Exercise. Derive as corrected macroscopic equation

$$\dot{y} = a_1\left(1 - \frac{1}{2a'_1}\right) - \frac{a_2 a''_1}{4a'_1} - \frac{a'''_1}{2a'_1 a''_1}\dot{y}^2 + \frac{\ddot{y}}{2a'_1}. \tag{8.14}$$

Exercise. There is no need to solve (8.12) and (8.9) simultaneously. Inasmuch as σ^2 appears in (8.12) as a correction it suffices to compute σ^2 from (8.9) using for y the uncorrected macroscopic value given by (8.6). Hence (8.12), (8.9) may be replaced with the following scheme [Kubo et al., loc. cit.]. Set $\langle y \rangle = y + u$, and solve

$$\dot{y} = a_1(y), \tag{8.15a}$$

$$\partial_t \sigma^2 = a_2(y) + 2a'_1(y)\sigma^2, \tag{8.15b}$$

$$\dot{u} = a'_1(y)u + \tfrac{1}{2}a''_1(y)\sigma^2. \tag{8.15c}$$

Compare (X.4.8).

Exercise. Show that this set of equations can be solved, i.e., it can be reduced to a number of integrations.

9. The adjoint equation

As a companion to the M-equation it is sometimes advantageous to consider the *adjoint* or *backward equation*

$$\dot{Q} = \widetilde{\mathbb{W}} Q, \tag{9.1}$$

where $\widetilde{\mathbb{W}}$ is again the transpose or adjoint of $\mathbb{W}$ introduced in (7.18). For a discrete range it may be written more explicitly

$$\dot{q}_n = \sum_{n'} (q_{n'} - q_n) W_{n'n}. \tag{9.2}$$

---

[*] E.G.D. Cohen, in: *Fundamental Problems in Statistical Mechanics* II (Proc. Second NUFFIC Summer Course; E.G.D. Cohen ed., North-Holland, Amsterdam 1968).

[**] N.G. van Kampen, in: *Stochastic Processes in Chemical Physics* (K.E. Shuler ed., Interscience, New York 1969).

The formal solution of (9.1) is

$$Q(t) = e^{t\tilde{\mathbb{W}}} Q(0). \tag{9.3}$$

Alternatively one may write

$$\tilde{Q}(t) = \tilde{Q}(0) \, e^{t\mathbb{W}}, \tag{9.4}$$

where the vector Q is now written in the transposed form, i.e., as a matrix of one row. It is clear that the solution of these equations is equivalent to solving the master equation itself. The significance of the adjoint equation appears from the following three features.

(i) Suppose one has a set of states n, and a system that is in any state n with probability $p_n(t)$, whose evolution is governed by an M-equation. Let Q be a quantity or property relating to the system, such that in each state n it has a value q_n. Then the average of Q at time t is, of course, $\Sigma_n q_n p_n(t)$. This can be expressed in the initial distribution:

$$\langle Q \rangle_t = \sum_{nn'} q_n (e^{t\mathbb{W}})_{nn'} p_{n'}(0). \tag{9.5}$$

According to (9.3) this equation can be interpreted as follows. Define a time-dependent vector $q_n(t)$ by stipulating that at $t = 0$ it has the components q_n, and for $t > 0$ it evolves according to (9.1). Then the average of Q at time t equals the average of $q_n(t)$ over the initial distribution:

$$\langle Q \rangle_t = \sum_n q_n p_n(t) = \sum_n q_n(t) p_n(0). \tag{9.6}$$

Thus one has formally transferred the time dependence from the probability distribution onto the observed quantity – in analogy with the quantum mechanical transformation from the Schrödinger representation to the Heisenberg representation. Accordingly one may define a time-dependent vector $Q(t)$ by setting

$$Q(t) = e^{t\tilde{\mathbb{W}}} Q, \tag{9.7}$$

so that $\langle Q \rangle_t = \langle Q(t) \rangle_0$.

(ii) Let $p_{n,m}(t)$ be the solution of an M-equation with $p_{n,m}(0) = \delta_{n,m}$. Then with respect to the subscript m it obeys the adjoint or "backward equation"

$$\dot{p}_{n,m}(t) = \sum_{m'} \tilde{\mathbb{W}}_{mm'} p_{n,m'}(t) = \sum_{m'} p_{n,m'}(t) \mathbb{W}_{m'm}. \tag{9.8}$$

The proof is evident from the fact that the solution of (9.8) is formally

$$p_{n,m}(t) = (e^{t\mathbb{W}})_{n,m},$$

which is identical with the solution (2.4) of the M-equation.

(iii) The following fact is also easily verified, but is important enough to rate as a theorem. *If $\mathbb{W}$ obeys detailed balance each solution $p_n(t)$ of the* master equation is associated with a solution $q_n(t)$ of the adjoint equation by

$$\frac{p_n(t)}{p_n^e} = q_n(t). \tag{9.9}$$

To put it differently, every right eigenvector ϕ_n is associated with a left eigenvector ψ_n (with the same eigenvalue) through

$$\frac{\phi_n}{p_n^e} = \psi_n. \tag{9.10}$$

This is useful to remember when attempting to solve a master equation, because the left eigenvectors are often simpler functions of n than the right eigenvectors. In fact, for $\lambda = 0$ one has $\phi_n = p_n^e$ whereas $\psi_n = 1$.

The most frequent use, however, of the adjoint equation is in connection with absorbing boundaries and first-passage problems, see chapter XII.

Exercise. Write the formulas of this section for continuous and multi-component variables.

Exercise. Convince yourself that the characteristic function $G(k, t)$ of the distribution $P_1(y, t)$ of a Markov process is a quantity of the type (9.7).

Exercise. Derive (9.8) in a similar way as the M-equation was derived in V.1 starting from the Chapman–Kolmogorov equation (IV.2.1).

Exercise. For Markov processes that are not stationary or homogeneous one also has a forward, or master equation and a backward equation,

$$\frac{\partial P_{1|1}(y, t | y_0, t_0)}{\partial t} = \int W_t(y | y') P_{1|1}(y', t | y_0, t_0) \, dy', \tag{9.11a}$$

$$-\frac{\partial P_{1|1}(y, t | y_0, t_0)}{\partial t_0} = \int P_{1|1}(y, t | y', t_0) W_{t_0}(y' | y_0) \, dy'. \tag{9.11b}$$

where $W_t(y | y')$ is a (possibly time-dependent) transition probability.

Remarkable Exercise. The following remark deserves more emphasis than as a mere Exercise. It is evident from (9.2) that the largest q_n decreases and the smallest one increases. Assuming a finite number of states one may conclude that all $q_n(t)$ tend to one and the same constant. Subsequently it is possible by invoking (9.6) to transfer this result to the $p_n(t)$ and in this way rederive the results of V.3. This derivation requires somewhat more thought but less algebra than the one in 3.

10. Other equations related to the master equation

Another equation, related to the M-equation, occurs in the problem of *stochastic monitoring*. Let X be a random quantity with some distribution $P(x)$. Suppose it is observed by means of an independent apparatus, which registers, rather than the actual value x, some other value u with probability $R(u|x)$. When the observer has read u, the *a posteriori* probability for the actual value to be x is according to Bayes' rule

$$P^u(x) = \frac{P(x)R(u|x)}{\int P(x')R(u|x')\,dx'}. \tag{10.1}$$

Note that, even when the observer fully knows his apparatus, i.e., the function $R(u|x)$, he is still not able to tell anything about the probable value of x, unless he knows the *a priori* distribution $P(x)$ – as mentioned before in the Exercise in I.5. The redeeming feature is that (10.1) is rather insensitive to the precise form of $P(x)$ when the apparatus is good, i.e., when $R(u|x)$ is sharply peaked around $u = x$.

Next consider a process $Y(t)$ observed by the apparatus. We ask the *a posteriori* probability $P^u(y, t)$ supposing that all values $u(t')$ from $t' = 0$ to $t' = t$ have been monitored. We derive an equation for $P^u(y, t)$ in the case that $Y(t)$ is a stationary Markov process governed by the M-equation (1.5).

$P^u(y, t)$ is a functional of $u(t')$ in $0 \leq t' \leq t$. Accordingly $P^u(y, t+\tau)$ depends on the same set of values of $u(t')$ and in addition on $u(t')$ in the interval $t \leq t' \leq t+\tau$. If τ is small and u is smooth one may replace the values in the latter interval by the single value $u(t+\tau)$, which we shall call u_2. Suppose that the value of Y at t equals y_1; the joint probability of y_2, u_2 at $t+\tau$, conditional on y_1, is

$$T_\tau(y_2|y_1)R(u_2|y_2).$$

T_τ is the transition probability (IV.3.1) of Y alone. Hence the conditional probability of Y at $t+\tau$, knowing u_2, is

$$P^u(y_2, t+\tau) = \int P^u(y_1, t)\,dy_1 \frac{T_\tau(y_2|y_1)R(u_2|y_2)}{\int T_\tau(y|y_1)R(u_2|y)\,dy}. \tag{10.2}$$

Now substitute (1.1). The numerator of the fraction in (10.2) becomes

$$\{1 - \tau a_0(y_1)\}\delta(y_1 - y_2)R(u_2|y_1) + \tau W(y_2|y_1)R(u_2|y_2).$$

The denominator becomes

$$\{1 - \tau a_0(y_1)\}R(u_2|y_1) + \tau \int W(y|y_1)R(u_2|y)\,dy.$$

Dividing and omitting all powers of τ beyond the first, one obtains for the

10. OTHER EQUATIONS RELATED TO THE MASTER EQUATION

fraction itself

$$\delta(y_1 - y_2)\left\{1 - \frac{\tau}{R(u_2|y_1)} \int W(y|y_1) R(u_2|y) \, dy\right\} + \tau W(y_2|y_1) \frac{R(u_2|y_2)}{R(u_2|y_1)}.$$

Substitute in (10.2), divide by τ, and go to the limit $\tau \to 0$:

$$\frac{\partial P^u(y_2, t)}{\partial t} = -\frac{P(y_2, t)}{R(u_2|y_2)} \int W(y|y_1) R(y_2|y) \, dy$$

$$+ R(u_2|y_2) \int W(y_2|y_1) \frac{P(y_1, t)}{R(u_2|y_1)} \, dy_1.$$

This can be written in the form of a master equation:

$$\frac{\partial P^u(y, t)}{\partial t} = \int \left\{ W^u(y|y') P^u(y', t) - W^u(y'|y) P^u(y, t) \right\} dy'. \tag{10.3}$$

The transition probability in this equation is

$$W^u(y|y') = R(u|y) W(y|y') \frac{1}{R(u|y')}, \tag{10.4}$$

and depends on time through $u(t)$.

Exercise. When $Y(t)$ is a Gaussian process and $R(u|y)$ is a Gauss distribution, then $P^u(y, t)$ is also Gaussian, provided that the initial $P^u(y, 0)$ is a delta function or Gaussian.

Exercise. In analogy with (V.2.1) we define

$$\mathbb{W}^u(y|y') = W^u(y|y') - \delta(y - y') \int W^u(y''|y) \, dy''.$$

Show that it is related to $\mathbb{W}(y|y')$ by

$$\mathbb{W}^u(y|y') = R(u|y) \mathbb{W}(y|y') \frac{1}{R(u|y')} - \delta(y - y') \int R(u|y'') \mathbb{W}(y''|y) \, dy'' \frac{1}{R(u|y)}.$$

Exercise. Chapter VIII will treat the case that $\mathbb{W}$ is a differential operator, namely

$$\mathbb{W}(y|y') = -\delta'(y - y') A(y') + \tfrac{1}{2} \delta''(y - y') B(y'). \tag{10.5}$$

Show that the corresponding equation for P^u is

$$\frac{\partial P^u(y, t)}{\partial t} = -\frac{\partial}{\partial y}\left\{ A(y) + \frac{d \log R(u|y)}{dy} B(y) \right\} P^u + \frac{1}{2} \frac{\partial^2}{\partial y^2} B(y) P^u. \tag{10.6}$$

A third related equation appears if one tries to express the Markov character in terms of the characteristic functional $G([k])$ defined in (III.4.4). It turns out that a more elaborate functional of $k(t)$ is needed. It is defined for $t_0 \leq t_1$ by

$$\Gamma(y_1, t_1 | y_0, t_0) = \left\langle \exp\left[i \int_{t_0}^{t_1} k(t') Y(t') \, dt'\right] \right\rangle_{y_0, t_0}^{y_1, t_1} \quad (10.7)$$

The notation is meant to indicate the following. From the ensemble of sample functions of $Y(t)$ extract the subensemble of those that obey $Y(t_0) = y_0$. From this subensemble consider those functions that at t_1 pass through the gate $y_1, y_1 + \Delta y_1$. For each of these make up the integral and put it in the exponent. Multiply with the probability with which each sample function occurs *in the subensemble*. The result is (10.7) (multiplied by Δy_1). The dependence of Γ on the test function $k(t)$ is not indicated.

This "curtailed characteristic functional" clearly obeys at equal times

$$\Gamma(y_1, t_0 | y_0, t_0) = \delta(y_1 - y_0). \quad (10.8)$$

For $t_1 \to \infty$ it produces the characteristic functional of the subensemble

$$\int dy_1 \, \Gamma(y_1, \infty | y_0, t_0) = G_{y_0, t_0}([k]). \quad (10.9)$$

The limit is taken for an arbitrary but fixed test function $k(t)$ that vanishes outside a finite support.

If $Y(t)$ is a Markov process one has the following equation of Chapman–Kolmogorov type. If $t_0 < t_1 < t_2$ one sees from the above definition:

$$\Gamma(y_2, t_2 | y_0, t_0) = \int \Gamma(y_2, t_2 | y_1, t_1) \, dy_1 \, \Gamma(y_1, t_1 | y_0, t_0).$$

Now take $t_2 = t_1 + \Delta t$ and use

$$\Gamma(y_2, t_1 + \Delta t_1 | y_1, t_1) = \exp[ik(t_1) y_2 \, \Delta t] T_{\Delta t}(y_2 | y_1). \quad (10.10)$$

The second factor is given in (1.1) and one obtains for $\Delta t \to 0$, after readjusting the notation,[*]

$$\frac{\partial \Gamma(y, t | y_0, t_0)}{\partial t} = \int W(y | y') \Gamma(y', t | y_0, t_0) \, dy' + ik(t) y \Gamma(y, t | y_0, t_0). \quad (10.11)$$

This equation together with (10.8) determines Γ and hence G.

Exercise. In (10.10) I tacitly used the notation for stationary $Y(t)$. Show that this restriction is unnecessary.

---

[*] D.A. Darling and A.J.F. Siegert, Proc. Natl. Acad. Sci. USA **42**, 525 (1956).

10. OTHER EQUATIONS RELATED TO THE MASTER EQUATION

Exercise. The definition (**10**.7) may also be written

$$P_1(y_0)\Gamma(y_1, t_1 | y_0, t_0) = \left\langle \delta[Y(t_0) - y_0] \delta[Y(t_1) - y_1] \exp\left[i \int_{t_0}^{t_1} k(t')Y(t') \, dt'\right] \right\rangle,$$

the average now being taken over the entire ensemble of functions $Y(t)$.

Exercise. To find $\langle Y(t_1) \rangle$ one has to solve (**10**.11) for the special choice $k(t) = \varepsilon\delta(t - t_1)$ to first order in ε. This is no harder than solving the M-equation itself. In the same way one finds $\langle Y(t_1)Y(t_2) \rangle$ and higher moments.

Exercise. For the radioactive decay in IV.6 the equation reads[*]

$$\dot{\Gamma}(n, t) = (n + 1)\Gamma(n + 1, t) - n\Gamma(n, t) + ik(t)n\Gamma(n, t).$$

Solve it and find G and hence $\langle n(t) \rangle$ and $\langle\langle n(t_1)n(t_2) \rangle\rangle$. Verify the above Markov condition.

Exercise. Same question for the Wiener process.

---

[*] N.G. van Kampen, Physics Letters A **76**, 104 (1980).

Chapter VI

ONE-STEP PROCESSES

The one-step or birth-and-death processes are a special class of Markov processes, which occur in many applications and can be analyzed in some detail.

1. Definition; the Poisson process

Many stochastic processes are of a special type called "birth-and-death processes" or "generation–recombination processes". We employ the less loaded name "one-step processes". This type is defined as a continuous time Markov process whose range consists of integers n and whose transition matrix $\mathbb{W}$ permits only jumps between adjacent sites,

$$\mathbb{W}_{nn'} = r_{n'}\delta_{n,n'-1} + g_{n'}\delta_{n,n'+1} \quad (n \neq n'). \tag{1.1}$$

The diagonal elements are, of course, $\mathbb{W}_{nn} = -(r_n + g_n)$, and the master equation is

$$\dot{p}_n = r_{n+1}p_{n+1} + g_{n-1}p_{n-1} - (r_n + g_n)p_n. \tag{1.2}$$

The coefficient r_n is the probability per unit time that, being at n, a jump occurs to $n - 1$, while g_n is the probability per unit time for a jump to $n + 1$, see fig. 11.[*]

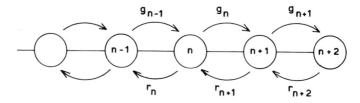

Fig. 11. The one-step process and its transition probabilities.

One-step processes occur whenever the stochastic process consists of the absorption or emission of photons or particles, the excitation and deexcitation of atoms or nuclei, or of electrons in semiconductors, the birth and

---

[*] The coefficients r_n and g_n derive their names from recombination and generation of charge carriers in semiconductors, and are often denoted by μ_n and λ_n in birth-and-death problems.

1. DEFINITION; THE POISSON PROCESS

death of individuals, or the arrival and departure of customers. The name does not imply that it is impossible for n to jump by two or more units in a time Δt, but only that the probability for this to happen is $\mathcal{O}(\Delta t^2)$, or at most $o(\Delta t)$. This is not true when births occur in twins or litters, or when neutrons diffuse in fissionable material, each fission producing several neutrons.

The one-step processes may be subdivided into the following three subclasses according to their range.

(a) The range consists of all integers $-\infty < n < \infty$.
(b) The range is half-infinite; say $n = 0, 1, 2, \ldots$.
(c) The range is finite, $n = 0, 1, 2, \ldots, N$.

If the range consists of several intervals separated by gaps there cannot be any transitions across the gaps, so that the process decomposes into several independent processes of either class (b) or class (c).

If $n = 0$ is a boundary, equation (**1.2**) is meaningless for $n = 0$ and has to be replaced with

$$\dot{p}_0 = r_1 p_1 - g_0 p_0. \tag{1.3}$$

Alternatively it is possible, and sometimes convenient, to declare (**1.2**) valid for $n = 0$ with the added definition

$$r_0 = g_{-1} = 0. \tag{1.4}$$

Similarly an upper boundary N requires a special equation:

$$\dot{p}_N = g_{N-1} p_{N-1} - r_N p_N, \tag{1.5}$$

or alternatively an additional definition:

$$r_{N+1} = g_N = 0. \tag{1.6}$$

Another subdivision of one-step processes is based on the coefficients r_n, g_n.

(α) The coefficients are constant, independent of n, except perhaps at the boundaries. This type will be referred to as "random walks" (although that term has also been used for more general cases). These processes are treated in the next section.

(β) The coefficients are linear functions of n, not both constant. This type will be called "linear one-step processes" and the general solution is given in **6**. Note that there must be at least one boundary since otherwise negative values for the transition probabilities would occur.

(γ) All other cases will be called "nonlinear", and are the subject of chapter X. It should be clear that the master equation is by definition always a linear equation in the unknown p_n, and that the term "nonlinear" here refers to the coefficients.

An important example of a one-step process with constant transition probabilities is the *Poisson process*, defined by

$$r_n = 0, \qquad g_n = q, \qquad p_n(0) = \delta_{n,0}, \tag{1.7}$$

where q is a constant parameter. The master equation is

$$\dot{p}_n = q(p_{n-1} - p_n). \tag{1.8}$$

It is a random walk over the integers $n = 0, 1, 2, \ldots$ with steps to the right alone, but at random times. The relation to chapter II becomes more clear by the following alternative definition. Every random set of events can be treated in terms of a stochastic process Y by defining $Y(t)$ to be the number of events between some initial time $t = 0$ and t. Each sample function consists of unit steps and takes only integral values $n = 0, 1, 2, \ldots$ (fig. 5). In general this Y is not Markovian, but if the events are independent (in the sense of II.2) there is a probability $q(t) \, dt$ for a step to occur between t and $t + dt$, regardless of what happened before. If, moreover, q does not depend on time, Y is a Poisson process.

Exercise. Find r_n and g_n of the decay process in IV.6, and determine to which subclass this process belongs.

Exercise. Show that for the Poisson process

$$p_n(t) = \frac{(qt)^n}{n!} e^{-qt},$$

in agreement with (II.2.6). Note that the Poisson *process* is not stationary, even though the random set of events is, according to the definition of "stationary" for random events. The reason is that $Y(t)$ is the total number of events *from an initial time*. Compare the definition in IV.2.

Exercise. For independent non-stationary events, as defined by (II.2.1), equation (1.8) still holds although q now depends on time ("time-dependent shot noise"). Find again $p_n(t)$.

2. Random walk with continuous time

Consider the unbounded *symmetrical random walk*: r_n and g_n constant and equal to one another. This constant may then be absorbed into the time unit, so that the master equation is as in (V.2.11)

$$\dot{p}_n = p_{n+1} + p_{n-1} - 2p_n \quad (-\infty < n < \infty). \tag{2.1}$$

This random walk differs from those considered in I.4 and IV.5 in that the time varies continuously. The transition probabilities are now taken per unit time. This simple example often suffices to illustrate more complicated processes; in particular it contains the essential features of a physical diffusion process.*⁾

As initial distribution we take

$$p_n(0) = \delta_{n,0}. \tag{2.2}$$

This is sufficient because, owing to the invariance for a shift in n, this also covers the case $p_n(0) = \delta_{n,m}$ for every m, and by a suitable superposition of these any initial distribution can be reproduced.

There are many ways of solving the differential–difference equation (2.1); we choose a method that can also be used in more general cases. The essential tool is the probability generating function $F(z, t)$ defined in I.2:

$$F(z, t) = \sum_{-\infty}^{\infty} z^n p_n(t), \tag{2.3}$$

where z is an auxiliary variable. Since $\sum_n p_n(t) = 1$ and $p_n(t) \geq 0$ this function certainly exists for $|z| = 1$. Some properties are

$$F(1, t) = 1, \quad F'(1, t) = \langle n(t) \rangle, \quad F''(1, t) = \langle n(t)^2 \rangle - \langle n(t) \rangle. \tag{2.4}$$

The prime indicates differentiation with respect to z. An alternative form of the last two equations is often convenient,

$$\left[\frac{\partial \log F}{\partial z}\right]_{z=1} = \langle n(t) \rangle, \tag{2.5a}$$

$$\left[\frac{\partial^2 \log F}{\partial z^2}\right]_{z=1} = \langle n(t)^2 \rangle - \langle n(t) \rangle^2 - \langle n(t) \rangle. \tag{2.5b}$$

Exercise. Establish the relation between the probability generating function and the characteristic function. Derive the identities (2.4) and (2.5) from the known properties of the latter.

Exercise. Derive for all n

$$p_n(t) = \frac{1}{2\pi i} \oint z^{-n-1} F(z, t) \, dt. \tag{2.6}$$

The integral is taken around the unit circle.

Exercise. In the discrete-time random walk the steps were taken at fixed time points. Suppose now that the times of the steps are randomly distributed as given by the Poisson process. Show that this is identical to the situation described by (2.1).**⁾

---

*⁾ For a review of simple and more complicated random walks, see G.H. Weiss and R.J. Rubin, in: *Advances in Chemical Physics* **52** (Wiley-Interscience, New York 1983). Many applications are given by J.W. Haus and K.W. Kehr, Physics Reports **150**, 203 (1987), with a depressing list of references.

**⁾ FELLER II, p. 177; COX AND MILLER, p. 239.

Equation (2.1) may now be written as an equation for $F(z, t)$. Multiply (2.1) by z^n and sum over all n:

$$\frac{\partial F(z, t)}{\partial t} = \left(z + \frac{1}{z} - 2\right) F(z, t). \tag{2.7}$$

This differential equation can be integrated; the general solution is

$$F(z, t) = \Omega(z) \exp\left[t\left(z + \frac{1}{z} - 2\right)\right]. \tag{2.8}$$

The solution contains an arbitrary function $\Omega(z)$, which will now be adjusted to the initial condition (2.2). This condition in terms of F states $F(z, 0) = 1$. Substitution in (2.8) yields

$$1 = F(z, 0) = \Omega(z).$$

Thus the final solution for the generating function is

$$F(z, t) = \exp\left[t\left(z + \frac{1}{z} - 2\right)\right]. \tag{2.9}$$

From this result one may find the $p_n(t)$ by expanding in powers of z,

$$F(z, t) = e^{-2t} \sum_{k,l=0}^{\infty} \frac{t^{k+l}}{k!\, l!} z^{k-l}.$$

Hence for $-\infty < n < \infty$

$$p_n(t) = e^{-2t} \sum_{l} \frac{t^{2l+n}}{(l+n)!\, l!}, \tag{2.10}$$

where the sum extends over all integral values of l for which both l and $n+l$ are nonnegative.

Exercise. It is often more convenient to use (2.9) rather than (2.10). Derive with the aid of (2.5)

$$\langle n(t) \rangle = 0, \qquad \langle n(t)^2 \rangle = 2t.$$

Thus the mean square of the distance covered between 0 and t grows proportionally with t, just as in the discrete-time random walk. How could this have been shown *a priori* (without explicitly solving the M-equation)?

Exercise. Show that (2.10) may also be written

$$p_n(t) = e^{-2t} I_{|n|}(2t), \tag{2.11}$$

where I_n is the n-th modified Bessel function. Verify directly that (2.11) obeys (2.1). Also write the solution of (2.1) with arbitrary $p_n(0)$ in this way.

Exercise. From the known properties of I_n follows the asymptotic expression for large t (more precisely: $t \to \infty$, $n \to \infty$ with n^2/t fixed),

$$p_n(t) = \frac{1}{\sqrt{4\pi t}} \exp[-n^2/4t]. \tag{2.12}$$

Derive this result by means of (2.6) and the method of steepest descent.

Exercise. Repeat the calculation in the text with $G(k, t)$ in lieu of $F(z, t)$.

Exercise. The *asymmetric random walk* on an infinite lattice is governed by the master equation

$$\dot{p}_n = \alpha p_{n+1} + \beta p_{n-1} - (\alpha + \beta) p_n \quad (\alpha \neq \beta). \tag{2.13}$$

It describes diffusion in the presence of an external force. Solve this equation by means of the transformation

$$p_n(t) = q_n(t)(\beta/\alpha)^{\frac{1}{2}n} \exp[-(\alpha + \beta - 2\sqrt{\alpha\beta})t].$$

(A more direct method will result from the general treatment in **6**.) Find $\langle n(t) \rangle$ and $\langle\langle n(t) \rangle\rangle$.

Exercise. Let Y_0 be the Wiener process and $Y_1, Y_2, \ldots, Y_r$ random walks with different step sizes and transition probabilities. Show that $Y_0 + Y_1 + Y_2 + \cdots + Y_r$ is a process with independent increments, see (IV.4.7), and find its transition probability.

3. General properties of one-step processes

Much writing is saved by introducing the "step operator" $\mathbb{E}$, defined by its effect on an arbitrary function $f(n)$:

$$\mathbb{E}f(n) = f(n+1), \qquad \mathbb{E}^{-1}f(n) = f(n-1). \tag{3.1}$$

If n ranges from $-\infty$ to $+\infty$ it is possible to regard $\mathbb{E}$ as an actual operator in function space, but if there are one or two boundaries it is better to regard $\mathbb{E}$ simply as an abbreviating symbol. Most of its properties are immediately evident, but we note in particular

$$\sum_{n=0}^{N-1} g(n)\mathbb{E}f(n) = \sum_{n=1}^{N} f(n)\mathbb{E}^{-1}g(n), \tag{3.2}$$

for any pair of test functions f, g. The difference in the limits of the summations in both members is awkward. Fortunately, in practice it is often immaterial because the missing terms are zero anyway.

With the aid of this symbol the master equation for one-step processes may be written

$$\dot{p}_n = (\mathbb{E} - 1)r_n p_n + (\mathbb{E}^{-1} - 1)g_n p_n. \tag{3.3}$$

This expression remains valid at the boundary points if one stipulates (1.4) and (1.6). As an example of (3.2) we compute

$$\frac{d}{dt}\langle n\rangle = \sum n(\mathbb{E}-1)r_n p_n + \sum n(\mathbb{E}^{-1}-1)g_n p_n$$
$$= \sum r_n p_n(\mathbb{E}^{-1}-1)n + \sum g_n p_n(\mathbb{E}-1)n$$
$$= -\langle r_n\rangle + \langle g_n\rangle. \tag{3.4}$$

The result is intuitively clear and is the same as (V.8.3).

Remark. The coefficients r_n and g_n cannot be negative but may be zero. If for some k both r_k and g_{k-1} are zero, it means that the process decomposes into two separate ones. If for some k one has $r_k > 0$, $g_{k-1} = 0$ transitions from k to $k-1$ are possible, but no jumps in the reverse direction can occur. One expects that the states with $n \geq k$ will be depleted, and that all probability will end up in the states $n < k$. Thus the set of states $n < k$ is "absorbing" in the sense of V.2. In fact, we shall find that in this case $p_n^s = 0$ for $n \geq k$. The case $r_k = 0$, $g_{k-1} > 0$ is, of course, analogous.

Exercise. Apply these considerations to the M-equation for radioactive decay and find in this way the absorbing states.

Exercise. Write the right-hand side of (3.3) in the form of a $\mathbb{W}$-matrix acting on the vector $\{p_n\}$. Show that the two cases distinguished in the above remark correspond to a decomposable and reducible matrix.

Exercise. Derive

$$\frac{d}{dt}\langle n^2\rangle = 2\langle n(g_n - r_n)\rangle + \langle g_n + r_n\rangle. \tag{3.5}$$

Exercise. Find a differential equation for the generating function $F(z,t)$ supposing that r_n and g_n are polynomials in n.

Exercise. In the case of a random walk (symmetric or asymmetric) the equations (3.4) and (3.5) can be solved. Find the solutions and compare them with those obtained in the previous section.

Exercise. Also find the solutions of (3.4) and (3.5) for the general linear one-step process. Why can they not be solved in the nonlinear case?

We shall now find a general expression for the stationary solution of a one-step process. From (3.3) one has

$$0 = (\mathbb{E}-1)r_n p_n^s + (\mathbb{E}^{-1}-1)g_n p_n^s$$
$$= (\mathbb{E}-1)\{r_n p_n^s - \mathbb{E}^{-1}g_n p_n^s\}.$$

This equation states that $\{\ \}$ is independent of n,

$$r_n p_n^s - \mathbb{E}^{-1}g_n p_n^s = -J. \tag{3.6}$$

The constant has been called $-J$ because it represents the net flow of probability from n to $n-1$.

3. GENERAL PROPERTIES OF ONE-STEP PROCESSES

First consider subclass (c), i.e., $n = 0, 1, 2, \ldots, N$. On substituting in (3.6) $n = 0$ and remembering (1.3) or (1.4) one finds $J = 0$. Hence (3.6) states

$$r_n p_n^s = g_{n-1} p_{n-1}^s. \tag{3.7}$$

By applying this relation repeatedly,

$$p_n^s = \frac{g_{n-1} g_{n-2} \cdots g_1 g_0}{r_n r_{n-1} \cdots r_2 r_1} p_0^s. \tag{3.8}$$

This determines all p_n^s in terms of p_0^s, which is subsequently fixed by the normalization condition

$$\frac{1}{p_0^s} = 1 + \sum_{n=1}^N \frac{g_0 g_1 \cdots g_{n-1}}{r_1 r_2 \cdots r_n}. \tag{3.9}$$

The same result (3.8) also applies to the case of a half-infinite range $n = 0, 1, 2, \ldots$. In the two-sided infinite range one has (3.8) for $n \geq 0$, and for $n \leq 0$ one obtains from (3.7)

$$p_n^s = \frac{r_{n+1} r_{n+2} \cdots r_{-1} r_0}{g_n g_{n+1} \cdots g_{-2} g_{-1}} p_0^s. \tag{3.10}$$

However, in this case it is no longer possible to prove that $J = 0$. One cannot exclude the possibility of a constant flow from $-\infty$ to $+\infty$, as in the asymmetric random walk. Such solutions would describe, for instance, diffusion in an open system, such as diffusion through a medium between two reservoirs with different densities. The stationary solution is no longer unique, but depends on the current J, which depends on additional information concerning the physical problem one is dealing with. See Exercise.

Exercise. Find the explicit stationary distribution in the case $r_n = \alpha n^2$, $g_n = \beta(n+1)$ with constant α, β.
Exercise. Same question for $r_n = \alpha n^2$, $g_n = \beta$.
Exercise. The result (3.8) tacitly assumes that all r_n are different from zero. Discuss the case that $r_k = 0$ for some k. Find p_n^s for that case.
Exercise. Find the stationary distribution for the asymmetric random walk with reflecting boundary described for $n \geq 1$ by (2.13) with $\alpha > \beta$, together with the special boundary equation

$$\dot{p}_0 = \alpha p_1 - \beta p_0. \tag{3.11}$$

(This is a model for diffusion of heavy particles in a gravitational field or in a centrifuge; compare VIII.3.)
Exercise. Two volumes V_1 and V_2 communicate through an aperture and contain a gas of N non-interacting molecules. The M-equation for the number n of molecules

in V_1 is

$$\dot{p}_n = \alpha(\mathbb{E} - 1)np_n + \beta(\mathbb{E}^{-1} - 1)(N - n)p_n. \tag{3.12}$$

Find the stationary distribution. [See (I.1.5).]

Exercise. Find the stationary distribution for the radioactive decay process described by the master equation (V.1.7).

Exercise. Show that the general solution of (3.6) with $J \neq 0$ is given for $n \geq 0$ by

$$p_n^s = -\frac{J}{r_n}\left[1 + \frac{g_{n-1}}{r_{n-1}} + \frac{g_{n-1}g_{n-2}}{r_{n-1}r_{n-2}} + \cdots + \frac{g_{n-1}g_{n-2}\cdots g_1}{r_{n-1}r_{n-2}\cdots r_1}\right] + \text{const.}\, \frac{g_{n-1}g_{n-2}\cdots g_0}{r_n r_{n-1}\cdots r_1}. \tag{3.13}$$

Also find the expression for p_n with $n < 0$.

Exercise. In the case of a half-infinite interval the stationary distribution is given by (3.8) and (3.9) with $N = \infty$. Study the condition on r_n, g_n for (3.9) to converge. Find an example in which these conditions are not satisfied. What is the physical consequence of the lack of convergence?

Suppose we have a closed, isolated physical system whose non-equilibrium behavior is adequately described by a one-step master equation for a single quantity. Then, supposing the quantity is an even function, we know that detailed balance holds, which for a one-step process reads

$$r_n p_n^e = g_{n-1} p_{n-1}^e. \tag{3.14}$$

This equation has the same form as (3.7); for an isolated system the stationary solution of the master equation p^s is identical with the thermodynamic equilibrium p^e.

Yet it is necessary for a clear understanding to distinguish between the logical status of (3.7) and (3.14). Equation (3.7) is merely the master equation with left-hand side set equal to zero; it owes its simple form to the restriction to one-step processes. It has no physical content and applies to open systems as well, and even to non-physical systems, such as populations. On the other hand (3.14) states a physical principle: p^e is regarded as known from equilibrium statistical mechanics and the equation provides a connection between the transition probabilities r_n, g_n, which must hold if the system is closed and isolated.

Exercise. Apply detailed balance to (3.12) and find a relation between α and β. Show that they are equal when $V_1 = V_2$ and conclude: thermodynamics precludes the existence of one-way apertures, or one-way mirrors.

Exercise. In the preceding model let V_2 and N go to infinity with finite density $N/V_2 \equiv \rho$. Write the master equation and show that the stationary solution is a Poisson distribution. (Could this have been known *a priori*?) What does detailed balance tell us in this case?

Exercise. For one-step processes $\mathbb{W}$ is a tridiagonal matrix. With the aid of (3.8) a similarity transformation can be constructed which makes the matrix symmetric, as in (V.6.15). Prove in this way that any finite one-step process has a complete set of eigenfunctions, and that its autocorrelation function consists of a sum of exponentials.[*]

Exercise. From this it can be concluded that no two eigenvalues coincide. [Hint: Show that the eigenvector belonging to any eigenvalue is unique by successively solving the equations for its components.]

4. Examples of linear one-step processes

These have been defined in section 1 as one-step processes in which r_n and g_n are linear functions of n, not both constant. Some examples have already been encountered: the decay process in IV.6 and the density fluctuations of a gas in an Exercise of the previous section. We here list some other examples, leaving it to the reader to fill in the details, even if they are not formally registered as Exercises.

(i) *Quantized harmonic oscillator* interacting with a radiation field. Let $n = 0, 1, 2, \ldots$ be the states of the oscillator, having energies $h\nu(n + \tfrac{1}{2})$. The transition probabilities are proportional to the matrix elements of the dipole moment, which vanish except between adjacent states; hence this is a one-step process. The matrix element between $n - 1$ and n is proportional to n. The probability per unit time for a jump from $n - 1$ to n is $g_{n-1} = \beta n$, where β is a factor which will depend on the radiation density ρ at frequency ν, but not on n. The probability for a jump from n to $n - 1$ is $r(n) = \alpha n$ with a similar factor α. The M-equation is (see XVII.3)

$$\dot{p}_n = \alpha(\mathbb{E} - 1)n p_n + \beta(\mathbb{E}^{-1} - 1)(n + 1)p_n. \tag{4.1}$$

The stationary solution is found to be a Pascal distribution

$$p_n^s = \text{const.} \, (\beta/\alpha)^n. \tag{4.2}$$

Let the total system of harmonic oscillator plus radiation field come to equilibrium. Then we know from equilibrium statistical mechanics

$$p_n^e = \text{const.} \, \exp\left[-\frac{nh\nu}{kT}\right]. \tag{4.3}$$

Since this must be the same as (4.2)

$$\frac{\beta}{\alpha} = \exp\left[-\frac{h\nu}{kT}\right]. \tag{4.4}$$

---

[*] For a more general theorem concerning tridiagonal matrices, see J.H. Wilkinson, *The Algebraic Eigenvalue Problem* (Clarendon, Oxford 1965) pp. 335 and 336.

It is possible to say more. As g_n must be proportional to the density ρ of photons present, one may write $\beta = A\rho$, where A may depend on ν but not on ρ. As there is spontaneous and stimulated emission one may write $\alpha = B + C\rho$. Substitution in (4.4) yields

$$\rho = \frac{B}{A\,e^{h\nu/kT} - C}.$$

This is the celebrated Einstein derivation of Planck's law; to complete it one takes into account that for large T the distribution must become identical with the Rayleigh–Jeans law.[*]

(ii) *Maser amplification.*[**] Let n be the number of quanta in a given mode of the electromagnetic field in a volume containing excited atoms or molecules. The quanta disappear through the walls at a rate $r_n = \alpha n$ but are produced by stimulated emission and through input, $g_n = \beta n + \gamma$. From (3.4) follows that $\langle n(t) \rangle$ grows exponentially when $\beta > \alpha$, but for $\beta < \alpha$ tends to

$$\langle n(\infty) \rangle = \frac{\gamma}{\alpha - \beta}.$$

Thus in the stationary state the input γ is amplified by the factor $(\alpha - \beta)^{-1}$. From (3.5) one finds for the fluctuations in the stationary state

$$\sigma^2(\infty) = \frac{\alpha\gamma}{(\alpha - \beta)^2}.$$

The stationary distribution itself is a negative binomial or Pólya distribution. There is no thermodynamic p^e, since the system is subject to a continuous input.

(iii) *Adsorption.* A large reservoir contains a gas of molecules with practically constant density ρ. They can be adsorbed on a small surface having N sites available. When the number of adsorbed molecules is n, one easily sees $r_n = \alpha n$, $g_n = \beta(N - n)$. The M-equation turns out to be the same as (3.12). The reader therefore knows already that the stationary solution is the binomial distribution (I.1.5):

$$p_n^s = \text{const.} \binom{N}{n} \left(\frac{\beta}{\alpha}\right)^n. \tag{4.5}$$

Comparison with p_n^e of equilibrium statistical mechanics tells us

$$\frac{\beta}{\alpha} = \rho \frac{\zeta}{(2\pi m kT)^{3/2}}, \tag{4.6}$$

where ζ is the internal partition function of an adsorbed molecule.

---

[*] A. Einstein, Physik. Zeits. **18**, 121 (1917).
[**] K. Shimoda, H. Takahashi, and C.H. Townes, J. Phys. Soc. Japan **12**, 686 (1957).

(iv) *Chemical reactions* give rise to a large variety of M-equations, most of them nonlinear. The general treatment is given in VII. We here take the simple reaction scheme

$$A \underset{k'}{\overset{k}{\rightleftarrows}} X, \tag{4.7}$$

where A is supposed to be so abundant that its amount n_A is practically constant. The number n of molecules X has a probability per unit time $g_n = kn_A$ to increase by one; and $r_n = k'n$ to decrease (k and k' are reaction constants):

$$\dot{p}_n = k'(\mathbb{E} - 1)np_n + kn_A(\mathbb{E}^{-1} - 1)p_n. \tag{4.8}$$

The stationary solution is a Poisson distribution. The reason is that the molecules X are created and annihilated independently of each other. In reaction schemes in which several molecules X react together they are no longer independent and deviations from the Poisson distribution will occur (see VII.3).

(v) *Growth of a population.* n is the number of individuals in a population of a certain species such as bacteria. Each individual has a probability α per unit time to die and a probability β to produce an additional individual. α and β are assumed fixed, independent of the age of the individual, otherwise the process would not be Markovian. Thus $r_n = \alpha n$ and $g_n = \beta n$. According to (3.4)

$$\frac{d}{dt}\langle n \rangle = (\beta - \alpha)\langle n \rangle,$$

which is Malthus' law of exponential growth (when $\beta > \alpha$). The variance obeys, according to (3.5),

$$\frac{d\sigma^2}{dt} = 2(\beta - \alpha)\sigma^2 + (\beta + \alpha)\langle n \rangle \tag{4.9}$$

and hence also grows exponentially. Note that the observation of $\langle n \rangle$ only determines $\beta - \alpha$, but observation of the deviations from the average also yields $\beta + \alpha$ and hence α and β separately.

(vi) *Cosmic ray cascades.* When a cosmic ray electron enters an absorbing material, like lead, it creates additional electrons through Bremsstrahlung followed by pair creation. Let t be the traversed thickness, n the number of electrons (both positive and negative). Bhabha and Heitler[*] described this cascade by a one-step process with $r_n = 0, g_n = \beta$. The M-equation is the

---

[*] H.J. Bhabha and W. Heitler, Proc. Roy. Soc. A **159**, 432 (1937).

same as (1.8) but the initial value is $p_n(0) = \delta_{n,1}$, so that

$$p_n(t) = \frac{(\beta t)^{n-1}}{(n-1)!} e^{-\beta t} \quad (n = 1, 2, 3, \ldots).$$

The mean square fluctuations are $\langle n \rangle - 1$, which is less than observed. Furry[*] improved the model by taking $r_n = 0$, $g_n = \gamma n$ and found

$$p_n(t) = e^{-\gamma t}(1 - e^{-\gamma t})^{n-1}.$$

This is a Pascal distribution with $\langle n \rangle = e^{\gamma t}$ and variance $\sigma_n^2(t) = e^{\gamma t}(e^{\gamma t} - 1)$. The reason why the fluctuations are so large is that the cascade process not only magnifies the average number of electrons, but also the fluctuations about that average. Later work took into account the absorption of electrons, their distribution over various energies, and the photons as a separate entity.[**]

Exercise. Supposing that the harmonic oscillator (i) starts out from a state n_0, find $\langle n \rangle$ and σ_n^2 as functions of t and verify that they tend to their equilibrium values.

Exercise. Write the M-equation for example (ii) and use it to find $\langle n \rangle$ and σ_n^2 as functions of t.

Exercise. Write the M-equation for the reactions

$$A + X \to 2X, \quad X \rightleftharpoons B.$$

The first reaction is autocatalytic and its reverse is neglected. Solve the equation. [GARDINER, p. 274.]

Exercise. In example (v) take $\beta < \alpha$ and find the stationary distribution.

Exercise. Write the master equation for the Furry model of cosmic ray cascades. Notice that it differs from that for the population growth, but gives the same equation for $\langle n \rangle$ although a different one for $\langle n^2 \rangle$.

Exercise. Ulam has considered the motion of a single particle in one dimension, which bounces back and forth between one fixed wall and one wall that oscillates with velocity $\pm \frac{1}{2}$ (see fig. 12). The speed (absolute value of the velocity) v of the particle is changed into $v + 1$ by a head-on collision and into $v - 1$ when it overtakes the receding wall. Hammersley replaced this dynamical problem with a stochastic one by introducing the following Stosszahlansatz (which is physically reasonable when the fixed wall is far away, but sidesteps the real issue, viz., how do the stochastic aspects arise from the underlying deterministic dynamics?). The probability for a collision to occur in dt is taken to be $v\,dt$ and the ratio of head-on to receding collisions is $AC/BC = (v + \frac{1}{2})/(v - \frac{1}{2})$. Moreover the initial v is taken $\frac{1}{2}$, so that the only accessible values for v are $\frac{1}{2} + n$ ($n = 0, 1, 2, \ldots$). Show that the M-equation has

---

[*] W.H. Furry, Phys. Rev. **52**, 569 (1937). This has been called the Furry process.

[**] See BHARUCHA-REID; and N. Arley, *On The Theory of Stochastic Processes and their Application to the Theory of Cosmic Radiation* (Wiley, New York 1943); J. Nishimura, in: *Encyclopedia of Physics* **46**/2 (S. Flügge ed., Springer, Berlin 1967).

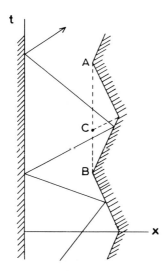

Fig. 12. Hammersley's model of a bouncing particle.

the form (4.1) with $\alpha = \beta = \frac{1}{2}$. Show that the average speed grows indefinitely, and that its fluctuations grow equally fast.[*]

5. Natural boundaries

In the examples of the previous section, and in most other applications, the coefficients r_n and g_n are not just a collection of numbers, but are given as simple analytic functions $r(n)$, $g(n)$ of the variable n. If that were not so, there could be no hope to find explicit solutions (unless the number of states is very small). However, it also implies that the special equations (1.3) and (1.5) at the boundaries are to be taken seriously and cannot be incorporated in the general equation by the simple trick described in (1.4) and (1.6) without spoiling the analytic character. Hence it is necessary, in the case of two boundaries, to write the master equation in three separate lines,

$$\dot{p}_n = r(n+1)p_{n+1} + g(n-1)p_{n-1} - \{r(n) + g(n)\}p_n \qquad (5.1)$$

for $n = 1, 2, \ldots, N-1$ and

$$\dot{p}_0 = r(1)p_1 - g(0)p_0, \qquad (5.2)$$

$$\dot{p}_N = g(N-1)p_{N-1} - r(N)p_N. \qquad (5.3)$$

---

[*] The articles of Ulam and Hammersley are in the Proceedings of the Fourth Berkeley Symposium on Mathematical Statistics and Probability, Vol. III (J. Neyman ed., University of California Press, Berkeley 1961).

Note. It may happen that the boundary extends its influence over several sites, so that several additional lines have to be written. An example is the model for diffusion-controlled chemical reactions in section 7. Also "impurities" may occur, i.e., internal sites at which the r_n and g_n deviate from the analytic expression used in (5.1). Then the equations for the p_n of two or more internal sites have to be written separately. For the time being we exclude such complications.

There is a special type of boundary, however, for which this exceptional character can be rendered harmless. Consider class (b) of section 3, i.e., $n = 0, 1, 2, \ldots$. We shall call the boundary $n = 0$ "natural" if
   (i) the general equation (5.1) is valid down to $n = 1$ and only the boundary equation (5.2) has to be supplied;
   (ii) and moreover $r(0) = 0$.
In that case the following device can be utilized.

Declare (5.1) valid for all n from $-\infty$ to $+\infty$ but only consider solutions of this extended equation whose *initial* values $p_n(0)$ are zero for $n < 0$. It is then clear that the $p_n(t)$ for $n < 0$ remain zero at all $t > 0$, because the condition $r(0) = 0$ guarantees that no transitions occur from the state 0 into the state -1. Consequently the general equation (5.1) reduces to (5.2) at $n = 0$. Thus the boundary merely restricts the possible choice of initial condition, but no longer occurs in the equation itself – which greatly facilitates its solution.

The reason why the boundaries in physical problems are often natural becomes obvious by looking at the simple example of radioactive decay in IV.6. The probability for an emission to take place is proportional to the number n of radioactive nuclei, and therefore automatically vanishes at $n = 0$. The same consideration applies when n is the number of molecules of a certain species in a chemical reaction, or the number of individuals in a population. Whenever by its nature n cannot be negative any reasonable master equation should have $r(0) = 0$. However, this does not exclude the possibility that something special happens at low n by which the analytic character of $r(n)$ is broken, as in the example of diffusion-controlled reactions. A boundary that is not natural will be called "artificial" in section 7.

Exercise. Show that the coefficients in (5.2) and (5.3) cannot be altered without violating conservation of probability.

Exercise. Suppose the boundary at $n = 0$ requires *two* special equations, for $\dot{p}_0$ and $\dot{p}_1$. How many additional coefficients enter when probability must be conserved? And for k special equations?

Exercise. What is the condition for the upper boundary $n = N$ to be natural?

Exercise. Check that all boundaries in section 4 are natural, but that the boundary (3.11) is not.

Exercise. Prove for linear one-step problems, by using (3.4), that a natural boundary has the effect that $\langle n \rangle$ cannot cross it.

6. Solution of linear one-step processes with natural boundaries

The equation to be solved may be written

$$\dot{p}_n = a(\mathbb{E} - 1)(r + n)p_n + b(\mathbb{E}^{-1} - 1)(g + n)p_n, \tag{6.1}$$

with constants a, b, r, g. At first we ignore boundaries and treat it as a difference equation for $-\infty < n < \infty$, with initial condition $p_n(0) = \delta_{nm}$. Moreover we assume $a \neq 0$, $b \neq 0$, $a \neq b$. From the general solution found in this way the various special cases can be obtained by substituting special values or taking appropriate limits.

We use again the probability generating function (VI.2.3). Multiply (6.1) by z^n and sum over all n,

$$\frac{\partial F(z, t)}{\partial t} = a \sum (z^{n-1} - z^n)(r + n)p_n + b \sum (z^{n+1} - z^n)(g + n)p_n$$

$$= ar\left(\frac{1}{z} - 1\right)F + a(1 - z)\frac{\partial F}{\partial z} + bg(z - 1)F + b(z^2 - z)\frac{\partial F}{\partial z}$$

$$= (1 - z)(a - bz)\frac{\partial F}{\partial z} + (1 - z)\left(\frac{ar}{z} - bg\right)F. \tag{6.2}$$

This linear partial differential equation for F can be solved by the standard method of characteristics.[*] The characteristic curves in the (z, t)-plane are determined by

$$-dt = \frac{dz}{(1 - z)(a - bz)}.$$

Integration yields the equation for these curves (fig. 13)

$$\frac{1 - z}{a - bz} e^{(b - a)t} = C,$$

where C is an integration constant, by which the various curves are identified. The variation of F along each separate characteristic curve is determined by

$$-\frac{dz}{a - bz} = \frac{d \log F}{ar/z - bg},$$

---

[*] I.N. Sneddon, *Elements of Partial Differential Equations* (McGraw-Hill, New York 1957) ch. 2.

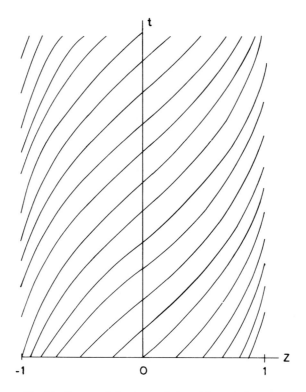

Fig. 13. Characteristic curves of (6.2) with $a = 1$, $b = -1$.

which yields

$$F(z, t) = \text{const.} \; z^{-r}(a - bz)^{r-g}.$$

The constant may be chosen differently for each characteristic and is therefore an arbitrary function $\Omega(C)$. Thus the general solution of (6.2) is

$$F(z, t) = z^{-r}(a - bz)^{r-g} \Omega\left(\frac{1-z}{a-bz} e^{(b-a)t}\right). \tag{6.3}$$

This function Ω will now be determined so as to satisfy the initial condition $p_n(0) = \delta_{nm}$. In terms of F this initial condition reads

$$F(z, 0) = z^m.$$

Substitution in (6.3) gives

$$\Omega\left(\frac{1-z}{a-bz}\right) = z^{m+r}(a-bz)^{g-r}.$$

6. ONE-STEP PROCESSES WITH NATURAL BOUNDARIES

Introduce temporarily the variable ζ by

$$\zeta = \frac{1-z}{a-bz}, \qquad z = \frac{a\zeta - 1}{b\zeta - 1},$$

Then our equation for Ω tells

$$\Omega(\zeta) = \left(\frac{a\zeta - 1}{b\zeta - 1}\right)^{m+r} \left(\frac{b-a}{b\zeta - 1}\right)^{g-r}.$$

Inserting this expression for Ω into (6.3) one finally arrives at

$$F(z,t) = z^m \left[\frac{a\varepsilon - b + a(1-\varepsilon)z^{-1}}{a-b}\right]^{m+r} \left[\frac{a - b\varepsilon - b(1-\varepsilon)z}{a-b}\right]^{-m-g}, \qquad (6.4)$$

where $\varepsilon = e^{(b-a)t}$.

Exercise. Obtain the M-equation for the random walk as a limiting case of (6.1) and find in this way the solution (2.9).

Exercise. Check that $F(1,t) = 1$ and find the first two moments of n as functions of t. Verify that they satisfy the equations (3.4) and (3.5).

Exercise. If I substitute in (6.1) $n = -n'$ another M-equation of the same type results. What are the coefficients a', b', r', g' of this new equation? Show that a half-infinite range extending to $-\infty$ can always be transformed to the range $(0, \infty)$.

As mentioned before there must be at least one boundary to prevent negative values of r_n or g_n. Without loss of generality this boundary may be taken as a lower boundary at $n = 0$. Then (6.4) makes sense only with $m \geq 0$. It is also necessary, however, that it contains no negative powers of z, otherwise it is merely a solution of the unbounded difference equation (6.1), but does not provide a solution of the actual master equation with the boundary. This condition is satisfied if $r = 0$, that is, the boundary must be natural for our solution to work. Then F may be written

$$F(z,t) = (a-b)^g [a(1-\varepsilon) + (a\varepsilon - b)z]^m [a - b\varepsilon - b(1-\varepsilon)z]^{-m-g}. \qquad (6.5)$$

It is now necessary to distinguish various cases. *First* suppose the range is $(0, \infty)$. Then $g_n = b(n+g)$ must be nonnegative for all $n \geq 0$; that implies

| either | $b > 0$, | $g \geq 0$; | (6.6) |
| or | $b = 0$, | $bg = \beta \geq 0$. | (6.7) |

In the former case (6.5) applies as it stands. In the latter case one has to take the limit $b \to 0$ with $bg = \beta$, with the result

$$F(z,t) = [1 - \varepsilon + \varepsilon z]^m \exp\left[-\frac{\beta}{a}(1-\varepsilon)(1-z)\right], \qquad (6.8)$$

where $\varepsilon = e^{-at}$.

Secondly, let the range be finite, $0 \leq n \leq N$. Then both possibilities (6.6) and (6.7) are allowed and moreover

$$b < 0, \qquad g \leq -N. \tag{6.9}$$

In order that (6.5) is a solution it must be a polynomial in z of order N, for every $m = 0, 1, 2, \ldots, N$. Hence $-g = N$, so that $g_n = b'(N-n)$ with $b' = -b > 0$. Thus the upper boundary must also be natural and F becomes

$$F(z, t) = (a + b')^{-N} [a(1 - \varepsilon) + (a\varepsilon + b')z]^m [a + b'\varepsilon + b'(1 - \varepsilon)z]^{N-m}, \tag{6.10}$$

where $\varepsilon = \exp[-(a + b')t]$.

The various categories for which solutions have been found are summarized in the following table.

(i) Range $(-\infty, \infty)$ with constant coefficients (random walk).
(ii) Range $(0, \infty)$ with natural boundary, hence $r_n = an$, while g_n is given by (6.6) or (6.7).
(iii) Range $(0, N)$ with two natural boundaries, $r_n = an$ and $g_n = b'(N - n)$.
(iv) The exceptional case mentioned in the next Exercise.

Exercise. The condition $r = 0$ is not the only way of avoiding negative powers in (6.5): one might have $a = 0$. Discuss this case.
Exercise. Solve the M-equation (3.12).
Exercise. Show that all examples of section **4** belong to the categories (i)–(iv), and solve their master equations.
Exercise. For example (v) of section **4** show

$$p_0(t) = \left[\frac{\alpha - \alpha e^{-(\beta-\alpha)t}}{\beta - \alpha e^{-(\beta-\alpha)t}} \right]^m. \tag{6.11}$$

Conclude that there is a positive probability that the population dies out in an early stage and never becomes large. It follows that σ^2 must be at least of order $\langle n \rangle^2$.
Exercise. For category (ii) suppose $a > b$ and take the limit $F(z, \infty)$. What happens for $b > a$?
Exercise. A limiting case in category (ii) is obtained for $a = b$. Show that this yields

$$F(z, t) = [at + (1 - at)z]^m [1 + at - atz]^{-m-g}.$$

Study the corresponding master equation and its explicit solution $p_n(t)$.
Exercise. Find the equilibrium distribution for category (iii).
Exercise. A long-lived radioactive substance A decays into B through two short-lived intermediaries. $A \to X \to Y \to B$. When the amount of A is supposed constant, the joint distribution of the numbers n, m of nuclei X, Y obeys the bivariate master equation

$$\dot{p}_{nm} = \alpha(\mathbb{E}_n^{-1} - 1)p_{nm} + \beta(\mathbb{E}_m - 1)mp_{nm} + \gamma(\mathbb{E}_n \mathbb{E}_m^{-1} - 1)np_{nm},$$

where $\mathbb{E}_n, \mathbb{E}_m$ are the step operators for n, m, respectively. Solve this equation with

$p_{nm}(0) = \delta_{n,0}\delta_{m,0}$ and show that, after a sufficently long time, p_{nm} is the product of two Poisson distributions.

Exercise. As a generalization of the preceding example consider the chemical chain reaction involving r variable compounds

$$B \xrightarrow{\beta} X_1, \quad X_1 \xrightarrow{\gamma_1} X_2, \quad X_2 \xrightarrow{\gamma_2} X_3, \quad \ldots, \quad X_{r-1} \xrightarrow{\gamma_{r-1}} X_r,$$

$$X_1 \xrightarrow{\alpha_1} A_1, \quad X_2 \xrightarrow{\alpha_2} A_2, \quad \ldots, \quad X_{r-1} \xrightarrow{\alpha_{r-1}} A_{r-1}, \quad X_r \xrightarrow{\alpha_r} A_r.$$

Show that the master equation can, in principle, be solved by means of an r variable generating function. In particular, prove that all solutions tend to a stationary joint Poisson distribution.[*]

7. Artificial boundaries

Artificial boundaries have been defined as boundaries at which the occupation probability of one or more sites obeys special equations, not covered by the analytic expression $r(n)$ and $g(n)$ that apply to the other n. The variety of possible artificial boundaries is, of course, endless. A restricted class are the *pure boundaries*, defined as those in which only the end site requires a special equation. Another subdivision is in *reflecting boundaries*: those which conserve total probability; and *absorbing boundaries*, at which probability disappears. The latter definition requires comment.

As an example consider the set of equations for p_n with $n = 0, 1, 2, \ldots$,

$$\dot{p}_n = p_{n+1} + p_{n-1} - 2p_n \quad (n = 1, 2, \ldots), \tag{7.1a}$$

$$\dot{p}_0 = p_1 - 2p_0. \tag{7.1b}$$

The site $n = 0$ is a pure, artificial boundary. The equations can be interpreted as a random walk on $-\infty < n < \infty$ in which the transitions from -1 to 0 are impossible: a drunkard's walk with a bottomless pit on one side. The total probability is not conserved,

$$\frac{d}{dt} \sum_{n=0}^{\infty} p_n = -p_0. \tag{7.2}$$

The reader who is shocked by the heresy of a probability distribution that is not normalized to unity can be pacified in two ways. First he may interpret p_n as the density of an ensemble of independent particles, each conducting a random walk until it disappears into the pit. The non-conservation (7.2) then simply tells us that the total number of remaining particles decreases.

---

[*] This scheme has been used as a model for carcinogenesis by H.G. Tucker, in the Proceedings of the Fourth Berkeley Symposium on Mathematical Statistics and Probability, Vol. IV (J. Neyman ed., University of California Press, Berkeley 1961) p. 387.

Alternatively one may always introduce an additional state, which we shall call a "limbo state" and label by a star. The probability p_* of being in this state is by definition

$$p_* = 1 - \sum_{n=0}^{\infty} p_n, \qquad (7.3)$$

and obeys according to (7.2)

$$\dot{p}_* = p_0. \qquad (7.4)$$

The set (7.1) supplemented with (7.4) contributes a proper probability-conserving master equation in the extended space $n = *, 0, 1, 2, \ldots$. Its matrix $\mathbb{W}$ has the limbo state $*$ as an absorbing state – in this case the bottomless pit.

The introduction of a limbo state is a purely formal device, because (7.1) by itself is a closed set of equations for the p_n ($n = 0, 1, 2, \ldots$). We shall therefore still call it a master equation, although its matrix is not a $\mathbb{W}$-matrix. Having solved it one may afterwards determine $p_*(t)$ from (7.3) or (7.4). Note that the average time that the walker survives before entering the limbo state is given by

$$\int_0^\infty t\dot{p}_* \, dt = -\sum_{n=0}^{\infty} \int_0^\infty t\dot{p}_n \, dt = \sum_{n=0}^{\infty} \int_0^\infty p_n \, dt. \qquad (7.5)$$

The average is taken over an ensemble of independent walkers.

We shall now list some examples of artificial boundaries. One example with a reflecting pure boundary has already been encountered in (3.11).

(i) *A volatile gas is dissolved in an inert solvent.* The gas molecules diffuse until they reach the surface and evaporate. Replacing the diffusion of each molecule by a one-dimensional random walk one has the set of equations (see fig. 14)

$$\dot{p}_n = p_{n+1} + p_{n-1} - 2p_n \quad (n = 1, 2, \ldots), \qquad (7.6a)$$

$$\dot{p}_0 = p_1 - (1+c)p_0. \qquad (7.6b)$$

At the boundary site $n = 0$ the molecule has a normal probability per unit time to return to site 1, but also a probability c per unit time to evaporate.

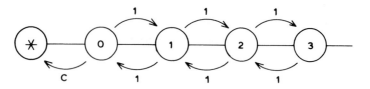

Fig. 14. Random walk with absorbing boundary.

7. ARTIFICIAL BOUNDARIES

This is a pure absorbing boundary (when $c > 0$). The limbo state represents the evaporated molecules. For $c = 0$ the boundary is reflecting.

For $c = 1$ one obtains (7.1b) as a special case, distinguished by the fact that the probability for stepping into limbo is the same as for the other steps to the left. We shall refer to this special case of an absorbing pure boundary as *totally absorbing*. It is particularly important in connection with first-passage problems, see chapter XII, but it is by no means the only possible absorbing boundary.

(ii) *Diffusion-controlled chemical reaction.*[*] A molecule A wanders about in a solvent until it happens to meet a fixed molecule B to which it can be bound, but, of course, the molecule AB may dissociate again. This picture can be schematized as a random walk on the half-lattice $n = 0, 1, 2, \ldots$, where $n = 0$ represents the bound state. The master equation is

$$\dot{p}_n = p_{n+1} + p_{n-1} - 2p_n \quad (n = 2, 3, \ldots), \tag{7.7a}$$

$$\dot{p}_1 = p_2 + \gamma p_0 - 2p_1, \tag{7.7b}$$

$$\dot{p}_0 = p_1 - \gamma p_0. \tag{7.7c}$$

This is an example of a reflecting impure boundary.

The stationary solution is

$$p_n = \gamma p_0 \quad \text{for } n \geqslant 1.$$

p_0 is an arbitrary factor, which cannot be found from the master equation (7.7), but is available to normalize the distribution. It is clear, however, that it cannot be normalized as a probability distribution since $\sum p_n = \infty$. The physical reason is, of course, that our molecule A roams about in an infinite space. To remedy this, one must introduce a gas of independent molecules A with a certain density ρ. More precisely, ρ is the average number of molecules A at a far-away site $n \gg 1$. This situation corresponds to the normalization $\gamma p_0 = \rho$. The detailed balance equation (V.4.2) can now be applied,

$$\gamma p_0^e = p_1^e, \qquad \gamma = \rho\, e^{-\chi/kT}, \tag{7.8}$$

where χ is the binding energy. Note that for $\gamma = 0$ equations (7.7) reduce to (7.1): infinite binding energy turns the bound state into an absorbing boundary.

(iii) In deference to the fact that artificial boundaries are quite common in *queuing problems* we include one example of that type, although it is

---

[*] A.M. North, *The Collision Theory of Chemical Reaction in Liquids* (Methuen, London 1964); H. Eyring, in: *Physical Chemistry, An Advanced Treatise*, Vol. VII (H. Eyring et al. eds., Academic Press, New York 1975).

outside the scope indicated by the title of this book.*) A telephone exchange is able to handle N connections simultaneously; the number of connections at any moment is $n = 0, 1, 2, \ldots, N$. Connections are made at random time points as long as the exchange can handle them,

$$g_n = \text{const.} = \beta \quad \text{for } n = 0, 1, 2, \ldots, N-1; \qquad g_N = 0.$$

Once a connection is made it is discontinued at a random moment, $r_n = \alpha n$. (This somewhat unrealistic assumption is needed for the Markov character.) The master equation is

$$\dot{p}_0 = \alpha p_1 - \beta p_0, \tag{7.9a}$$

$$\dot{p}_n = \alpha(n+1)p_{n+1} + \beta p_{n-1} - (\alpha n + \beta)p_n \quad (n = 1, 2, \ldots, N-1), \tag{7.9b}$$

$$\dot{p}_N = \beta p_{N-1} - \alpha N p_N. \tag{7.9c}$$

There is a natural boundary at $n = 0$ and an artificial reflecting pure boundary at $n = N$.

Exercise. In the radioactive decay process the state $n = 0$ is an absorbing state. Show that the equations for the remaining p_n ($n = 1, 2, \ldots$) constitute a master equation with absorbing boundary. The state $n = 0$ functions as limbo state.

Exercise. For the same decay process show that for any $N \geq 1$ the states $n \geq N$ obey a master equation with absorbing boundary. Its limbo state consists of all states with $n < N$.

Exercise. A vessel contains a gas of unstable molecules, which dissociate as described by the decay process. At random moments additional molecules enter the vessel or are produced inside it. The M-equation is

$$\dot{p}_n = (\mathbb{E} - 1)np_n + a(\mathbb{E}^{-1} - 1)p_n \quad (n = 1, 2, \ldots), \tag{7.10a}$$

$$\dot{p}_0 = p_1 - ap_0. \tag{7.10b}$$

Is this boundary natural or artificial?

Exercise. For the general one-step process (5.1) show that the information: "$n = 0$ is an absorbing pure boundary" determines the equation for p_0 up to one positive constant, as in (7.6b).

Exercise. For the infinite symmetric random walk an explicit solution of an absorbing boundary problem can be obtained by the *reflection principle*. Solve the M-equation with initial condition $p_n(0) = \delta_{nm} - \delta_{n,-m}$. The solution $p_n(t)$ for $n > 0$ obeys

$$\dot{p}_n = p_{n+1} + p_{n-1} - 2p_n \quad (n > 1), \tag{7.11a}$$

$$\dot{p}_1 = p_2 - 2p_1. \tag{7.11b}$$

---

*) D.R. Cox and W.L. Smith, *Queues* (Methuen, London and Wiley, New York 1961); J.W. Cohen, *The Single Server Queue* (North-Holland, Amsterdam 1969).

This is the master equation for a random walk with an absorbing boundary at $n = 0$.

Exercise. Solve by the same method the M-equation

$$\dot{p}_n = p_{n+1} + p_{n-1} - 2p_n \quad (n > 1), \tag{7.12a}$$

$$\dot{p}_1 = p_2 + 2p_0 - 2p_1, \tag{7.12b}$$

$$\dot{p}_0 = p_1 - 2p_0, \tag{7.12c}$$

which describes a random walk with an impure reflecting boundary.[*]

Exercise. Solve the M-equation with pure reflecting boundary

$$\dot{p}_n = p_{n+1} + p_{n-1} - 2p_n \quad (n > 0), \tag{7.13a}$$

$$\dot{p}_0 = p_1 - p_0. \tag{7.13b}$$

8. Artificial boundaries and normal modes

Only a few problems with artificial boundaries can be treated by the reflection principle. In this section the method of normal modes is expounded, which in principle is able to deal with artificial boundaries of all types. Rather than develop this method in full generality we demonstrate it on the example (ii) of section 7: the model for diffusion-controlled chemical reactions.

In order to solve the set (7.7) one first seeks normal modes. Knowing since V.7 that the eigenvalues are real non-positive we set

$$p_n(t) = \phi_n e^{-\lambda t}.$$

ϕ_n must be an eigenvector of the master equation (7.7), i.e.,

$$(2 - \lambda)\phi_n = \phi_{n+1} + \phi_{n-1} \quad (n = 2, 3, \ldots), \tag{8.1a}$$

$$(2 - \lambda)\phi_1 = \phi_2 + \gamma\phi_0, \tag{8.1b}$$

$$(\gamma - \lambda)\phi_0 = \phi_1. \tag{8.1c}$$

The first line is a difference equation with constant coefficients and can therefore be solved in the standard way by the Ansatz $\phi_n = z^n$, where z is to be solved from

$$2 - \lambda = z + \frac{1}{z}. \tag{8.2}$$

This equation has two solutions z_1, z_2 with $z_1 z_2 = 1$. Hence we have found

$$\phi_n = C_1 z_1^n + C_2 z_2^n. \tag{8.3}$$

---

[*] For a more sophisticated application of the reflection principle, see M. Schwarz and D. Poland, J. Chem. Phys. **63**, 557 (1975).

In order that (**8.1a**) is satisfied for all n indicated, this form of ϕ_n must be true for all ϕ_n occurring in it, i.e., for $n = 1, 2, 3, \ldots$. The remaining ϕ_0 and the constants C_1, C_2 are available to satisfy the other two equations,

$$(2 - \lambda)(C_1 z_1 + C_2 z_2) = C_1 z_1^2 + C_2 z_2^2 + \gamma \phi_0, \tag{8.4a}$$

$$(\gamma - \lambda)\phi_0 = C_1 z_1 + C_2 z_2. \tag{8.4b}$$

Eliminating ϕ_0 and using (**8.2**) one obtains

$$-\frac{C_1}{C_2} = \frac{z_1 + (1-\gamma)z_2 + \gamma - 2}{z_2 + (1-\gamma)z_1 + \gamma - 2}.$$

A restriction on the admissible values of λ is obtained by studying the behavior of ϕ_n for large n. If the two roots of (**8.2**) are real, at least one must be in absolute value greater than unity (excluding the special cases $z_1 = z_2 = \pm 1$, which are covered by the next paragraph). Then (**8.3**) will grow exponentially for large n. *Physically* that means an exponentially increasing density of molecules A, which may not be an impossible situation, but is not applicable to diffusion-controlled chemical reactions. The *mathematical* reason for rejecting these solutions is that the remaining ones will turn out to constitute a complete set, i.e., they suffice for reproducing any initial state that we shall be interested in.

Accordingly we only allow those values of λ for which the roots of (**8.2**) are complex. Since they must lie on the unit circle we may put

$$z_1 = e^{iq}, \qquad z_2 = e^{-iq}, \qquad 0 \leqslant q \leqslant \pi. \tag{8.5}$$

Consequently

$$\lambda = 2 - 2\cos q = 4\sin^2 \tfrac{1}{2} q \tag{8.6}$$

is real and $0 \leqslant \lambda \leqslant 4$. Moreover

$$\frac{C_1}{C_2} = -\frac{2 - 2\cos q - \gamma + \gamma e^{-iq}}{2 - 2\cos q - \gamma + \gamma e^{iq}} \tag{8.7}$$

has unit modulus and may therefore be written $e^{2i\eta}$ with real $\eta(q)$. Substitution in (**8.3**) yields

$$\phi_n^{(q)} = C \cos[qn + \eta(q)] \quad (n = 1, 2, 3, \ldots). \tag{8.8}$$

This is a plane wave with phase shift η determined by the boundary. The component for $n = 0$ is obtained from (**8.1c**):

$$\phi_0^{(q)} = -C \frac{\cos(q + \eta)}{2 - 2\cos q - \gamma}. \tag{8.9}$$

Summarizing: for each q in the interval (**8.5**) we have found one eigenfunction. Its components for $n > 0$ are given by (**8.8**), while the component for

8. ARTIFICIAL BOUNDARIES AND NORMAL MODES

$n = 0$ is given by (8.9). Of course, it is determined up to a constant factor C. The associated eigenvalue is (8.6).

Exercise. The stationary solution mentioned in 7 is not among the solutions found here, although it obviously satisfies (8.1). Where did it get lost?

Exercise. Verify that the zeros in the denominator of (8.9) are cancelled by zeros in the numerator. Derive the alternative expression

$$\phi_0^{(q)} = \frac{C}{\gamma} \cos \eta. \tag{8.10}$$

Exercise. The value of η is only defined up to a multiple of π. Show that it can be made unique by choosing $\eta(0) = 0$ and continuing analytically. Show that the analytic function $\eta(q)$ obtained in this way is odd:

$$\eta(-q) = -\eta(q). \tag{8.11}$$

Exercise. Show that the function $S(q) \equiv e^{2i\eta(q)}$ defined by (8.7) has the following properties, reminiscent of the S-matrix:[*]
(i) $S(q)^* = S(q)^{-1} = S(-q)$ for all real q;
(ii) $S(q)$ is holomorphic in the upper half-plane and tends to a finite value in the limit Im $q \to +\infty$.

The remaining task is to show that the set of normal modes found is complete. For this purpose we choose the normalization $C = \sqrt{2/\pi}$ and prove the completeness relation. First we take $m > 0, n > 0$ and prove

$$\frac{2}{\pi} \int_0^\pi \cos[qm + \eta(q)] \cos[qn + \eta(q)] \, dq = \delta_{nm}. \tag{8.12}$$

Using (8.11) one may write for the left-hand side

$$\frac{1}{2\pi} \int_{-\pi}^\pi \{e^{iq(m+n)} e^{2i\eta(q)} + e^{iq(m-n)}\} \, dq.$$

The second term of the integrand is the desired δ_{mn}. The first term can be shown to vanish by considering the closed contour in the complex q-plane in fig. 15. The function $e^{2i\eta(q)}$ is periodic and according to the above Exercise, holomorphic and bounded in the upper half of the complex q-plane. Hence the contribution of the dotted line vanishes when it is shifted to infinity owing to the factor $e^{iq(n+m)}$, while the two vertical lines cancel because of the periodicity. This proves (8.12) for $n, m > 0$.

---

[*] N.G. van Kampen and I. Oppenheim, J. Mathem. Phys. **13**, 842 (1972). This use of the S-matrix for describing the effect of a boundary has been applied to hydrodynamics by P.G. Wolynes, Phys. Rev. A **13**, 1235 (1976) and to gas dynamics by K. Nakazato, J. Phys. Soc. Japan **43**, 1154 (1977).

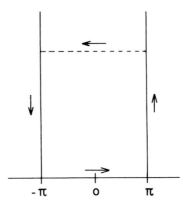

Fig. 15. The integration contour in the complex q-plane.

A separate proof is needed when n, or m, or both, are zero. The calculations are similar to the one above and are left to the reader.

The solution of the master equation (7.7) with the initial value $p_n(0) = \delta_{nm}$ is now readily seen to be for $n > 0$

$$p_n(t) = \int_0^\pi \phi_n^{(q)} \phi_m^{(q)} e^{-t(2 - 2\cos q)} \, dq. \tag{8.13a}$$

This follows from the general formula (V.7.11) since in the present case the weight factor Φ_0 in the denominator of (V.7.11) is constant for $n > 0$. For $n = 0$, however, it is equal to $1/\gamma$, so that

$$p_0(t) = \gamma \int_0^\pi \phi_0^{(q)} \phi_m^{(q)} e^{-t(2 - 2\cos q)} \, dq. \tag{8.13b}$$

Equations (8.13) give the probability distribution of the molecule at $t > 0$ when it is known to start out from m at $t = 0$. For instance, if the molecule is initially in the bound state $m = 0$, the probability that at time t it is still *or again* bound is

$$\gamma \int_0^\pi (\phi_0^{(q)})^2 e^{-t(2 - 2\cos q)} \, dq = \frac{2}{\pi\gamma} \int_0^\pi \cos^2 \eta(q) \, e^{-4t \sin^2 \frac{1}{2} q} \, dq. \tag{8.14}$$

Exercise. Find the analog of (8.12) for $n = m = 0$.

Exercise. What is the probability for a molecule in the bound state to be *still* there at t, without having left it?

Exercise. Show that, in contrast, the *average time* corresponding to (8.14) is infinite.

Exercise. Solve the absorbing boundary problem (7.6). Show that for $c = 1$ the solution coincides with the one found by the reflection principle (7.12).

Exercise. Solve (7.13) by the method of this section.

Exercise. As a model for diffusion in a gravitational field take the asymmetric random walk (2.13) for $n = 0, 1, 2, \ldots$ with a reflecting pure boundary.

Exercise. Solve the following master equation for a random walk between two reflecting boundaries

$$\dot{p}_0 = p_1 - p_0,$$
$$\dot{p}_n = p_{n+1} + p_{n-1} - 2p_n \quad (n = 1, 2, \ldots, N-1),$$
$$\dot{p}_N = p_{N-1} - p_N.$$

As the range is now bounded on both sides one arrives at an eigenvalue equation for q, which can be solved. Make sure that all $N + 1$ eigenfunctions are found.

Exercise. The motion of a defect in a crystal has been described by eq. (7.7a), combined with an absorbing boundary:

$$\dot{p}_1 = p_2 - (1 + \gamma)p_1.$$

Solve this set of equations.*)

9. Nonlinear one-step processes

When $r(n)$ and $g(n)$ are both linear in n it is usually impossible**) to give an explicit solution of the master equation other than the stationary solution. An approximate treatment is given in chapter VIII and a systematic approximation method will be developed in chapter X. We here merely list a few typical examples.†)

(i) *Intrinsic semiconductor.* The valence band can accommodate N electrons, the conduction band M. When $M \geqslant N$ there may be $n = 0, 1, 2, \ldots, N$ excited electrons. Each of them can recombine when it meets one of the n holes, hence $r(n) = \alpha n^2$. Each of the $N - n$ electrons in the valence band can be excited into one of the $M - n$ holes, hence $g(n) = \beta(N - n)(M - n)$. The boundaries at $n = 0$ and at $n = N$ are natural. The master equation is

$$\dot{p}_n = \alpha(\mathbb{E} - 1)n^2 p_n + \beta(\mathbb{E}^{-1} - 1)(N - n)(M - n)p_n. \tag{9.1}$$

---

*) C.A. Condat, Phys. Rev. A **39**, 2112 (1989).
**) That is, the known methods do not work, as shown explicitly in N. Dubin, *A Stochastic Model for Immunological Feedback in Carcinogenesis* (Lecture Notes in Biomathematics Nr. 9; Springer, Berlin 1976).
†) The literature abounds with examples; see, e.g., S. Karlin, *A First Course in Stochastic Processes* (Academic Press, New York 1966).

The stationary solution (3.8) is

$$p_n^s = \text{const.} \binom{N}{n}\binom{M}{n}\left(\frac{\beta}{\alpha}\right)^n, \tag{9.2}$$

Note the relation to the hypergeometric distribution (I.1.6). This p_n^s coincides with p_n^e provided that α and β are connected by the detailed balance relation (4.4), where $h\nu$ is the energy gap, energy differences inside each band being neglected.

In the case of a photoconductor β is increased by a constant γ proportional to the incident light intensity. The system is no longer closed and the new β is no longer connected with α by detailed balance. The stationary solution (9.2) is no longer identical with the thermodynamic equilibrium. Another remark is, that it is possible to represent the effect of the incident photons in this simple way of adding γ to the generation probability only if the arrival times of the photons are uncorrelated (shot noise). When they are correlated the number n is no longer a Markov process, and a more sophisticated description is needed, see XV.3.

(ii) *Dissociation and recombination of a two-atomic gas.* N_X atoms X and N_Y atoms Y in a volume Ω form n molecules XY. Supposing that there are enough inert molecules present to collide with, the probability per unit time for a dissociation to occur is $r(n) = \alpha n$. For recombination an atom X has to meet an atom Y, hence $g(n) = \beta(N_X - n)(N_Y - n)$. Natural boundaries at $n = 0$ and at N_X or N_Y, whichever is smaller. One has

$$p_n^s = \text{const.} \binom{N_X}{n}\binom{N_Y}{n} n! \left(\frac{\beta}{\alpha}\right)^n. \tag{9.3}$$

This is the same distribution as the p_n^e that one finds from ordinary statistical mechanics, provided that

$$\frac{\beta}{\alpha} = \frac{1}{\Omega}\left(\frac{m_X + m_Y}{m_X m_Y}\right)^{3/2} (2\pi kT)^{-3/2} \frac{\zeta_{XY}}{\zeta_X \zeta_Y},$$

where ζ denotes the internal partition function.

(iii) *A chemical reaction.* Consider the auto-catalytic reaction

$$A + X \underset{k'}{\overset{k}{\rightleftarrows}} 2X, \tag{9.4}$$

where A is again so abundant that its amount n_A may be taken constant. If Ω is the volume, n_A/Ω is the concentration of A and one has $g_n = k(n_A/\Omega)n$, where k is the reaction constant as usually defined in chemical kinetics. Similarly

$$r_n = (k'/\Omega)n(n-1),$$

9. NONLINEAR ONE-STEP PROCESSES

because two molecules X have to collide for one to disappear. The master equation is

$$\dot p_n = \frac{k'}{\Omega}(\mathbb{E} - 1)n(n - 1)p_n + \frac{k}{\Omega}n_A(\mathbb{E}^{-1} - 1)np_n. \tag{9.5}$$

Its stationary solution is the equilibrium distribution

$$p_n^e = \frac{1}{n!}\left(\frac{kn_A}{k'}\right)^n \exp\left[-\frac{kn_A}{k'}\right]. \tag{9.6}$$

This is the Poisson distribution. Its average

$$\langle n \rangle^e = kn_A/k'$$

is identical with the macroscopic value as given by the law of mass action. More general chemical reactions are treated in chapter VII.

(iv) *Growth of a competitive population.* Let n be the number of individuals in a population of a certain species of bacteria. Each individual has a natural death rate α and a probability β per unit time to produce a second one by fission. α and β are assumed fixed, independent of the age of the individual; otherwise the process would not be Markovian. Moreover, competition gives rise to an additional death rate $\gamma(n - 1)$, proportional to the number of other individuals present. In the macroscopic rate equation one may replace $n - 1$ with n:

$$\dot n = (\beta - \alpha)n - \gamma n^2, \tag{9.7}$$

which is called the *Malthus–Verhulst equation*. The stochastic formulation of this model is a one-step process with

$$g(n) = \beta n, \qquad r(n) = \alpha n + \gamma n(n - 1). \tag{9.8}$$

The master equation is therefore

$$\dot p_n = \alpha(\mathbb{E} - 1)np_n + \beta(\mathbb{E}^{-1} - 1)np_n + \gamma(\mathbb{E} - 1)n(n - 1)p_n. \tag{9.9}$$

This equation is further studied in XIII.3.

(v) In *superradiance* a very highly excited collective state of a set of atoms decays through successive photon emissions. The probability for being in the state $n = 0, 1, 2, \ldots$ obeys the master equation[*]

$$\dot p_n = (\mathbb{E} - 1)r(n)p_n \quad (n = 0, 1, 2, \ldots, N - 1), \tag{9.10a}$$

$$\dot p_N = -r(N)p_N. \tag{9.10b}$$

Observe that the steps only go in one direction, as in radioactive decay ("pure death process"). As a consequence it is, in principle, possible to obtain

---

[*] G.H. Weiss, J. Statist. Phys. **6**, 179 (1972).

explicit expressions for the eigenfunctions and eigenvalues (although a complication arises when $r(n)$ is such that it takes the same value for two different values n).

(vi) "Alkemade's diode"[*] is the model of fig. 16. The whole is in thermal equilibrium, but the work functions W_1, W_2 of both diode plates are different, resulting in a contact potential V_c. Accordingly the equilibrium charge on the condenser is V_c/C. Let n be the excess of electrons on the left condenser plate above this equilibrium value and $V = -en/C$ the corresponding excess potential. The electrons that jump from left to right face the potential threshold W_1, so that r_n is a constant, A, given by Richardson's formula. The electrons jumping from right to left face $W_2 + V_c + V = W_1 + V$, hence

$$r_n = A, \qquad g_n = A \exp\left[-\frac{e^2}{kTC}n\right]. \qquad (9.11)$$

This remains true for $n < 0$ provided that V does not exceed V_c.

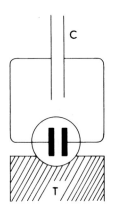

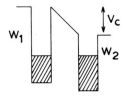

Fig. 16. Alkemade's diode.

---

[*] C.T.J. Alkemade, Physica **24**, 1029 (1958); N.G. van Kampen, Physica **26**, 585 (1960); J. Mathem. Phys. **2**, 592 (1961).

9. NONLINEAR ONE-STEP PROCESSES

Remark. The distinction between linear and nonlinear one-step processes has more physical significance than appears from the mathematical distinction between linear and nonlinear functions $r(n)$ and $g(n)$. In many cases n stands for a number of individuals, such as electrons, quanta, or bacteria. The master equation for p_n is linear in n when these individuals do not interact, but follow their own individual random history regardless of the others. A nonlinear term in the equation means that the fate of each individual is affected by the total number of others present, as is particularly clear in example (iv) above. Thus linear master equations play a role similar to the ideal gas in gas theory. This state of affairs is described more formally in VII.6.

Exercise. An n-type semiconductor, not heavily doped, can be described by (9.1), with $M \gg N$. Take the appropriate limit $M \to \infty$, find the corresponding master equation and equilibrium distribution.

Exercise. For low excitation (low temperature) one may take in (9.1) the limit $M \to \infty$, $N \to \infty$. The resulting M-equation is (with a new β)

$$\dot{p}_n = \alpha(\mathbb{E} - 1)n^2 p_n + \beta(\mathbb{E}^{-1} - 1)p_n. \tag{9.12}$$

Derive this, find the equilibrium distribution and the relation between α and β.

Exercise. In example (ii) take the limit $N_Y \to \infty$. The resulting M-equation is linear and can be solved.

Exercise. Show that the average of (9.6) is indeed identical with the value given by the law of mass action.

Exercise. Show that the eigenvalues of (9.10) are $\lambda_n = r(n)$ and find the corresponding eigenfunctions when no degeneracy occurs. In the model of Weiss $r(n) = n(N - n)$, so that the eigenvalues are two-fold degenerate. How does one overcome this difficulty?

Exercise. The following modifications of Ehrenfest's urn model is nonlinear.[*] Two urns each contain a mixture of black and white balls. Every second I draw with one hand a ball from one urn and with the other a ball from the other urn, and transfer both. Write the difference equation for the probability $p_n(t)$ of having n white balls in the left urn.

---

[*] P.S. de Laplace, *Théorie analytique des probabilités* (3rd Ed., Courcier, Paris 1820) p. 292.

Chapter VII

CHEMICAL REACTIONS

Chemical reactions constitute an ample field for applications of stochastic methods, and have already provided us with several examples. This chapter slightly deviates from the main line in order to provide a firm basis for these applications. The last two sections deal with topics of more general interest, which, however, can best be formulated in a chemical context.

1. Kinematics of chemical reactions

Consider a *closed* volume Ω containing a mixture of chemical compounds X_j ($j = 1, 2, \ldots, J$). Let n_j be the number of molecules X_j. It is convenient to represent the set $\{n_j\}$ geometrically by a vector $\boldsymbol{n}$ in a J-dimensional "state space". The integral values of the n_j constitute a lattice. Every lattice point in the "octant" of nonnegative values corresponds to a state of the mixture and vice versa (fig. 17).

The state of the mixture changes when a chemical reaction occurs. A typical reaction is determined by a set of stoichiometric coefficients s_j, r_j in the form

$$s_1 X_1 + s_2 X_2 + \cdots \rightarrow r_1 X_1 + r_2 X_2 + \cdots . \tag{1.1}$$

Both sides can be written as a sum over all j when zero values of s_j, r_j are admitted. If for any k one has $s_k = r_k \neq 0$ the corresponding X_k is a catalyst. If $r_k > s_k > 0$ then X_k is an autocatalyst. So far the s_j, r_j are defined up to a

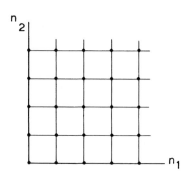

Fig. 17. The state space of a binary mixture.

1. KINEMATICS OF CHEMICAL REACTIONS

common factor, but we may take for s_j the actual number of molecules needed for a reactive collision. A reaction that proceeds through intermediate steps (chain reaction) then has to be written as a sequence of single-collision reactions, the intermediate products being included as separate items among the X_j. As three-body collisions are rare one meets in practice only reactions with $\sum s_j$ equal to 1 or 2; or possibly 3 if a catalyst is involved. But the following theoretical considerations are not subject to this restriction.[*]

Each reactive collision of type (1.1) changes the state $\{n_j\}$ of the mixture into $\{n_j + r_j - s_j\}$. In the geometrical representation this means that it changes the state vector $\boldsymbol{n}$ by adding to it a vector $\boldsymbol{v}$ with components $v_j = r_j - s_j$. As the reaction proceeds the state vector runs over a sequence of lattice points lying on a straight line. This line cannot extend to infinity and must therefore end on one of the boundaries of the physical octant.

With the reaction (1.1) is associated the reverse reaction

$$\sum r_j X_j \to \sum s_j X_j. \tag{1.2}$$

It has the effect of *subtracting* $\boldsymbol{v}$ from the state vector. Thus, starting from an initial state $\boldsymbol{n}^0$, both reactions together cause the state vector to move over a discrete chain of lattice points lying on a straight line between two boundaries of the physical octant. The accessible points are

$$\boldsymbol{n} = \boldsymbol{n}^0 + \xi \boldsymbol{v}, \tag{1.3}$$

where ξ takes all integer values between an upper and a lower bound.

Now suppose another reaction s'_j, r'_j and its reverse are possible. Starting from $\boldsymbol{n}^0$ a second chain of lattice points becomes accessible. Together with the previous reaction, a network of points can now be reached, viz.,

$$\boldsymbol{n}^0 + \xi \boldsymbol{v} + \xi' \boldsymbol{v}' \quad (\xi, \xi' = \ldots, -1, 0, 1, \ldots). \tag{1.4}$$

When in this way *all* possible reactions are taken into account a sublattice is generated of points accessible from $\boldsymbol{n}^0$. It certainly cannot cover the whole octant, because $\sum_j n_j$ is bounded.

As the reactions take place in a closed vessel, there is no other way by which the n_j can vary. Hence this bounded sublattice is the set of accessible states of the system. The physical octant decomposes in such sublattices and the system is confined to that sublattice on which its initial state $\boldsymbol{n}^0$ happens to lie. In open systems there are additional possibilities for $\boldsymbol{n}$ to change; they will be considered in section 3.

It is possible to parametrize the accessible sublattice in the way indicated by (1.4). Each possible reaction ρ has a vector $\boldsymbol{v}^{(\rho)}$. All lattice points accessible

---

[*] Our description is a drastic idealization of real chemical reactions, as described, e.g., by J.B. Anderson, in: *Advances in Chemical Physics* **41** (Wiley, New York 1980).

from $\boldsymbol{n}^0$ are by construction

$$\boldsymbol{n} = \boldsymbol{n}^0 + \sum_\rho \xi_\rho \boldsymbol{v}^{(\rho)}. \tag{1.5}$$

Each parameter ξ_ρ takes the integral values $\ldots, -2, -1, 0, 1, 2, \ldots$ and is called "degree of advancement", because it indicates how far the reaction ρ has advanced.[*] The values of the $\{\xi_\rho\}$ are limited by the requirement $n_j \geq 0$ for all j, but the next section will show that this limitation creates no difficulties.

First suppose that the representation (1.5) is unique, in the sense that for given $\boldsymbol{n}^0$ each accessible point $\boldsymbol{n}$ is represented by a single set of values $\{\xi_\rho\}$. Then (1.5) maps the accessible sublattice onto the integral value lattice in the space with coordinates ξ_ρ. Each lattice point in the accessible part of this space corresponds to one and only one state of the mixture. Each reactive collision corresponds to a unit step parallel to one of the coordinate axes ξ.

Unfortunately, there is no reason why (1.5) should be unique. There may well be two different sets of ξ_ρ that lead from $\boldsymbol{n}^0$ to the same $\boldsymbol{n}$. That implies that there is a set of integers ζ_ρ, not all zero, such that

$$\sum_\rho \zeta_\rho \boldsymbol{v}^{(\rho)} = 0. \tag{1.6}$$

If that is so, it is still possible to find a smaller set of lattice vectors $\boldsymbol{w}^{(\sigma)}$, such that each point of the accessible sublattice is uniquely represented by

$$\boldsymbol{n} = \boldsymbol{n}^0 + \sum_\sigma \eta_\sigma \boldsymbol{w}^{(\sigma)}, \tag{1.7}$$

with integers η_σ. Then again each lattice point in the space with coordinates η_σ corresponds to one and only one state of the mixture. But while in ξ-space reactions correspond to unit steps, in η-space they do not. Hence not much has been gained with respect to the original representation in the space of state vectors $\boldsymbol{n}$.

The reactions that are possible in a closed vessel are restricted by conservation laws for the atoms involved. Let α label the various kinds of atoms and suppose X_j contains m_j^α atoms of kind α, where $m_j^\alpha = 0, 1, 2, \ldots$. Then the stoichiometric coefficients of (1.1) obey for each α

$$\sum_j s_j m_j^\alpha = \sum r_j m_j^\alpha \quad \text{or} \quad \boldsymbol{v} \cdot \boldsymbol{m}^\alpha = 0.$$

---

[*] Introduced by Th. De Donder, *L'affinité* (Gauthier-Villars, Paris 1927). Other names have sprung up: "progress variable" [DE GROOT AND MAZUR, p. 199], "extent of the reaction" [G.R. Gavalas, *Nonlinear Differential Equations of Chemically Reacting Systems* (Springer, Berlin 1968)], and "reaction parameter" [O.J. Heilmann, Kong. Danske Vidensk. Selsk. Mat.-fys. Medd. **38**, no. 11 (1972)].

As this holds for all reactions, the accessible sublattice lies entirely on the intersection of hyperplanes given by

$$\boldsymbol{n} \cdot \boldsymbol{m}^\alpha = A^\alpha, \tag{1.8}$$

where A^α is the total number of available atoms α.

The conservation laws (1.8) need not all be independent. If a group of different atoms stays together through all reactions, it gives rise to a single conservation law. For example, the reaction[*]

$$2NO + Cl_2 \rightleftharpoons 2NOCl$$

involves three kinds of molecules, but the conservation laws for N and O coincide. On the other hand, there may be other conservation laws, in addition to those expressing conservation of atoms. For instance, if X_k occurs only as a catalyst, n_k is conserved by itself.

All conservation laws together define a linear subspace of lattice points in the total state space. The *accessible* subspace lies in this subspace and usually is identical with it, but not necessarily so. A counterexample would be

$$2X \rightleftharpoons 2Y,$$

in which two molecules X by colliding may change into a different modification Y. The conservation law $n_X + n_Y = A$ defines a straight line in the two-dimensional state space (see fig. 18), but only every other lattice point is accessible from a given $\boldsymbol{n}^0$.

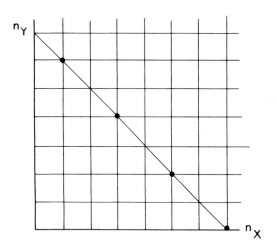

Fig. 18. Accessible states for $2X \rightleftharpoons 2Y$ with $A = 7$.

---

[*] K.J. Laidler, *Chemical Kinetics* (3rd Ed., Harper and Row, New York 1987).

Example. Let X_1 be a monomer and X_j the polymer molecule consisting of j such monomers. We suppose that the maximum length is J, and that subject to this restriction all polymers can combine and decompose:

$$X_i + X_j \rightleftharpoons X_{i+j} \quad (i, j \geq 1, i + j \leq J).$$

There are $[J/2](J - [J/2])$ such reaction pairs.[*)] On the other hand, there are J compounds, subject to one conservation law,

$$\sum_{j=1}^{J} j n_j = A.$$

The accessible lattice lies on a $(J-1)$-dimensional hyperplane in the J-dimensional state space. Clearly the ξ_ρ cannot be independent when

$$[J/2](J - [J/2]) > J - 1, \quad \text{i.e. } J \geq 4.$$

In particular, for $J = 4$ there are four possible reaction pairs,

$$X_1 + X_1 \rightleftharpoons X_2, \quad X_1 + X_2 \rightleftharpoons X_3, \quad X_1 + X_3 \rightleftharpoons X_4, \quad X_2 + X_2 \rightleftharpoons X_4.$$

Suppose that initially only X_1 is present, so that $\boldsymbol{n}^0 = \{n_1^0, 0, 0, 0\}$. Then

$$n_1 = n_1^0 - 2\xi_1 - \xi_2 - \xi_3, \quad n_2 = \xi_1 - \xi_2 - \xi_4, \quad n_3 = \xi_2 - \xi_3, \quad n_4 = \xi_3 + \xi_4.$$

Thus the four variables n_j are determined by the four ξ_ρ, but the converse is not true since the n_j obey identically

$$n_1 + 2n_2 + 3n_3 + 4n_4 = n_1^0.$$

Exercise. Argue that in (1.1) one must have $r_i < s_i$ for at least one i, and that therefore the reaction must come to a stop at the side $n_i = 0$ of the octant (unless it has been stopped before by one of the other sides).

Exercise. Sketch the state space and the accessible lattice for the reaction

$$H_2 + I_2 \rightleftharpoons 2HI. \tag{1.9}$$

Indicate in the picture the conservation laws and the parameter ξ.

Exercise. Analyze in the same way (although without sketch) the following reaction scheme for $H_2 + Br_2 \rightleftharpoons 2HBr$,

$$Br_2 + M \rightleftharpoons 2Br + M,$$

$$Br + H_2 \rightleftharpoons HBr + H,$$

$$H + Br_2 \rightleftharpoons HBr + Br.$$

M stands for some inert catalyst.

Exercise. If X_1 occurs exclusively as a catalyst it may be eliminated by considering the state space of $X_2, X_3, \ldots$ alone. Show that this does not affect the ξ_ρ nor the conservation laws.

[*)] $[J/2]$ is the integral part of $\tfrac{1}{2}J$, i.e., $[J/2] = \tfrac{1}{2}J$ when J is even and $[J/2] = \tfrac{1}{2}(J-1)$ when J is odd.

Exercise. Prove the assertion that any sublattice can be represented by (1.7) for a suitable choice of the $w^{(\sigma)}$. [Hint: Choose as $w^{(\sigma)}$ the smallest possible lattice vectors and show that the lattice they generate coincides with the sublattice.]

2. Dynamics of chemical reactions

In chemical kinetics it is customary to write rate equations in terms of the densities or "concentrations" $c_j = n_j/\Omega$. For the rate at which the reaction (1.1) occurs one takes, following Van 't Hoff,

$$k_+ \prod_{j=1}^{J} c_j^{s_j}. \qquad (2.1)$$

Here k_+ is a constant, which involves the cross-section for a collision of the required molecules, times the probability for the collision to result in a reaction. More precisely, (2.1) is the number of collisions per unit time per unit volume in which $\{n_j\} \to \{n_j + r_j - s_j\}$. The *rate equations* are therefore

$$\frac{dn_i}{dt} = \Omega k_+ (r_i - s_i) \prod_{j=1}^{J} \left(\frac{n_j}{\Omega}\right)^{s_j}. \qquad (2.2)$$

Remark. This equation is, of course, not a universal truth, but holds when the following physical requirements are satisfied.

(i) The mixture must be homogeneous in order that the density at each point of Ω equals n_j/Ω. For sufficiently slow reactions homogeneity can be achieved or approximated by stirring. Departures from homogeneity are the subject of chapter XIV.

(ii) The elastic, non-reactive collisions must be sufficiently frequent to ensure that the Maxwell velocity distribution is maintained. Otherwise the collision frequency could not be proportional to the product of densities, but more details of the velocity distribution would enter. This requirement will be satisfied in the presence of a solvent or an inert gas, but is actually better obeyed than might be expected.

(iii) The internal degrees of freedom of the molecules are also supposed to be in thermal equilibrium, with the same temperature T as the velocities. Otherwise the fraction of collisions that result in a reaction would depend on the details of the distribution over internal states, and not just on the concentrations. Long-lived excited states, however, may be taken into account by listing them among the X_j as a separate species[*], but a clear-cut difference in time scales is indispensable.

(iv) The temperature must be constant in space and time in order that one may treat the reaction rate coefficients as constants even though they depend strongly on temperature.

---

[*] See, e.g., J.H. Gibbs and P.D. Fleming, J. Statist. Phys. **12**, 375 (1975), but note that their "master equation" is our "rate equation".

These assumptions may not be very realistic in many actual chemical reactions, but they do not violate any physical law and their validity can therefore be approximated to any desired accuracy in suitable experiments. They ensure that the state of the mixture is fully described by the set of numbers $\{n_j\}$.

The expression (2.1) is not the precise number of reactive collisions, but the average. The actual number fluctuates around it and we want to find the resulting fluctuations in the n_j around the macroscopic values determined by (2.2). In order to describe them one needs the joint probability distribution $P(\boldsymbol{n}, t)$. Although it is written as a function of all n_j, it is defined on the accessible sublattice alone. Alternatively one may regard it as a distribution over the whole physical octant, which is zero on all points that are not accessible.

The same assumptions on which the macroscopic equation (2.2) is based lead to a definite form of the master equation for P, apart from a small but essential modification.[*] In (2.1) the probability for a collision involving s_j molecules X_j is taken proportional to $n_j^{s_j}$; more precisely this factor should be

$$n_j(n_j - 1)(n_j - 2) \cdots (n_j - s_j + 1) = \frac{n_j!}{(n_j - s_j)!}. \tag{2.3}$$

We abbreviate this expression by $((n_j))^{s_j}$. Then the terms in the master equation corresponding to (1.1) and its reverse (1.2) are readily seen to be

$$\dot{P}(\boldsymbol{n}, t) = k_+ \Omega \left(\prod_i \mathbb{E}_i^{s_i - r_i} - 1 \right) \prod_j \left\{ \frac{((n_j))^{s_j}}{\Omega^{s_j}} \right\} P$$

$$+ k_- \Omega \left(\prod_i \mathbb{E}_i^{r_i - s_i} - 1 \right) \prod_j \left\{ \frac{((n_j))^{r_j}}{\Omega^{r_j}} \right\} P, \tag{2.4}$$

where k_- is the reaction rate constant for the reverse reaction. The total M-equation including all reactions ρ is a sum of such terms, each with its own s_j^ρ, r_j^ρ, k_+^ρ, k_-^ρ.

Exercise. Determine the first jump moment $a_1^{(j)}(\{n\})$ of n_j. Write the macroscopic equation for n_j.

Exercise. Show that the boundaries are natural in the sense of VI.5. That is, the M-equation guarantees that P vanishes as soon as one n_j is negative, provided this is true for the initial P. Note that the substitution of $((n_j))^{s_j}$ in place of $n_j^{s_j}$ is essential.

---

[*] Basic articles for chemical master equations are D.A. McQuarrie, J. Appl. Prob. **4**, 413 (1967) and E.W. Montroll in: *Energetics in Metallurgical Phenomena III* (W.M. Mueller ed., Gordon and Breach, New York 1967).

3. THE STATIONARY SOLUTION

Exercise. Justify (2.3) as follows. Let ω be a volume element with diameter of the order of the interaction range between molecules. The probability for having in ω the set $\{s_j\}$ of molecules needed for a reaction is, according to (I.1.5),

$$\prod_j \binom{n_j}{s_j} \left(\frac{\omega}{\Omega}\right)^{s_j} \left(1 - \frac{\omega}{\Omega}\right)^{n_j - s_j}.$$

For $\omega \ll \Omega$ this yields (2.3), apart from a constant factor.

Exercise. When only a single reaction is possible, the master equation (2.4) represents a one-step process on the chain of accessible lattice points. Exhibit this fact more clearly by writing the master equation in terms of the degree of advancement ξ.

Exercise. The process in which two incident protons are absorbed to produce one outgoing photon has been described by the equation[*]

$$\dot{P}(n, m, t) = (\mathbb{E}_n^2 \mathbb{E}_m^{-1} - 1)n(n-1)mP(n, m, t).$$

Reduce this equation to a single-variable one-step process of pure birth type. [Compare the observation in connection with (VI.9.10).]

3. The stationary solution

The grand canonical distribution of an ideal mixture of chemical compounds is

$$P^g(\mathbf{n}) = \prod_j \frac{(\Omega z_j)^{n_j}}{n_j!} e^{-\Omega z_j} \quad (n_j = 0, 1, 2, \ldots). \tag{3.1}$$

Here z_j is the partition function of one molecule j in a unit volume:

$$z_j = \left(\frac{2\pi m}{\beta}\right)^{3/2} \sum_\nu e^{-\varepsilon_\nu/kT}. \tag{3.2}$$

The sum extends over all internal vibrational, rotational and electronic states. According to (3.1) the n_j are independent Poisson variables with averages $\langle n_j \rangle = \Omega z_j$, and z_j is the equilibrium value of the concentration.

One expects P^g to be a stationary solution of our master equation (2.4). Substitution shows that this is actually so, provided that

$$\frac{k_+}{k_-} = \prod_j z_j^{r_j - s_j}. \tag{3.3}$$

Once more the known equilibrium distribution leads to a relation between the rates of a transition and its reverse. Relation (3.3) is called the "law of mass action".

Yet the grand canonical distribution (3.1) does not correctly describe the fluctuations of the n_j in a closed vessel. This is obvious because it assigns a

---
[*] H.D. Simaan, J. of Physics (London) A **11**, 1799 (1979).

non-zero probability to each point in the octant, whereas actually only the points of a certain sublattice can be reached. The correct distribution is proportional to (3.1) on that sublattice, and zero outside it:

$$P^e(\mathbf{n}) = C \prod_j \frac{(\Omega z_j)^{n_j}}{n_j!} e^{-\Omega z_j} \Delta(\mathbf{n}, \mathbf{n}^0). \tag{3.4}$$

Here $\Delta(\mathbf{n}, \mathbf{n}^0) = 1$ if $\mathbf{n}$ is accessible from $\mathbf{n}^0$, and $\Delta(\mathbf{n}, \mathbf{n}^0) = 0$ if $\mathbf{n}$ is inaccessible. To show that (3.4) is actually a solution of the master equation observe that each term (2.4) of the total master equation vanishes on substituting (3.1) and therefore also on substituting (3.4), because (3.4) is proportional to (3.1) for all values of $\mathbf{n}$ that occur in the master equation.

Example. Dissociation of a diatomic molecule

$$X_2 \underset{k_-}{\overset{k_+}{\rightleftharpoons}} 2X_1. \tag{3.5}$$

The stoichiometric coefficients are

$$s_1 = 0, \qquad s_2 = 1, \qquad r_1 = 2, \qquad r_2 = 0.$$

As initial state we take $n_1 = 0, n_2 = n_2^0$. The conservation of atoms is expressed by $n_1 + 2n_2 = 2n_2^0$. The accessible lattice points are indicated in fig. 19. In equilibrium the probability for the mixture to be in the state represented by such a point is given by (3.4). The probability for the values n_2 is

$$P^s(n_2) = C \frac{(\Omega z_2)^{n_2}}{n_2!} \frac{(\Omega z_1)^{2n_2^0 - 2n_2}}{(2n_2^0 - 2n_2)!}$$

$$= C' \frac{1}{n_2!(2n_2^0 - 2n_2)!} \left(\frac{z_2}{\Omega z_1^2}\right)^{n_2}. \tag{3.6}$$

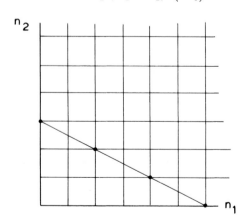

Fig. 19. The state space for the dissociation reaction (3.5) with $n_2^0 = 3$.

Evidently this is not a Poisson distribution. Alternatively, in terms of the degree of advancement

$$n_1 = 2\xi, \qquad n_2 = n_2^0 - \xi \quad (\xi = 0, 1, 2, \ldots, n_2^0),$$

one has

$$P^s(\xi) = C'' \frac{1}{(n_2^0 - \xi)!\,(2\xi)!} \left(\frac{\Omega z_1^2}{z_2}\right)^\xi.$$

In the language of V.2 the total master equation is decomposable. The state space of vectors **n** decomposes into sublattices between which no transition is possible. A separate master equation applies to each sublattice. According to V.3 in each sublattice the probability tends to a unique stationary solution. Any solution of the total master equation tends to that stationary solution on each sublattice, but the weights allotted to the several sublattices are fixed by the initial condition. If the initial condition is $P(\mathbf{n}, 0) = \delta(\mathbf{n}, \mathbf{n}^0)$ all sublattices other than that accessible from $\mathbf{n}^0$ have zero weight, as expressed by (3.4). If one wants to describe an ensemble of vessels, with various $\mathbf{n}^0$, the weights will be different and depend on the initial ensemble. In particular, (3.1) may occur for a suitably chosen ensemble; but it does *not* describe the fluctuations of **n** as they occur in a single closed vessel in the course of time.

Exercise. Under which conditions can one write for the Δ in (3.4)

$$\Delta(\mathbf{n}, \mathbf{n}^0) = \prod_\alpha \delta(\mathbf{n} \cdot \mathbf{m}^\alpha, \mathbf{n}^0 \cdot \mathbf{m}^\alpha),$$

where $\delta(\cdot, \cdot)$ is a Kronecker delta and α runs over all conservation laws?

Exercise. Consider the reaction between two isomers

$$X \rightleftharpoons Y.$$

Determine the probability distribution of n_1 in equilibrium, including its normalizing constant. Also find the variance and show that it does not agree with a Poisson distribution.

Exercise. Compute from (3.6) for the variance

$$\frac{\langle (\Delta n_1)^2 \rangle}{\langle n_1 \rangle} = \left(1 + 16 \frac{z_2}{z_1^2} \frac{n_2^0}{\Omega}\right)^{-1/2} + \mathcal{O}(\Omega^{-1}).$$

Note that the deviation from Poisson does *not* vanish in the thermodynamic limit.

Exercise. Verify that the vanishing of the probability outside the physical octant $n_j \geqslant 0$ is automatically taken into account if one agrees to set $n! \equiv \Gamma(n+1) = \infty$ for negative integer n.

Exercise. The way in which the stationary solution of (3.4) was found in the text was by noting that (3.1) does obey (2.4). A systematic derivation is also possible, however, along the following lines. Write equation (2.4) for the stationary solution in the form

$$0 = \left(\prod_i \mathbb{E}_i^{s_i - r_i} - 1\right)\left[k_+ \prod_j \left\{\frac{((n_j))^{s_j}}{\Omega^{s_j}}\right\} P - k_- \left(\prod_i \mathbb{E}_i^{r_i - s_i}\right) \prod_j \left\{\frac{((n_j))^{r_j}}{\Omega^{r_j}}\right\} P\right].$$

Deduce that [ ] = 0 and subsequently find P^s.

Exercise. The method in the previous Exercise is made possible by the fact that (2.4) is virtually a one-step M-equation, as mentioned in an earlier Exercise. Write the equation in terms of ξ and apply (VI.3.8) to find the stationary solution

4. Open systems

A much larger variety of phenomena can be described as stationary states of *open* chemical systems, i.e., systems in which molecules can be injected and from which molecules can be extracted. The simplest possibility for injection of molecules X is at a constant rate b. Extraction of X can only be done as long as there are some X present; the rate of extraction must therefore vanish when $n = 0$; the simplest possible choice for it is an. Then the macroscopic equation for the number n of molecules X has the form

$$\dot{n} = b - an + \cdots, \tag{4.1}$$

where the dots stand for the internal chemical reactions examined in the preceding sections.

These macroscopic terms do not yet specify the injection and extraction mechanisms on the mesoscopic level. The simplest case is that the molecules are injected independently at random moments (shot noise) and extracted independently. That corresponds to terms in the transition probability of the form

$$W(n|n') = b\,\delta_{n,n'+1} + an'\,\delta_{n,n'-1} + \cdots.$$

In the master equation they give rise to the terms

$$\dot{p}_n(t) = b(\mathbb{E}^{-1} - 1)p_n + a(\mathbb{E} - 1)np + \cdots. \tag{4.2}$$

This is not the only possible way of adding terms to the master equation that give rise to the macroscopic terms (4.1). The molecules might be injected, e.g., in clusters. Such a different choice for the mesoscopic description would affect the fluctuations in n. In general, *whenever a system is subject to an external force or agency, one cannot compute the fluctuations if that force is merely known macroscopically, one must also know its stochastic properties.*[*]

---

[*] The quantum mechanical description of open systems is discussed by W.R. Frensley, Rev. Mod. Phys. **62**, 745 (1990).

4. OPEN SYSTEMS

For chemical reactions there is a convenient and natural way to specify the noise properties of the injection mechanism, which has already been used in some examples. One supposes that the molecules X are produced from a compound B, which is present in large amount and slowly decays into X. The production is then practically constant and the reverse reaction negligible, just as the production of helium by uranium may be regarded as constant. In other words, one describes the open system as a limiting case of a closed system that is not in equilibrium. The role of B is reduced to that of a reservoir.

To describe this limit more precisely consider

$$B \underset{k'}{\overset{k}{\rightleftarrows}} X.$$

The M-equation for the joint distribution is

$$\partial_t P(n_X, n_B, t) = k(\mathbb{E}_B \mathbb{E}_X^{-1} - 1) n_B P + k'(\mathbb{E}_B^{-1} \mathbb{E}_X - 1) n_X P.$$

Now let $n_B \to \infty$, $k \to 0$ with constant $k n_B = b$. Then

$$k \mathbb{E}_B n_B = k n_B + k \to b.$$

Since n_B no longer occurs as a variable one may sum over it to obtain the marginal distribution for n_X alone,

$$\sum_{n_B} P(n_X, n_B, t) = P(n_X, t),$$

$$\partial_t P(n_X, t) = b(\mathbb{E}_X^{-1} - 1) P + k'(\mathbb{E}_X - 1) n_X P.$$

Finally take the limit $k' \to 0$ and the first term of (4.2) results.

The extraction term in (4.2) can also be obtained as a limit of a reaction in a closed vessel. Consider

$$X \underset{k'}{\overset{k}{\rightleftarrows}} A,$$

with master equation

$$\partial_t P(n_X, n_A, t) = k(\mathbb{E}_X \mathbb{E}_A^{-1} - 1) n_X P + k'(\mathbb{E}_X^{-1} \mathbb{E}_A - 1) n_A P.$$

Take the limit $k' \to 0$ and sum over n_A; the result is the second term in (4.2). The rate coefficient k has to be kept constant equal to a; hence, since according to (3.3)

$$k'/k = z_X/z_A \to 0,$$

A must have an extremely large internal partition function in order to act as a sink.

Exercise. Give a few other master equations compatible with (4.1).

Exercise. Write the M-equation for the reaction

$$A \rightleftharpoons X, \qquad 2X \rightleftharpoons B,$$

where A and B act as reservoirs. For certain values of the parameters the stationary state is such that the net transport between A and B vanishes. Show that in that case the stationary distribution of X is Poissonian.

Exercise. Generally, in any open system with reservoirs A, B, C, ... and reactants X, Y, Z, ..., if the stationary state does not transport matter among the reservoirs, the stationary distribution is multi-Poissonian.

5. Unimolecular reactions

An entirely different aspect of chemical reactions gives rise to stochastic problems of a different type. Consider a gaseous reaction in which a single molecule dissociates,

$$XY \to X + Y, \tag{5.1}$$

e.g., C_2H_5Cl into $C_2H_4 + HCl$. If one starts with pure XY a reaction can occur only when an XY molecule collides with another one, and the rate should therefore initially be proportional to the *square* of the density. But experiment tells us that the rate is actually proportional to the density itself, at least if the density is not too low. Following a suggestion by F.A. Lindemann this is explained by the following picture.[*] A collision of a molecule XY does not have the effect of either dissociating XY or not. Rather it leaves XY in an activated state (XY)*, which is metastable and will ultimately dissociate if left alone. In the meantime, however, another collision may occur by which the molecule returns to the ground state. Thus the reaction (5.1) should be written

$$XY \rightleftharpoons (XY)^*,$$
$$(XY)^* \to X + Y. \tag{5.2}$$

When the density n of XY molecules is sufficiently high, the reaction on the first line maintains a supply of (XY)* molecules proportional to n. Each (XY)* has a certain probability per unit time to decompose according to the second line. Hence the overall reaction proceeds at a rate proportional to n rather than n^2. See the Exercise below.

---

[*] N.B. Slater, *Theory of Unimolecular Reactions* (Cornell University Press, Ithaca, NY 1959); K.J. Laidler, op. cit.

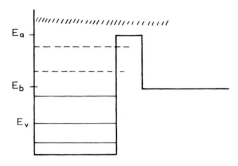

Fig. 20. Bound states, approximately bound states and decaying states.

It is not realistic to distinguish merely one stable state and one activated state. The molecule XY has several degrees of freedom corresponding to the rotational, vibrational and electronic motion of its constituents. Hence there is a large number of eigenstates v with energies E_v ($v = 0, 1, 2, ...$). Strictly speaking the energy spectrum is continuous above the binding energy E_b. However, even above E_b there exist approximately bound states v with approximately defined energies $E_v > E_b$, as is known from Gamow's theory of alpha decay.[*] (fig. 20.) They have a very small probability per unit time for X and Y to separate. Only above a certain energy E_a, called *activation energy*, is the dissociation probability appreciable. One expects therefore the dissociation rate to involve the very small factor $\exp[-E_a/kT]$, called the *Arrhenius factor*.

A collision leaves the XY molecule in a state v with probability p_v. Between collisions p_v is constant in time when $E_v < E_b$, but the activated states can dissociate:

$$\dot{p}_v = -\Gamma_v p_v \quad (E_v > E_a). \tag{5.3a}$$

Collisions will change p_v and, since there is a certain probability per unit time to undergo a collision of a specified type, one has a master equation

$$\dot{p}_v = \sum_\mu (W_{v\mu} p_\mu - W_{\mu v} p_v). \tag{5.3b}$$

As emphasized in IV.1 this can be true only on a coarse time scale on which the collision is instantaneous. Combining (5.3a) and (5.3b)

$$\dot{p}_v = \sum_\mu W_{v\mu} p_\mu - \Gamma_v p_v \quad (v = 0, 1, 2, ...; \Gamma_v = 0 \text{ for } E_v < E_a). \tag{5.4}$$

This equation does not conserve probability. However, probability conservation can be restored by adding an equation for the probability p_* that the

---

[*] J.M. Blatt and V.F. Weisskopf, *Theoretical Nuclear Physics* (Wiley, New York 1952) p. 383.

molecule is dissociated:

$$p_* = 1 - \sum p_v, \qquad \dot{p}_* = \sum \Gamma_v p_v. \tag{5.5}$$

Equations (5.4) and (5.5) together are a probability conserving master equation. The dissociated state $*$ is an absorbing state and consequently $p_*(\infty) = 1$, $p_v(\infty) = 0$. However, the chemist wants to know how fast this dissociation occurs. Suppose one has solved (5.4) with initial condition $p_v(0) = \delta_{v0}$; i.e., the molecule starts in its ground state. Then the probability that it dissociates between t and $t + dt$ is $\dot{p}_*(t)\,dt$. The average dissociation time is, compare (VI.7.5),

$$\int_0^\infty t\dot{p}_*(t)\,dt. \tag{5.6}$$

In order to obtain numerical results it is necessary, on the one hand to know $W_{v\mu}$ and Γ_v, on the other hand to solve (5.4). It is not our task to describe the various theories and approximations that concern the first problem, but it is fair to say that at best they lead to qualitative results. As a consequence there is a lot of leeway in choosing for $W_{v\mu}$ and Γ_v expressions that facilitate the handling of the second problem. We mention two approaches, but it has to be admitted that they are more interesting as exercises in stochastic processes than useful for actually calculating unimolecular reaction rates.

Montroll and Shuler[*] took for $W_{v\mu}$ a one-step transition matrix, cut off by an absorbing boundary, which amounts to setting

$$\Gamma_v = 0 \quad \text{for} \quad v = 0, 1, 2, \cdots, N - 1; \quad \Gamma_N > 0.$$

In order to obtain an analytic solution they subsequently took the $W_{v\mu}$ belonging to a quantized harmonic oscillator, which is not unreasonable for the vibrations of a diatomic molecule[**]. Thus their M-equation has an artificial boundary:

$$\dot{p}_v = \alpha(\mathbb{E} - 1)v p_v + \beta(\mathbb{E}^{-1} - 1)(v+1)p_v \quad (v = 0, 1, 2, \ldots, N-1),$$
$$\dot{p}_N = -\alpha N p_N + \beta N p_{N-1} - \Gamma_N p_N,$$
$$\dot{p}_* = \Gamma_N p_N. \tag{5.7}$$

The rather laborious solution proceeds through the determination of eigenvalues and eigenfunctions as in VI.8, which turn out to be the somewhat

---

[*] E.W. Montroll and K.E. Shuler, J. Chem. Phys. **26**, 454 (1957); *Advances in Chemical Physics* **1** (Interscience, New York 1958) p. 361; S.K. Kim, J. Chem. Phys. **28**, 1057 (1958).
[**] L. Landau and E. Teller, Physik. Z. Sovjetunion **10**, 34 (1936).

unfamiliar Gottlieb polynomials.*⁾ Unfortunately the reaction rate, obtained with reasonable values of α, β, N, Γ_N, is several orders of magnitude smaller than the observed one.

Secondly, Kramers took for $W_{\nu\mu}$ a differential operator, which led to the celebrated Kramers escape problem, treated in XIII.6.

Exercise. Write the rate equations for (5.2) and show that the dissociation rate is indeed proportional to the density of XY when the second line is much slower than the first.

Exercise. In the case of two states (5.4) may be written

$$\dot p_0 = -\beta p_0 + \alpha p_1, \qquad (5.8a)$$

$$\dot p_1 = \beta p_0 - \alpha p_1 - \gamma p_1. \qquad (5.8b)$$

Find for the average survival time $(\alpha + \beta + \gamma)/\beta\gamma$. Also show that the decrease of $p_0 + p_1$ is exponential for large t, but the rate is *not* given by this survival time, unless γ is very small.

Exercise. Dissociation and recombination in the presence of an inert diluent is described by the reaction scheme

$$X_2 + M \rightleftharpoons X + X + M.$$

If the molecule X_2 has internal levels v the rate equation is**⁾

$$\dot c_v = \sum_\mu (w_{v\mu} c_\mu - w_{\mu v} c_v) + u_v c^2 - v_v c_v,$$

where c is the concentration of atoms X. Construct the M-equation.

Exercise. Construct the M-equation for ionization of multi-level hydrogen atoms in a plasma of protons and electrons†⁾

$$H + e \rightleftharpoons p + 2e.$$

Exercise. The molecules X have vibrational levels $v = 0, 1, 2, \ldots$. On colliding they may exchange a vibrational quantum by the process††⁾

$$X_{v+1} + X_{v-1} \rightleftharpoons 2X_v \quad (v = 1, 2, \ldots).$$

Construct the M-equation and obtain from it the rate equation and the equilibrium distribution (for fixed particle number and energy).

Exercise. An example of (5.4) is the equation

$$\dot p_n = \alpha(\mathbb{E} - 1)np_n + \beta(\mathbb{E}^{-1} - 1)(n+1)p_n - (a + bn)p_n \quad (0 \leqslant n < \infty),$$

---

*⁾ S.H. Northrup and J.T. Hynes, J. Chem. Phys. **73**, 2700 and 2715 (1980), simplified the problem by setting $p_v = p \exp[-\beta E_v]$ for the bound states, $E_v < E_a$; this amounts to assuming a single bound state.

⁾ C.A. Brau, J. Chem. Phys. **47, 1153 (1967) calls this a "nonlinear master equation".

†⁾ W.L. Hogarth and D.L.S. McElwain, Proc. Roy. Soc. A **345**, 251 and 265 (1975); A.W. Yau and H.O. Pritchard, Proc. Roy. Soc. A **362**, 113 (1978).

††⁾ T.A. Bak and P.G. Sørensen, in: *Advances in Chemical Physics* **15** (Interscience, New York 1969).

which has been used as a model for dissociation[*]. Solve it with initial condition $p_n(0) = \delta_{n,0}$ and find the dissociation as a function of time. For which initial conditions is the dissociation exponential?

6. Collective systems

The molecules of an ideal gas provide a concrete example for the following consideration, which, however, is of more general validity. Consider a system which can be in different states labelled n, and whose evolution is described by a master equation

$$\dot{p}_n = \sum_{n'} (W_{nn'} p_{n'} - W_{n'n} p_n). \tag{6.1}$$

We shall refer to this system as a "molecule" and to the states n as "levels" in order to distinguish them from the system and states to be introduced now.

Consider a large system, consisting of a collection of N similar molecules, *independent of one another* and each governed by (6.1). The precise state of this collective system is specified by a set of $n_1, n_2, n_3, \ldots, n_N$, being the labels of the levels in which the several molecules reside. The probability for the collective system to be in this precise state is the product of the probabilities of the individual molecules

$$p_{n_1} p_{n_2} \cdots p_{n_N}. \tag{6.2}$$

Now suppose one is merely interested in the number of molecules occupying each level, regardless of their identity. That is, one is merely concerned with the gross state defined by the set of occupation numbers $\{N\} = N_1, N_2, \ldots, N_n, \ldots$. The probability for the collective system to be in this gross state is

$$P(\{N\}) = \sum p_{n_1} p_{n_2} \cdots p_{n_N}, \tag{6.3}$$

where the sum extends over all values of $n_1, n_2, \ldots, n_N$ that are consistent with the prescribed occupation numbers.

Remark. It should be clear that this transition to an occupation number description is a purely algebraic step. In this respect it is similar to what in quantum mechanics is denoted by the misleading term "second quantization". The only difference is that here we deliberately eliminate the information about the identity of the "molecules", whereas in quantum mechanical applications (e.g., to photons or

---

[*] K.F. Freed and D.F. Heller, J. Chem. Phys. **61**, 3942 (1974); V. Kenkre and V. Seshadri, Phys. Rev. **A 15**, 197 (1977).

electrons) that information is absent from the start, owing to the indistinguishability of the particles. We still count the states according to Boltzmann statistics.

The probability (6.3) of the gross state changes whenever one of the molecules jumps from one level n' to another level n. The probability that one of the $N_{n'}$ molecules in level n' makes such a jump during Δt is $W_{nn'} N_{n'} \Delta t$. The probability that two or more jump is of order $(\Delta t)^2$. Hence P obeys the master equation

$$\dot{P}(\{N\}, t) = \sum_{n,n'} W_{nn'}(\mathbb{E}_n^{-1}\mathbb{E}_{n'} - 1) N_{n'} P(\{N\}, t). \tag{6.4}$$

This master equation is uniquely determined by the one-molecule master equation (6.1), because the physics of both is the same. We shall now show how the solutions of (6.4) can be expressed in terms of those of (6.1).

It is sufficient to examine the solution of (6.4) in which initially all N molecules are in one and the same level m:

$$P(\{N\}, 0) = \delta(N_m, N) \prod_{n \neq m} \delta(N_n, 0), \tag{6.5}$$

where the deltas stand for Kronecker delta symbols. Let $p_{n,m}(t)$ be the solution of (6.1) with $p_{n,m}(0) = \delta(n, m)$. At $t > 0$ each of the molecules has a probability $p_{n,m}(t)$ to occupy level n. Hence according to (I.3.11) the probability for the occupation numbers $N_1, N_2, \ldots$ to be realized is

$$P(\{N\}, t) = \binom{N}{N_1 N_2 \cdots} \{p_{1,m}(t)\}^{N_1} \{p_{2,m}(t)\}^{N_2} \cdots. \tag{6.6}$$

It is understood that the multinominal coefficient vanishes unless all N_n are nonnegative and add up to N. This is the solution of (6.4) with initial condition (6.5).

The distribution (6.6) is the multivariate generalization of the binomial distribution. Now consider an ensemble of similar systems in which the total N is not constant but distributed according to Poisson with average $\langle N \rangle$. Then the probability distribution in this grand ensemble is

$$P^g(\{N\}, t) = \sum_{N=0}^{\infty} \frac{\langle N \rangle^N}{N!} e^{-\langle N \rangle} P(\{N\}, t)$$

$$= \frac{\langle N_1 \rangle^{N_1}}{N_1!} e^{-\langle N_1 \rangle} \frac{\langle N_2 \rangle^{N_2}}{N_2!} e^{-\langle N_2 \rangle} \cdots,$$

where $\langle N_1 \rangle = \langle N \rangle p_{1,m}(t)$, etc. This is a product of independent Poisson distributions.

Exercise. Verify by direct substitution that (6.6) obeys (6.4).

Exercise. Deduce from (6.4) that the averages $\langle N_n \rangle$ obey the same equation (6.1)

$$\partial_t \langle N_n \rangle = \sum_{n'} \{W_{nn'} \langle N_{n'} \rangle - W_{n'n} \langle N_n \rangle\}$$

and therefore

$$\langle N_n \rangle_t / N = p_{n,m}(t). \tag{6.7}$$

This fact confirms that our collection of molecules may serve as an ensemble to visualize the probabilities p_n.

Exercise. Find the solution of (6.4) when initially the molecules are distributed over the levels in some arbitrary way.

Exercise. Show that (6.4) can be interpreted as the M-equation of an appropriately chosen chemical reaction scheme.

Exercise. Suppose I group the levels together into "phase cells", each consisting of one or more levels. Then the probability distribution of the occupation numbers of the cells is again given by (6.6), where $p_{n,m}(t)$ is the probability for a molecule starting from *level* m to be in *cell* n at time t.

Exercise. It is also possible to deduce from (6.4) equations for the second moments. The most condensed way of expressing the result is in terms of the factorial cumulants (I.3.13):

$$\partial_t [N_n N_k] = \sum_{n'} \mathbb{W}_{nn'} [N_{n'} N_k] + \sum_{k'} \mathbb{W}_{kk'} [N_n N_{k'}]. \tag{6.8}$$

This formula is utilized in chapter XIV.

Take the special case that there are only two levels $n = 1, 2$. Then (6.4) reduces to

$$\dot{P}(N_1, N_2, t) = W_{12}(\mathbb{E}_1^{-1} \mathbb{E}_2 - 1) N_2 P + W_{21}(\mathbb{E}_2^{-1} \mathbb{E}_1 - 1) N_1 P.$$

Since $N_2 = N - N_1$ this may be written as

$$\dot{P}(N_1, t) = \beta(\mathbb{E}_1^{-1} - 1)(N - N_1) P + \alpha(\mathbb{E}_1 - 1) N_1 P. \tag{6.9}$$

This is a linear one-step master equation of the type studied in VI.6. Such equations can therefore be interpreted as describing an ensemble of two-level systems, each obeying the simple M-equation

$$\dot{p}_1 = -\alpha p_1 + \beta p_2, \qquad \dot{p}_2 = \alpha p_1 - \beta p_2. \tag{6.10}$$

As the solution of this M-equation is trivial, it follows that the solutions of (6.9) can immediately be found in the form of (6.6). This explains why we were able to solve linear one-step master equations explicitly (although not all of them can be interpreted in this way). On the other hand, when the coefficients in a one-step master equation are nonlinear functions of n (as in VI.9) it means that the molecules interact, and the solution is essentially more difficult.

Exercise. Show that the solution of the decay process in IV.6 is an example of the reduction of a linear M-equation to an equation of type (6.10).

Exercise. The quantized harmonic oscillator in example (i) of VI.4 can be regarded as a collection of photons, each of which is either bound or free. Why is it nevertheless not possible to interpret the M-equation (VI.4.1) in the way described here?

Exercise. Reformulate Ehrenfest's urn model (IV.5.4) by noticing that each ball has two states[*].

To a chemist the entropy of a system is a *macroscopic* state function, i.e., a function of the thermodynamic variables of the system. In statistical mechanics, entropy is a *mesoscopic* quantity, i.e., a functional of the probability distribution, viz., the functional given by (V.5.6) and (V.5.7). It is never a *microscopic* quantity, because on the microscopic level there is no irreversibility.[**]

The mesoscopic entropy of a physical system with states n is, according to V.5,

$$s(t) = -k \sum_n p_n \log \frac{p_n}{p_n^e} + s^e, \qquad (6.11)$$

where s^e is constant for a closed and isolated system. The entropy of a system consisting of N independent replicas is, according to V.5,

$$S(t) = -k \sum_{\{N\}} P(\{N\}) \log \frac{P(\{N\})}{P^e(\{N\})} + S^e.$$

Because of (6.3) this is equal to

$$S(t) = -k \sum_{\{N\}} P(\{N\}) \log \prod_n \left(\frac{p_n}{p_n^e}\right)^{N_n} + S^e$$

$$= -k \sum_n \langle N_n \rangle \log \frac{p_n}{p_n^e} + S^e. \qquad (6.12)$$

On account of (6.7) this equals $Ns(t)$ if one disregards the additional constant. Normally one defines the equilibrium entropy S^e in such a way that $S^e = Ns^e$, but when the replicas consist of gases occupying the same volume, a different choice is appropriate[†].

---

[*] F.G. Hess, Amer. Math. Monthly **61**, 323 (1954).
[**] This point was forcefully argued by P. and T. Ehrenfest in their famous article quoted in V.4. Yet it is still often ignored and then leads to the "paradox" that the entropy is a constant of motion.
[†] This is called the "Gibbs paradox", see *Essays in Theoretical Physics in Honour of Dirk ter Haar* (W. Parry ed.; Pergamon, Oxford 1984) p. 303.

186   VII. CHEMICAL REACTIONS

Exercise. A volume V is subdivided into two compartments V_1, V_2 communicating through a small hole. It contains an ideal gas; each molecule may be in V_1 and in V_2 and has therefore two "levels". Apply (6.10) to find the solution of (6.9) and relate (6.11) via (6.12) to the thermodynamic entropy. [Compare (I.1.5).]

Exercise. A donor in a crystal is surrounded by z lattice sites labelled j, each of which has a probability $e^{-\alpha_j}$ to be an acceptor. For each acceptor there is a probability β_j per unit time for the excitation of the donor to jump into the acceptor. Suppose at $t = 0$ there are N such donors, independent from each other. What is the probability distribution of the number of surviving donors at any later time?[*]

7. Composite Markov processes

In various applications the following model has been used, which is of more general interest. Consider a molecule having a number of internal states or "levels" i. From each i it can jump to any other level j with a fixed transition probability $\gamma_{j,i}$ per unit time. Moreover the molecule is embedded in a solvent in which it diffuses with a diffusion constant depending on its state i. The probability at time t for finding it in level i at the position $\mathbf{r}$ with margin $d^3 r$ is $P_i(\mathbf{r}, t) d^3 r$. While the molecule resides in i the probability obeys

$$\dot{P}_i(\mathbf{r}, t) = D_i \nabla^2 P_i(\mathbf{r}, t). \tag{7.1}$$

This model has been used for the explanation of a number of observed transport properties.[**]

The master equation for $P_j(\mathbf{r}, t)$ is obtained by adding to (7.1) the rate of change due to the jumps between levels,

$$\dot{P}_i(\mathbf{r}, t) = D_i \nabla^2 P_i(\mathbf{r}, t) + \sum_j \{\gamma_{i,j} P_j(\mathbf{r}, t) - \gamma_{j,i} P_i(\mathbf{r}, t)\}. \tag{7.2}$$

The equation assumes that during the jumps the position $\mathbf{r}$ does not change, and that the jump probabilities are constants independent of $\mathbf{r}$.

The scheme can be generalized by replacing the diffusion operator with a general operator $\mathbb{F}_i$ acting on the $\mathbf{r}$-dependence and depending parametrically on i. For instance, the physics of chromatography has been explained[†] by considering independent molecules, which can reside in the liquid ($i = 1$) or be absorbed on a surface ($i = 2$). While in the liquid they are carried along

---

[*] A. Blumen and J. Manz, J. Chem. Phys. **71**, 4694 (1979).

[**] K.S. Singwi and A. Sjölander, Phys. Rev. **119**, 863 (1960); H.L. Friedman and A. Ben-Naim, J. Chem. Phys. **48**, 120 (1968); J.W. Haus and K.W. Kehr, op. cit. in VI.2; M.O. Caceres, H. Schnörer, and A. Blumen, Phys. Rev. A **42**, 4462 (1990).

[†] J.C. Giddings and H. Eyring, J. Physical Chem. **59**, 416 (1955); J.C. Giddings, J. Chemical Phys. **26**, 169 (1957).

7. COMPOSITE MARKOV PROCESSES

by its flow velocity v, so that

$$\mathbb{F}_1 = -(v \cdot \nabla), \qquad \mathbb{F}_2 = 0. \tag{7.3}$$

It is, of course, possible to add refinements, such as diffusion in the liquid, or several types of absorption sites.

It is therefore useful to study master equations of the form

$$\dot{P}_i(r, t) = \mathbb{F}_i P_i(r, t) + \sum_j \{\gamma_{i,j} P_j(r, t) - \gamma_{j,i} P_i(r, t)\}. \tag{7.4}$$

Solving this equation means determining the probability to find at $t > 0$ the system in (i, r) when it was at $t = 0$ in (i_0, r_0). This problem decomposes into two successive steps: first find how the molecule jumps among the levels regardless of r, and subsequently add on the behavior in r. This is the reason why we use the name "composite Markov process" for any random process obeying a master equation of the type (7.4).[*]

Take a particular realization of the former process, i.e., suppose that one knows the sequence of times τ_σ at which a jump took place and also the levels that were successively occupied:

$$
\begin{array}{ccccccc}
0 < \tau_1 < \tau_2 < & \cdots & < \tau_s < t \\
\big| & \big| & \big| & & \big| & \big| \\
i_0 & i_1 & i_2 & \cdots & i_{s-1} & i_s.
\end{array}
\tag{7.5}
$$

The probability for being at r under this condition is

$$[P_{i_s}(r, t)]_{\text{cond}} = \exp[(t - \tau_s)\mathbb{F}_{i_s}] \exp[(\tau_s - \tau_{s-1})\mathbb{F}_{i_{s-1}}]$$
$$\cdots \exp[(\tau_2 - \tau_1)\mathbb{F}_{i_1}] \exp[\tau_1 \mathbb{F}_{i_0}] P_{i_0}(r, 0). \tag{7.6}$$

The solution of (7.4) is obtained by multiplying (7.6) with the probability that the realization (7.5) occurs and summing over all realizations.

To find that probability, first note that the probability for any level i to survive at least a time τ is according to (IV.6.3)

$$e^{-\gamma_i \tau}, \qquad \text{where} \quad \gamma_i = \sum \gamma_{j,i}.$$

The probability that a jump takes place from i to j at a time between τ and

---

[*] By a slight generalization it is also possible to treat certain composite non-Markov processes, see G.H. Weiss, J. Statist. Phys. **8**, 221 (1973); K.J. Lindenberg and R. Cukier, J. Chem. Phys. **67**, 568 (1977); N.G. van Kampen, Physica A **96**, 435 (1979).

$\tau + d\tau$ is therefore

$$\gamma_{j,i} e^{-\gamma_i \tau} d\tau.$$

Hence the probability density for (7.5) to be realized is

$$\gamma_{i_1,i_0} \exp[-\gamma_{i_0}\tau_1] \gamma_{i_2,i_1} \exp[-\gamma_{i_1}(\tau_2 - \tau_1)]$$
$$\cdots \gamma_{i_s,i_{s-1}} \exp[-\gamma_{i_{s-1}}(\tau_s - \tau_{s-1})] \exp[-\gamma_{i_s}(t - \tau_s)].$$

Consequently

$$P_i(\mathbf{r}, t) = \sum_{i_0} \left[\delta_{ii_0} \exp[-\gamma_{i_0} t] \exp[t \mathbb{F}_{i_0}] \right.$$
$$+ \sum_{s=1}^{\infty} \sum_{i_1,\ldots,i_{s-1}} \int d\tau_1 \, d\tau_2 \cdots d\tau_s \exp[-\gamma_i(t-\tau_s)] \exp[(t-\tau_s)\mathbb{F}_i]$$
$$\times \gamma_{i,i_{s-1}} \exp[-\gamma_{i_{s-1}}(\tau_s - \tau_{s-1})] \exp[(\tau_s - \tau_{s-1})\mathbb{F}_{i_{s-1}}]$$
$$\left. \cdots \gamma_{i_1,i_0} \exp[-\gamma_{i_0}\tau_1] \exp[\tau_1 \mathbb{F}_{i_0}] \right] P_{i_0}(\mathbf{r}, 0). \quad (7.7)$$

The integration extends over all values of $\tau_1, \tau_2, \ldots, \tau_s$ with the limits given by (7.5), and $\gamma_{i,i} = 0$.

This formidable expression can be written more compactly. Regard $P_i(\mathbf{r}, t)$ as the components of a vector $P(\mathbf{r}, t)$ and define the matrices

$$X_{ij}(t) = \gamma_{i,j} e^{-\gamma_j t + t\mathbb{F}_j}, \qquad Y_{ij}(t) = \delta_{ij} e^{-\gamma_j t + t\mathbb{F}_j}.$$

Then (7.7) reduces to

$$P(\mathbf{r}, t) = \left[Y(t) + \sum_{s=1}^{\infty} \int_0^t d\tau_s \int_0^{\tau_s} d\tau_{s-1} \int_0^{\tau_{s-1}} d\tau_{s-2} \cdots \int_0^{\tau_2} d\tau_1 \right.$$
$$\left. \times Y(t - \tau_s) X(\tau_s - \tau_{s-1}) X(\tau_{s-1} - \tau_{s-2}) \cdots X(\tau_1) \right] P(\mathbf{r}, 0). \quad (7.8)$$

Next introduce Laplace transforms such as

$$\int_0^\infty e^{-\lambda t} P(\mathbf{r}, t) \, dt = \hat{P}(\mathbf{r}, \lambda).$$

Then the convolutions in (7.8) become products,

$$\hat{P}(\mathbf{r}, \lambda) = \left[\hat{Y}(\lambda) + \sum_{s=1}^{\infty} \hat{Y}(\lambda)\{\hat{X}(\lambda)\}^s \right] P(\mathbf{r}, 0)$$
$$= \hat{Y}(\lambda)\{1 - \hat{X}(\lambda)\}^{-1} P(\mathbf{r}, 0). \quad (7.9)$$

Hence the solution of (7.4) is

$$P(r, t) = \frac{1}{2\pi i} \int_{c-i\infty}^{c+i\infty} e^{t\lambda}\, \hat{Y}(\lambda)\{1 - \hat{X}(\lambda)\}^{-1}\, d\lambda \cdot P(r, 0), \qquad (7.10)$$

where c is taken to the right of all singularities in the integrand.

Example. As a model for two-stage diffusion take $i = 1, 2$ and $\mathbb{F}_i$ as in (7.1). Then $\gamma_{1,2} = \gamma_2$ and $\gamma_{2,1} = \gamma_1$. For computing the cross-section for neutron scattering one needs to know the probability density $G_s(r, t)$ that a molecule that, at $t = 0$, was at $r = 0$ will, at time t, be at r. The differential cross-section is its double Fourier transform $G_s(k, \omega)$. It is convenient to apply the Fourier transformation in space right away to (7.4) so that both operators $\mathbb{F}_i$ reduce to factors,

$$\mathbb{F}_i = -D_i k^2.$$

One then has

$$\hat{X}_{ij}(\lambda) = \gamma_{i,j} \int_0^\infty e^{t(\mathbb{F}_j - \gamma_j - \lambda)}\, dt = \frac{\gamma_{i,j}}{\lambda + \gamma_j + D_j k^2},$$

$$\hat{Y}_{ij}(\lambda) = \frac{\delta_{ij}}{\lambda + \gamma_j + D_j k^2}.$$

Finally we take for the initial distribution a delta function at $r = 0$:

$$P_i(k, 0) = (2\pi)^{-3} g_i,$$

where g_i is the equilibrium probability of level i.

Substitution in (7.9) yields

$$\begin{pmatrix} \hat{P}_1(k, \lambda) \\ \hat{P}_2(k, \lambda) \end{pmatrix} = \begin{pmatrix} \hat{Y}_{11} & 0 \\ 0 & \hat{Y}_{22} \end{pmatrix} \frac{1}{1 - \hat{X}_{12}\hat{X}_{21}} \begin{pmatrix} 1 & \hat{X}_{12} \\ \hat{X}_{21} & 1 \end{pmatrix} \begin{pmatrix} g_1 \\ g_2 \end{pmatrix} (2\pi)^{-3}.$$

As neutron scattering does not distinguish between the molecular levels, one wants to know the probability density in space $G_s(r, t) = P_1(r, t) + P_2(r, t)$. Its Fourier–Laplace transform is

$$\hat{G}_s(k, \lambda) = \hat{P}_1(k, \lambda) + \hat{P}_2(k, \lambda) = \frac{g_1 \hat{Y}_{11} + g_2 \hat{Y}_{22} + g_1 \hat{Y}_{22}\hat{X}_{21} + g_2 \hat{Y}_{11}\hat{X}_{12}}{(2\pi)^3 (1 - \hat{X}_{12}\hat{X}_{21})}$$

$$= \frac{\lambda + \gamma_1 + \gamma_2 + (g_1 D_2 + g_2 D_1) k^2}{(2\pi)^3 \{(\lambda + \gamma_1 + D_1 k^2)(\lambda + \gamma_2 + D_2 k^2) - \gamma_1 \gamma_2\}}. \qquad (7.11)$$

The quantity actually observed is

$$G_s(k, \omega) = \hat{G}_s(k, i\omega) + \hat{G}_s(k, -i\omega). \qquad (7.12)$$

Exercise. Show that in this example $g_1 = \gamma_2(\gamma_1 + \gamma_2)^{-1}$ and $g_2 = \gamma_1(\gamma_1 + \gamma_2)^{-1}$.

Exercise. The behavior of $G_s(r, t)$ for large t is determined by the pole of (7.10) with the largest real part. Show in this way that $G_s(r, t)$ for large t obeys a diffusion equation with renormalized diffusion constant $D' = g_1 D_1 + g_2 D_2$. How large does t have to be?

Exercise. Solve in the same way the chromatography model specified by (7.3). Show that the average drift velocity is $v' = g_1 v$.

Exercise. For the process defined in connection with (II.1.8) derive for $t_1 > 0$

$$f_1(t_1) = \frac{1}{2\pi i} \int_{-i\infty}^{i\infty} \frac{\hat{w}(\lambda)}{1 - \hat{w}(\lambda)} e^{\lambda t} \, d\lambda.$$

Also show that

$$f_n(t_1, t_2, \ldots, t_n) = f_1(t_1) f_1(t_2 - t_1) \cdots f_1(t_n - t_{n-1}).$$

Exercise. The solution of the ordinary M-equation for non-composite Markov processes, such as (V.1.5), can also be written as a sum over realizations. The result is, in analogy with (7.10),

$$P(y, t \mid y_0, 0) = \frac{1}{2\pi i} \int_{c-i\infty}^{c+i\infty} e^{t\lambda} \hat{Y}(\lambda) \{1 - \hat{X}(\lambda)\}^{-1} \, d\lambda.$$

Derive this formula directly and show that its poles are the eigenvalues of $\mathbb{W}$.

Application. Consider a Markov process with discrete sites and governed by the M-equation

$$\dot{p}_i = \sum_j (\gamma_{i,j} p_j - \gamma_{j,i} p_i) \equiv \sum_j \Gamma_{ij} p_j.$$

Query: if it starts from i_0, what is, after time t, the average time θ_k spent at site k? Consider the joint characteristic function

$$G(\{\varepsilon\}) = \langle \exp[-\varepsilon_1 \theta_1 - \varepsilon_2 \theta_2 - \cdots] \rangle.$$

This is the same expression as (7.6) if each $\mathbb{F}_i$ is replaced with $-\varepsilon_i$. Subsequently one has to sum over all histories, which in this case includes a sum over the final site i_s. Hence the result is obtained from (7.10):

$$G = \frac{1}{2\pi i} \int e^{t\lambda} \, d\lambda \sum_i \frac{1}{\lambda + \gamma_i + \varepsilon_i} \{1 - \hat{X}(\lambda)\}^{-1}_{ii_0},$$

$$\hat{X}_{ij}(\lambda) = \frac{\gamma_{ij}}{\lambda + \gamma_j + \varepsilon_j}.$$

It remains to compute the reciprocal.

We confine ourselves to the averages $\langle \theta_k \rangle$ and therefore require only the first orders in the ε_k,

$$(1 - \hat{X})_{ij} = \delta_{ij} - \frac{\gamma_{i,j}}{\lambda + \gamma_j} + \frac{\gamma_{i,j} \varepsilon_j}{(\lambda + \gamma_j)^2}$$

$$= \left\{ \lambda \delta_{ij} - \Gamma_{ij} + \frac{\gamma_{i,j} \varepsilon_j}{\lambda + \gamma_j} \right\} \frac{1}{\lambda + \gamma_j},$$

$$(1 - \hat{X})^{-1}_{ij} = (\lambda + \gamma_i) \left[(\lambda - \Gamma)^{-1}_{ij} - \sum_{h,k} (\lambda - \Gamma)^{-1}_{ih} \frac{\gamma_{h,k} \varepsilon_k}{\lambda + \gamma_k} (\lambda - \Gamma)^{-1}_{kj} \right].$$

7. COMPOSITE MARKOV PROCESSES

$(\lambda - \Gamma)^{-1}$ is the resolvent of Γ and can be expressed in the right and left eigenvectors ξ_n, η_n of Γ,

$$(\lambda - \Gamma)^{-1} = \sum_n \frac{\xi_n \eta_n}{\lambda + \lambda_n}.$$

All eigenvalues $-\lambda_n$ are negative except $\lambda_0 = 0$, so that

$$(\lambda - \Gamma)_{ij}^{-1} = \frac{1}{\lambda} p_i^s + \text{terms regular for Re } \lambda \geq 0.$$

This is enough if we are interested only in large t and omit exponentially decreasing terms.

Collecting these results we have obtained

$$G = \frac{1}{2\pi i} \int e^{t\lambda} \, d\lambda \sum_i \frac{\lambda + \gamma_i}{\lambda + \gamma_i + \varepsilon_i} \left(\frac{p_i^s}{\lambda} - \frac{p_i^s}{\lambda} \sum_{h,k} \frac{\gamma_{h,k} \varepsilon_k}{\lambda + \gamma_k} \frac{p_k^s}{\lambda} \right).$$

The zeroth order in ε gives $\sum p_i^s = 1$, as it should. The next order gives as coefficient of ε_k

$$\langle \theta_k \rangle = \frac{1}{2\pi i} \int e^{t\lambda} \, d\lambda \sum_{i,h} \frac{p_i^s}{\lambda} \frac{\gamma_{h,k}}{\gamma_k} \frac{p_k^s}{\lambda} = t p_k^s. \tag{7.13}$$

Conclusion. For large t the fraction of the time the particle spends on site k equals p_k^s. This is the *ergodic property*, announced in IV.5. For the present case of Markov processes it is less profound than for the Hamilton systems in statistical mechanics, for which it was originally formulated.

Exercise. Verify that all omitted terms are of higher order in ε or less singular in λ. The terms with λ^{-1} represent the transient effect of the initial state.

Exercise. Show that, having found the average sojourn times, one obtains immediately the solution of the chromatography problem.

Exercise. Derive

$$\langle\langle \theta_k \theta_l \rangle\rangle = t p_k^s p_l^s \left(\frac{1}{\gamma_k} + \frac{1}{\gamma_l} \right).$$

Remark. The essential feature of our composite process is that i is an independent process by itself, while the transition probabilities of r are governed by i. This situation can be formulated more generally.[*] Take a Markov process $Y(t)$, discrete or continuous, having an M-equation with kernel

$$\Gamma(y \mid y') = \gamma(y \mid y') - \gamma(y) \delta(y - y').$$

Let X be a second variable with transition probabilities $\mathbb{W}_y(x \mid x')$, which depend parametrically on the value of $y(t)$. Then the joint variable $\{X, Y\}$ constitutes a

---

[*] N.G. van Kampen, loc. cit.; M.A. Rodriguez, L. Pesquera, M. San Miguel, and J.M. Sancho, J. Statist. Phys. **40**, 669 (1985).

Markov process with M-equation

$$\dot{P}(x, y, t) = \int \mathbb{W}_y(x\,|\,x')P(x', y, t)\,\mathrm{d}x' + \int \Gamma(y\,|\,y')P(x, y', t)\,\mathrm{d}y'. \tag{7.14}$$

The stochastic function $X(t)$ by itself is not Markovian. This is an example of the fact discussed in IV.1: If one has an r-component Markov process and one ignores some of the components, the remaining $s < r$ components are still a stochastic process but in general not Markovian. Conversely, it is often possible to study non-Markovian processes by regarding them as the projection of a Markov process with more components[*]. We return to this point in IX.7.

---

[*] For a recent example, see L. Schimansky-Geier and Ch. Zülicke, Z. Phys. B **79**, 453 (1990).

Chapter VIII

THE FOKKER–PLANCK EQUATION

The Fokker–Planck equation is a special type of master equation, which is often used as an approximation to the actual equation or as a model for more general Markov processes. Its elegant mathematical properties should not obscure the fact that its application in physical situations requires a physical justification, which is not always obvious, in particular not in nonlinear systems.

1. Introduction

The Fokker–Planck equation is a master equation in which $\mathbb{W}$ is a differential operator of second order

$$\frac{\partial P(y,t)}{\partial t} = -\frac{\partial}{\partial y} A(y)P + \frac{1}{2}\frac{\partial^2}{\partial y^2} B(y)P. \tag{1.1}$$

The range of y is necessarily continuous, and for the present time is supposed to be $(-\infty, +\infty)$. The coefficients $A(y)$ and $B(y)$ may be any real differentiable functions with the sole restriction $B(y) > 0$. The equation can be broken up into a continuity equation for the probability density

$$\frac{\partial P(y,t)}{\partial t} = -\frac{\partial J(y,t)}{\partial y}, \tag{1.2}$$

where $J(y,t)$ is the probability flux, and a "constitutive equation"

$$J(y,t) = A(y)P - \frac{1}{2}\frac{\partial}{\partial y} B(y)P. \tag{1.3}$$

We shall meet more general Fokker–Planck equations; the special form (1.1) is also called "Smoluchowski equation", "generalized diffusion equation", or "second Kolmogorov equation". The first term on the right-hand side has been called "transport term", "convection term", or "drift term"; the second one "diffusion term" or "fluctuation term". Of course, these names should not prejudge their physical interpretation. Some authors distinguish between Fokker–Planck equations and master equations, reserving the latter name to the jump processes considered hitherto.

Let $P(y,t|y_1,t_1)$ for $t \geq t_1$ be that solution of (1.1) that at t_1 reduces to $\delta(y-y_1)$. According to (IV.2.2) one may construct a Markov process whose transition probability is $P(y_2, t_2 | y_1, t_1)$ and whose one-time distribution $P_1(y_1, t_1)$ may still be chosen arbitrarily at one initial time t_0. If one chooses for P_1 the stationary solution of (1.1),

$$P^s(y) = \frac{\text{const.}}{B(y)} \exp\left[2 \int_0^y \frac{A(y')}{B(y')} dy' \right], \tag{1.4}$$

the resulting Markov process is stationary. This is only possible, however, when this P^s is integrable so that it can be normalized to represent a probability distribution.

Markov processes whose master equation has the form (1.1) have been called "continuous", because it can be proved that their sample functions are continuous (with probability 1). This name has sometimes led to the erroneous idea that all processes with a *continuous range* are of this type and must therefore obey (1.1).

By definition the Fokker–Planck equation is always a linear equation for P.[*] Hence the adjective "linear" is available for use in a different sense. We shall call the Fokker–Planck equation (1.1) *linear* if A is a linear function of y and B is constant:

$$\frac{\partial P(y, t)}{\partial t} = -\frac{\partial}{\partial y}(A_0 + A_1 y)P + \tfrac{1}{2} B_0 \frac{\partial^2 P}{\partial y^2}. \tag{1.5}$$

If $A_1 < 0$ the stationary solution (1.4) is Gaussian. In fact, in that case it is possible by shifting y and rescaling, to reduce (1.5) to (IV.3.20), so that one may conclude: *the stationary Markov process determined by the linear Fokker–Planck equation is the Ornstein–Uhlenbeck process.* For $A_1 \geq 0$ there is no stationary probability distribution.

Exercise. Let $P(y, t | y_0, t_0)$ be solution of (1.1). Take $t = t_0 + \Delta t$ and compute for small Δt the moments of $y - y_0 \equiv \Delta y$. Show that for $\Delta t \to 0$

$$\frac{\langle \Delta y \rangle}{\Delta t} = A(y_0), \quad \frac{\langle (\Delta y)^2 \rangle}{\Delta t} = B(y_0), \quad \frac{\langle (\Delta y)^v \rangle}{\Delta t} = 0 \quad (v \geq 3). \tag{1.6}$$

Exercise. Find the singular kernel $\mathbb{W}(y | y')$ corresponding to the differential operator in (1.1). Calculate its jump moments (V.8.2) and compare the result with (1.6). [Compare (V.10.5).]

Exercise. Show that the differential operator in (1.1) has the properties required of a $\mathbb{W}$-matrix: it conserves probability and positivity. Also, when $B(y) > 0$ it is

---

[*] The Landau equation in plasma theory is a nonlinear variant, but there P is a particle density rather than a probability. L.D. Landau, Physik. Z. Sovjetunion **10**, 154 (1963) = *Collected Papers* (D. ter Haar ed., Pergamon, Oxford 1965) p. 163. The same is true for the "nonlinear Fokker–Planck equation" in M. Shiino, Phys. Rev. A **36**, 2393 (1987).

1. INTRODUCTION

irreducible and all solutions tend to the stationary solution (1.4) (provided that it is normalizable).

Exercise. Show that any linear second order differential operator with these properties must have the form (1.1).

Exercise. Solve (1.1) for $B(y) = 2$, $A(y) = -y + 1/y$, $y > 0$. Also for $B =$ constant $\neq 2$.[*]

Exercise. Give the explicit solution of (1.5).

We have introduced the Fokker–Planck equation as a special kind of M-equation. Its main use, however, is as an approximate description for any Markov process $Y(t)$ whose individual jumps are small. In this sense the linear Fokker–Planck equation was used by Rayleigh[**], Einstein[†], Smoluchowski[††], and Fokker[‡], for special cases. Subsequently Planck[‡‡] formulated the general nonlinear Fokker–Planck equation from an arbitrary M-equation assuming only that the jumps are small. Finally Kolmogorov[§] provided a mathematical derivation by going to the limit of infinitely small jumps.

As an approximate substitute for the general master equation (V.1.5) the Fokker–Planck equation (1.1) has two alluring features. First it is a differential equation rather than a differentio-integral equation. Even though it can still not be solved explicitly except for a few special cases, it is easier to handle. More important is the second feature, namely, the fact that it does not require the knowledge of the entire kernel $W(y \mid y')$ but merely of the functions $A(y)$ and $B(y)$. For any actual stochastic process these can be determined with a minimum of detailed knowledge about the underlying mechanism. The way this is done is as follows.

Suppose the physics of some system suggests that a quantity y should be approximately a Markov process. One picks a time Δt so small that y cannot change very much during Δt, but large enough for the Markov assumption to apply. Then compute for this short time the average change $\langle \Delta y \rangle_y$ and its mean square $\langle (\Delta y)^2 \rangle_y$ to first order in Δt. According to the equations (1.6) this is enough to find $A(y)$ and $B(y)$ and hence to set up the Fokker–

---

[*] STRATONOVICH, Section IV.4. Other solvable Fokker–Planck equations are given by R.I. Cukier, K. Lakatos-Lindenberg, and K.E. Shuler, J. Statist. Phys. **9**, 137 (1973).

[**] Lord Rayleigh, Philos. Mag. **32**, 424 (1891) = *Scientific Papers* III (University Press, Cambridge 1902) p. 473.

[†] A. Einstein, Ann. Physik (4) **17**, 549 (1905); **19**, 371 (1906).

[††] M. von Smoluchowski, Ann. Physik (4) **21**, 756 (1906); Physik. Zeits. **17**, 557 and 585 (1916).

[‡] A.D. Fokker, Thesis (Leiden 1913); Ann. Physik (4) **43**, 810 (1914).

[‡‡] M. Planck, Sitzungsber. Preuss. Akad. Wissens. (1917) p. 324 = *Physikalische Abhandlungen und Vorträge* II (Vieweg, Braunschweig 1958) p. 435.

[§] A. Kolmogorov, Mathem. Annalen **104**, 415 (1931).

Planck equation. In this way the actual equations of motion need be solved only during Δt, which can be done by some perturbation theory. The Fokker–Planck equation then serves to find the long-time behavior. This separation between short-time behavior and long-time behavior is made possible by the Markov assumption.

There is also an alternative, more phenomenological way of finding the functions A and B. From (1.1) one has

$$\partial_t \langle y \rangle = \langle A(y) \rangle. \tag{1.7}$$

If one neglects the fluctuations one has $\langle A(y) \rangle = A(\langle y \rangle)$ and one obtains a differential equation for $\langle y \rangle$ alone

$$\partial_t \langle y \rangle = A(\langle y \rangle).$$

This equation is identified with the macroscopic equation of motion for the system, which is supposedly known. Thus the function $A(y)$ is obtained from the knowledge of the macroscopic behavior. Subsequently one obtains $B(y)$ by identifying (1.4) with the equilibrium distribution, which at least for closed physical systems is known from ordinary statistical mechanics. Thus the knowledge of the macroscopic law and of equilibrium statistical mechanics suffices to set up the Fokker–Planck equation and therefore to compute the fluctuations.

This phenomenological identification of A and B has been utilized by Einstein and others with great success (section 3), but *only for linear Fokker–Planck equations*. If the macroscopic law is nonlinear a difficulty arises, first pointed out by D.K.C. MacDonald.[*] The flaw in the argument lies in the identification of the coefficient $A(y)$ with the macroscopic law. The two may well differ by a term of the same order as the fluctuations: once one neglects the fluctuations such a term is invisible anyway. The consequence was that different authors obtained different, but equally plausible expressions for noise in nonlinear systems. This difficulty led to the more fundamental approach in chapter X.

The present chapter, however, is concerned with the traditional approach. We shall carefully stay within the limits of validity, although these limits will only become clear in chapter X. The reader is advised that any doubts he may have must be postponed.

Exercise. Compute the quantities (1.6) for the random walk (VI.2.1) and construct the Fokker–Planck equation. Show that it is satisfied by the asymptotic distribution (VI.2.12).

---

[*] Philos Mag. **45**, 63 and 345 (1954); Phys. Rev. **108**, 541 (1957). For a detailed review see N.G. van Kampen, in: *Fluctuation Phenomena in Solids* (R.E. Burgess ed., Academic Press, New York 1965).

2. DERIVATION OF THE FOKKER–PLANCK EQUATION

Exercise. Same question for the asymmetric random walk (VI.2.13).

Exercise. In an RC circuit with non-Ohmic resistance $R(V)$ the Fokker–Planck equation for the voltage V would be, according to the above phenomenological argument,

$$\frac{\partial P(V,t)}{\partial t} = \frac{\partial}{\partial V} \frac{V}{CR(V)} P + \frac{1}{2} \frac{\partial^2}{\partial V^2} B(V) P, \tag{1.8a}$$

$$B(V) = e^{CV^2/2kT} \left[\text{const.} - 2 \int_0^V \frac{V'}{CR(V')} e^{-CV'^2/2kT} \, dV' \right]. \tag{1.8b}$$

See, however, the discussion in X.2.

Exercise. Apply to (1.1) the nonlinear transformation $\bar{y} = \phi(y)$ and show that the transformed density $\bar{P}(\bar{y}, t)$ obeys a Fokker–Planck equation with coefficients

$$\bar{B}(\bar{y}) = \{\phi'(y)\}^2 B(y), \qquad \bar{A}(\bar{y}) = \phi'(y) A(y) + \tfrac{1}{2} \phi''(y) B(y). \tag{1.9}$$

Exercise. The criterion (V.6.1) for detailed balance cannot readily be applied to (1.1), but the equivalent formulation (V.7.5) is suitable for differential master operators.[*] Show that (1.1) obeys detailed balance if and only if

$$A(y) = \frac{1}{2 P^e(y)} \frac{d}{dy} P^e(y) B(y). \tag{1.10}$$

In that case the equation may be written

$$\frac{\partial P(y,t)}{\partial t} = \frac{1}{2} \frac{\partial}{\partial y} P^e(y) B(y) \frac{\partial}{\partial y} \frac{P(y,t)}{P^e(y)}. \tag{1.11}$$

Exercise. For the "linear" Fokker–Planck equation (1.5), the equation (1.7) is a closed equation for the mean value $\langle y \rangle$, which can be solved. Show that in this way all moments $\langle y^n \rangle$ can be obtained successively.

Exercise. The same is true when B_0 is replaced with $B(y) = B_0 + B_1 y + B_2 y^2$. Moreover it is also possible, but much harder, to find the moments of[**]

$$\frac{\partial P(y,t)}{\partial t} = \frac{\partial}{\partial y} (-\alpha y + y^3) P + \frac{1}{2} \frac{\partial^2}{\partial y^2} y^2 P \quad (0 < y < \infty). \tag{1.12}$$

2. Derivation of the Fokker–Planck equation

Planck derived the Fokker–Planck equation as an approximation to the master equation (V.1.5) in the following way. First express the transition probability W as a function of the size r of the jump and of the starting

---

[*] S.R. de Groot and N.G. van Kampen, Physica **21**, 39 (1954).
[**] R. Graham and A. Schenzle, Phys. Rev. A **25**, 1731 (1982); N. Banai and L. Brenig, Physica A **119**, 512 (1983).

point:
$$W(y|y') = W(y'; r), \quad r = y - y'. \tag{2.1}$$

The general M-equation (V.1.5) then reads

$$\frac{\partial P(y, t)}{\partial t} = \int W(y - r; r) P(y - r, t) \, dr - P(y, t) \int W(y; -r) \, dr. \tag{2.2}$$

The *basic assumption* is that only small jumps occur, i.e., $W(y'; r)$ is a sharply peaked function of r but varies slowly with y'. Somewhat more precisely, there exists a $\delta > 0$ such that

$$W(y'; r) \approx 0 \qquad \text{for } |r| > \delta, \tag{2.3a}$$
$$W(y' + \Delta y; r) \approx W(y'; r) \quad \text{for } |\Delta y| < \delta. \tag{2.3b}$$

The *second assumption* is that the solution $P(y, t)$ one is interested in also varies slowly with y, in the same sense as (2.3b). It is then possible to deal with the shift from y to $y - r$ in the first integral in (2.2) by means of a Taylor expansion up to second order:

$$\frac{\partial P(y, t)}{\partial t} = \int W(y; r) P(y, t) \, dy - \int r \frac{\partial}{\partial y} \{W(y; r) P(y, t)\} \, dr$$
$$+ \frac{1}{2} \int r^2 \frac{\partial^2}{\partial y^2} \{W(y; r) P(y, t)\} \, dr - P(y, t) \int W(y; -r) \, dr.$$

Note that the dependence of $W(y; r)$ on its second argument r is fully maintained; an expansion with respect to this argument is not allowed as W varies rapidly with r. The first and fourth terms cancel. The other two terms can be written with the aid of the jump moments

$$a_\nu(y) = \int_{-\infty}^{\infty} r^\nu W(y; r) \, dr \tag{2.4}$$

already defined in (V.8.2). The result is

$$\frac{\partial P(y, t)}{\partial t} = -\frac{\partial}{\partial y} \{a_1(y) P\} + \frac{1}{2} \frac{\partial^2}{\partial y^2} \{a_2(y) P\}. \tag{2.5}$$

Thus we have derived the Fokker–Planck equation (1.1) from the master equation and in doing so obtained expressions for the coefficients in terms of the transition probability W. These expressions are the same as (1.6) in slightly different notation.

Remark. It is not hard to include all terms of the Taylor expansion of (2.2):

$$\frac{\partial P(y, t)}{\partial t} = \sum_{\nu=1}^{\infty} \frac{(-1)^\nu}{\nu!} \left(\frac{\partial}{\partial y}\right)^\nu \{a_\nu(y) P\}. \tag{2.6}$$

2. DERIVATION OF THE FOKKER–PLANCK EQUATION

This is called the *Kramers–Moyal expansion*.[*] Formally (2.6) is identical with the master equation itself and is therefore not easier to deal with, but it suggests that one may break off after a suitable number of terms. The Fokker–Planck approximation assumes that all terms after $v = 2$ are negligible. Kolmogorov's proof is based on the assumption that $a_v = 0$ for $v > 2$. This, however, is never true in physical systems[**]. In the next chapter we shall therefore expand the M-equation systematically in powers of a small parameter and find that the successive orders do *not* simply correspond to the successive terms in the Kramers–Moyal expansion.

Exercise. For the decay process in (IV.6) construct the Fokker–Planck equation using (1.6). Show that it gives the first and second moments correctly, but not P^s.

Exercise. Find the higher a_v for the random walk and note that they are not small. How can it be explained that nevertheless the Fokker–Planck equation gave the correct result (VI.2.12)?

Exercise. A high-speed particle traverses a medium in which it encounters randomly located scatterers, which slightly deflect it with differential cross-section $\sigma(\theta)$. Find the Fokker–Planck equation for the total deflection, supposing that it is small.

Exercise. Derive from (2.5)

$$\partial_t \langle y \rangle = \langle a_1(y) \rangle,$$

$$\partial_t \langle\langle y^2 \rangle\rangle = 2\langle (y - \langle y \rangle) a_1(y) \rangle + \langle a_2(y) \rangle. \qquad (2.7)$$

Show also that these are exact consequences of the M-equation itself. But the analogous equations for the higher moments are not correctly reproduced by the Fokker–Planck equation. (Compare V.8.)

Exercise. Consider the process with independent increments defined in (IV.4.7), whose M-equation was solved in an Exercise of V.1. Investigate how the Fokker–Planck approximation modifies the general solution. Conclude that for $P(y, t | y_0, t_0)$ the approximation is bad when $t - t_0 < a_4/a_2^2$, owing to the fact that the second assumption above is not satisfied.

To make the Fokker–Planck equation exact, rather than an approximation, one has to allow the coefficients in W to depend on a parameter ε in such a way that the assumptions made are exact in the limit $\varepsilon \to 0$.[†] We demonstrate this approach for the asymmetric random walk, whose master equation (VI.2.13) is

$$\dot{p}_n = \alpha(p_{n+1} - p_n) + \beta(p_{n-1} - p_n). \qquad (2.8)$$

First the steps are scaled down by setting

$$\varepsilon n = x, \qquad p_n(t) = \varepsilon P(x, t).$$

---

[*] J.E. Moyal, J. Roy. Statist. Soc. B **11**, 150 (1949); H.A. Kramers, Physica **7**, 284 (1940).

[**] N.G. van Kampen, in: *Thermodynamics and Kinetics of Biological Systems* (I. Lamprecht and A.I. Zotin eds., W. de Gruyter, Berlin 1983).

[†] FELLER I, p. 323.

This would have the effect that the whole process bogs down in a small interval of x, and in order to compensate for the smallness of the steps one has to scale up the transition rates α, β. To keep the drift on the original scale one must put

$$\beta - \alpha = \frac{A}{\varepsilon},$$

with a constant A independent of ε. Similarly to keep the spreading from disappearing one must put

$$\beta + \alpha = \frac{B}{\varepsilon^2}.$$

Then (2.8) becomes

$$\frac{\partial P(x, t)}{\partial t} = \left(\frac{B}{2\varepsilon^2} - \frac{A}{2\varepsilon}\right)\left\{\varepsilon \frac{\partial P}{\partial x} + \tfrac{1}{2}\varepsilon^2 \frac{\partial^2 P}{\partial x^2} + \cdots\right\}$$
$$+ \left(\frac{B}{2\varepsilon^2} + \frac{A}{2\varepsilon}\right)\left\{-\varepsilon \frac{\partial P}{\partial x} + \tfrac{1}{2}\varepsilon^2 \frac{\partial^2 P}{\partial x^2} - \cdots\right\}$$
$$= -A \frac{\partial P}{\partial x} + \tfrac{1}{2}B \frac{\partial^2 P}{\partial x^2} + \mathcal{O}(\varepsilon^2). \tag{2.9}$$

Thus one obtains the Fokker–Planck equation as an exact result in the limit $\varepsilon \to 0$. Yet this derivation is unsatisfactory, because in actual applications one does not have a parameter ε that goes to zero. The question is: for *given* α, β (or more generally for given W), how good an approximation is provided by the Fokker–Planck equation? This question is better answered by Planck's derivation, and in a more systematic way by the derivation in chapter X.

Exercise. Verify that the coefficients in (2.9) are indeed the same as in (1.6).
Exercise. Show that a different dependence of α and β on ε will not lead to (2.9).
Exercise. The master equation for a Malthusian population is

$$\dot{p}_n = \alpha(\mathbb{E} - 1)np_n + \beta(\mathbb{E}^{-1} - 1)np_n.$$

Derive the approximation (2.5) directly and also by an appropriate ε limit.

3. Brownian motion

The physical description of the Brownian particle was given in IV.1. It led to the conclusion that its coordinate X may be treated on a coarse time scale as a Markov process. Accordingly we have the picture of a particle

3. BROWNIAN MOTION

that makes random jumps back and forth over the X-axis. The jumps may have any length, but the probability for large jumps falls off rapidly. Moreover the probability is symmetrical and independent of the starting point. Hence

$$a_1 = \frac{\langle \Delta X \rangle_X}{\Delta t} = 0, \qquad a_2 = \frac{\langle (\Delta X)^2 \rangle_X}{\Delta t} = \text{const.}$$

The Fokker–Planck equation for the transition is therefore

$$\frac{\partial P(X, t)}{\partial t} = \frac{a_2}{2} \frac{\partial^2 P(X, t)}{\partial X^2}. \tag{3.1}$$

Even without solving this equation one can draw an important conclusion. It has the same form as the diffusion equation (IV.2.8) and in fact it *is* the diffusion equation for the Brownian particles in the fluid. Consequently $\tfrac{1}{2} a_2$ is identical with the phenomenological diffusion constant D. On the other hand, a_2 is expressed in microscopic terms by (2.4) or by (1.6). This establishes Einstein's relation

$$D = \frac{\langle (\Delta X)^2 \rangle_X}{2 \Delta t}. \tag{3.2}$$

It connects the macroscopic constant D with the microscopic jumps of the particle.

Consider an ensemble of Brownian particles which at $t = 0$ are all at $X = 0$. Their positions at $t \geq 0$ constitute a stochastic process $X(t)$, which is Markovian by assumption and whose transition probability is determined by (3.1). That is, just the Wiener process defined in IV.2. Their density at $t > 0$ is given by the solution of (3.1) with initial condition $P(X, 0) = \delta(X)$, which is given by (IV.2.5):

$$P(X, t) = \frac{1}{\sqrt{4\pi D t}} \exp\left[-\frac{X^2}{4Dt}\right]. \tag{3.3}$$

This is a Gaussian with maximum at the origin and whose width grows with a square root of time:

$$\sqrt{\langle X^2(t) \rangle} = \sqrt{2Dt}.$$

Next consider the same Brownian particle, subject to an additional constant force, say a gravitational field Mg in the direction of $-X$. If we write $M\gamma$ for the friction of the particle in the surrounding fluid, it will now receive an average drift velocity $-g/\gamma$. This is superimposed on the Brownian motion, so that now

$$a_1 = \frac{\langle \Delta X \rangle_x}{\Delta t} = -\frac{g}{\gamma}, \qquad a_2 = 2D. \tag{3.4}$$

The resulting Fokker–Planck equation is

$$\frac{\partial P(X,t)}{\partial t} = \frac{g}{\gamma}\frac{\partial P}{\partial X} + D\frac{\partial^2 P}{\partial X^2}. \tag{3.5}$$

It is physically obvious that this equation has no stationary solution when X is allowed to range from $-\infty$ to $+\infty$. However, if one imagines a reflecting bottom at $X = 0$, the equation has to be solved for $X > 0$ only, with the boundary condition that the flow vanishes:

$$\frac{g}{\gamma} P + D\frac{\partial P}{\partial X} = 0 \quad \text{at } X = 0. \tag{3.6}$$

With this modification there will be a stationary solution. Now one knows from equilibrium statistical mechanics that this stationary solution is nothing but the barometric density formula,

$$P^e(X) = \text{const. exp}\left[-\frac{Mg}{kT}X\right]. \tag{3.7}$$

It is easily seen that this function does, indeed, satisfy (3.5) and (3.6) provided that

$$D = \frac{kT}{M\gamma}. \tag{3.8}$$

This is the other relation of Einstein. Together with (3.2) it connects the damping coefficient γ with the mean square of the fluctuations

$$\frac{\langle (\Delta X)^2 \rangle_x}{\Delta t} = \frac{2kT}{M\gamma}. \tag{3.9}$$

Exercise. In (3.4) it is assumed that a_2 is not affected by the presence of the field. In order to justify this compare the displacement ΔX with field with the average displacement $\Delta_0 X$ without field,

$$\Delta X = \Delta_0 X - (g/\gamma)\Delta t,$$

and compute $\langle \Delta X^2 \rangle/\Delta t$.

Exercise. Solve equation (3.5) for $-\infty < X < \infty$, when $P(X, 0)$ is given, by means of the characteristic function.

Exercise. Solve the same equation by means of the substitution

$$P(X,t) = R(X,t)\exp\left[-\frac{Mg}{2kT}X - \frac{Mg^2}{4\gamma kT}t\right].$$

Exercise. Solve the same equation for $0 < X < \infty$ with boundary conditions (3.6) and arbitrary $P(X, 0)$.

Exercise. Consider the right-hand side of (3.5) as a linear operator $\mathbb{W}$ acting on the space of functions $P(X)$ defined for $0 < X < \infty$ and obeying (3.6). Verify that it has the symmetry property (V.7.5) and is negative semi-definite, the only eigenfunction with zero eigenvalue being (3.7).

Exercise. Argue that an overdamped particle subject to an external force with potential $U(X)$ is described by[*]

$$\frac{\partial P(X, t)}{\partial t} = \frac{1}{M\gamma}\left[\frac{\partial}{\partial X} U'(X)P + kT\frac{\partial^2 P}{\partial X^2}\right]. \quad (3.10)$$

"Overdamped" refers to the assumption that γ is so large that the *velocity* may be taken proportional to the force. [Compare (XI.2.4).]

Exercise. For a pendulum in a potential $U(\theta)$ and subject to a constant torque τ this equation is

$$\frac{\partial P(\theta, t)}{\partial t} = \frac{1}{Ml^2\gamma}\left[\frac{\partial}{\partial \theta}\{U'(\theta) - \tau\}P + kT\frac{\partial^2 P}{\partial \theta^2}\right] \quad (0 \leqslant \theta < 2\pi). \quad (3.11)$$

Since θ is bounded the stationary solution does not have zero flow as in (3.6), but instead one has the condition that P must be a periodic function of θ. Find the stationary solution and the corresponding flow and derive from it the average angular velocity $\langle \dot\theta \rangle^s$.

Exercise. Let $E(t) = \exp[i\omega_0 t + i\phi(t)]$ represent a wave with random phase ϕ, whose probability obeys

$$\frac{\partial P(\phi, t)}{\partial t} = D\frac{\partial^2 P}{\partial \phi^2}.$$

The output of a detector with frequency response ψ is

$$U = \left|\frac{1}{2\pi}\int_{-\infty}^{\infty}\psi(\omega)\,d\omega\int_{-\infty}^{\infty}e^{-i\omega t}E(t)\,dt\right|^2.$$

Show that

$$\langle U \rangle = \int_{-\infty}^{\infty}|\psi(\omega)|^2\frac{D/\pi}{(\omega - \omega_0)^2 + D^2}\,d\omega. \quad (3.12)$$

Thus the random phase gives rise to a Lorentz broadening of the spectral line.

Exercise. Two particles diffuse independently. Show that their mutual distance obeys the diffusion equation, with a diffusion constant equal to the sum of the diffusion constants of the separate particles.[**]

---

[*] It was applied to diffusion-controlled reactions and quenching by E.W. Montroll, J. Chem. Phys. **14**, 202 (1946).

[**] M. von Smoluchowski, Z. Phys. Chemie **92**, 129 (1917).

4. The Rayleigh particle

The Rayleigh particle is the same particle as the Brownian particle, but studied on a finer time scale. Time differences Δt are regarded that are small compared to the time in which the *velocity* relaxes, but, of course, still large compared to the duration of single collisions with the gas molecules. Thus the stochastic function to be considered is the velocity rather than the position. It is sufficient to confine the treatment to one dimension; this is sometimes emphasized by the name "Rayleigh piston".[*]

The macroscopic law for the velocity V is the linear damping law

$$\dot{V} = -\gamma V, \tag{4.1}$$

when V is not too large. Hence

$$a_1(V) = \frac{\langle \Delta V \rangle_V}{\Delta t} = -\gamma V.$$

The second jump moment a_2 must be positive even at $V=0$ and may therefore be taken constant when V is not too large,

$$a_2(V) = a_{2,0} + \mathcal{O}(V^2) \approx a_{2,0}. \tag{4.2}$$

Thus the Fokker–Planck equation for $P(V,t)$ is

$$\frac{\partial P(V,t)}{\partial t} = \gamma \frac{\partial}{\partial V} VP + \frac{a_{2,0}}{2} \frac{\partial^2 P}{\partial V^2}. \tag{4.3}$$

One knows from equilibrium statistical mechanics that the stationary solution must be

$$P^e(V) = \left(\frac{M}{2\pi kT}\right)^{1/2} \exp\left[-\frac{M}{2kT} V^2\right], \tag{4.4}$$

where T is the temperature of the gas. Substitution in (4.3) yields

$$\frac{a_{2,0}}{2} = \gamma \frac{kT}{M}. \tag{4.5}$$

The result is Rayleigh's equation for the probability density of the velocity of a heavy particle,

$$\frac{\partial P(V,t)}{\partial t} = \gamma \left\{ \frac{\partial}{\partial V} VP + \frac{kT}{M} \frac{\partial^2 P}{\partial V^2} \right\}. \tag{4.6}$$

---

[*] Reviewed by M.R. Hoare in: *Advances in Chemical Physics* **20** (Wiley–Interscience, New York 1971) and M.R. Hoare, S. Ravel and M. Rahman, Philos. Trans. Roy. Soc. A **305**, 383 (1982); W. Driessler; J. Statist. Phys. **24**, 595 (1981).

This is a linear Fokker–Planck equation. Apart from constants which can be scaled away, it is identical with the equation (IV.3.20) obeyed by the transition probability of the Ornstein–Uhlenbeck process. The stationary solution of (4.6) is the same as the P_1 given in (IV.3.10). Thus, in equilibrium $V(t)$ is the Ornstein–Uhlenbeck process.

Exercise. Deduce directly from (4.6) that for fixed $V(0) = V_0$

$$\langle V(t) \rangle_{V_0} = V_0 \, e^{-\gamma t}, \tag{4.7}$$

$$\langle V(t)^2 \rangle_{V_0} = V_0^2 \, e^{-2\gamma t} + \frac{kT}{M}(1 - e^{-2\gamma t}). \tag{4.8}$$

Therefore

$$\langle\langle V(t)V(t+\tau) \rangle\rangle^e = \frac{kT}{M} e^{-\gamma \tau}. \tag{4.9}$$

Exercise. Show that the solution of (4.6) with initial condition $P(V, 0) = \delta(V - V_0)$ is

$$P(V, t) = \left[\frac{2\pi kT}{M}(1 - e^{-2\gamma t}) \right]^{-1/2} \exp\left[-\frac{M}{2kT} \frac{(V - V_0 \, e^{-\gamma t})^2}{1 - e^{-2\gamma t}} \right]. \tag{4.10}$$

Exercise. Find this solution by guessing that it should have the Gaussian form $P(V, t) = \exp[-AV^2 - BV - C]$ and determining A, B, C as functions of t.

Exercise. Solve (4.6) systematically by transforming it into an equation for the characteristic function $G(k, t)$ of P, which is of first order and can be solved by the method of VI.6.

Exercise. The rotation of an ellipsoidal particle suspended in a fluid obeys the macroscopic equation of motion

$$I_i \dot{\omega}_i = \sum_{j,k} \varepsilon_{ijk} I_j \omega_j \omega_k - C_i \omega_i.$$

I_i are the three principal moments of inertia, ω_i the components of the angular velocity along the principal axes, ε_{ijk} is Levi-Civita's completely anti-symmetric tensor, and C_i are friction coefficients. The corresponding Fokker–Planck equation is[*]

$$\frac{\partial P(\omega, t)}{\partial t} = -\sum_i \frac{1}{I_i} \frac{\partial}{\partial \omega_i} \left\{ \sum_{j,k} \varepsilon_{ijk} I_j \omega_j \omega_k - C_i \omega_i \right\} P + \sum_i D_i \frac{\partial^2 P}{\partial \omega_i^2}. \tag{4.11}$$

Find the relation between C_i and D_i.

Having found the description of the particle on a fine time scale we ought to be able to deduce the coarser description of **3** from it. Consider an

---

[*] P.S. Hubbard, Phys. Rev. A **15**, 329 (1977); G.W. Ford, J.T. Lewis, and J. McConnell, Phys. Rev. A **19**, 907 (1979); G. Wyllie, Physics Reports **61**, 327 (1980).

ensemble of identical but independent Brownian particles, which at $t = 0$ all have the position $X = 0$ with velocities distributed according to the equilibrium distribution. Their velocities constitute an Ornstein–Uhlenbeck process and we have to study the random process $X(t)$ defined by

$$X(t) = \int_0^t V(t')\, dt'.$$

According to this formula $X(t)$ is a sum of Gaussian variables and is therefore itself Gaussian. It has zero mean because

$$\langle X(t) \rangle = \int_0^t \langle V(t') \rangle \, dt' = 0.$$

The averages are, of course, taken over the ensemble specified above. We have to compute

$$\langle X(t_1) X(t_2) \rangle = \int_0^{t_1} dt' \int_0^{t_2} dt'' \, \langle V(t') V(t'') \rangle$$

$$= \langle V^2 \rangle \int_0^{t_1} dt' \int_0^{t_2} dt'' \, e^{-\gamma|t' - t''|}.$$

The integral is elementary. As the left-hand side is manifestly symmetric in t_1, t_2 it suffices to compute it for $0 \leq t_1 \leq t_2$; the result is

$$\langle X(t_1) X(t_2) \rangle = \frac{kT}{M} \left[\frac{2}{\gamma} t_1 - \frac{1}{\gamma^2} + \frac{1}{\gamma^2} \{ e^{-\gamma t_1} + e^{-\gamma t_2} - e^{-\gamma(t_2 - t_1)} \} \right]. \quad (4.12)$$

The process $X(t)$ is now fully specified since it is Gaussian and the first two moments are known. But it is not the same as the Wiener process determined by (3.1), because the autocorrelation function is more complicated than (IV.2.7a). In fact, $X(t)$ is not even Markovian, owing to the fact that it is still described on the fine time scale belonging to the Rayleigh particle. On the coarse time scale only time differences much larger than the damping time $1/\gamma$ of the velocity are admitted,

$$t_1 \gg 1/\gamma, \qquad t_2 - t_1 \gg 1/\gamma.$$

In that approximation (4.12) reduces to (IV.2.7a) and $X(t)$ is indistinguishable from the solution of (3.1). Note that the value of the constant a_2 of (3.1) has now been found without the detour involving a gravitational field.

Exercise. Take an ensemble of Rayleigh particles whose position at $t = 0$ is the origin and whose initial velocities have a Maxwell distribution. Show

$$\lim_{t \to \infty} \frac{\langle X(t)^2 \rangle}{t} = 2D. \quad (4.13)$$

(In kinetic theory this is often taken as the *definition* of the diffusion constant, in spite of the obvious fact that, strictly speaking, in a finite vessel the limit is zero.)

Exercise. Construct the equation for $P(V, t)$ in the presence of a constant force and deduce from it (3.5).

Exercise. The transition probability per unit time for the velocity V of a Rayleigh piston in a gas is

$$W(V'|V) = vA\left(\frac{M+m}{2m}\right)^2 |V' - V| F\left(\frac{M+m}{2m}V' - \frac{M-m}{2m}V\right). \tag{4.14}$$

M is the mass of the piston, A its cross section, m the mass of the gas molecules, v their number density and F their velocity distribution.[*]

Exercise. Compute from (4.14) the jump moments, taking for F the Maxwell distribution. Show that (4.1) and (4.2) hold when V is small compared to the average speed of the gas molecules, and can therefore be used to describe equilibrium fluctuations if $M \gg m$.

Exercise. Formulate the process described by the V and X of a Rayleigh particle as a composite Markov process in the sense of VII.7.

5. Application to one-step processes

Suppose one is faced with a one-step problem in which the coefficients r_n and g_n are nonlinear but can be represented by smooth functions $r(n), g(n)$. "Smooth" means not only that $r(n)$ and $g(n)$ should be continuous and a sufficient number of times differentiable, but also that they vary little between n and $n+1$. Suppose furthermore that one is interested in solutions $p_n(t)$ that can similarly be represented by a smooth function $P(n, t)$. It is then reasonable to approximate the problem by means of a description in which n is treated as a continuous variable. Moreover, since the individual steps of n are small compared to the other lengths that occur, one expects that the master equation can be approximated by a Fokker–Planck equation. The general scheme of section 2 provides the two coefficients, but we shall here use an alternative derivation, particularly suited to one-step processes.

The M-equation to be solved is

$$\dot{P}(n, t) = (\mathbb{E} - 1)r(n)P(n, t) + (\mathbb{E}^{-1} - 1)g(n)P(n, t). \tag{5.1}$$

Since the step operator $\mathbb{E}$ here only acts on smooth functions it may be replaced with a Taylor expansion

$$\mathbb{E} = 1 + \frac{\partial}{\partial n} + \frac{1}{2}\frac{\partial^2}{\partial n^2} + \frac{1}{3!}\frac{\partial^3}{\partial n^3} + \cdots. \tag{5.2}$$

---

[*] Various aspects of the corresponding master equation have been studied, see C.T.J. Alkemade, N.G. van Kampen, and D.K.C. MacDonald, Proc. Roy. Soc. A **271**, 449 (1963); M.R. Hoare and C.H. Kaplinsky, Physica A **81**, 349 (1975).

Hence, omitting all derivatives beyond the second,

$$\dot{P}(n, t) = \frac{\partial}{\partial n}\{r(n) - g(n)\}P + \frac{1}{2}\frac{\partial^2}{\partial n^2}\{r(n) + g(n)\}P. \tag{5.3}$$

This is the Fokker–Planck approximation for the one-step process.

As an example take Burgess' treatment of the fluctuations in semi-conductors.[*] n is the number of charge carriers in the conduction band, $g(n)$ and $r(n)$ are the generation and recombination probabilities per unit time. There is no need to specify them explicitly, it suffices to know that $r(n)$ will be monotonically increasing and $g(n)$ decreasing with n, as is physically reasonable. In general it is not possible to obtain the exact solution of (5.3); nor would it be wise because its physical relevance cannot be trusted owing to the approximations already made to arrive at (5.3). The following limited conclusions, however, are still correct, as will be shown in X.3 and X.4.

The macroscopic equation associated with (5.3) is

$$\dot{n} = -r(n) + g(n). \tag{5.4}$$

The stationary value of n is therefore determined by

$$r(n^s) = g(n^s). \tag{5.5}$$

There is one solution, see fig. 21. We want to study the fluctuations in the stationary state. For this purpose set $n = n^s + x$, substitute this in (5.3), expand the coefficients in x and retain only the lowest non-zero terms:

$$\frac{\partial P(x, t)}{\partial t} = \{r'(n^s) - g'(n^s)\}\frac{\partial}{\partial x}xP + \frac{r(n^s) + g(n^s)}{2}\frac{\partial^2 P}{\partial x^2}. \tag{5.6}$$

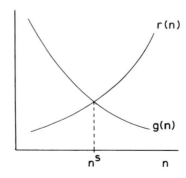

Fig. 21. The stationary state as a balance between generation and recombination.

---

[*] A. van der Ziel, *Noise* (Prentice, Englewood Cliffs, NJ 1954); K.M. van Vliet and J.R. Fasset, in: *Fluctuation Phenomena in Solids* (R.E. Burgess ed., Academic Press, New York 1965).

5. APPLICATION TO ONE-STEP PROCESSES

Thus our additional approximation for the neighborhood of n^s leads to a linear Fokker–Planck equation of the same form as (4.6). The fluctuations in the stationary state are therefore again an Ornstein–Uhlenbeck process. It will be shown in X.4 that (5.6) is a consistent approximation.*

Exercise. Derive from (5.6)

$$\langle x^2 \rangle = \langle\langle n^2 \rangle\rangle^s = \frac{1}{2} \frac{r(n^s) + g(n^s)}{r'(n^s) - g'(n^s)}, \tag{5.7}$$

$$\langle\langle n(t)n(t+\tau)\rangle\rangle^s = \langle\langle n^2 \rangle\rangle^s \, e^{-\tau/\tau_0}, \tag{5.8}$$

$$1/\tau_0 = r'(n^s) - g'(n^s), \tag{5.9}$$

$$S_n(\omega) = \frac{2}{\pi} \langle\langle n^2 \rangle\rangle^s \, \frac{\tau_0}{1+\omega^2 \tau_0^2}. \tag{5.10}$$

Exercise. Show that the same results are obtained by simply linearizing $r(n)$ and $g(n)$ around n^s.

Exercise. Show that the results (5.5) and (5.7) are in agreement with the explicit expression (VI.3.8) for $P^s(n)$.

Exercise. Write the Kramers–Moyal expansion for one-step processes using (5.2).

Exercise. Construct the Fokker–Planck approximation for the M-equation (VI.9.12) and use it to find $\langle n \rangle^s$ and $\langle\langle n^2 \rangle\rangle^s$.

So far the Fokker–Planck approximation has only been formulated for cases where there is no boundary, or where the boundary is too far away to bother about it. The question now is how a boundary with certain physical properties is to be translated into a boundary condition for the differential equation. In the case of a reflecting boundary the answer is clear: the probability flow (1.3) has to vanish, as in (3.6),

$$a_1(y)P(y,t) - \frac{1}{2}\frac{\partial}{\partial y} a_2(y)P(y,t) = 0 \quad \text{at the boundary.} \tag{5.11}$$

In the case of an absorbing boundary, however, the answer is not evident. It will be obtained by utilizing the connection with one-step processes.

Take a one-step process with analytic $r(n), g(n)$ but with an absorbing boundary at $n = 0$. As pointed out in VI.7 this does not at all uniquely specify how the r_n, g_n near the boundary differ from the analytic expressions $r(n), g(n)$. Even if the boundary is pure, so that only r_1 can be modified, the requirement that the boundary should be absorbing merely implies $r_1 > 0$. The boundary absorbs as soon as there is some probability for a transition

---

* A numerical comparison is given by C.F. Clement and M.H. Wood, Proc. Roy. Soc. London A **371**, 553 (1980).

from the ultimate state into limbo. One special choice is $r_1 = r(1)$; this is the totally absorbing boundary, which is relevant to first-passage problems, see ch. XII. We shall here adopt this type and therefore start from the master equation

$$\dot{p}_n = r(n+1)p_{n+1} + g(n-1)p_{n-1} - \{r(n) + g(n)\}p_n \quad (n > 1), \quad (5.12\text{a})$$

$$\dot{p}_1 = r(2)p_2 - \{r(1) + g(1)\}p_1. \quad (5.12\text{b})$$

It is immediately clear that (5.12b) can be subsumed in (5.12a) if one defines $p_0 = 0$. (This is merely a formal device; p_0 is not the same as the probability p_* attached to the limbo state.) With this definition we may declare (5.12a) valid for all $n \geq 1$ with the additional stipulation $p_0 = 0$. If one now approximates (5.12a) by a differential equation the boundary condition for $P(x, t)$ becomes $P(0, t) = 0$. Although this conclusion was reached for a special type of absorbing boundary it can be shown to be general: in the continuous description the different absorbing boundaries encountered in (VI.7) all reduce to the same boundary condition $P(0, t) = 0$.[*]

6. The multivariate Fokker–Planck equation

The generalization of the Fokker–Planck equation (1.1) to the case that there are r variables y_i is

$$\frac{\partial P(y, t)}{\partial t} = - \sum_{i=1}^{r} \frac{\partial}{\partial y_i} A_i(y)P + \tfrac{1}{2} \sum_{i,j=1}^{r} \frac{\partial^2}{\partial y_i \, \partial y_j} B_{ij}(y)P. \quad (6.1)$$

The coefficients A_i and B_{ij} may be any real differentiable functions with the sole restriction that the matrix B_{ij} is taken to be symmetric and must be positive definite. More precisely, at each point y of the r-dimensional space, $B_{ij}(y)$ must be nonnegative semi-definite:

$$\sum_{i,j=1}^{r} B_{ij}(y) x_i x_j \geq 0 \quad \text{for any vector } \{x_i\}. \quad (6.2)$$

The physical interpretation of the coefficients is similar to (1.6): for $\Delta t \to 0$,

$$\frac{\langle \Delta y_i \rangle_y}{\Delta t} = A_i(y), \quad \frac{\langle \Delta y_i \, \Delta y_j \rangle_y}{\Delta t} = B_{ij}(y). \quad (6.3)$$

The nonlinear multivariate Fokker–Planck equation (6.1) will be derived and studied in chapter X.

---

[*] N.G. van Kampen and I. Oppenheim, J. Mathem. Phys. 13, 842 (1972).

6. THE MULTIVARIATE FOKKER–PLANCK EQUATION

The linear multivariate Fokker–Planck equation is

$$\frac{\partial P(y,t)}{\partial t} = -\sum_{i,j} A_{ij} \frac{\partial}{\partial y_i} y_j P + \tfrac{1}{2} \sum_{ij} B_{ij} \frac{\partial^2 P}{\partial y_i \, \partial y_j}, \quad (6.4)$$

where A_{ij} and B_{ij} are now constant matrices, B_{ij} symmetric and obeying (6.2), but A_{ij} unrestricted. We shall solve this equation with the initial condition

$$P(y,0) = \prod_{i=1}^{r} \delta(y_i - y_{i0}) - \delta^r(y - y_0). \quad (6.5)$$

It will be found that the solution is a Gaussian, and consequently fully determined by the first and second moments. Hence, as a preliminary step, we first compute these moments.

Multiply (6.4) with y_k and integrate over all y:

$$\partial_t \langle y_k \rangle = \sum_j A_{kj} \langle y_j \rangle. \quad (6.6)$$

The solution with initial condition (6.5) is, written in matrix notation,

$$\langle y \rangle_t = e^{tA} y_0. \quad (6.7)$$

Next multiply (6.4) with $y_k y_l$ and integrate,

$$\partial_t \langle y_k y_l \rangle = \sum_i A_{ki} \langle y_i y_l \rangle + \sum_j A_{lj} \langle y_k y_j \rangle + B_{kl}. \quad (6.8)$$

It is convenient to consider instead of these moments the covariances

$$\langle\langle y_k y_l \rangle\rangle = \langle y_k y_l \rangle - \langle y_k \rangle \langle y_l \rangle \equiv \Xi_{kl}.$$

With the aid of (6.6) one finds that they obey the same equation (6.8), but they vanish at $t=0$. In matrix notation

$$\partial_t \Xi = A\Xi + \Xi \tilde{A} + B. \quad (6.9)$$

$\tilde{A}$ is the transpose of A.

To solve the equation, transform Ξ to a kind of interaction representation Ξ^* by setting

$$\Xi(t) = e^{tA} \, \Xi^*(t) \, e^{t\tilde{A}}.$$

Substitution in (6.9) yields

$$\partial_t \Xi^*(t) = e^{-tA} B \, e^{-t\tilde{A}}.$$

Integrating with the initial value $\Xi^*(0) = 0$,

$$\Xi^*(t) = \int_0^t e^{-t'A} B \, e^{-t'\tilde{A}} \, dt',$$

$$\Xi(t) = \int_0^t e^{(t-t')A} B \, e^{(t-t')\tilde{A}} \, dt'. \quad (6.10)$$

Having found the moments and covariance we are now able to construct the corresponding Gaussian distribution

$$P(y, t) = (2\pi)^{-\frac{1}{2}r}(\text{Det } \Xi)^{-\frac{1}{2}} \exp[-\tfrac{1}{2}(\tilde{y} - \langle \tilde{y} \rangle)\Xi^{-1}(y - \langle y \rangle)]. \qquad (6.11)$$

In order to prove that this Gaussian is, indeed, a solution one must substitute it into (6.4) and merely verify that it obeys the equation. This, however, is not a trivial calculation and is therefore left for an Exercise. The following systematic solution of (6.4), without Gaussian guess, is actually simpler.

To solve (6.4) systematically[*] we transform it into an equation for the characteristic function $G(k, t)$:

$$\frac{\partial G(k, t)}{\partial t} = \sum_{i,j} A_{ij} k_i \frac{\partial G}{\partial k_j} - \tfrac{1}{2} \sum_{i,j} B_{ij} k_i k_j G. \qquad (6.12)$$

This partial difference equation is of the first order, due to the fact that the coefficients in (6.4) were linear in the y. The characteristics of (6.12) are determined by

$$\dot{k}_j = - \sum_i k_i A_{ij}.$$

The solution is, in vector notation,

$$k = e^{-t\tilde{A}} c,$$

where c is a vector of r constants, which parametrize the characteristics. Along these curves G varies according to

$$\frac{d \log G}{dt} = -\tfrac{1}{2} \tilde{k} B k = -\tfrac{1}{2} \tilde{c} e^{-t A} B e^{-t\tilde{A}} c.$$

Hence we have, with arbitrary $\Omega(c)$,

$$G = \Omega(c) \exp\left[-\tfrac{1}{2} \int_0^t \tilde{c} e^{-t'A} B e^{-t'\tilde{A}} c \, dt' \right].$$

Re-expressing the c in the actual variables k, one has

$$G(k, t) = \Omega(e^{t\tilde{A}} k) \exp\left[-\tfrac{1}{2} \int_0^t \tilde{k} e^{(t-t')A} B e^{(t-t')\tilde{A}} k \, dt' \right].$$

The prefactor is determined by the initial value

$$G(k, 0) = \Omega(k) = \exp[i k \cdot y_0].$$

That gives us finally

$$G(k, t) = \exp\left[i\tilde{k} e^{tA} y_0 - \tfrac{1}{2} \tilde{k} \cdot \int_0^t e^{(t-t')A} B e^{(t-t')\tilde{A}} dt' \cdot k \right]. \qquad (6.13)$$

---

[*] M. Lax, Rev. Mod. Phys. **32**, 25 (1960).

6. THE MULTIVARIATE FOKKER–PLANCK EQUATION

This is the characteristic function of the solution of (6.4). It is quadratic in the auxiliary variables k and corresponds therefore to a Gaussian distribution of the y. The averages are seen to be the same as previously found in (6.7) and the correlations are also the same as given by (6.10). Hence we have now found (6.11) in a constructive fashion.

Exercise. The matrix Ξ is by definition symmetric and positive definite (or at least semi-definite). Show that Ξ^* has the same properties, and that they are preserved by equation (6.9).

Exercise. A Brownian particle in a harmonic potential is described by a joint distribution $P(x, v, t)$ for its position and velocity, which obeys (see next section)

$$\frac{\partial P(x, v, t)}{\partial t} = -v\frac{\partial P}{\partial x} + x\frac{\partial P}{\partial v} + \gamma\left(\frac{\partial}{\partial v}vP + \frac{\partial^2 P}{\partial v^2}\right), \qquad (6.14)$$

with constant damping coefficient γ. Find $P(x, v, t)$ with the initial condition $P(x, v, 0) = \delta(x - x_0)\,\delta(v - v_0)$.

Exercise. Suppose that A is such that all solutions of (6.6) tend to zero as t goes to infinity. Show that (6.11) tends to a limiting distribution, which depends on A and B but not on the initial y_0.

Exercise. In particular this situation obtains if the equation refers to a closed thermodynamic system. The corresponding matrix $\Xi(\infty) = \Xi^e$ is known from equilibrium statistical mechanics. This makes it possible to establish a correlation between the diffusion matrix B and the convection matrix A:

$$B = -A\Xi^e - \Xi^e \tilde{A},$$

which is the fluctuation–dissipation theorem, see (3.8).

Exercise. Verify by direct substitution that (6.11) satisfies (6.4) [as was done in the first edition of this book].

In chapter X we shall encounter a linear Fokker–Planck equation of the form (6.4) whose coefficients A_{ij}, B_{ij} are given functions of time. The solution is again Gaussian, and can be obtained in much the same way as before.

First the equation (6.6) for the averages $\langle y_k \rangle_t$ has to be solved. As A_{kj} now depends on t the solution no longer has the form (6.7) but it still has the general form

$$\langle y \rangle_t = Y(t) y_0, \qquad (6.15)$$

where $Y(t)$ is the "evolution matrix" or "propagator" determined by the matrix equation

$$\dot{Y}(t) = A(t) Y(t), \qquad Y(0) = 1.$$

Similarly, in equation (6.9) for the covariances the matrices A and B are now time-dependent. We define the interaction representation by setting

$$\Xi(t) = Y(t)\Xi^*(t)\tilde{Y}(t)$$

and find instead of (6.10)

$$\Xi(t) = \int_0^t Y(t)Y^{-1}(t')B(t')\tilde{Y}^{-1}(t')\tilde{Y}(t)\,\mathrm{d}t'.$$

The Gaussian (6.11) with this Ξ and with the averages (6.15) constitutes the solution of the linear Fokker–Planck equation (6.4) with time-dependent coefficients.[*]

Exercise. Find the explicit solution of the one-variable time-dependent Fokker–Planck equation

$$\frac{\partial P(y,t)}{\partial t} = -A(t)\frac{\partial}{\partial y}yP + \tfrac{1}{2}B(t)\frac{\partial^2 P}{\partial y^2}.$$

Exercise. Solve the equation

$$\frac{\partial P(y,t)}{\partial t} = -\sum_i C_i(t)\frac{\partial P}{\partial y_i} + \tfrac{1}{2}\sum_{i,j}B_{ij}(t)\frac{\partial^2 P}{\partial y_i\,\partial y_j}.$$

Exercise. The most general Fokker–Planck equation is

$$\frac{\partial P(y,t)}{\partial t} = -\sum_i C_i\frac{\partial P}{\partial x_i} - \sum_{i,j}A_{ij}\frac{\partial}{\partial x_i}x_j P + \tfrac{1}{2}\sum_{i,j}B_{ij}\frac{\partial^2 P}{\partial x_i\,\partial x_j}, \qquad (6.16)$$

with time-dependent coefficients. Solve this equation by reducing it to (6.4) with the aid of the substitution $x_i = y_i + u_i(t)$, where $u_i(t)$ is a solution of $\dot{u} = A(t)u + C(t)$.

Exercise. A chain of r rotators, interacting harmonically through the potential

$$U(\vartheta_1, \vartheta_2, \ldots, \vartheta_r) = \tfrac{1}{2}\kappa\sum_{j=1}^r (\vartheta_{j+1} - \vartheta_j)^2, \qquad \vartheta_{r+1}\equiv\vartheta_1.$$

They are strongly damped, subject to Brownian motion and to an external torque τ,

$$\frac{\partial P(\vartheta,t)}{\partial t} = \frac{1}{Ml^2\gamma}\sum_{j=1}^r\left[\frac{\partial}{\partial\vartheta_j}\left(\frac{\partial U}{\partial\vartheta_j} - \tau\right)P + kT\frac{\partial^2 P}{\partial\vartheta_j^2}\right], \qquad (6.17)$$

compare (3.11). Find the steady solution and derive from it the average velocity with which the whole string rotates. Also find the time-dependent solution with $P(\vartheta, 0) = \delta(\vartheta)$. [For a nonlinear modification of this system, see S.E. Trullinger et al., Phys. Rev. Letters **40**, 202 (1978).]

Exercise. Let $P(x,t)$ be some time-dependent Gaussian distribution with r variables,

$$P(x,t) = (2\pi)^{-\frac{1}{2}r}(\mathrm{Det}\,M)^{\frac{1}{2}}\exp\left[-\tfrac{1}{2}\sum M_{ij}(x_i - \mu_i)(x_j - \mu_j)\right],$$

with given $\mu_i(t)$ and symmetric, positive definite $M_{ij}(t)$. Show that it obeys an equation

---

[*] S. Chandrasekhar, Rev. Mod. Phys. **15**, 1 (1943); reprinted in WAX.

of the form (**6.16**) and determine $B_{ij}(t)$, $A_{ij}(t)$, $C_i(t)$. The last two are not unique; in fact one may take $A_{ij} = 0$.

Exercise. From the previous Exercise it follows that even for non-Markovian Gaussian processes the conditional probability $P(x, t | x_0, t_0)$ obeys an equation of type (**6.16**). However, this is *not* a master equation, as is betrayed by the fact that the coefficients depend on t_0. Compare V.1. Indicate this dependence in the "Fokker–Planck equation" used by S.A. Adelman, J. Chem. Phys. **64**, 124 (1976).

7. Kramers' equation

Consider a Brownian particle subject to a force $F(X)$ depending on the position. The obvious generalization of the Fokker–Planck equation (3.5) is*[)]

$$\frac{\partial P(X, t)}{\partial t} = -\frac{\partial}{\partial X} \frac{F(X)}{M\gamma} P + D \frac{\partial^2 P}{\partial X^2}. \tag{7.1}$$

We shall call this a *quasilinear* Fokker–Planck equation, to indicate that it has the form (**1.1**) with constant B but nonlinear A.**[)] It is clear that this equation can only be correct if $F(X)$ varies so slowly that it is practically constant over a distance in which the velocity is damped. On the other hand, the Rayleigh equation (**4.6**) involves only the velocity and cannot accommodate a spatial inhomogeneity. It is therefore necessary, if F does not vary sufficiently slowly for (**7.1**) to hold, to describe the particle by the joint probability distribution $P(X, V, t)$. We construct the bivariate Fokker–Planck equation for it.

To find the coefficients of the first order derivatives note

$$\langle \Delta X \rangle_{X,V} = V \Delta t, \qquad \langle \Delta V \rangle_{X,V} = \left\{ \frac{F(X)}{M} - \gamma V \right\} \Delta t. \tag{7.2}$$

There are three second order derivatives and their coefficients are

$$\frac{\langle (\Delta X)^2 \rangle_{X,V}}{\Delta t} = V^2 \Delta t \to 0, \tag{7.3a}$$

$$\frac{\langle \Delta X \, \Delta V \rangle_{X,V}}{\Delta t} = V \left\{ \frac{F(X)}{M} - \gamma V \right\} \Delta t \to 0, \tag{7.3b}$$

$$\frac{\langle (\Delta V)^2 \rangle_{X,V}}{\Delta t} = \gamma \frac{kT}{M}. \tag{7.3c}$$

---

*[)] Examples have been encountered in (3.10) and (6.17).
**[)] In analogy with "quasilinear" as used for partial differential equations.

The last one is taken equal to the coefficient in (4.6) because the external force does not affect the collisions with the gas molecules. Thus we have found the *Kramers equation*[*)]

$$\frac{\partial P(X, V, t)}{\partial t} + V\frac{\partial P}{\partial X} + \frac{F(X)}{M}\frac{\partial P}{\partial V} = \gamma \left\{\frac{\partial}{\partial V}VP + \frac{kT}{M}\frac{\partial^2 P}{\partial V^2}\right\}. \qquad (7.4)$$

It is a bivariate quasilinear Fokker–Planck equation of the form (6.1) with semi-definite B_{ij}.

We shall now solve the Kramers equation (7.4) approximately for large γ by means of a systematic expansion in powers of γ^{-1}. Straightforward perturbation theory is not possible because the time derivative occurs among the small terms. This makes it a problem of singular perturbation theory, but the way to handle it can be learned from the solution method invented by Hilbert and by Chapman and Enskog for the Boltzmann equation.[**)] To simplify the writing I eliminate the coefficient kT/M by rescaling the variables,

$$V = v\sqrt{kT/M}, \qquad X = x\sqrt{kT/M}, \qquad F(X) = f(x)\sqrt{kT/M}.$$

The equation may then be written

$$\frac{\partial}{\partial v}\left(vP + \frac{\partial P}{\partial v}\right) = \frac{1}{\gamma}\left(\frac{\partial}{\partial t} + v\frac{\partial}{\partial x} + f(x)\frac{\partial}{\partial v}\right)P. \qquad (7.5)$$

It will be solved by setting

$$P(x, v, t) = P^{(0)} + \gamma^{-1} P^{(1)} + \gamma^{-2} P^{(2)} + \cdots$$

and satisfying the equation to each order in γ^{-1}.

Terms of order γ^0 are

$$\frac{\partial}{\partial v}\left(vP^{(0)} + \frac{\partial P^{(0)}}{\partial v}\right) = 0. \qquad (7.6)$$

This is solved by taking

$$P^{(0)}(x, v, t) = e^{-\frac{1}{2}v^2}\phi(x, t) \qquad (7.7)$$

with so far arbitrary $\phi(x, t)$.

[*)] Also called "Klein–Kramers equation"; see O. Klein, Arkiv Mat. Astr. Fys. **16**, No. 5 (1922). It was used by H.A. Kramers for describing chemical reactions, see XIII.6.

[**)] S. Chapman and T.G. Cowling, *The Mathematical Theory of Non-Uniform Gases* (University Press, Cambridge 1939). Kramers used a slightly different method, which was criticized and improved by H.C. Brinkman, Physica **22**, 29 (1956). For the present application, see G. Wilemski, J. Statist. Phys. **14**, 153 (1976); U.M. Titulaer, Physica A **91**, 321 (1978); **100**, 234 and 251 (1980); N.G. van Kampen, Physics Reports **124**, 69 (1985).

Terms of order γ^{-1} *yield*

$$\left\{\frac{\partial}{\partial v}v + \frac{\partial^2}{\partial v^2}\right\}P^{(1)} = \left(\frac{\partial \phi}{\partial t} + v\frac{\partial \phi}{\partial x} - vf\phi\right)e^{-\frac{1}{2}v^2}. \qquad (7.8)$$

This has to be solved for $P^{(1)}$. However, the operator $\{\ \}$ acting on $P^{(1)}$ cannot be inverted, because it has an eigenvalue zero. The corresponding left eigenvector is a constant, and the right-hand side of (7.8) must be orthogonal to it. In fact, by integrating (7.8) over v one obtains

$$0 = \frac{\partial \phi}{\partial t}.$$

Thus the condition that (7.8) can be solved for $P^{(1)}$ states that the so far arbitrary ϕ (and therefore $P^{(0)}$) must be independent of time.

Having satisfied the solubility condition we can now proceed to solve $P^{(1)}$ from

$$\frac{\partial}{\partial v}\left(v + \frac{\partial}{\partial v}\right)P^{(1)} = \left(\frac{d\phi}{dx} - f\phi\right)v\,e^{-\frac{1}{2}v^2}. \qquad (7.9)$$

One easily finds that this is obeyed by

$$P^{(1)}(x, v, t) = v\left(f\phi - \frac{d\phi}{dx}\right)e^{-\frac{1}{2}v^2} + \psi(x, t)\,e^{-\frac{1}{2}v^2} \qquad (7.10)$$

with arbitrary $\psi(x, t)$.

The terms of order γ^{-2} *become with the use of* (7.10)

$$\left\{\frac{\partial}{\partial v}v + \frac{\partial^2}{\partial v^2}\right\}P^{(2)} = v^2\frac{d}{dx}\left(f\phi - \frac{d\phi}{dx}\right)e^{-\frac{1}{2}v^2} + (1-v^2)f\left(f\phi - \frac{d\phi}{dx}\right)e^{-\frac{1}{2}v^2}$$

$$+ \left(\frac{\partial \psi}{\partial t} + v\frac{\partial \psi}{\partial x} - vf\psi\right)e^{-\frac{1}{2}v^2}.$$

Again integration over v leads to the solubility condition

$$\frac{d}{dx}\left(f\phi - \frac{d\phi}{dx}\right) + \frac{\partial \psi}{\partial t} = 0. \qquad (7.11)$$

At this point we stop and collect the result,

$$P(x, v, t) = e^{-\frac{1}{2}v^2}\left[\phi(x) + \gamma^{-1}v\left(f\phi - \frac{d\phi}{dx}\right) + \gamma^{-1}\psi(x, t) + \mathcal{O}\gamma^{-2}\right], \qquad (7.12)$$

where ψ obeys (7.11). We are interested in the distribution over x and therefore integrate over v to obtain the marginal distribution,

$$P(x, t) = \sqrt{2\pi}\,[\phi(x) + \gamma^{-1}\psi(x, t) + \mathcal{O}\gamma^{-2}]. \qquad (7.13)$$

Equation (7.11) now states

$$\frac{\partial P(x,t)}{\partial t} + \gamma^{-1}\left(\frac{d}{dx}fP - \frac{d^2P}{dx^2}\right) = \mathcal{O}\gamma^{-2}. \qquad (7.14)$$

On restoring the original variables this becomes identical to (7.1).

Exercise. Solve the Kramers equation for constant F.

Exercise. (7.7) is not the only solution of (7.6), nor is (7.10) the only one of (7.8). Find the other solutions and show that they are inadmissible.

Exercise. It is possible to solve (7.4) explicitly for the special case of an harmonic potential: $F(x) = -a^2 x$. The result is contained in (6.12), but the explicit evaluation requires a number of elementary integrations. Having found $P(x, v, t)$ one can determine the marginal distribution $P(x, t)$ and verify that it obeys (7.1) when γ is large. (More directly one can first determine the marginal distribution from the general expression (6.12), which reduces the number of integrations needed.)

Exercise. Show that (7.14) remains true in the next order.[*]

Exercise. Reduce (7.1) to an eigenvalue problem by setting $P(X, t) = \Phi_n(X) e^{-\lambda_n t}$. On the other hand, consider a Schrödinger equation

$$\phi''(X) + \{E - V(X)\}\phi(X) = 0.$$

If ϕ_0 is the ground state, take in (7.1)

$$\frac{F(X)}{M\gamma} = 2D\frac{\phi'_0(X)}{\phi_0(X)}.$$

Then both problems are equivalent and one has[**]

$$\Phi_n(X) = \phi_0(X)\phi_n(X), \qquad \lambda_n = (E_n - E_0)D. \qquad (7.15)$$

Remark. Equation (7.14) is strictly correct in the sense that it is the first term of a mathematically well-defined asymptotic expansion in $1/\gamma$. To estimate whether it is also a good approximation we require that $\gamma^{-1}P^{(1)} \ll P^{(0)}$. The differential operator on the left-hand side of (7.9) has eigenvalues of order 1, the zero eigenvalue having been extracted by the solubility condition. Thus $P^{(1)}$ is comparable with the right-hand side, i.e., of order $dP^{(0)}/dx - fP^{(0)}$. Our requirement amounts therefore to

$$\frac{1}{\gamma}\left|\frac{d\log P^{(0)}}{dx} - f(x)\right| \ll 1.$$

If $P^{(0)}$ is sharply peaked the first term is large and the approximation is bad. If $P^{(0)}$ is the equilibrium distribution both terms are equal. For intermediate situations a reasonable estimate is $\gamma \gg f$, or

$$\frac{F}{\gamma\sqrt{MkT}} \ll 1. \qquad (7.16)$$

In the same way one sees that higher orders involve higher powers of this quantity.

---

[*] For higher orders, see J.L. Skinner and P.G. Wolynes, Physica A **96**, 561 (1979); RISKEN, ch. 10.

[**] H. Risken and H.D. Vollmer, Z. Phys. **204**, 240 (1967); L.F. Favella, Ann. Inst. Henri Poincaré **A7**, 77 (1967).

Chapter IX

THE LANGEVIN APPROACH

The Langevin approach is widely used for the purpose of finding the effect of fluctuations in macroscopically known systems. The fluctuations are introduced by adding random terms to the equations of motion, called "noise sources". This approach is popular because it gives a more concrete picture than the Fokker–Planck equation, but it is mathematically equivalent to it. In nonlinear cases it is subject to the same difficulties, and some additional ones.

1. Langevin treatment of Brownian motion

After the work of Einstein and Smoluchowski an alternative treatment of Brownian motion was initiated by Langevin.[*] Consider the velocity of the Brownian particle, as in VIII.4. When the mass is taken to be unity it obeys the equation of motion

$$\dot{V} = -\gamma V + L(t). \qquad (1.1)$$

The right-hand side is the force exerted by the molecules of the surrounding fluid. This force is unknown, but the following three physically plausible properties concerning it are postulated.

(i) The force consists of a damping term linear in V with a constant coefficient γ, plus a remaining term $L(t)$. This $L(t)$ is irregular and unpredictable, but its averaged properties over an ensemble of similar systems are simple. The ensemble may consist of many particles in the same field, provided their mutual distances are so large that they do not influence each other. Or it may consist of a series of successive observations of the same particle, provided the time intervals between them are so large that they do not influence each other, as discussed in III.2. Thus $L(t)$ can be treated as a stochastic process.

(ii) The stochastic properties of $L(t)$ are given regardless of V, so that $L(t)$ acts as an external force. Its average vanishes,

$$\langle L(t) \rangle = 0. \qquad (1.2)$$

---

[*] P. Langevin, Comptes Rendus Acad. Sci. (Paris) **146**, 530 (1908). The present treatment is due to G.E. Uhlenbeck and L.S. Ornstein, Phys. Rev. **36**, 823 (1930), reprinted in WAX.

(iii) The force $L(t)$ is caused by the collisions of the individual molecules of the surrounding fluid and varies rapidly. This is expressed by postulating for its autocorrelation function

$$\langle L(t)L(t')\rangle = \Gamma\, \delta(t - t'), \tag{1.3}$$

where Γ is a constant. The idea is that each collision is practically instantaneous and that successive collisions are uncorrelated. Actually the right-hand side should be a function of $|t - t'|$ with a sharp peak of width equal to the duration of a single collision. As long as this is shorter than all other relevant times one may use a delta function for convenience.

A term having the properties (i), (ii), (iii) is called a Langevin force, and (1.1) is the Langevin equation. We emphatically include in that name these three properties: equation (1.1) without any specification of $L(t)$ is, of course, void. Note that (1.2) and (1.3) do not completely specify the stochastic process but merely its first two moments. A complete specification will be given in section 3, but it is not needed for the following calculation.

The Langevin equation is the prototype of a stochastic differential equation, i.e., a differential equation whose coefficients are random functions of time with given stochastic properties, see chapter XVI. It defines $V(t)$ as a stochastic process when an initial condition is supplied, say $V(0) = V_0$. Then (1.1) can be solved explicitly:

$$V(t) = V_0\, e^{-\gamma t} + e^{-\gamma t} \int_0^t e^{\gamma t'}\, L(t')\, dt. \tag{1.4}$$

Average this equation over a subensemble of Brownian particles all having the same initial V_0. It is allowed to use (1.2) for this subensemble, because $L(t')$ for $t' > t$ is independent of V_0. Hence

$$\langle V(t)\rangle_{V_0} = V_0 e^{-\gamma t}. \tag{1.5}$$

Moreover, after squaring (1.4) and utilizing (1.3)

$$\langle \{V(t)\}^2 \rangle_{V_0} = V_0^2\, e^{-2\gamma t} + \Gamma\, e^{-2\gamma t} \int_0^t dt' \int_0^t dt''\, e^{\gamma(t'+t'')}\langle L(t')L(t'')\rangle$$

$$= V_0^2\, e^{-2\gamma t} + \frac{\Gamma}{2\gamma}\left(1 - e^{-2\gamma t}\right). \tag{1.6}$$

The hitherto unknown Γ can now be identified by employing the fact that for $t \to \infty$ the mean square velocity must have the known thermal value,

$$\langle \{V(\infty)\}^2 \rangle = \frac{\Gamma}{2\gamma} = kT. \tag{1.7}$$

This equation relates Γ, the size of the fluctuating term, to the damping constant γ. It is the simplest form of the general fluctuation–dissipation theorem. It is due to this theorem that the Langevin approach is so successful: the noise is fully expressed in the macroscopic damping constant, together with the temperature. Just as the Einstein relation in VIII.3, one obtains an identity because one knows beforehand the size of the equilibrium fluctuations from ordinary equilibrium statistical mechanics. The physical picture is that the random kicks tend to spread out V, while the damping term tries to bring V back to zero. The balance between these two opposing tendencies is the equilibrium distribution.

Exercise. Restore the mass M of the Brownian particle in the equations and compute from (1.4) for $t_1 > 0, t_2 > 0$

$$\langle V(t_1)V(t_2)\rangle = \left(V_0^2 - \frac{kT}{M}\right)e^{-\gamma(t_1+t_2)} + \frac{kT}{M}e^{-\gamma|t_1-t_2|}. \tag{1.8}$$

Exercise. The autocorrelation function of V in equilibrium follows from (1.8) in the limit $t_1 \to \infty, t_2 \to \infty, t_1 - t_2 = \tau$. Find it also directly from (1.5) and (1.7).

Exercise. Obtain the mean square displacement

$$\langle\{X(t) - X(0)\}^2\rangle = \left(V_0^2 - \frac{kT}{M}\right)\left(\frac{1-e^{-\gamma t}}{\gamma}\right)^2 + \frac{2kT}{M\gamma}\left(t - \frac{1-e^{-\gamma t}}{\gamma}\right). \tag{1.9}$$

Deduce from it the Einstein relation (VIII.3.8).

Exercise. Consider the harmonically bound Brownian particle

$$\ddot{X} + \gamma\dot{X} + \omega^2 X = L(t). \tag{1.10}$$

Find $\langle X(t)\rangle$ and $\langle\{X(t)\}^2\rangle$ for given $X(0), \dot{X}(0)$. Show that for $t \gg \gamma^{-1}$ the value of $\langle\{X(t)\}^2\rangle$ is the one known from the equipartition law, provided that Γ is again given by (1.7).

Remark. On taking the average of (1.1) term by term and using (1.2) one obtains

$$\partial_t\langle V\rangle = -\gamma\langle V\rangle.$$

This is a closed equation for the average, from which one finds directly (1.5). Next multiply (1.1) with V and take the average so as to get an equation for $\langle V^2\rangle$,

$$\partial_t\langle V^2\rangle = -2\gamma\langle V^2\rangle + 2\langle VL(t)\rangle.$$

This is not a closed equation because the last term is unknown. That is the reason for the more elaborate calculation of Uhlenbeck and Ornstein that led to (1.6). In fact, using their result we now find in hindsight

$$\langle VL(t)\rangle = \tfrac{1}{2}\Gamma.$$

2. Applications

The fluctuation–dissipation theorem (1.7) tells us that wherever there is damping there must be fluctuations. They are small from the macroscopic point of view because of the factor kT in (1.7). In practice they appear only when they are magnified in some way. For instance, the velocity fluctuations (1.6) of the Brownian particle are not seen, but they build up a mean square displacement (1.9), which can be observed under a microscope. In electronics fluctuations are important because of the large amplifications used.

Consider an RC-circuit with linear resistance R. The charge Q on the condenser obeys macroscopically an equation with damping due to R, and must therefore be supplemented with a noise term,

$$\frac{dQ}{dt} = -\frac{Q}{RC} + L(t). \tag{2.1}$$

In thermal equilibrium the mean electrostatic energy on the condenser is

$$\langle Q^2 \rangle / 2C = \tfrac{1}{2} kT. \tag{2.2}$$

Hence the constant Γ in (1.3) must be equal to $2kT/R$. Here T should be the temperature of the entire circuit, but it may be taken to be that of the resistor, since that is where the noise is produced; the temperatures of condenser and wires are irrelevant.

The fact that the fluctuations in (2.1) are fully determined by the parameters R and T of the resistor alone, has led to the picture of a "noise source" located in the resistor. The source may be regarded as producing a fluctuating current δI, to be added to the macroscopic current. From (2.1) one sees $\delta I(t) = -L(t)$, and therefore

$$\langle \delta I(t)\, \delta I(t') \rangle = (2kT/R)\, \delta(t - t'). \tag{2.3}$$

Alternatively one may say that the source produces a random voltage $\delta V(t) = -RL(t)$, so that

$$\langle \delta V(t)\, \delta V(t') \rangle = (2kTR)\, \delta(t - t'). \tag{2.4}$$

These are merely other ways of describing the same facts, but in practice it is convenient to be able to insert the noise sources into a network, just like any other source, without having to write first the equations for the whole network.

Consider a general linear circuit whose instantaneous electrical state is described by a set of currents and voltages. If we denote all these variables by v_ν the network equations have the general form

$$\dot v_\nu = \sum_\mu A_{\nu\mu} v_\mu + F_\nu + L_\nu(t). \tag{2.5}$$

Here $A_{\nu\mu}$ is a constant matrix; the external input F_ν will also be taken constant, and $L_\nu(t)$ represents the noise sources, which in general obey

$$\langle L_\nu(t) \rangle = 0, \qquad \langle L_\nu(t) L_\mu(t') \rangle = \Gamma_{\nu\mu}\, \delta(t - t'). \tag{2.6}$$

2. APPLICATIONS

It is easy to transform the F_ν away by setting

$$v_\nu + \sum_\mu (A^{-1})_{\nu\mu} F_\mu = u_\nu.$$

This amounts to subtracting the stationary values. The deviations u_ν from the stationary values obey

$$\dot{u}_\nu = \sum_\mu A_{\nu\mu} u_\mu + L_\nu(t). \tag{2.7}$$

This equation together with (2.6) is the general multivariate Langevin equation.

Exercise. The spectral density of the fluctuations (2.3), (2.4) is

$$S_I(\omega) = \frac{2}{\pi}\frac{kT}{R}, \quad S_V(\omega) = \frac{2}{\pi}kTR \quad \text{or} \quad S_I(f) = \frac{4kT}{R}, \quad S_V(f) = 4kTR.$$

Exercise. Construct the Langevin equation of a Brownian particle in three dimensions with gravity. Find the correlation matrix $\langle\langle v_i(t)v_j(0)\rangle\rangle$ of its velocity components.

Exercise. Find from (2.7) – in matrix notation, the tilde indicates the transpose –

$$\langle u(t)\tilde{u}(0)\rangle = \int_0^\infty e^{(t+\tau)A}\, \Gamma\, e^{\tau\tilde{A}}\, d\tau.$$

The condition is that all eigenvalues of A are negative so that the system is dissipative.

Exercise. When F in (2.5) depends on time it is still possible to reduce that equation to (2.7).

Exercise. An LRC-circuit is given by

$$L\frac{dI}{dt} + RI + \frac{Q}{C} = V(t) + L(t), \tag{2.8}$$

where $V(t)$ is an externally imposed potential. Derive

$$\langle\langle I(t)I(0)\rangle\rangle = \frac{kT}{L} e^{-\gamma t}\left(\cos\omega t + \frac{\gamma}{\omega}\sin\omega t\right),$$

where

$$\gamma = \frac{R}{2L}, \quad \omega^2 = \frac{1}{LC} - \gamma^2.$$

Exercise. The velocity v of a charged particle in a constant magnetic field and random electric field obeys

$$\dot{v} = v \wedge B - \beta v + E(t), \tag{2.9}$$

where $\langle E_i(t)E_j(t')\rangle = C\delta_{ij}\delta(t-t')$. Find its mean square velocity in equilibrium and its mean square displacement.[*]

---

[*] B. Korsunoğlu, Annals Phys. **17**, 259 (1962). R. Czopnik and P. Garbaczewski, Phys. Rev. E **63**, 021105 (2001).

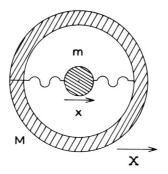

Fig. 22. Simplified itinerant oscillator model.

Exercise. A polymer in solution has been described by[*]

$$\dot{x}_n = x_{n+1} + x_{n-1} - 2x_n + L_n(t),$$

$$\langle L_n(t) L_{n'}(t') \rangle = \Gamma \delta_{nn'} \delta(t - t').$$

Solve this equation and find $\langle \{x_n(t) - x_n(0)\}^2 \rangle$ in equilibrium.

Exercise. Consider the following simplified version of the itinerant oscillator model[**]. A body moves in a fluid and contains in its interior a damped oscillator (fig. 22). The equations of motion are

$$m\ddot{x} + \beta(\dot{x} - \dot{X}) + a^2(x - X) = K(t),$$

$$M\ddot{X} + \beta(\dot{X} - \dot{x}) + \gamma \dot{X} + a^2(X - x) = -K(t) + L(t),$$

where K and L are independent Langevin forces. Find their constants Γ.

Exercise. Delta functions do not occur in nature. In any physical application $L(t)$ has an autocorrelation time $\tau_c > 0$; for a Brownian particle τ_c is at least as large as the duration of an individual collision. It is therefore more physical to write instead of the delta function in (1.3) some sharply peaked function $\phi(t - t')$ of width τ_c. Show that this leads to the same results provided that $\gamma \tau_c \ll 1$.

3. Relation to Fokker–Planck equation

One expects that the Langevin equation (1.1) is equivalent to the Fokker–Planck equation (VIII.4.6). This cannot be literally true, however, because the Fokker–Planck equation fully determines the stochastic process $V(t)$, whereas the Langevin equation does not go beyond the first two moments. The reason is that the postulates (i), (ii), (iii) in section 1 say nothing about

---

[*] G. Ronca, J. Chem. Phys. **67**, 4965 (1977).
[**] N.E. Hill, Proc. Phys. Soc. (London) **82**, 723 (1963); J.H. Calderwood and W.T. Coffey, Proc. Roy. Soc. A **356**, 269 (1977).

the higher moments of $L(t)$. One therefore customarily supplements them with the following postulate.

(iv) $L(t)$ is Gaussian; i.e., all odd moments vanish and the even moments are given by the same rule (I.6.11) that holds for the Gaussian distribution. For instance,

$$\langle L(t_1)L(t_2)L(t_3)L(t_4)\rangle = \langle L(t_1)L(t_2)\rangle \langle L(t_3)L(t_4)\rangle + \cdots$$
$$= \Gamma^2 \{\delta(t_1 - t_2)\,\delta(t_3 - t_4) + \delta(t_1 - t_3)\,\delta(t_2 - t_4)$$
$$+ \delta(t_1 - t_4)\,\delta(t_2 - t_3)\}. \tag{3.1}$$

Alternatively one may postulate that all higher *cumulants* are zero. This specifies all stochastic properties of $L(t)$ in terms of the single parameter Γ. The $L(t)$ defined in this way is called *Gaussian white noise*. From the mathematical point of view it does not really exist as a stochastic function (no more than the delta function exists as a function); and in physics it never really occurs but serves as a model for any rapidly fluctuating force.

With the aid of this fourth postulate the solution $V(t)$ of (1.1), together with the initial condition $V(0) = V_0$, is fully determined. We first prove that this process $V(t)$ is Gaussian. One sees from (1.4) that $V(t)$ is a linear combination of all values that $L(t')$ takes for $0 \leq t' \leq t$. Since the joint distribution of all $L(t')$ is Gaussian it follows that the value of V at time t has a Gaussian distribution. By the same token, the joint distribution of $V(t_1), V(t_2), \ldots$ is Gaussian. Hence V is uniquely determined by its first and second moments, which were computed in (1.5) and (1.8). Thus the process $V(t)$ with all its moments is now known explicitly.

Exercise. Show that the property of $L(t)$ to be Gaussian white noise is expressed by the following identity of its characteristic functional:

$$\left\langle \exp\left[i \int k(t)L(t)\,dt\right]\right\rangle = \exp\left[-\tfrac{1}{2}\Gamma \int \{k(t)\}^2\,dt\right]. \tag{3.2}$$

Exercise. Prove that

$$W(t) = \int_0^t L(t')\,dt' \tag{3.3}$$

is the Wiener process.

Exercise. A rod-like molecule rotates freely in a plane, but is subjected to a Langevin force due to the surroundings:

$$\ddot{\phi} + \gamma\dot{\phi} = L(t).$$

Find $\langle \cos\phi(t_1)\cos\phi(t_2)\rangle^s$, which is the autocorrelation function of the x-component of the dipole moment of the molecule. [W.T. Coffey and A. Morita, J. Physics (London) D **9**, L17 (1976).]

Exercise. The Ornstein–Uhlenbeck process (IV.3.10), (IV.3.11) satisfies the "generalized Langevin equation with memory kernel"

$$\dot{y}(t) = -(1 - 2\alpha)y(t) + 2\alpha(1 - \alpha) \int_{-\infty}^{t} e^{-\alpha(t - t')} y(t') \, dt' + f(t),$$

where $f(t)$ is Gaussian white noise. Hence it is not possible to conclude from such an equation that y cannot be Markovian, in spite of the integral over previous history. [Hint: Determine the characteristic functional of the stochastic function $f(t)$ defined by this equation, using Fourier transformation.]

The equivalence of the Langevin equation (1.1) to the Fokker–Planck equation (VIII.4.6) for the velocity distribution of our Brownian particle now follows simply by inspection. The solution of (VIII.4.6) was also a Gaussian process, see (VIII.4.10), and its moments (VIII.4.7) and (VIII.4.8) are the same as the present (1.5) and (1.6). Hence the autocorrelation function (1.8) also applies to both, so that both solutions are the same process. Q.E.D.

Conclusion. The Langevin equation

$$\dot{y} = -\gamma y + L(t) \tag{3.4}$$

with Gaussian white noise specified by

$$\langle L(t)L(t')\rangle = \Gamma \, \delta(t - t') \tag{3.5}$$

represents the same Markov process y as

$$\frac{\partial P(y, t)}{\partial t} = \gamma \frac{\partial}{\partial y} yP + \frac{\Gamma}{2} \frac{\partial^2 P}{\partial y^2}. \tag{3.6}$$

This is the master equation for the Ornstein–Uhlenbeck process. This result is readily extended to more variables, but we emphasize that the equations must be linear.

Exercise. Insert into (1.1) a constant force and derive in this way (VIII.3.5).
Exercise. Find the Fokker–Planck equation for the *LRC* circuit (2.8).
Exercise. Same question for the charged particle (2.9).
Exercise. Same question for the itinerant oscillator in fig. 22.
Exercise. Construct the Langevin equation that corresponds to the Kramers equation (VIII.7.4).
Exercise. The generating functional of the solution of (1.1) can be found explicitly with the aid of (1.4) and (3.2). Show in this way that $V(t)$ is the Ornstein–Uhlenbeck process that is given by (VIII.4.6).
Exercise. Another type of itinerant oscillator is given by two angles ϕ_1, ϕ_2 obeying

$$I_1 \ddot{\phi}_1 + \zeta_1 \dot{\phi}_1 + V'(\phi_1 - \phi_2) + \mu_1 E \sin \phi_1 = L_1(t),$$

$$I_2 \ddot{\phi}_2 + \zeta_2 \dot{\phi}_2 - V'(\phi_1 - \phi_2) + \mu_2 E \sin \phi_2 = L_2(t).$$

I_1 and I_2 are moments of inertia; ζ_1, ζ_2 damping coefficients; μ_1, μ_2 dipole moments;

E is an electric field and V the interaction potential. When L_1, L_2 are independent Langevin forces, write the corresponding Fokker–Planck equation. Find the equilibrium solution and determine Γ_1 and Γ_2.[*]

Exercise. The Langevin equation for a Brownian particle driven by a periodic force is, in suitable units,

$$\dot{V} = -\gamma V + L(t) + a\sin(t+\eta). \tag{3.7}$$

a and η are constants. We are interested in the case that η, although constant in time, is a random number with constant probability in $(0, 2\pi)$. Show that the joint probability distribution of V and $\vartheta = \eta + t$ obeys

$$\frac{\partial P(V,\vartheta,t)}{\partial t} = \gamma \frac{\partial}{\partial V} VP + \frac{\Gamma}{2}\frac{\partial^2 P}{\partial V^2} - a\sin\vartheta \frac{\partial P}{\partial V} - \frac{\partial P}{\partial \vartheta}.$$

Exercise. Derive the stationary solution of this equation, viz.,

$$P^s(V,\vartheta) = (2\pi)^{-3/2} \exp[-(\gamma/\Gamma)\{V - u(\vartheta)\}^2].$$

On multiplying this solution by $e^{in\vartheta}$ one obtains an infinite sequence of eigenvalues ni.[**]

Remark. In equation (3.7) suppose η is not stochastic, but merely a constant parameter. This provides a simple example of a wide class of stochastic systems involving a deterministic, time dependent driving force. Other examples are treated in the book by COFFEY ET AL. Another example is the Kramers' particle (VIII.7.4), in which the force $F(X)$ derives from a bistable potential (as in Fig. 37 on page 335) and in addition contains a periodic driving force $A\cos\omega t$. The transition rate between the two minima will be determined by this deterministic force and the stochastic diffusion together. It may happen that the stochastic term facilitates the transition, which gave rise to the term "stochastic resonance". Numerous examples have been studied, ranging from biological systems to ice ages[†]. Unfortunately they are impenetrable to analytic approaches and have to be studied numerically. I can only refer the reader to the surveys in J. Statist. Physics **70** (January 1993); Phys. Reports **234**/4–5 (1993); Rev. Mod. Phys. **70**, 223 (1998); Fluctuation and Noise Letters **4**/2, L 247 (2004).

4. The Langevin approach

The approach used in section **2** may be generally formulated as follows. Suppose one has a system whose macroscopic behavior is known and one also knows that there must be fluctuations. In order to describe these fluctuations one then proceeds by the following three steps.

---

[*] W.T. Coffey, J.K. Vij, and P.M. Corcoran, Proc. Roy. Soc. London A **425**, 169 (1989).

[**] P. Jung and P. Hänggi, Europhys. Letters **8**, 505 (1989). For a similar case, see P. Jung, Z. Phys. B **76**, 521 (1989).

[†] P. Hoyng, D. Schmitt, M.A.J.H. Ossendrijver, Phys. Earth and Planet. Interiors **130**, 143 (2002).

(1) Write the deterministic macroscopic equations of motion of the system.
(2) Add a Langevin force with the properties (i)–(iv) stipulated above.
(3) Adjust the constant Γ so that the stationary solution reproduces the correct mean square fluctuations as known from statistical mechanics, or find Γ from other considerations, see, e.g., (**5**.1) below.

These three steps constitute what I call "the Langevin approach". It is widely utilized, also when the fluctuations are not due to thermal motion or to discreteness of particles, and even when they are of unspecified or unknown provenance. The applications in many branches of physics, chemistry and biology are far too numerous to list. The approach has been highly successful whenever the deterministic equations are linear, but for nonlinear cases it leads to difficulties, which are analyzed in this section. The purpose is to shield the reader from the labyrinthine literature and to convince him of the necessity of the more firmly based treatment in the next chapter.

We begin with an innocuous case. Consider a pendulum suspended in air and consequently subject to damping accompanied by a Langevin force. This force is, of course, the same as the one in equation (**1**.1) for the Brownian particle, because the collisions of the air molecules are the same. They depend on the instantaneous value of V, but they are insensitive to the fact that there is a mechanical force acting on the particle as well. Hence for small amplitudes the motion is governed by the linear equation (**1**.10). For larger amplitudes the equation becomes nonlinear:

$$\ddot{\phi} + \gamma\dot{\phi} + \omega^2 \sin\phi = l^{-1}L(t). \tag{4.1}$$

l is the length of the pendulum, the mass equals unity, and $L(t)$ is the same as for the free particle. This equation, although nonlinear, is perfectly well established thanks to the fact that we know the physical cause of the fluctuations.[*] In the same way the insertion of Langevin terms into the hydrodynamic Navier–Stokes equations[**] can be justified because the nonlinearity is due to the term $(v\cdot\nabla)v$, which is purely mechanical, or rather kinematical.

However, the structure of equation (**4**.1) differs essentially from that of (**1**.1) or (**1**.2). If one averages (**4**.1) term by term the result is

$$\partial_t^2 \langle\phi\rangle + \gamma\, \partial_t\langle\phi\rangle + \omega^2 \langle\sin\phi\rangle = 0. \tag{4.2}$$

One sees that, in contrast to the linear case, it is *not* true that the average obeys the deterministic macroscopic equation from which one started. In fact, (**4**.2) is

---

[*] In particular the anharmonic Brownian oscillator has been studied extensively by J.B. Morton and S. Corrsin, J. Mathem. Phys. **10**, 361 (1969); K.S.J. Nordholm and R. Zwanzig, J. Statist. Phys. **11**, 143 (1974); R.F. Rodríguez and N.G. van Kampen, Physica A **85**, 347 (1976); M.F. Dimentberg, Intern. J. Nonlinear Mech. **11**, 83 (1976); R.C. Desai and R. Zwanzig, J. Statist. Phys. **19**, 1 (1978).

[**] L.D. Landau and E.M. Lifshitz, *Fluid Mechanics* (Pergamon, Oxford 1959) ch. 17; R.D. Mountain, Advances in Molecular Relaxation Processes **9**, 255 (1977).

4. THE LANGEVIN APPROACH

not a closed equation for $\langle\phi\rangle$ at all; rather it involves all higher moments, as can be seen from

$$\langle \sin\phi\rangle = \sin\langle\phi\rangle - \tfrac{1}{2}\langle(\phi-\langle\phi\rangle)^2\rangle\cos\langle\phi\rangle + \cdots .$$

The situation is analogous to the one described by (V.8.5). Moreover, it follows that the equation for $\langle\phi\rangle$ involves the higher moments of $L(t)$ as well, and they have been fixed rather artificially by postulate (iv) in the previous section. Of course, $\phi(t)$ is no longer Gaussian.

More generally, suppose one has a physical system with a nonlinear equation of motion $\dot y = A(y)$ and, following the Langevin approach, one adds a Langevin term to describe the fluctuations in the system,

$$\dot y = A(y) + L(t). \tag{4.3}$$

We call this equation "quasilinear" to express the fact that the coefficient of $L(t)$ is still a constant.[*] Although its solution cannot be given explicitly it can still be argued that it is equivalent to the quasilinear Fokker–Planck equation

$$\frac{\partial P(x,t)}{\partial t} = -\frac{\partial}{\partial y}A(y)P + \frac{\Gamma}{2}\frac{\partial^2 P}{\partial y^2}. \tag{4.4}$$

First it is clear that for each sample function L equation (4.3) uniquely determines $y(t)$ when $y(0)$ is given. Since the values of L at different times are stochastically independent, it follows that y is Markovian. Hence it obeys an M-equation, which may be written in the Kramers–Moyal form (VIII.2.6). We compute the successive coefficients (VIII.2.4).

An exact consequence of (4.3) is

$$\Delta y = \int_t^{t+\Delta t} A(y(t'))\,dt' + \int_t^{t+\Delta t} L(t')\,dt'.$$

Hence the average with fixed $y(t)$ is

$$\langle \Delta y\rangle = A(y(t))\,\Delta t + \mathcal{O}(\Delta t)^2.$$

That yields the first term in (4.4). Next

$$\langle(\Delta y)^2\rangle = \left\langle\left\{\int_t^{t+\Delta t} A(y(t'))^2\,dt\right\}^2\right\rangle$$

$$+ 2\int_t^{t+\Delta t} dt' \int_t^{t+\Delta t} dt'' \langle A(y(t'))L(t'')\rangle$$

$$+ \int_t^{t+\Delta t} dt' \int_t^{t+\Delta t} dt'' \langle L(t')L(t'')\rangle.$$

---

[*] This equation is commonly characterized by the term "additive noise", but a transformation of the variable y changes the "additive noise" into the "multiplicative noise" of (4.5), compare XVI.1.

The first line is of order $(\Delta t)^2$ and does therefore not contribute to a_2. The last line equals $\Gamma \Delta t$ as before, and agrees with (4.4). In the second line expand $A(y(t'))$:

$$2A(y(t)) \Delta t \int_t^{t+\Delta t} dt'' \langle L(t'') \rangle$$

$$+ 2A'(y(t)) \int_t^{t+\Delta t} dt' \int_t^{t+\Delta t} dt'' \langle \{y(t') - y(t)\} L(t'') \rangle + \cdots .$$

The first term vanishes and the second one is $o(\Delta t)$, because it is a double integral and $\{y(t') - y(t)\}$ contains no delta functions. By a similar argument one sees that the higher terms are $o(\Delta t)$, and also that $\langle (\Delta y)^\nu \rangle = o(\Delta t)$ for $\nu > 2$. This proves the equivalence of (4.3) and (4.4).

Finally consider the fully nonlinear Langevin equation

$$\dot{y} = A(y) + C(y)L(t). \tag{4.5}$$

It can be reduced to the preceding case on dividing by $C(y)$ and transforming to a variable $\bar{y}$:

$$\bar{y} = \int \frac{dy}{C(y)}, \qquad \frac{A(y)}{C(y)} = \bar{A}(\bar{y}), \qquad \bar{P}(\bar{y}) = P(y)C(y). \tag{4.6}$$

This has been shown above to be equivalent to

$$\frac{\partial \bar{P}(\bar{y}, t)}{\partial t} = -\frac{\partial}{\partial \bar{y}} \bar{A}(\bar{y}) \bar{P} + \frac{\Gamma}{2} \frac{\partial^2 \bar{P}}{\partial \bar{y}^2}. \tag{4.7}$$

Transforming back to the original y,

$$\frac{\partial P(y, t)}{\partial t} = -\frac{\partial}{\partial y} A(y) P + \frac{\Gamma}{2} \frac{\partial}{\partial y} C(y) \frac{\partial}{\partial y} C(y) P. \tag{4.8}$$

Thus the Langevin equation (4.5) is equivalent to the Fokker–Planck equation (4.8); or alternatively[*]

$$\frac{\partial P(y, t)}{\partial t} = -\frac{\partial}{\partial y} [A(y) + \tfrac{1}{2} \Gamma C(y) C'(y)] P + \frac{\Gamma}{2} \frac{\partial^2}{\partial y^2} [C(y)]^2 P. \tag{4.9}$$

However, there is a problem concerning (4.5). This equation, as it stands, has no well-defined meaning. To see this, note that $L(t)$ can be visualized as a sequence of delta peaks arriving at random times (as shown explicitly in the next section). According to (4.5) each delta function in $L(t)$ causes a jump in $y(t)$. Hence the value of y at the time that the delta function arrives is undetermined, and therefore also the value of $C(y)$. The equation does not specify whether one should insert in $C(y)$ the value of y before the jump, after the jump, or

---

[*] Another derivation is given by M. Lax, Rev. Mod. Phys. **38**, 541 (1966).

perhaps the mean of both. These different options lead to different Fokker–Planck equations and hence to different results. (The Fokker–Planck equation, of course, does not suffer from poly-interpretability.)[*]

Stratonovich[**] opted for the mean value. He read (**4.5**) as

$$y(t+\Delta t) - y(t) = A(y(t))\Delta t + C\left(\frac{y(t)+y(t+\Delta t)}{2}\right)\int_t^{t+\Delta t} L(t')\,dt'. \quad (4.10)$$

It can be proved that this choice leads to (**4.8**).[†] This shows that our naive use of the transformation of variables amounted to opting for the Stratonovich interpretation.

Itô[††] opted for the value of y before the arrival of the delta peak. He read (**4.5**) as

$$y(t+\Delta t) - y(t) = A(y(t))\Delta t + C(y(t))\int_t^{t+\Delta t} L(t')\,dt' \quad (4.11)$$

and proved that this is equivalent to

$$\frac{\partial P(y,t)}{\partial t} = -\frac{\partial}{\partial y}A(y)P + \frac{\Gamma}{2}\frac{\partial^2}{\partial y^2}[C(y)]^2 P. \quad (4.12)$$

Apparently this interpretation is not compatible with the familiar way of transforming variables; new transformation laws have to be formulated. They are given in (**4.14**).

Exercise. Itô finds for the last term in (**4.5**) that $\langle C(y)L(t)\rangle = 0$, but according to Stratonovich $\langle C(y)L(t)\rangle = \frac{1}{2}C(y)C'(y)$.

Exercise. Under a transformation $\bar{y}=\phi(y)$ the coefficients of (**4.8**) transform into

$$\bar{A}(\bar{y}) = A(y)\frac{d\phi}{dy}, \qquad \bar{C}(\bar{y}) = C(y)\frac{d\phi}{dy}. \quad (4.13)$$

If one applies formally the transformation to the meaningless equation (**4.5**) one also obtains (**4.13**). Hence the connection between (**4.5**) and (**4.8**) is invariant for nonlinear transformations of the variable y.

---

[*] For this discussion, see also A.H. Gray and T.K. Caughy, J. Math. and Phys. **4**, 288 (1965); R.E. Mortensen, J. Statist. Phys. **1**, 271 (1969); N.G. van Kampen, J. Statist. Phys. **24**, 175 (1981). The same dilemma arose in quantum mechanics: M. Blume, J. Mathem. Phys. **19**, 2004 (1978); H. Gzyl, J. Mathem. Phys. **20**, 1714 (1979).

[**] R.L. Stratonovich, SIAM J. Control **4**, 362 (1966); STRATONOVICH, ch. 4, sect. 8.

[†] J.L. Doob, *Stochastic Processes* (Wiley, New York 1953) ch. 6, sect. 3.

[††] K. Itô, Proc. Imp. Acad. Tokyo **20**, 519 (1944); Mem. Amer. Mathem. Soc. **4**, 51 (1951).

Exercise. The same transformation applied to (**4.12**) yields, compare (VIII.1.9),

$$\bar{A}(\bar{y}) = A(y)\frac{d\phi}{dy} + \tfrac{1}{2}\{C(y)\}^2 \frac{d^2\phi}{dy^2}, \qquad \bar{C}(\bar{y}) = C(y)\frac{d\phi}{dy}. \qquad (4.14)$$

Hence *in the Itô interpretation of* (**4.5**) *these are the proper transformation formulas of the coefficients*, rather than (**4.13**), which one might expect naively.[*]

Exercise. The damping of a Brownian particle is described in terms of the energy by $\dot{E} = -2\gamma E$. Construct a Langevin force which gives the correct fluctuations in the Stratonovich interpretation. Find another Langevin force that is correct in the Itô interpretation.

Exercise. Any equation of the form (**4.5**) with Itô prescription is equivalent with another equation with Stratonovich prescription. Find the connection between the coefficients in both equations. The difference between their coefficients A is sometimes called the "spurious drift".

Exercise. The multivariate version of (**4.5**)

$$\dot{y}_\nu = A_\nu(y) + C_\nu(y)L(t)$$

suffers from the same dilemma unless

$$\sum_\mu C_\mu(y)\frac{\partial C_\nu(y)}{\partial y_\mu} = 0.$$

Find also the corresponding condition for

$$\dot{y}_\nu = A_\nu(y) + \sum_j C_{\nu,j}(y)L_j(t),$$

where $L_j(t)$ are mutually independent Langevin processes.

5. Discussion of the Itô–Stratonovich dilemma

Before embarking on this discussion one fact must be established. In many applications the autocorrelation function of $L(t)$ is not really a delta function, but merely sharply peaked with a small $\tau_c > 0$. Accordingly $L(t)$ is a proper stochastic function, not a singular one. Then (**4.5**) is a well-defined stochastic differential equation (in the sense of chapter XVI) with a well-defined solution. *If one now takes the limit* $\tau_c \to 0$ *in this solution, it becomes a solution of the Stratonovich form* (**4.8**) *of the Fokker–Planck equation.* This theorem has been proved officially[**], but the result can also be seen as follows.

---

[*] K. Itô, Nagoya Mathem. J. **1**, 35 (1950) and **3**, 55 (1951); L. Arnold, *Stochastische Differentialgleichungen* (Oldenburg, Munich 1973).

[**] E. Wong and M. Zakai, Annals Mathem. Statistics **36**, 1560 (1965).

5. DISCUSSION OF THE ITÔ–STRATONOVICH DILEMMA

As long as $L(t)$ is not singular one is free to apply any transformation of y using the familiar rules of calculus; in particular one may transform (4.5) into the quasilinear equation (4.3). In the latter the limit $\tau_c \to 0$ gives no problem and the result is (4.4). Transforming back to the original variables gives (4.8), as we found before. Hence, whenever the delta function stands for a sharp, but not infinitely sharp peak the Stratonovich interpretation is appropriate. The Itô interpretation cannot even be formulated unless τ_c is strictly zero.

Nonetheless the mathematicians prefer the Itô interpretation of (4.5); they call it the "Itô equation" and write it in the form

$$\mathrm{d}y = A(y)\,\mathrm{d}t + C(y)\,\mathrm{d}W(t),$$

where $W(t)$ stands for the Wiener process, see (3.3).[*] We repeat, however, that the two forms of the equation can be transformed into each other at the expense of adding or subtracting a correction term to $A(t)$, see the above Exercise about the "spurious drift".

The confusion among the physicists is caused not only by an undue respect for mathematics, but also by their indiscriminate application of the Langevin approach to nonlinear systems. They would write (4.5), with $A(y)$ borrowed from the macroscopic equation, and add the fluctuating term with a $C(y)$ determined by some other physical considerations. This would be done regardless of the physical mechanism responsible for the fluctuations – as if stochasticity is part and parcel of the mathematics and requires no physical cause. Having postulated (4.5) on such shaky grounds, one is then faced with the dilemma how to interpret it. Different interpretations lead to different convection terms in the corresponding Fokker–Planck equation, see (4.9) and (4.12). In order to discuss this dilemma it is necessary to distinguish between external and internal noise.[**]

External noise denotes fluctuations created in an otherwise deterministic system by the application of a random force, whose stochastic properties are supposed to be known. Examples are: a noise generator inserted into an electric circuit, a random signal fed into a transmission line, the growth of a species under influence of the weather, random loading of a bridge, and most other stochastic problems that occur in engineering. In all these cases clearly (4.5) holds if one inserts for $A(y)$ the deterministic equation of motion for the isolated system, while $L(t)$ is approximately but never completely white. Thus for external noise the Stratonovich result (4.8) and (4.9) applies, in which $A(y)$ represents the dynamics of the system with the noise turned off.

---

[*] Occasionally and confusingly the term "stochastic differential equations" is used exclusively for equations of this form.

[**] H. Mori, Prog. Theor. Phys. **53**, 1617 (1975).

Remark. The white noise limit is not sufficiently defined by just saying $\tau_c \to 0$. We have to construct a sequence of processes which in this limit reduce to Gaussian white noise. For that purpose take a long time interval $(0, T)$ and a Poisson distribution of time points τ_σ in it with density ν. To each τ_σ attach a random number c_σ; they are independent and identically distributed, with zero mean. Consider the process

$$Y(t) = \beta \sum_{\sigma=1}^{s} c_\sigma \tau_c^{-1} \psi\left(\frac{t - \tau_\sigma}{\tau_c}\right),$$

where $\psi(t)$ is a fixed function, positive and with finite support or rapidly falling off. For $\tau_c \to 0$ the functions $\tau_c^{-1}\psi$ will reduce to a delta peak. We compute the characteristic functional of Y.

$$G([k]) = \sum_{s=0}^{\infty} \frac{(\nu T)^s}{s!} e^{-\nu T} \int_0^T \frac{d\tau_1}{T} \cdots \int_0^T \frac{d\tau_s}{T}$$

$$\times \left\langle \exp\left[i\beta \sum_{\sigma=1}^{s} c_\sigma \int k(t)\psi\left(\frac{t-\tau_\sigma}{\tau_c}\right) \frac{dt}{\tau_c}\right] \right\rangle.$$

The $\langle\ \rangle$ indicate the average over all the c_σ. As these are independent the average factorizes into s individual averages, and the summation can be performed.

$$G([k]) = \exp\left[\nu \int_0^T d\tau \left\{\left\langle \exp i\beta c \int k(t)\psi\left(\frac{t-\tau}{\tau_c}\right)\frac{dt}{\tau_c}\right\rangle - 1\right\}\right]$$

$$= \exp\left[\nu \sum_{m=1}^{\infty} \frac{(i\beta)^m}{m!} \langle c^m \rangle \int_0^T d\tau \left\{\int k(\tau + \tau_c \theta)\psi(\theta)\, d\theta\right\}^m\right].$$

As we have assumed $\langle c \rangle = 0$ the series starts with $m = 2$. Now take the limit

$$\tau_c \to 0, \qquad \beta \to 0, \qquad \nu \to \infty, \qquad \nu\beta = \lambda.$$

The peaks become sharp, and weak, but so dense that their net effect does not vanish but gives

$$G([k]) = \exp\left[-\tfrac{1}{2}\lambda\langle c^2\rangle \int_0^T \{k(\tau)\}^2\, d\tau\right].$$

This is the generating function of Gaussian white noise, see (3.2).

Exercise. Derive the expression for $G([k])$ with the aid of (II.5.7), thereby avoiding the introduction of T. [Hint: compare (II.3.8).]

Exercise. Formulate the white noise limit for a random matrix process.

Internal or intrinsic noise is caused by the fact that the system itself consists of discrete particles; it is inherent in the very mechanism by which the system evolves, as described in III.2. All our examples concerning chemical reactions, emission and absorption of light, growth of populations, etc., were of this type. Internal noise cannot be switched off and it is therefore impossible to identify $A(y)$ as the evolution equation for the system in isolation. One usually identifies it

5. DISCUSSION OF THE ITÔ–STRATONOVICH DILEMMA

with the deterministic macroscopic equation, see V.8. However, the macroscopic equation is merely the approximation that consists in neglecting the fluctuations and is therefore not precise enough to be used in equation (4.5), which purports to describe the fluctuations. A correction term added to $A(y)$ which like the one in (4.9) is of the same order as the fluctuations would not show up in the macroscopic approximation. It is not possible in this way to determine $A(y)$ with the accuracy needed to distinguish between (4.9) and (4.12); nor can one exclude the possibility of other forms. Hence the Langevin approach does not work for internal noise.

Conclusion. *For internal noise one cannot just postulate a nonlinear Langevin equation or a Fokker–Planck equation and then hope to determine its coefficients from macroscopic data.*[*] The more fundamental approach of the next chapter is indispensable.[**]

Brillouin's paradox.[†] The *I-V* characteristic of a metal-oxide rectifier or a p-n junction is

$$I = A\left\{\exp\left[\frac{eV}{kT}\right] - 1\right\}.$$

A simple model is the diode circuit with two metals with known work functions, as in fig. 16 in VI.9. The whole is in thermal equilibrium at temperature T ("Alkemade's diode")[††]. The charge Q on the condenser obeys the phenomenological equation

$$\dot{Q} = -A\left\{\exp\left[\frac{eQ}{kTC}\right] - 1\right\}.$$

If one adds to this equation a Langevin term to describe the noise produced by the rectifier one obtains for the average of Q

$$\frac{d}{dt}\langle Q \rangle = -A\left\{\left\langle\exp\left[\frac{eQ}{kTC}\right]\right\rangle - 1\right\}$$
$$= -A\left\{\frac{e}{kTC}\langle Q \rangle + \frac{1}{2}\left(\frac{e}{kTC}\right)^2\langle Q \rangle^2 + \cdots\right\}.$$

It follows that in equilibrium

$$\langle Q \rangle^e = -\frac{1}{2}\frac{e}{kTC}\langle Q^2 \rangle^e = -\tfrac{1}{2}e.$$

---

[*] This conclusion has been explicitly demonstrated for a generalized Rayleigh particle model by C.T.J. Alkemade, N.G. van Kampen, and D.K.C. MacDonald, Proc. Roy. Soc. (London) A **271**, 449 (1963).

[**] Other *postulated* nonlinear modifications of Langevin are studied in M. Gitterman, Physica A **352**, 309 (2005).

[†] L. Brillouin, Phys. Rev. **78**, 627 (1950).

[††] C.T.J. Alkemade, Physica **24**, 1029 (1958).

Thus the diode in thermal equilibrium maintains a charge and hence a voltage on the condenser: the rectifier rectifies its own fluctuations!

However small the effect, it violates the second law of thermodynamics.[*] An array of 10^{12} diodes would be able to run a flashlight.[**] The answer to this paradox is that one cannot trust the phenomenological equation to the extent that one may use it for deducing a result that is itself of the order of the fluctuations. This example demonstrates the danger of adding a Langevin term to a nonlinear equation.[†]

Although we have found that for internal noise the Itô–Stratonovich dilemma is undecidable for lack of a precise $A(t)$ there are cases in which the Itô equation seems the more appropriate option. As an example we take the decay process defined in IV.6; the M-equation is (V.1.7) and the average obeys the radioactive decay law (V.1.9). As the jumps are relatively small one may hope to describe the process by means of a Langevin equation. Following the Langevin approach we guess

$$\dot{n} = -\gamma n + \sqrt{n}\, L(t), \tag{5.1}$$

because we expect the fluctuations in the decay rate to be proportional to the square root of the number of active nuclei present. To endow this equation with a meaning one has to supply an interpretation rule. Itô's rule gives, according to (4.12), the following Fokker–Planck equation:

$$\frac{\partial P(n,t)}{\partial t} = \gamma \frac{\partial}{\partial n} nP + \frac{\Gamma}{2} \frac{\partial^2}{\partial n^2} nP. \tag{5.2}$$

The Stratonovich equation gives according to (4.8)

$$\frac{\partial P(n,t)}{\partial t} = \gamma \frac{\partial}{\partial n} nP + \frac{\Gamma}{2} \frac{\partial}{\partial n} \sqrt{n} \frac{\partial}{\partial n} \sqrt{n}\, P$$

$$= \frac{\partial}{\partial n}(\gamma n - \tfrac{1}{2}\Gamma)P + \frac{\Gamma}{2} \frac{\partial^2}{\partial n^2} nP. \tag{5.3}$$

One sees that the Itô equation reproduces the correct equation (V.1.9) for the average and the Stratonovich equation does not. The reason is that in this process the jump probability from n to $n-1$ is proportional to the number n of available nuclei *before* the jump, just as was assumed by Itô in (4.11). The same thing holds for chemical reactions, emission of photons, etc. But the higher moments give difficulties. At any rate these jump processes are better treated by means of the master equation – to be solved by means of the expansion developed in the next chapter.

---

[*] A. Marek, Physica **25**, 1358 (1959).

[**] R. McFee, Amer. J. Phys. **39**, 814 (1971).

[†] This particular model has been solved exactly in J. Mathem. Phys. **2**, 592 (1961).

Exercise. Show that the Brillouin paradox remains the same when the Fokker–Planck approach of VIII.5 is used.

Exercise. Why is it not possible to find Γ in (5.2) with the aid of the fluctuation–dissipation theorem? On the other hand, the expressions (VIII.1.6) give directly $\Gamma = \gamma$.

Exercise. Show that (5.2) with $\Gamma = \gamma$ gives the correct second moment but the wrong third moment.

Exercise. Solve (5.1) by setting $n = y^2$. Show explicitly that the moments are the same as those given by (5.3) – as they should.

Exercise. Write (5.1) in the form $\dot{n} + n = f(t)$, thereby defining the stochastic process f. Using the results from IV.6 one can find the properties of f. From

$$n(t) - n_0 e^{-t} = e^{-t} \int_0^t e^{t'} f(t') \, dt'$$

one obtains immediately $\langle f(t) \rangle = 0$, and after squaring

$$\langle f(t_1) f(t_2) \rangle = n_0 e^{-t_1} \delta(t_1 - t_2).$$

Derive also, θ being the Heaviside step function,

$$\langle f(t_1) f(t_2) f(t_3) \rangle = -n_0 e^{-t_1} \delta(t_1 - t_2) \delta(t_1 - t_3) + n_0 e^{-t_1} \delta(t_1 - t_2) \theta(t_1 - t_3)$$
$$+ n_0 e^{-t_1} \delta(t_1 - t_3) \theta(t_1 - t_2) + n_0 e^{-t_2} \delta(t_2 - t_3) \theta(t_3 - t_1).$$

How could it have been seen *a priori* that f cannot be Gaussian?

6. Non-Gaussian white noise

We consider a process $\Lambda(t)$ which shares with the Langevin force $L(t)$ the properties (1.2) and (1.3) but not the Gaussian property (3.1). The higher cumulants of Λ do not vanish but they are supposed to be delta-correlated

$$\langle\langle \Lambda(t_1) \Lambda(t_2) \cdots \Lambda(t_m) \rangle\rangle = \Gamma_m \, \delta(t_1 - t_2) \, \delta(t_1 - t_3) \cdots \delta(t_1 - t_m) \qquad (6.1)$$

for $m \geqslant 1$. The Γ_m are constants, Γ_2 is the same as Γ in 1, and $\Gamma_1 = \langle \Lambda(t) \rangle$ not necessarily zero. We take the Γ_m independent of time so that the process is stationary, if it exists.

The generating functional of $\Lambda(t)$ is

$$\langle e^{i \int k(t) \Lambda(t) dt} \rangle = \exp\left[\sum_{m=1}^{\infty} \frac{i^m}{m!} \int k(t_1) \cdots k(t_m) \langle\langle \Lambda(t_1) \cdots \Lambda(t_m) \rangle\rangle \, dt_1 \cdots dt_m \right]$$

$$= \exp\left[\sum_{m=1}^{\infty} \frac{i^m}{m!} \Gamma_m \int \{k(t)\}^m \, dt \right]. \qquad (6.2)$$

Its logarithm

$$\int dt \sum_{m=1}^{\infty} \frac{i^m}{m!} \Gamma_m \{k(t)\}^m$$

is a sum of terms referring to individual times t, which shows that the values of $\Lambda(t)$ at different times are statistically independent. [This resembles the "completely random process" (IV.3.2), but our present Λ has no well-defined distribution at a single time t since (6.1) is singular.]

To bring these singularities down to earth we consider the integrated process

$$Z(t) = \int_0^t \Lambda(t')\, dt' \quad (t > 0). \tag{6.3}$$

Evidently the increment $z = Z(t_2) - Z(t_1)$ for $t_2 > t_1$ is independent of the previous values of Z. Hence $Z(t)$ is a process with independent increments as defined in (IV.4.7). The probability distribution of the increment z is the transition probability $T_{t_2-t_1}(z)$ introduced in (IV.4.7). Its characteristic function is obtained from (6.2) on inserting $k(t) = k\,\theta(t - t_1)\,\theta(t_2 - t)$:

$$\int e^{ikz}\, T_\tau(z)\, dz = \exp\left[\tau \sum_{m=1}^\infty \frac{(ik)^m}{m!}\, \Gamma_m\right]. \tag{6.4}$$

It has the form (IV.4.8), as it should. The coefficients are the Γ_m that characterized the Λ. Vice versa, it is now possible to construct a white noise process by taking any process with independent increments and differentiating it; its Γ_m are given by (6.4).

Exercise. The Poisson process (IV.2.6) has independent increments. Show that it gives rise to a white noise with $\Gamma_m = 1$ for all m.

Exercise. The random walk (IV.2.2) gives $\Gamma_m = 0$ for odd m and $\Gamma_m = 2$ for even m.

Exercise. The Wiener process has independent increments, see (IV.2.4). The resulting $\Lambda(t)$ is the familiar Gaussian white noise.

A process with independent increments can be generated by compounding Poisson processes in the following way.[*] Take a random set of dots on the time axis forming shot noise as in (II.3.14); the density f_1 will now be called ρ. Define a process $Z(t)$ by stipulating that, at each dot, Z jumps by an amount z (positive or negative), which is random with probability density $w(z)$. Clearly the increment of Z between t and $t + \tau$ is independent of previous history and its probability distribution has the form (IV.4.7). It is easy to compute.

The number n of dots in the interval τ is Poissonian with average $\rho\tau$. The increment z is the sum of the n jumps. Hence the distribution of z is

$$T_\tau(z) = \sum_{n=0}^\infty \frac{(\rho\tau)^n}{n!}\, e^{-\rho\tau} w(z) * w(z) * \cdots * w(z),$$

---

[*] D.L. Snyder, *Random Point Processes* (Wiley–Interscience, New York 1975) ch. 3.

6. NON-GAUSSIAN WHITE NOISE

where the product is the convolution of n factors w,

$$\int e^{ikz} T_\tau(z)\, dz = \sum_{n=0}^{\infty} \frac{(\rho\tau)^n}{n!}\, e^{-\rho\tau} \left\{ \int e^{ikz} w(z)\, dz \right\}^n$$

$$= \exp\left[\rho\tau \int (e^{ikz} - 1)w(z)\, dz\right]. \tag{6.5}$$

On comparing this expression with (**6.4**) one sees that the compound Poisson process defined here is the same as $Z(t)$ in (**6.3**) if

$$\Gamma_m = \rho \int z^m\, w(z)\, dz. \tag{6.6}$$

Summary. A non-Gaussian white noise $\Lambda(t)$ is characterized by the constants Γ_m. Every $\Lambda(t)$ is the derivative of a process with independent increments $Z(t)$ and every $Z(t)$ has a derivative $\Lambda(t)$. The process $Z(t)$ is characterized by its transition probability $T_\tau(z)$ and the connection with $\Lambda(t)$ is given by (**6.4**). On the other hand, the $Z(t)$ is a compound Poisson process.[*] The latter is characterized by ρ and $w(z)$, and the connection is (**6.5**). The connection between the noise and the Poisson process is (**6.6**).

Exercise. Argue that our "compound Poisson process" may be regarded as a superposition of simple Poisson processes as defined in IV.2.
Exercise. Show that the random walk is a compound Poisson process.
Exercise. Show that the Wiener process can be obtained as the limiting case of a compound Poisson process when the jumps become infinitely small, but infinitely dense on the time axis. Compare the Remark in **5**.
Exercise. Derive the M-equation

$$\frac{\partial T_\tau(z)}{\partial \tau} = \rho \int w(z-z')T(z')\, dz' - \rho T_\tau(z)$$

$$= \rho \int w(\zeta)\{T_\tau(z-\zeta) - T_\tau(z)\}\, d\zeta. \tag{6.7}$$

Let us now consider the Langevin-like equation[**]

$$\dot{y} = A(y) + \Lambda(t). \tag{6.8}$$

For fixed initial value this defines a stochastic process, which is clearly Markovian. As in **4** we integrate over Δt,

$$y(t+\Delta t) - y(t) = A(y)\, \Delta t + Z(t+\Delta t) - Z(t) + \mathcal{O}(\Delta t)^2.$$

---

[*] We have only shown that a compound Poisson process is at the same time a process with independent increments. The converse is proved in FELLER II, p. 204.

[**] "Langevin-like" because $\Lambda(t)$ is not a Langevin force as defined in **1**.

Accordingly the probability distribution of y varies by a drift term and by the change of $Z(t)$, as given in (**6.**7):

$$\frac{\partial P(y,t)}{\partial t} = -\frac{\partial}{\partial y}A(y)P(y,t) + \rho \int w(\zeta)\{P(y-\zeta,t) - P(y,t)\}\,\mathrm{d}\zeta. \qquad (6.9)$$

This translates the Langevin-like equation (**6.**8) into a master equation.

Exercise. When the stationary solution $P^s(y)$ of (**6.**9) is known all Γ_m are determined.[*]
[Hint: Fourier transform (**6.**9).]

Exercise. If it happens that $A(y) = c(\mathrm{d}/\mathrm{d}y)\log P^s(y)$, then $\Lambda(t)$ is Gaussian white noise.[*]

Exercise. A damped harmonic oscillator with delta-correlated fluctuations in the frequency is given by the Langevin equation[**]

$$\dot{x} = p, \qquad \dot{p} = -\gamma p - [\Omega^2 + \Lambda(t)]x.$$

Show that this is equivalent to the M-equation

$$\frac{\partial P(x,p,t)}{\partial t} = -p\frac{\partial P}{\partial x} + \frac{\partial}{\partial p}(\gamma p + \Omega^2 x)P + \rho \int w(\zeta)\left\{P(x,p+\zeta x,t) - P(x,p,t)\right\}\,\mathrm{d}\zeta.$$

Exercise. Consider the Langevin-like equation

$$\dot{y} = -\gamma y + \Lambda(t), \qquad y(0) = 0, \quad t \geqslant 0.$$

Find the characteristic functional of $y(t)$ and conclude that $y(t)$ is not Gaussian.

7. Colored noise

This term refers to equations that look like Langevin equations

$$\dot{y} = A(y) + C(y)\xi(t), \qquad (7.1)$$

but in which the noise $\xi(t)$ is not white. Rather, $\xi(t)$ is supposed to be a given stationary process with

$$\langle \xi(t) \rangle = 0, \qquad \langle \xi(t)\xi(t') \rangle = \kappa(t-t'). \qquad (7.2)$$

The autocorrelation function κ is not a delta function, and $\xi(t)$ is some well-defined, non-singular stochastic function, so that (7.1) is also well-defined. Such Langevin-like equations have been amply discussed in the literature.[†] The fact

---

[*] P. Mazur and D. Bedeaux, Physica A **173**, 155 (1991).

[**] B.J. West, K. Lindenberg, and V. Seshadri, Physica A **102**, 470 (1980); N.G. van Kampen, ibid. p. 489.

[†] For references, see N.G. van Kampen, J. Statist. Phys. **54**, 1289 (1989); J. Casademunt and J.M. Sancho, Phys. Rev. A **39**, 4915 (1989).

that $\xi(t)$ is not white creates some profound differences with the genuine Langevin equation (4.5). Equations of this type are stochastic differential equations, which will be treated more fully in chapter XVI. Here only a few comments are made.

a. It is permissible to use the transformation (4.6) in order to reduce the coefficient $C(y)$ to unity. In practice this has the drawback that approximate solution methods then often lead to moments of the transformed variable $\bar{y}$ rather than of y. Also, for a set of coupled equations,

$$\dot{y}_i = A_i(y) + \sum_r C_{ir}(y)\xi_r(t),$$

such a reduction of $C_{ir}(y)$ is not generally possible.[*]

b. As the autocorrelation time τ_c of $\xi(t)$ is not zero, $y(t)$ is not a Markov process. This is rather obvious, but the explicit argument runs as follows. Take any t_0 and consider the ensemble of all solutions $y(t)$ with fixed $y(t_0)$. The probability distribution of $y(t_1)$ at some $t_1 > t_0$ depends on the values of $\xi(t)$ for $t_0 < t < t_1$. These values are correlated with the values taken by $\xi(t)$ at times $t < t_0$, and those were involved in determining $y(t)$ at $t < t_0$. Hence the values of y at $t > t_0$ with fixed $y(t_0)$ are not statistically independent of the values of y before t_0.

c. This fact has the consequence that some familiar ideas must be revised.

(i) *There is no longer a uniquely defined stochastic $y(t)$ that obeys* (7.1) *and has a given initial value $y(t_0) = y_0$*. There is a unique solution with this initial condition if one also supposes that y and ξ were disconnected before t_0 and that the coupling is switched on at t_0. Any previous interaction, however, would result in a correlation between $y(t_0)$ and $\xi(t_0)$, so that for given y_0 the $\xi(t_0)$ is biased.

(ii) Under certain conditions there is a unique *stationary* solution of (7.1). It has an autocorrelation function

$$\langle\langle y(t_1)y(t_2)\rangle\rangle^s = K(|t_1 - t_2|).$$

To compute it one may take the solution $y(t)$ with the above initial condition, compute the corresponding $\langle\langle y(t_1)y(t_2)\rangle\rangle$, and take the limit $t_1 \to \infty, t_2 \to \infty$ with fixed $t_1 - t_2$. See XVII, section **6**.

(iii) A difficulty with the first-passage problem for a non-Markov process will be considered in XII.**6**.

d. Analytic solutions are, of course, rare and most treatments deal with approximation methods. Almost all are based on the assumption that the autocorrelation time τ_c of $\xi(t)$ is small. A systematic expansion in powers of τ_c gives in lowest order the white noise result (4.5) in the Stratonovich interpretation (4.8). This approximation to the true process is Markovian.

---

[*] R. Graham, Z. Phys. B **36**, 397 (1977).

Higher approximations give corrections to this Fokker–Planck equation, but it remains of the form of an M-equation. Thus the expansion in τ_c produces to all orders an approximation in the form of a Markov process, see XVI.5.

e. The following device can be used to study colored noise even when τ_c is not small. Take for $\xi(t)$ in (7.1) some Markov process, governed by an M-equation, say

$$\dot{\Pi}(\xi, t) = \int \left\{ \gamma(\xi|\xi')\Pi(\xi', t) - \gamma(\xi'|\xi)\Pi(\xi, t) \right\} d\xi' \equiv \mathbb{W}\Pi. \qquad (7.3)$$

Then the bivariate process $\{y, \xi\}$ is Markovian and the joint probability $\mathscr{P}$ obeys an M-equation

$$\frac{\partial \mathscr{P}(y, \xi, t)}{\partial t} = -\frac{\partial}{\partial y} \left\{ A(y) + C(y)\xi \right\} \mathscr{P} + \mathbb{W}\mathscr{P}. \qquad (7.4)$$

The joint process is a composite Markov process as defined in VII.7.

A convenient choice for ξ is the Ornstein–Uhlenbeck process. Then (7.3) is given by (IV.3.20) and (7.4) becomes

$$\frac{\partial \mathscr{P}(y, \xi, t)}{\partial t} = -\frac{\partial}{\partial y} \left\{ A(y) + C(y)\xi \right\} \mathscr{P} + \gamma \frac{\partial}{\partial \xi} \xi \mathscr{P} + \frac{\Gamma}{2} \frac{\partial^2 \mathscr{P}}{\partial \xi^2}. \qquad (7.5)$$

Alternatively one may introduce the same process through the Langevin equation (1.1); then the joint process is described by two coupled equations:

$$\dot{y} = A(y) + C(y)\xi, \qquad (7.6a)$$

$$\dot{\xi} = -\gamma\xi + L(t). \qquad (7.6b)$$

The parameter $\gamma = 1/\tau_c$ serves to vary the color of the noise.

Another, even simpler choice for ξ is the two-valued Markov process (IV.2.3). The joint master equation reduces to two coupled equations for the pair of single-variable functions $\mathscr{P}(y, \pm 1, t) = P_\pm(y, t)$,

$$\frac{\partial P_+(y, t)}{\partial t} = -\frac{\partial}{\partial y} \left\{ A(y) + C(y) \right\} P_+ - \gamma_+ P_+ + \gamma_- P_-, \qquad (7.7a)$$

$$\frac{\partial P_-(y, t)}{\partial t} = -\frac{\partial}{\partial y} \left\{ A(y) - C(y) \right\} P_- + \gamma_+ P_+ - \gamma_- P_-. \qquad (7.7b)$$

Of course, equations (7.5) and (7.7) are still hard to solve and usually require approximate methods. Moreover it should be clear that this device does not solve the equation (7.1) for every given colored noise. It merely provides a model for investigating qualitatively the effect of positive τ_c. The conclusions obtained may

7. COLORED NOISE

be contingent on the type of noise one has chosen in the model. For instance, with the choice (7.5) the higher orders turn out to be especially simple owing to the vanishing of the higher cumulants in the Ornstein–Uhlenbeck process.[*]

Exercise. Construct the M-equation for

$$\ddot{x} + 2\beta\dot{x} + \omega^2 x = \xi(t) = \text{Ornstein–Uhlenbeck process}.$$

Exercise. Find the probability distribution of a free particle subject to dichotomic kicks:

$$\dot{x} = p, \qquad \dot{p} = \xi(t) = \text{two-valued Markov process},$$

when x and p are given initially.

Exercise. The same equations can be solved for arbitrary $\xi(t)$. Express the characteristic functional of x in that of ξ. Find that $\langle x(t)^2 \rangle \sim t^3$ whenever ξ has a finite autocorrelation time.

Exercise. Construct a Fokker–Planck equation for the process[**]

$$\dot{x} = \alpha x - x^3 + x\xi(t) + L(t),$$

where ξ is Ornstein–Uhlenbeck and L is Langevin. Why is there no Itô-Stratonovich dilemma?

Exercise. Write the M-equation for the joint variables y, ξ obeying

$$\dot{y} = -y + \xi, \qquad \dot{\xi} = -\gamma\xi + L(t). \tag{7.8}$$

Its solution with initial distribution $\delta(y - y_0)\, e^{-\frac{1}{2}\xi^2}/\sqrt{2\pi}$ establishes a marginal distribution $P(y, t)$ for y alone. Show that it obeys, if $\Gamma = 2\gamma$,

$$\frac{\partial P}{\partial t} = \frac{\partial}{\partial y} yP + \frac{1 - e^{-(1+\gamma)t}}{1+\gamma} \frac{\partial^2 P}{\partial y^2}.$$

[Hint: As P must be Gaussian it is sufficient to consider $\langle y \rangle$ and $\langle\langle y^2 \rangle\rangle$.]

Exercise. Consider equations (7.6) with[†]

$$A(y) = -\alpha \tanh y \quad \text{and} \quad C(y) = \beta[\cosh y]^{-1}.$$

Transform this problem into (7.8) and solve it.

---

[*] For an abundant list of references see I. Bena, R. Kawai, C. van den Broeck, K. Lindenberg, Fluctuations and Noise Letters **5/3**, L397 (2005).

[**] E. Peacock-López, F.J. de la Rubia, B.J. West, and K. Lindenberg, Phys. Letters A **136**, 96 (1989).

[†] M.-O. Hongler, Helv. Phys. Acta **52**, 280 (1979).

Chapter X

THE EXPANSION OF THE MASTER EQUATION

Internal noise is described by a master equation. When this equation cannot be solved exactly it is necessary to have a systematic approximation method – rather than the naive Fokker–Planck and Langevin approximations. Such a method will now be developed in the form of a power series expansion in a parameter Ω. In lowest order it reproduces the macroscopic equation and thereby demonstrates how a deterministic equation emerges from the stochastic description.

1. Introduction to the expansion

Only in rare cases is it possible to solve the master equation explicitly. For instance, we have seen that one-step master equations can be solved when the step probabilities r_n and g_n are constant or linear in n, but not otherwise. It is therefore important to find approximation methods, of which the Fokker–Planck approximation is the best known. Many other methods have been suggested in the literature, usually consisting of *ad hoc* prescriptions for cutting off higher moments of the fluctuations, and often determined by the needs or taste of the author rather than by logic. As a consequence they have led to the unreliable and contradictory results mentioned in VIII.1. One thing they have in common, however, namely the idea that somehow the fluctuations are small.

The situation can only be remedied by a systematic approximation method in the form of an expansion in powers of a small parameter. Only in that case does one have an objective measure for the size of the several terms. Our first task is therefore to select a suitable expansion parameter. It must be a parameter occurring in the M-equation, i.e., in the transition probability W. Furthermore, it must govern the size of the fluctuations, and therefore the size of the jumps. We denote this parameter by Ω and choose it in such a way that for *large* Ω the jumps are relatively *small*. In many cases Ω is simply the size of the system.

It will be helpful to demonstrate the expansion method first on a simple example before formulating it in full generality. Consider a volume Ω in which the following chemical reaction is in progress:

$$A \xrightarrow{k} X, \quad 2X \xrightarrow{k'} B. \tag{1.1}$$

1. INTRODUCTION TO THE EXPANSION

The concentration ϕ_A of molecules A is again taken constant and supposed so large that the reverse reactions may be neglected. Alternatively one may imagine that A is continually supplied and B is drained. The reaction scheme (1.1) therefore describes an *open* system, see VII.4. The chemical rate equation for the concentration ϕ of X is

$$\dot{\phi} = k\phi_A - 2k'\phi^2. \tag{1.2}$$

Now consider the reaction mesoscopically. The M-equation is according to (VII.2.4)

$$\dot{p}_n = k\phi_A \Omega(\mathbb{E}^{-1} - 1)p_n + (k'/\Omega)(\mathbb{E}^2 - 1)n(n-1)p_n. \tag{1.3}$$

It is nonlinear and cannot be solved explicitly by the method of VI.6. We shall therefore develop an approximate solution for large Ω. Note that the powers of Ω in the coefficients are written explicitly, so that the constants $k\phi_A$ and k' do not depend on Ω. For convenience we choose units in which $k' = \frac{1}{2}$ and $k\phi_A = 1$:

$$\dot{p}_n = \Omega(\mathbb{E}^{-1} - 1)p_n + (2\Omega)^{-1}(\mathbb{E}^2 - 1)n(n-1)p_n. \tag{1.4}$$

One expects that p_n will have a sharp maximum around the macroscopic value $n = \Omega\phi(t)$ determined by (1.2), with a width of order $n^{1/2} \sim \Omega^{1/2}$. To utilize this insight we set

$$n = \Omega\phi(t) + \Omega^{1/2}\xi. \tag{1.5}$$

Here $\phi(t)$ is a solution of (1.2) and ξ is the new variable replacing n. Accordingly the distribution p_n is now written as a function of ξ,

$$p_n(t) = \Pi(\xi, t). \tag{1.6}$$

The operator $\mathbb{E}$ changes n into $n + 1$ and therefore ξ into $\xi + \Omega^{-1/2}$, so that

$$\mathbb{E} = 1 + \Omega^{-1/2}\frac{\partial}{\partial\xi} + \frac{1}{2}\Omega^{-1}\frac{\partial^2}{\partial\xi^2} + \cdots. \tag{1.7}$$

The time derivative in (1.4) is taken with constant n; that means in the ξ, t plane along the direction given by $d\xi/dt = -\Omega^{1/2}d\phi/dt$. Hence

$$\dot{p}_n = \frac{\partial\Pi}{\partial t} - \Omega^{1/2}\frac{d\phi}{dt}\frac{\partial\Pi}{\partial\xi}. \tag{1.8}$$

Consequently the M-equation (1.4) in the new variable takes the form

$$\frac{\partial\Pi}{\partial t} - \Omega^{1/2}\frac{d\phi}{dt}\frac{\partial\Pi}{\partial\xi} = \Omega\left[-\Omega^{-1/2}\frac{\partial}{\partial\xi} + \frac{1}{2}\Omega^{-1}\frac{\partial^2}{\partial\xi^2} - \cdots\right]\Pi$$

$$+ \frac{1}{2}\Omega^{-1}\left[2\Omega^{-1/2}\frac{\partial}{\partial\xi} + 2\Omega^{-1}\frac{\partial^2}{\partial\xi^2} + \cdots\right]$$

$$\times (\Omega\phi + \Omega^{1/2}\xi)(\Omega\phi + \Omega^{1/2}\xi - 1)\Pi. \tag{1.9}$$

We are now in a position to collect the several powers of Ω. First there are a few large terms proportional to $\Omega^{1/2}$, which might make an expansion of Π in powers of $\Omega^{-1/2}$ illusory. However, each of them involves Π only in the form of a factor $\partial \Pi / \partial \xi$. As a result they cancel if

$$-\frac{d\phi}{dt} = -1 + \phi^2. \tag{1.10}$$

This is the macroscopic equation (1.2) and it is satisfied since we stipulated that for the function ϕ a macroscopic solution should be taken.

Next collect the terms in (1.9) of order Ω^0.

$$\frac{\partial \Pi}{\partial t} = 2\phi \frac{\partial}{\partial \xi} \xi \Pi + \tfrac{1}{2}(1 + 2\phi^2) \frac{\partial^2 \Pi}{\partial \xi^2}. \tag{1.11}$$

This is a linear Fokker–Planck equation, whose coefficients depend on t through $\phi(t)$. It has been solved in VIII.6 and the result was that Π is Gaussian. It therefore suffices to determine the first and second moments of ξ, which contain the most important information anyway. By the usual trick one obtains from (1.1)

$$\partial_t \langle \xi \rangle = -2\phi \langle \xi \rangle, \tag{1.12a}$$

$$\partial_t \langle \xi^2 \rangle = -4\phi \langle \xi^2 \rangle + 1 + 2\phi^2. \tag{1.12b}$$

Thus we have found the distribution of the fluctuations around the macroscopic value. They have been computed to order $\Omega^{-1/2}$ relative to the macroscopic value n, which will be called the *linear noise approximation*. In this order of Ω the noise is Gaussian even in time-dependent states far from equilibrium. Higher corrections are computed in X.6 and they modify the Gaussian character. However, they are of order Ω^{-1} relative to n and therefore of the order of a single molecule.

In particular let us take for ϕ the stationary solution of (1.10); $\phi = \phi^s = 1$. Then (1.11) reduces to a time-independent Fokker–Planck equation whose solution is the Ornstein–Uhlenbeck process. More directly one finds from (1.12b)

$$\langle\langle n^2 \rangle\rangle^s = \Omega \langle \xi^2 \rangle^s = \tfrac{3}{4}\Omega = \tfrac{3}{4}\langle n \rangle^s. \tag{1.13}$$

The factor $\tfrac{3}{4}$ shows that the stationary distribution is narrower than the Poisson distribution with the same average. This fact has caused some discussion[*], but can be understood as follows. The molecules X are created

---

[*] G. Nicolis and I. Prigogine, Proc. Nat'l Acad. Sci. USA **68**, 2102 (1971); A. Nitzan and J. Ross, J. Statist. Phys. **10**, 379 (1974); G. Nicolis, P. Allen, and A. Van Nypelseer, Prog. Theor. Phys. **52**, 1481 (1974); N. Saitô, J. Chem. Phys. **61**, 3644 (1974).

1. INTRODUCTION TO THE EXPANSION

independently of each other, but are annihilated in pairs; they are therefore not statistically independent from one another. If at any time there happen to be more X than on the average, their rate of annihilation is also enhanced, and for two reasons. First there are more candidates for annihilation, which is the normal reason for return to average in linear processes[*]. Secondly, however, the probability for each X to be annihilated is also greater than average, because there are more partners available to form a pair. This additional effect enhances the tendency to return to the average and thereby decreases the probability for large fluctuations. A similar argument applies when their number is below average.

Exercise. The function Π defined by (1.6) is not the probability density of ξ, but differs from it by a normalization factor. Find this factor and verify that no error is made by treating Π as probability density in (1.12).

Exercise. If $u = n/\Omega$ is the over-all density or concentration of X show that

$$\partial_t \langle u \rangle = 1 - \langle u \rangle^2 + \mathcal{O}(\Omega^{-1}), \tag{1.14a}$$

$$\partial_t \langle\langle u^2 \rangle\rangle = -4\langle u \rangle \langle\langle u^2 \rangle\rangle + 1 + \langle u \rangle^2 + \mathcal{O}(\Omega^{-1/2}). \tag{1.14b}$$

Exercise. Show for the reaction $A \to pX$, $qX \to B$ with arbitrary positive integers p, q

$$\langle\langle n^2 \rangle\rangle^s = \frac{p+q}{2q} \langle n \rangle^s.$$

For $p = q$ the result is the same as for a Poisson distribution. Verify that in this case p_n is in fact a Poisson distribution, even though the molecules X are not created and annihilated independently.

Exercise. Show that the exact stationary solution of (1.3) is

$$p_n^s = C \frac{\gamma^n}{n!} I_{n-1}(2\gamma), \tag{1.15}$$

where I_n denotes a modified Bessel function, and

$$\gamma^2 = k n_A \Omega / k', \qquad C^{-1} = \sqrt{2} I_1(2\gamma\sqrt{2}).$$

Exercise. Derive (1.13) from (1.15).
Exercise. Find $\langle\langle n^2 \rangle\rangle^s$ for the reaction

$$A \to pX, \qquad qX \to B + rX,$$

where $p \geqslant 1$, $q > r \geqslant 0$. Explain the result qualitatively.

---

[*] This effect is the one demonstrated by Ehrenfest's urn model (IV.5.4).

2. General formulation of the expansion method

Internal fluctuations are caused by the discrete nature of matter. The density of a gas fluctuates because the gas consists of molecules; fluctuations in a chemical reaction arise because the reaction consists of individual reactive collisions; current fluctuations exist because the current is made up of electrons; radioactive decay fluctuates owing to the individuality of the nuclei. Incidentally, this explains why the formulas for fluctuations in physical systems always contain atomic constants, such as Avogadro's number, the mass of a molecule, or the charge of an electron.

On the other hand, the macroscopic features are determined by all particles together. Thus one expects the importance of fluctuations to be relatively small when the system is large. In fact, this has been amply illustrated by the examples of linear systems treated so far. They led to the rule of thumb that in a collection of N particles the fluctuations are of order $N^{1/2}$. Their effect on the macroscopic properties will therefore be of order $N^{-1/2}$. Thus it is clear that the size of the system is a parameter that measures the relative importance of the fluctuations. We shall therefore introduce a size parameter Ω. The precise definition of Ω depends on the system, but we here formulate its general properties.

The presence of the parameter Ω creates a distinction between two scales. One scale is determined by the size of jumps and will be denoted by the variable X, so that when Ω varies the magnitude of the jumps measured in X remains the same. The other scale is the one on which macroscopic properties of the system are measured and is indicated by $x = X/\Omega$. One therefore expects that the probability for a transition to take place depends on x, i.e., when Ω varies this probability remains the same function of x. When Ω is the volume of the system, the distinction between X and x is the familiar one between extensive and intensive variable respectively. We now express this idea formally.

In the general form of the master equation (V.1.5) the y may be taken to be X or x; we here choose the former and also indicate the explicit dependence on Ω,

$$\dot{P}(X, t) = \int \{W_\Omega(X \mid X')P(X', t) - W_\Omega(X' \mid X)P(X, t)\} \, dX'. \quad (2.1)$$

As in (VIII.2.1) we write W as a function of starting point and jump length $r = X - X'$:

$$W_\Omega(X \mid X') = W_\Omega(X'; X - X') = W_\Omega(X'; r).$$

The dependence on r describes the relative probability of various jump lengths, while the dependence on X' determines the overall transition prob-

2. GENERAL FORMULATION OF THE EXPANSION METHOD

ability. Hence the idea described above is expressed by

$$W_\Omega(X'; X - X') = \Phi\left(\frac{X'}{\Omega}; X - X'\right) = \Phi(x'; r). \tag{2.2}$$

The formula means, of course, that Φ is a function of *two* variables $x' = X'/\Omega$ and r, and does not otherwise contain Ω. Note that

$$W_\Omega(X' \mid X) = \Phi(x; -r).$$

The stringent condition (2.2) can be relaxed. First it is clear that one may allow an arbitrary (positive) factor $f(\Omega)$, since it can always be absorbed in the time scale. It merely means that large systems may evolve more slowly than small systems. Secondly, as we are going to expand in Ω^{-1}, it does no harm when higher orders of Ω^{-1} occur in addition to (2.2). Thus the final form of our assumption is, rather than (2.2),

$$W_\Omega(X \mid X') = f(\Omega)\left\{\Phi_0\left(\frac{X'}{\Omega}; r\right) + \Omega^{-1}\Phi_1\left(\frac{X'}{\Omega}; r\right) + \Omega^{-2}\Phi_2 + \cdots\right\}. \tag{2.3}$$

It turns out that this assumption applies to almost all cases one meets in practice; we shall therefore refer to it as the *canonical form*. When W does not have this canonical form the expansion method does not work. Substitution of (2.3) in the master equation yields

$$\frac{\partial P(X, t)}{\partial t} = f(\Omega)\int\left\{\Phi_0\left(\frac{X-r}{\Omega}; r\right) + \Omega^{-1}\Phi_1\left(\frac{X-r}{\Omega}; r\right) + \cdots\right\}P(X - r, t)\,\mathrm{d}r$$

$$- f(\Omega)\int\left\{\Phi_0\left(\frac{X}{\Omega}; -r\right) + \Omega^{-1}\Phi_1\left(\frac{X}{\Omega}; -r\right) + \cdots\right\}\mathrm{d}r \cdot P(X, t). \tag{2.4}$$

Example. *The Rayleigh particle.* The velocity V is governed by an M-equation with $W(V|V')$ given by (VIII.4.14). The jumps in V are caused by collisions of the gas molecules and are therefore of order $(m/M)v$, where v is a typical velocity characterizing the velocity distribution F. They can be made small by choosing M large; accordingly we take $\Omega = M/m$. The variable in which the jumps remain of the same size is the momentum $P = MV$. This is our variable X, while mV serves as intensive variable x. The transition probability in the variable X is

$$W(X \mid X') = \frac{vA}{M}\left(\frac{M+m}{2m}\right)^2 \frac{|X - X'|}{M} F\left(\frac{M+m}{2mM}X - \frac{M-m}{2mM}X'\right),$$

$$W(X'; r) = vA\left(\frac{M+m}{2mM}\right)^2 |r| F\left(\frac{X'}{M} + \frac{M+m}{2mM}r\right)$$

$$= \frac{vA}{4m^2}(1 + \Omega^{-1})^2 |r| F\left(\frac{X'}{m\Omega} + \tfrac{1}{2}(1 + \Omega^{-1})\frac{r}{m}\right). \tag{2.5}$$

This has the form (2.3) with $f(\Omega) = 1$ and

$$\Phi_0(x; r) = \frac{vA}{4m^2} |r| F\left(\frac{x}{m} + \frac{r}{2m}\right),$$

etcetera.[*]

Exercise. For one-step processes the condition (2.3) for the canonical form is expressed by

$$r_n = f(\Omega) \left[\rho_0\left(\frac{n}{\Omega}\right) + \Omega^{-1} \rho_1\left(\frac{n}{\Omega}\right) + \cdots \right], \quad (2.6a)$$

$$g_n = f(\Omega) \left[\gamma_0\left(\frac{n}{\Omega}\right) + \Omega^{-1} \gamma_1\left(\frac{n}{\Omega}\right) + \cdots \right], \quad (2.6b)$$

where $\rho_0, \rho_1, \ldots$ and γ_0, γ_1 are functions of n/Ω not otherwise depending on Ω.

Exercise. Verify that (V.1.7) and (VI.9.5) have the canonical form (2.6).

Exercise. With respect to the master equation (VI.9.1) argue that N and M are proportional to Ω and α, β proportional to Ω^{-1}:

$$\dot{p}_n = \Omega a(\mathbb{E} - 1)\left(\frac{n}{\Omega}\right)^2 p_n + \Omega b(\mathbb{E}^{-1} - 1)\left(v - \frac{n}{\Omega}\right)\left(\mu - \frac{n}{\Omega}\right) p_n. \quad (2.7)$$

Verify that it has the canonical form (2.6).

Exercise. Show in the same way that the examples (ii), (iii), (iv) of VI.9 have the canonical form.

Exercise. Show that the master equation defined by (VI.9.11) has the canonical form with $\Omega = C$, or more elegantly $\Omega = CkT/e^2$.

Exercise. Treat the example of the Rayleigh particle by choosing $\Omega = (M + m)/2m$. With this choice $\Phi_1 = \Phi_2 = \cdots = 0$.

In order to proceed with the systematic expansion in $1/\Omega$ it is necessary to anticipate the way in which the solution $P(X, t)$ will depend on Ω. The initial condition is

$$P(X, 0) = \delta(X - X_0), \quad (2.8)$$

where in general X_0 will be of order Ω. One expects that at later times $P(X, t)$ is a sharp peak at some position of order Ω (in the X-scale), while its width will be of order $\Omega^{1/2}$ (fig. 23). To express this formally we set

$$X = \Omega\phi(t) + \Omega^{1/2}\xi. \quad (2.9)$$

The first term is macroscopic. The function $\phi(t)$ will have to be adjusted so as to follow the motion of the peak. It will then be found that $P(x, t)$,

---

[*] The problem is worked out in Can. J. Phys. **39**, 551 (1961), and again in R.F. Fox and M. Kac, BioSystems **8**, 197 (1977).

2. GENERAL FORMULATION OF THE EXPANSION METHOD

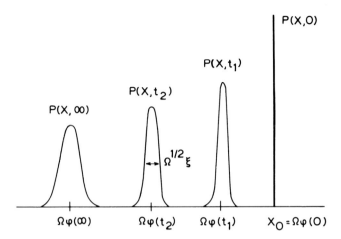

Fig. 23. The evolution of the probability density.

when expressed in ξ, does not depend on Ω to first approximation. That is the justification of the Ansatz (2.9).

This Ansatz is the essential step. The Ω-expansion is not just one out of a plethora of approximation schemes, to be judged by comparison with experimental or numerical results[*]. It is a systematic expansion in $\Omega^{-\frac{1}{2}}$ and is the basis for the existence of a macroscopic deterministic description of systems that are intrinsically stochastic. It justifies as a first approximation the standard treatment in terms of a deterministic equation with noise added, as in the Langevin approach. It will appear that in the lowest approximation the noise is Gaussian, as is commonly postulated. In addition, however, it opens up the possibility of adding higher approximations.

Equation (2.9) is a time-dependent transformation from the variable X to the new variable ξ, involving an as yet undetermined function $\phi(t)$. The function $P(X, t)$ transforms into a function $\Pi(\xi, t)$ of ξ according to

$$P(X, t) = P(\Omega\phi(t) + \Omega^{1/2}\xi, t) = \Pi(\xi, t). \qquad (2.10)$$

The transformation of the derivatives is, compare (1.8),

$$\frac{\partial^\nu \Pi}{\partial \xi^\nu} = \Omega^{\frac{1}{2}\nu} \frac{\partial^\nu P}{\partial X^\nu},$$

$$\frac{\partial \Pi}{\partial t} = \frac{\partial P}{\partial t} + \Omega \frac{d\phi}{dt} \frac{\partial P}{\partial X} = \frac{\partial P}{\partial t} + \Omega^{1/2} \frac{d\phi}{dt} \frac{\partial \Pi}{\partial \xi}.$$

---

[*] M. Gitterman and G.H. Weiss, Physica A **170**, 503 (1991).

X. THE EXPANSION OF THE MASTER EQUATION

The master equation (2.4) expressed in the new variables takes the form

$$\frac{\partial \Pi(\xi, t)}{\partial t} - \Omega^{1/2} \frac{d\phi}{dt} \frac{\partial \Pi}{\partial \xi}$$

$$= f(\Omega) \int \Phi_0(\phi(t) + \Omega^{-1/2}(\xi - \Omega^{-1/2}r); r) \Pi(\xi - \Omega^{-1/2}r, t) \, dr$$

$$+ \Omega^{-1} f(\Omega) \int \Phi_1(\phi(t) + \Omega^{-1/2}(\xi - \Omega^{-1/2}r); r) \Pi(\xi - \Omega^{-1/2}r, t) \, dr + \cdots$$

$$- f(\Omega) \int \Phi_0(\phi(t) + \Omega^{-1/2}\xi; -r) \, dr \cdot \Pi(\xi, t)$$

$$- \Omega^{-1} f(\Omega) \int \Phi_1(\phi(t) + \Omega^{-1/2}\xi; -r) \, dr \cdot \Pi(\xi, t) - \cdots . \qquad (2.11)$$

We are now in a position to repeat the expansion in VIII.2 with the difference that the size of all terms, as measured by their order in $\Omega^{-1/2}$, is manifest. In the first two terms on the right ξ is shifted by $-\Omega^{-1/2}r$. Using a Taylor expansion to take this shift into account one finds that the lowest term is cancelled by the last two lines and the result is

$$\frac{\partial \Pi(\xi, t)}{\partial t} - \Omega^{1/2} \frac{d\phi}{dt} \frac{\partial \Pi}{\partial \xi}$$

$$= -\Omega^{-1/2} f(\Omega) \frac{\partial}{\partial \xi} \int r \Phi_0(\phi(t) + \Omega^{-1/2}\xi; r) \, dr \cdot \Pi(\xi, t)$$

$$+ \tfrac{1}{2} \Omega^{-1} f(\Omega) \frac{\partial^2}{\partial \xi^2} \int r^2 \Phi_0(\phi(t) + \Omega^{-1/2}\xi; r) \, dr \cdot \Pi(\xi, t)$$

$$- \frac{1}{3!} \Omega^{-3/2} f(\Omega) \frac{\partial^3}{\partial \xi^3} \int r^3 \Phi_0(\phi(t) + \Omega^{-1/2}\xi; r) \, dr \cdot \Pi(\xi, t)$$

$$- \Omega^{-3/2} f(\Omega) \frac{\partial}{\partial \xi} \int r \Phi_1(\phi(t) + \Omega^{-1/2}\xi; r) \, dr \cdot \Pi(\xi, t) + \mathcal{O}(\Omega^{-2}). \qquad (2.12)$$

To simplify we define in analogy with (V.8.2) the rescaled jump moments

$$\alpha_{\nu, \lambda}(x) = \int r^\nu \Phi_\lambda(x; r) \, dr \qquad (2.13)$$

and rescale time by setting

$$\Omega^{-1} f(\Omega) t = \tau. \qquad (2.14)$$

2. GENERAL FORMULATION OF THE EXPANSION METHOD

Then the equation becomes

$$\frac{\partial \Pi(\xi, \tau)}{\partial \tau} - \Omega^{1/2} \frac{d\phi}{d\tau} \frac{\partial \Pi}{\partial \xi} = -\Omega^{1/2} \frac{\partial}{\partial \xi} \alpha_{1,0}(\phi(\tau) + \Omega^{-1/2} \xi) \cdot \Pi$$

$$+ \frac{1}{2} \frac{\partial^2}{\partial \xi^2} \alpha_{2,0}(\phi(\tau) + \Omega^{-1/2} \xi) \cdot \Pi$$

$$- \frac{1}{3!} \Omega^{-1/2} \frac{\partial^3}{\partial \xi^3} \alpha_{3,0}(\phi(\tau) + \Omega^{-1/2} \xi) \cdot \Pi$$

$$- \Omega^{-1/2} \frac{\partial}{\partial \xi} \alpha_{1,1}(\phi(\tau) + \Omega^{-1/2} \xi) \cdot \Pi + \mathcal{O}(\Omega^{-1}). \quad (2.15)$$

Expansion of the jump moments finally gives

$$\frac{\partial \Pi(\xi, \tau)}{\partial \tau} - \Omega^{1/2} \frac{d\phi}{d\tau} \frac{\partial \Pi}{\partial \xi}$$

$$= -\Omega^{1/2} \alpha_{1,0}(\phi) \frac{\partial \Pi}{\partial \xi} - \alpha'_{1,0}(\phi) \frac{\partial}{\partial \xi} \xi \Pi - \tfrac{1}{2} \Omega^{-1/2} \alpha''_{1,0}(\phi) \frac{\partial}{\partial \xi} \xi^2 \Pi$$

$$+ \tfrac{1}{2} \alpha_{2,0}(\phi) \frac{\partial^2 \Pi}{\partial \xi^2} + \tfrac{1}{2} \Omega^{-1/2} \alpha'_{2,0}(\phi) \frac{\partial^2}{\partial \xi^2} \xi \Pi$$

$$- \frac{1}{3!} \Omega^{-1/2} \alpha_{3,0}(\phi) \frac{\partial^3 \Pi}{\partial \xi^3}$$

$$- \Omega^{-1/2} \alpha_{1,1}(\phi) \frac{\partial \Pi}{\partial \xi} + \mathcal{O}(\Omega^{-1}). \quad (2.16)$$

Here ϕ stands for $\phi(\tau)$ and the primes indicate derivatives. *This is the systematic expansion of the master equation* and will serve as the starting point for the following sections.

Exercise. The relation between the old jump moments a_ν and the new $\alpha_{\nu, \lambda}$ is

$$a_\nu(X) = f(\Omega) \left[\alpha_{\nu, 0}\left(\frac{X}{\Omega}\right) + \frac{1}{\Omega} \alpha_{\nu, 1}\left(\frac{X}{\Omega}\right) + \cdots \right]. \quad (2.17)$$

Exercise. An alternative way of arriving at (2.16) is by starting from the Kramers–Moyal expansion (VIII.2.6) and carrying out the transformation (2.9), (2.10). Show that the result is the same. (It should be emphasized that the Kramers–Moyal expansion is by no means essential for the derivation of (2.16)!)

Exercise. Find the coefficients α in (2.16) explicitly for the Rayleigh particle using (2.5).

Exercise. The way in which (VI.9.12) depends on Ω is displayed by writing

$$\dot{p}_n = \frac{a}{\Omega}(\mathbb{E} - 1)n^2 p_n + b\Omega(\mathbb{E}^{-1} - 1)p_n. \tag{2.18}$$

Show that the exact stationary solution yields for large Ω

$$\langle\langle n^2 \rangle\rangle^s = \tfrac{1}{2}\langle n \rangle^s = \tfrac{1}{2}\Omega\sqrt{b/a}, \tag{2.19}$$

and agrees with the stationary solution of (2.16) for this case.

3. The emergence of the macroscopic law

At first sight, (2.16) is not a proper expansion for large Ω, because it contains two terms of order $\Omega^{1/2}$. That would mean that our Ansatz (2.9) is wrong[*]. On second thought, however, one notices that both these terms involve Π only through the factor $\partial\Pi/\partial\xi$; it is therefore possible to make them cancel each other by choosing ϕ such that

$$\frac{d\phi}{d\tau} = \alpha_{1,0}(\phi). \tag{3.1}$$

Since our aim is to solve the master equation with initial conditions (2.8), in which X_0 is of order Ω, the initial value of ϕ should be given by

$$\phi(0) = X_0/\Omega = x_0. \tag{3.2}$$

The function ϕ determined by (3.1) and (3.2) is the one to be used in the transformation (2.9). It determines the macroscopic part of X in such a way that the fluctuations around it are of order $\Omega^{1/2}$, as will be shown in the next section. Hence (3.1) is the macroscopic equation.

Remark. The equation (3.1) is not quite identical with (V.8.6), which was previously called the macroscopic equation, inasmuch as the latter also includes terms of order Ω^{-1}. The question which of the two is the correct one is moot, because terms of order Ω^{-1} (even of order $\Omega^{-1/2}$) can be transferred from the macroscopic part of (3.2) into the fluctuating part. To put it differently: the location of the peak $P(X, t)$ is not defined more precisely than permitted by its width, which is of order $\Omega^{1/2}$. Of course, one might agree to *define* the location as $\langle X \rangle$, or as the maximum of the peak, but there is no logical necessity for this and in the case of nonlinear processes it is awkward.

Normal usage, however, gives the edge to (3.1) rather than (V.8.6). In the case of a chemical reaction for instance, Ω is the volume; clearly the chemical rate equations

---

[*] It also means that we are faced with a singular perturbation problem: the time derivative is not among the largest terms. This explains why a straightforward perturbation calculation does not work but has to be replaced by the more sophisticated device described in this chapter.

3. THE EMERGENCE OF THE MACROSCOPIC LAW

apply to infinite volume and contain no terms of order Ω^{-1}. In the case of the Rayleigh particle one might be tempted to include higher orders in m/M in the macroscopic damping law, but their physical significance is doubtful, since they are much smaller than the fluctuations.

The stationary solutions of (3.1) are the roots of

$$\alpha_{1,0}(\phi) = 0. \tag{3.3}$$

There may be any number of them. *From the fact that the master equation (unless decomposable or splitting) has a single stationary solution P^s one cannot conclude that the macroscopic equation cannot have more than one stationary macroscopic state*, as will be seen in XIII.1.

There may also be various types of zeros of $\alpha_{1,0}$. They can best be reviewed in a hodogram, i.e., by plotting $\dot{\phi} \equiv \alpha_1(\phi)$ versus ϕ (fig. 24). Although the time is not represented one knows that ϕ must increase in the points above the ϕ-axis and decrease in the points below it, as indicated by the arrows. From fig. 24a it is clear that a stationary solution ϕ^s in which $\alpha_1'(\phi^s) < 0$ is

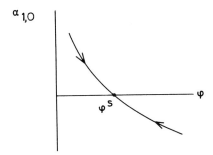

Fig. 24a. Stable stationary solution of the macroscopic equation.

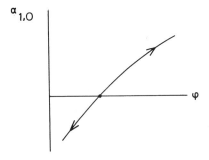

Fig. 24b. Unstable stationary solution.

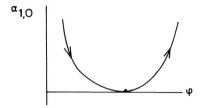

Fig. 24c. Another form of instability.

approached by all solutions in its neighborhood and is therefore *stable**[*]*. A stationary solution of the type of fig. 24b is clearly unstable. Other types occur when $\alpha'_{1,0}(\phi^s) = 0$, an example of which is shown in fig. 24c.

For the present discussion we suppose that (3.1) has only one solution and that it is stable. In fact we suppose more strongly that there is a positive constant h such that

$$\alpha'_{1,0}(\phi) \leqslant -h < 0 \quad \text{for all } \phi. \tag{3.4}$$

Then all solutions of (3.1) tend to ϕ^s at least as fast as $e^{-h\tau}$, see fig. 25. This stability condition is crucial for the present expansion. Chapters XI and XIII deal with cases where (3.4) does not hold.

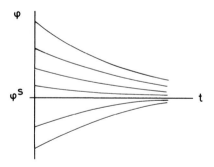

Fig. 25. Neighboring solutions tend to the (asymptotically) stable ϕ^s.

One other concept from the theory of differential equations will be needed. Suppose one has a solution $\phi(\tau)$ of (3.1) determined by some initial value $\phi(0)$. Consider a second solution with initial value $\phi(0) + \delta\phi(0)$, where $\delta\phi(0)$ is small. The difference $\delta\phi(\tau)$ between both solutions obeys, to first order

---

[*] Until further notice we use "stable" for what is technically called "asymptotically stable in the sense of Lyapunov"; see, e.g., J. La Salle and S. Lefshetz, *Stability by Liapunov's Direct Method* (Academic Press, New York 1961).

3. THE EMERGENCE OF THE MACROSCOPIC LAW

in $\delta\phi$,

$$\frac{d}{d\tau}\delta\phi = \alpha'_1(\phi(\tau))\,\delta\phi. \tag{3.5}$$

This linear equation for $\delta\phi$ can easily be solved when $\phi(\tau)$ is known. It is called the *linearized* or *variational equation* associated with (3.1). When it turns out that the solutions of (3.5) tend to zero as $\tau \to \infty$ it follows that this particular solution $\phi(\tau)$ of (3.1) is stable for small perturbations, or "locally stable". Clearly (3.5) cannot tell anything about global stability, i.e., the effect of large perturbations. One can only conclude from the local stability that $\phi(\tau)$ has a certain "domain of attraction": every solution starting inside this domain will tend to $\phi(\tau)$ for large τ. In this chapter, however, we postulate (3.4), which guarantees global stability.

Exercise. Find both versions (3.1) and (V.8.6) of the macroscopic equation belonging to the equation (VI.9.5). Check that the former is the familiar rate equation for the reaction.

Exercise. For the Rayleigh particle in the previous section the versions (3.1) and (V.8.6) of the macroscopic damping law coincide if one chooses $\Omega = (M + m)/2m$, but not if $\Omega = M/m$. In the latter case find the difference between $\alpha_1(V)$ and $\alpha_{1,0}(V)$.

Exercise. Show that (3.4) guarantees that all solutions of (3.1) are globally stable. Construct an example where all solutions are globally stable and yet (3.4) does not hold.

Exercise. Show that the autocatalytic chemical reaction[*]

$$A + 2X \rightleftharpoons 3X, \qquad X \rightleftharpoons B \tag{3.6}$$

has at least one stable macroscopically stationary state.

Exercise. Show that (VI.9.7) has one unstable solution. All other solutions are locally stable but not globally.

Exercise. Define a "potential function" $V(x)$ by setting $\alpha_1(\phi) = -V'(\phi)$. The condition (3.4) implies for $\phi \neq \phi^s$

$$V(\phi) > V(\phi^s) \quad \text{and} \quad \partial_t V(\phi) < 0.$$

Show that $V(\phi)$ is a Lyapunov function[**], which can be used to prove the global stability.

Exercise. V obeys even the stronger condition[†] $V''(\phi) > 0$. The properties of V together imply (3.4).

Exercise. For a one-step process the macroscopic equation is in the notation (2.6)

$$\dot\phi = \gamma_0(\phi) - \rho_0(\phi). \tag{3.7}$$

---

[*] F. Schlögl, Z. Phys. **253**, 147 (1972). See also H.K. Janssen, Z. Phys. **270**, 67 (1974); I. Matheson, D.F. Walls, and C.W. Gardiner, J. Statist. Phys. **12**, 21 (1975).
[**] La Salle and Lefshetz, loc. cit.
[†] F. Schlögl, Z. Phys. **343**, 303 (1971).

4. The linear noise approximation

Having taken care of the terms with $\Omega^{1/2}$ in (2.16) we are left with an equation for $\Pi(\xi, \tau)$ which is a proper expansion in $\Omega^{-1/2}$. The terms of order Ω^0 are

$$\frac{\partial \Pi(\xi, \tau)}{\partial \tau} = -\alpha'_{1,0}(\phi) \frac{\partial}{\partial \xi} \xi \Pi + \tfrac{1}{2} \alpha_{2,0}(\phi) \frac{\partial^2 \Pi}{\partial \xi^2}. \tag{4.1}$$

This is a linear Fokker–Planck equation whose coefficients depend on time through ϕ. This approximation was christened in section 1 "linear noise approximation". The solution of (4.1) was found in VIII.6 to be a Gaussian*), so that it suffices to determine the first and second moments of ξ. On multiplying (4.1) by ξ and ξ^2, respectively, one obtains

$$\partial_\tau \langle \xi \rangle = \alpha'_{1,0}(\phi) \langle \xi \rangle, \tag{4.2a}$$

$$\partial_\tau \langle \xi^2 \rangle = 2\alpha'_{1,0}(\phi) \langle \xi^2 \rangle + \alpha_{2,0}(\phi). \tag{4.2b}$$

Note that the variance obeys the same equation as $\langle \xi^2 \rangle$,

$$\partial_\tau \langle\langle \xi^2 \rangle\rangle = 2\alpha'_{1,0}(\phi) \langle\langle \xi^2 \rangle\rangle + \alpha_{2,0}(\phi). \tag{4.2c}$$

These equations determine both moments, provided their initial values are known. Our aim was to solve the master equation with initial delta distribution (2.8) and we took x_0 as initial value of the macroscopic part, see (3.2). It follows that the initial fluctuations vanish,

$$\langle \xi \rangle_0 = \langle \xi^2 \rangle_0 = \langle\langle \xi^2 \rangle\rangle_0 = 0. \tag{4.3}$$

Summary. The aim of solving the master equation with initial condition (2.1) has now been achieved in the linear noise approximation by the following three steps.

(i) Solve the macroscopic equation (3.1) with initial condition (3.2); call the solution $\phi(\tau | x_0)$.

(ii) Insert the $\phi(\tau | x_0)$ found into (4.2) and solve (4.2) with initial conditions (4.3).

(iii) Use the results to find the mean and variance of the original variable X,

$$\langle X \rangle_\tau = \Omega \phi(\tau | x_0) + \Omega^{1/2} \langle \xi \rangle_\tau, \tag{4.4a}$$

$$\langle\langle X^2 \rangle\rangle_\tau = \Omega \langle\langle \xi^2 \rangle\rangle_\tau, \tag{4.4b}$$

and take for $P(X, t)$ the Gaussian having this mean and variance.

---

*) The linear noise approximation may therefore also be called "the Gaussian approximation", but it should be clear that the Ω-expansion *derives* the Gaussian character rather than postulating it.

4. THE LINEAR NOISE APPROXIMATION

A few comments will clarify the situation.

a. From (4.2a), (4.3) one sees $\langle \xi \rangle_t = 0$ and hence, from (4.4a) and (3.1),

$$\partial_t \langle x \rangle = \alpha_{1,0}(\langle x \rangle) + \mathcal{O}(\Omega^{-1}). \tag{4.5}$$

Thus *in the linear noise approximation the average obeys the macroscopic law.*

b. The equations (3.1) and (4.2) can be solved, i.e., their solution can be obtained by a number of integrations as was previously mentioned in connection with (V.8.15). This is no longer true when there are more variables.

c. It is not necessary to take the initial value of ϕ identical with the location x_0 of the initial delta peak. A difference of order $\Omega^{-1/2}$ is permissible, since it can be taken into account as initial value of ξ. The equations (4.2) must then be solved with this new ϕ and the new initial values of $\langle \xi \rangle$ and $\langle \xi^2 \rangle$, although, of course, $\langle\langle \xi^2 \rangle\rangle_0$ remains zero.

d. This freedom is in particular useful when computing the autocorrelation function of the fluctuations in the stationary state, i.e.,

$$\langle\langle x(0)x(t) \rangle\rangle^s = \langle \{x_0 - x^s\}\{\langle x(t) \rangle_{x_0} - x^s\} \rangle^s. \tag{4.6}$$

In order to compute the conditional average $\langle x(t) \rangle_{x_0}$ in this expression it is not necessary to carry out the general scheme and solve (3.1) with initial condition $\phi(0) = x_0$. Rather one may take $\phi = \phi^s$ throughout and solve (4.2a) with initial value

$$\Omega^{-1/2}\langle \xi \rangle_0 = x_0 - x^s.$$

This is possible, because the values of x_0 that one needs in (4.6) differ from x^s by an amount of order $\Omega^{-1/2}$. The result is according to (4.2), for $t \geq 0$,

$$\langle\langle x(0)x(t) \rangle\rangle^s = \Omega^{-1}\langle \xi(0)\xi(t) \rangle^s$$
$$= \Omega^{-1}\langle \xi(0)^2 \rangle^s \exp[-|\alpha'_{1,0}(\phi^s)|\tau]$$
$$= \frac{1}{\Omega} \frac{\alpha_{2,0}(\phi^s)}{2|\alpha'_{1,0}(\phi^s)|} \exp[-|\alpha'_{1,0}(\phi^s)|\tau]. \tag{4.7}$$

This is the general formula (in the linear noise approximation) for the autocorrelation function of the fluctuations in a stable stationary state. *Hence it is possible to write down the fluctuation spectrum in an arbitrary system without solving any specific equations.* This fact is the basis of the customary noise theory.

e. The *a posteriori* justification of the Ansatz (2.9) consists in the observation that the resulting equation (4.1) does not contain Ω, so that the fluctuations are indeed of the order postulated in (2.9). If, however, the fluctuations grow with time, the result is applicable only during a limited period, until the fluctuations are of the same magnitude as the macroscopic part in spite of the fact that they are multiplied by the factor $\Omega^{1/2}$ rather than Ω. Our

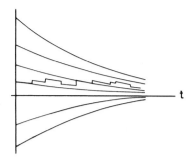

Fig. 26. The convergence of the macroscopic solution curves prevents the fluctuations from growing.

condition (3.4), however, which guarantees the stability of the macroscopic solutions, also guarantees that the solutions of (4.2) are bounded.

f. Figure 26 provides an intuitive explanation why fluctuations do not grow when the macroscopic solutions are stable. The macroscopic solutions converge; by a fluctuation the system may jump from one of the curves to a neighboring one, but the effect disappears in the course of time since the distance between both curves tends to zero.

In equilibrium, or in a stable state, the magnitude of the fluctuations is the outcome of the competition between the jumps and the macroscopic return to equilibrium. Both effects are represented by the second and first term, respectively, on the right of (4.2b). This is the basis of the Einstein relation (VIII.3.9) and of the fluctuation–dissipation theorem.

g. Equation (4.2a) for the average decrease of an initial fluctuation is identical with the variational equation (3.5) associated with the macroscopic equation. In linear approximation around equilibrium this means that the regression of fluctuations is governed by the macroscopic law. That is the assumption used by Onsager in his derivation of the reciprocal relations[*]. Thus (4.2a) is the generalization of the Onsager assumption to time-dependent states of nonlinear stable systems.

h. Our condition (3.4) ensured that there is just one ϕ^s, and that it is globally stable, i.e., all other solutions tend towards it. Some macroscopic laws possess a locally stable ϕ^s, that is, a solution of (3.3) having a limited domain of attraction, such that only solutions in this domain tend to ϕ^s. The points ϕ_a and ϕ_c in fig. 27 are such locally stable stationary solutions, as is clear from fig. 28. Well inside a domain of attraction the Ω-expansion can still be used for finding the fluctuations, in particular those around ϕ^s.

---

[*] L. Onsager, Phys. Rev. **37**, 405 and **38**, 2265 (1931); H.B.G. Casimir, Rev. Mod. Phys. **17**, 343 (1945); DE GROOT AND MAZUR.

4. THE LINEAR NOISE APPROXIMATION

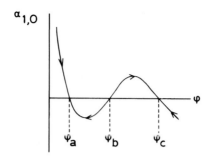

Fig. 27. Two locally stable stationary solutions separated by an unstable one.

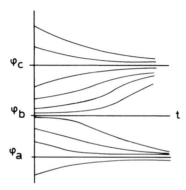

Fig. 28. All solutions other than ϕ_b itself tend to either ϕ_a or ϕ_c.

But there is also a small probability for a fluctuation to carry the system outside the region of attraction never to come back to ϕ^s. The probability for this to happen is typically of order $e^{-\Omega}$, and therefore usually very small, except near the boundary of the region of attraction. Clearly such a term can never be found in any series expansion in powers of $\Omega^{-1/2}$, and entirely different methods are needed (see chapter XIII). This obvious limitation has caused some undue misgivings about the validity of the Ω-expansion[*] and even of the master equation itself[**].

i. The correctness of the macroscopic law (3.1) has been proved rigorously in the following sense.[†] Choose a time interval $(0, T)$ and a permitted error $\delta > 0$; then the probability that for all $t \in (0, T)$ the actual value of x differs from $\phi(t)$ by no more than δ, tends to 1 as $\Omega \to \infty$. It was also proved that

---

[*] W. Horsthemke and L. Brenig, Z. Phys. **27**, 341 (1977); W. Horsthemke, M. Malek-Mansour, and L. Brenig, Z. Phys. B **28**, 135 (1977).

[**] M. Malek-Mansour and G. Nicolis, J. Statist. Phys. **13**, 197 (1975).

[†] T.G. Kurtz, J. Appl. Prob. **7**, 49 (1970); **8**, 344 (1971); J. Chem. Phys. **57**, 2976 (1972).

the error tends to a Gaussian, as given by the linear noise approximation. Note, however, that here *T is fixed before Ω goes to infinity*, which means that nothing is said about the behavior for long times. Hence the proofs equally apply to the unstable cases treated in the next two chapters, but then they merely describe the initial transient period (XI.1.3) rather than the behavior one is really interested in.

j. No nonlinear Fokker–Planck equation of the form (VIII.1.1) ever enters into the picture and hence also no nonlinear Langevin equation (IX.4.5) and no Itô–Stratonovich dilemma. A nonlinear Fokker–Planck equation will appear in the next chapter, when the stability condition (3.4) no longer applies.

Even in the present case, however, it is possible to manufacture a nonlinear Fokker–Planck equation, and a corresponding Langevin equation, which *as far as the linear noise approximation is concerned* reproduce the same results as found here.[*] But any features they contain beyond that approximation are spurious. For instance, one cannot conclude from this manufactured Fokker–Planck equation that the stationary distribution is given by (VIII.1.4).

Exercise. Write the linear noise approximation of the solution $P(X, t)$ with the delta initial value (2.8) explicitly in terms of ϕ and the solution of (4.2).

Exercise. Apply the linear noise approximation to the M-equation (VI.3.12).[**]

Exercise. Verify that *the linear noise approximation always leads to an Ornstein–Uhlenbeck process for the fluctuations in a stable stationary state.*

Exercise. Apply the Ω-expansion to (2.18) and derive

$$\frac{d}{dt}\frac{\langle n\rangle}{\Omega} = b - a\left(\frac{\langle n\rangle}{\Omega}\right)^2 + \mathcal{O}(\Omega^{-1}),$$

$$\frac{d}{dt}\frac{\langle\langle n^2\rangle\rangle}{\Omega} = -4a\frac{\langle n\rangle}{\Omega}\frac{\langle\langle n^2\rangle\rangle}{\Omega} + a\left(\frac{\langle n\rangle}{\Omega}\right)^2 + b + \mathcal{O}(\Omega^{-1/2}),$$

$$\langle\langle n(0)n(t)\rangle\rangle^s = \tfrac{1}{2}\langle n\rangle^s \exp[-2\sqrt{ab}\,t].$$

Note that experimentally one cannot determine a and b by measuring only $\langle n\rangle^s$ and $\langle\langle n^2\rangle\rangle^s$, but one has to observe the fluctuation spectrum as well.

Exercise. Solve (4.2) explicitly.

Exercise. Show that the terms $\Phi_1, \Phi_2, \ldots$ in (2.4) do not contribute to the linear noise approximation. Hence the distinction between (3.1) and (V.8.6) is irrelevant in this approximation.

Exercise. To solve the master equation with initial condition (2.8), take as initial value for the macroscopic value $\phi(0) = x_0 - c\Omega^{-1/2}$. Show that the resulting $\langle x\rangle_t$ is the same as when taking $\phi(0) = x_0$.

---

[*] Z.A. Akcasu, J. Statist. Phys. **16**, 33 (1977); N.G. van Kampen, J. Statist. Phys. **25**, 431 (1981).

[**] C. Schat, G. Abramson, and H.S. Wio, Amer. J. Phys. **59**, 357 (1991).

5. EXPANSION OF A MULTIVARIATE MASTER EQUATION

Exercise. The spreading of an epidemic in a population of Ω individuals, of which n are infected and $\Omega - n$ are not, may be described in the simplest case by $r(n) = 0$ (no cure) and $g(n) = \beta n(1 - n/\Omega)$. Find $\langle n \rangle$ and $\langle\langle n^2 \rangle\rangle$ as functions of time.

Exercise. Prove that (V.8.12) with (V.8.9) – or the alternative (V.8.15) – is a consistent approximation; in fact one has

$$\dot{y} = a_1(y) + \tfrac{1}{2}\sigma_y^2 a_1''(y) + \mathcal{O}(\Omega^{-1/2}), \tag{4.8a}$$

$$\partial_t \sigma_y^2 = 2a_1'(y)\sigma_y^2 + a_2(y) + \mathcal{O}(\Omega^{1/2}). \tag{4.8b}$$

Exercise. Verify that (VIII.5.6) is nothing but the linear noise approximation of (VIII.5.1).

5. Expansion of a multivariate master equation

In the case of more than one fluctuating quantity the Ω-expansion can be carried out in much the same way as for one variable. The main difference is that the macroscopic equations are more involved and cannot normally be solved explicitly in terms of a quadrature. Rather than give the general formulation we shall demonstrate the multivariate Ω-expansion on a suitable example.

Consider the following chemical reaction in an open system, involving two reactants X and Y,

$$A \xrightarrow{\alpha} X,$$

$$2X \xrightarrow{\gamma} Y,$$

$$Y \xrightarrow{\beta} B. \tag{5.1}$$

The first line describes production of atoms X from a fixed reservoir A at a constant rate $\alpha\Omega$. The macroscopic equations are

$$\dot{n}_X = \alpha\Omega - 2(\gamma/\Omega)n_X^2, \tag{5.2a}$$

$$\dot{n}_Y = (\gamma/\Omega)n_X^2 - \beta n_Y. \tag{5.2b}$$

We are interested in the fluctuations around these macroscopic values and therefore introduce the joint probability distribution $P(n_X, n_Y, t)$. It obeys the M-equation

$$\dot{P} = \alpha\Omega(\mathbb{E}_X^{-1} - 1)P + (\gamma/\Omega)(\mathbb{E}_X^2 \mathbb{E}_Y^{-1} - 1)n_X^2 P + \beta(\mathbb{E}_Y - 1)n_Y P. \tag{5.3}$$

$\mathbb{E}_X$ and $\mathbb{E}_Y$ are, of course, the step operators acting on n_X and n_Y. The equation is nonlinear, which makes an Ω-expansion necessary.

The transformation (2.9) is now extended to

$$n_X = \Omega\phi(t) + \Omega^{1/2}\xi, \qquad n_Y = \Omega\psi(t) + \Omega^{1/2}\eta, \qquad P(n_X, n_Y, t) = \Pi(\xi, \eta, t).$$

Substitution in the master equation yields, to the required order,

$$\frac{\partial\Pi}{\partial t} - \Omega^{1/2}\frac{d\phi}{dt}\frac{\partial\Pi}{\partial\xi} - \Omega^{1/2}\frac{d\psi}{dt}\frac{\partial\Pi}{\partial\eta}$$

$$= \alpha\Omega\left\{-\Omega^{-1/2}\frac{\partial}{\partial\xi} + \tfrac{1}{2}\Omega^{-1}\frac{\partial^2}{\partial\xi^2}\right\}\Pi$$

$$+ \frac{\gamma}{\Omega}\left\{\Omega^{-1/2}\left(2\frac{\partial}{\partial\xi} - \frac{\partial}{\partial\eta}\right) + \tfrac{1}{2}\Omega^{-1}\left(2\frac{\partial}{\partial\xi} - \frac{\partial}{\partial\eta}\right)^2\right\}(\Omega\phi + \Omega^{1/2}\xi)^2\Pi$$

$$+ \beta\left\{\Omega^{-1/2}\frac{\partial}{\partial\eta} + \tfrac{1}{2}\Omega^{-1}\frac{\partial^2}{\partial\eta^2}\right\}(\Omega\psi + \Omega^{1/2}\eta)\Pi. \tag{5.4}$$

Again the successive powers of Ω may be treated separately.

The terms of order $\Omega^{1/2}$ are all either proportional to $\partial\Pi/\partial\xi$ or to $\partial\Pi/\partial\eta$. The conditions for the terms of each type to vanish are

$$-\dot\phi = -\alpha + 2\gamma\phi^2, \tag{5.5a}$$

$$-\dot\psi = -\gamma\phi^2 + \beta\psi. \tag{5.5b}$$

These are the two macroscopic equations (5.2). They can be solved explicitly and one easily finds that all solutions tend to

$$\phi^s = \sqrt{\alpha/2\gamma}, \qquad \psi^s = \alpha/2\beta. \tag{5.6}$$

The solution curves in the (ϕ, ψ)-plane are sketched in fig. 29. Evidently all solutions tend to the stationary point, so that they are globally stable.

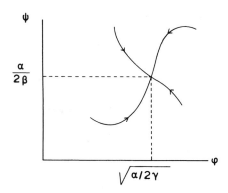

Fig. 29. Solution curves of (5.5).

5. EXPANSION OF A MULTIVARIATE MASTER EQUATION

The terms of order Ω^0 in (5.4) yield

$$\frac{\partial \Pi}{\partial t} = 2\gamma\phi\left(2\frac{\partial}{\partial \xi} - \frac{\partial}{\partial \eta}\right)\xi\Pi + \beta\frac{\partial}{\partial \eta}\eta\Pi$$
$$+ \tfrac{1}{2}\alpha\frac{\partial^2 \Pi}{\partial \xi^2} + \tfrac{1}{2}\gamma\phi^2\left(2\frac{\partial}{\partial \xi} - \frac{\partial}{\partial \eta}\right)^2\Pi + \tfrac{1}{2}\beta\psi\frac{\partial^2 \Pi}{\partial \eta^2}. \tag{5.7}$$

This is a multivariate linear Fokker–Planck equation of the type solved in VIII.6. We use it to determine the moments of ξ and η,

$$\partial_t\langle\xi\rangle = -4\gamma\phi\langle\xi\rangle, \tag{5.8a}$$
$$\partial_t\langle\eta\rangle = 2\gamma\phi\langle\xi\rangle - \beta\langle\eta\rangle. \tag{5.8b}$$

These equations are again the same as the variational equations associated with the macroscopic equations (5.5).

There are three moments of the second degree and they obey three coupled equations:

$$\partial_t\langle\xi^2\rangle = -8\gamma\phi\langle\xi^2\rangle + \alpha + 4\gamma\phi^2, \tag{5.9a}$$
$$\partial_t\langle\eta^2\rangle = 4\gamma\phi\langle\xi\eta\rangle - 2\beta\langle\eta^2\rangle + \gamma\phi^2 + \beta\psi, \tag{5.9b}$$
$$\partial_t\langle\xi\eta\rangle = -4\gamma\phi\langle\xi\eta\rangle + 2\gamma\phi\langle\xi^2\rangle - \beta\langle\xi\eta\rangle - 2\gamma\phi^2. \tag{5.9c}$$

They determine the variances and the covariance of the fluctuations of n_X and n_Y around the values of the macroscopic solution of (5.5). Rather than study the general time-dependent state, however, we concentrate on the stationary state.

To find the fluctuations in the stationary state one has to substitute for ϕ and ψ their stationary values (5.6). For simplicity we use the freedom in time unit to put $\sqrt{2\alpha\gamma} = 1$. Then the equations (5.8) for the first moments become

$$\partial_t\langle\xi\rangle = -2\langle\xi\rangle, \tag{5.10a}$$
$$\partial_t\langle\eta\rangle = \langle\xi\rangle - \beta\langle\eta\rangle. \tag{5.10b}$$

The equations for the second moments are conveniently written in matrix form,

$$\frac{d}{dt}\begin{bmatrix}\langle\xi^2\rangle \\ \langle\eta^2\rangle \\ \langle\xi\eta\rangle\end{bmatrix} = \begin{bmatrix}-4 & 0 & 0 \\ 0 & -2\beta & 2 \\ 1 & 0 & -2-\beta\end{bmatrix}\begin{bmatrix}\langle\xi^2\rangle \\ \langle\eta^2\rangle \\ \langle\xi\eta\rangle\end{bmatrix} + \begin{bmatrix}3\alpha \\ \alpha \\ -\alpha\end{bmatrix}. \tag{5.11}$$

As a first conclusion from these equations one finds

$$\langle \xi^2 \rangle^s = \tfrac{3}{4}\alpha, \quad \langle \eta^2 \rangle^s = \frac{3+2\beta}{4\beta(2+\beta)}\alpha, \quad \langle \xi\eta \rangle^s = \frac{-\alpha}{4(2+\beta)}. \quad (5.12)$$

Rewritten in the original quantities this tells us

$$\langle\langle n_X^2 \rangle\rangle^s = \tfrac{3}{4}\langle n_X \rangle^s, \quad (5.13\text{a})$$

$$\langle\langle n_Y^2 \rangle\rangle^s = \frac{3+2\beta}{4+2\beta}\langle n_Y \rangle^s, \quad (5.13\text{b})$$

$$\langle\langle n_X n_Y \rangle\rangle^s = -\frac{1}{4(2+\beta)}\langle n_X \rangle^s. \quad (5.13\text{c})$$

The first line shows that the fluctuations of X are the same as in the reaction (1.1), compare (1.13). This could have been predicted, because as far as X is concerned the present reaction scheme is the same. The third line shows that there is a negative correlation between X and Y, an obvious consequence of the fact that each time when two atoms X associate, n_X decreases and n_Y increases. Finally (5.13b) shows that the variance of n_Y is intermediate between the value belonging to a Poisson distribution and the value belonging to n_X.

As a second conclusion from the moment equations we compute the autocorrelation matrix. Using (5.10a) and (5.12) one first finds

$$\langle \xi(0)\xi(t) \rangle^s = \langle \xi^2 \rangle^s e^{-2t} = \tfrac{3}{4}\alpha\, e^{-2t},$$

$$\langle \eta(0)\xi(t) \rangle^s = \langle \eta\xi \rangle^s e^{-2t} = -\frac{\alpha}{4(2+\beta)} e^{-2t}.$$

Secondly, the solution of (5.10b),

$$\langle \eta \rangle_t = \langle \eta \rangle_0 e^{-\beta t} + \langle \xi \rangle_0 \frac{e^{-2t} - e^{-\beta t}}{\beta - 2},$$

leads to

$$\langle \eta(0)\eta(t) \rangle^s = \frac{(3+2\beta)\alpha}{4\beta(2+\beta)} e^{-\beta t} - \frac{\alpha}{4(2+\beta)} \frac{e^{-2t} - e^{-\beta t}}{\beta - 2},$$

$$\langle \xi(0)\eta(t) \rangle^s = \frac{-\alpha}{4(2+\beta)} e^{-\beta t} + \tfrac{3}{4}\alpha \frac{e^{-2t} - e^{-\beta t}}{\beta - 2}.$$

These correlations contain two relaxation times. The first one is $1/2$ and is due to the rate at which fluctuations of X disappear. The second one, $1/\beta$, is due to the rate at which the molecules Y disappear.

Exercise. Construct the hodogram of fig. 29 by drawing in the (ϕ, ψ)-plane the curves on which $\dot\phi = 0$ and $\dot\psi = 0$ and indicating the direction of the solution in the four parts of the plane.

Exercise. Find an analytic expression for the solution curves of (5.5).

Exercise. The solution of (5.11) requires the diagonalization of a 3×3 matrix (which in this example was trivial). However, the eigenvalues can be found from those of (5.10). How?

Exercise. Investigate the special case $\beta = 2$. Find in particular the spectral density of the fluctuations in n_Y.

Exercise. Perform the same calculation for the chemical reaction

$$A \to X, \quad X \to Y, \quad X + Y \to B.$$

Exercise. Also for the reaction[*]

$$A \to X, \quad X + Y \to 2Y, \quad Y \to B.$$

Exercise. A population is infected by an epidemic. It has m healthy individuals and n infectives. The rate equations are

$$\dot m = \beta\Omega - \gamma mn/\Omega, \quad \dot n = \gamma mn/\Omega - \alpha n.$$

Interpret these equations. Construct the M-equation. Find the variances and covariance in the stationary state[**].

Exercise. Same questions for a population of m males and n females with the macroscopic equations

$$\dot m = \beta n - \alpha m - \gamma m(m+n)/\Omega, \quad \dot n = \beta n - \alpha n - \gamma n(m+n)/\Omega.$$

Exercise. For the bivariate, doubly autocatalytic reaction

$$A + X \to X + Y, \quad X + Y \to Y$$

the time-dependent equation can be solved explicitly.

6. Higher orders

In this section we examine the higher orders beyond the linear noise approximation. They add terms to the fluctuations that are of order Ω^{-1} relative to the macroscopic quantities, i.e., of the order of a single particle. They also modify the macroscopic equation by terms of that same order, as has been anticipated in (V.8.12) and (4.8). These effects are obviously unimportant for most practical noise problems, but cannot be ignored in two cases. First, they tell us how equilibrium fluctuations are affected by the presence of nonlinear terms in the macroscopic equation, in particular how

---

[*] R.F.Fox, Proc. Nat'l. Acad. Sci. USA **76**, 2114 (1979).
[**] The macroscopic solutions are analyzed by H.W. Hethcote, in: *Mathematical Problems in Biology* (Lecture Notes in Biomathematics **2**; P.van den Driessche ed., Springer, Berlin 1974).

the noise in an electronic device is affected by the curvature of the I–V characteristic. Secondly, they are essential when single-particle events are actually observed, e.g., in the recombination current in photoconductors (see XV.2). Unless the reader is interested in these topics we advise him to skip this section, because the algebra is rather formidable.

Consider the equation (2.16) for the evolution of $\Pi(\xi, \tau)$. The terms of order $\Omega^{1/2}$ were taken care of in **3** by stipulating that ϕ obeys the equation (3.1). The terms of order Ω^0 were given in (4.1), and we now add the higher orders, writing explicitly all terms to order Ω^{-1}.

$$\frac{\partial \Pi}{\partial \tau} = -\frac{\partial}{\partial \xi}\left[\alpha'_{1,0}\xi + \tfrac{1}{2}\Omega^{-1/2}\alpha''_{1,0}\xi^2 + \frac{1}{3!}\Omega^{-1}\alpha'''_{1,0}\xi^3 + \cdots\right.$$

$$\left. + \Omega^{-1/2}\alpha_{1,1} + \Omega^{-1}\alpha'_{1,1}\xi + \cdots\right]\Pi$$

$$+ \frac{1}{2}\frac{\partial^2}{\partial \xi^2}[\alpha_{2,0} + \Omega^{-1/2}\alpha'_{2,0}\xi + \tfrac{1}{2}\Omega^{-1}\alpha''_{2,0}\xi^2 + \cdots + \Omega^{-1}\alpha_{2,1} + \cdots]\Pi$$

$$- \frac{1}{3!}\frac{\partial^3}{\partial \xi^3}[\Omega^{-1/2}\alpha_{3,0} + \Omega^{-1}\alpha'_{3,0}\xi + \cdots]\Pi$$

$$+ \frac{1}{4!}\frac{\partial^4}{\partial \xi^4}[\Omega^{-1}\alpha_{4,0} + \cdots]\Pi. \tag{6.1}$$

All coefficients are functions of the argument $\phi(\tau)$ and the primes indicate derivatives with respect to that argument.

While the linear noise approximation led to the linear Fokker–Planck equation (4.1) we now see that the higher powers of $\Omega^{-1/2}$ give rise to three modifications.

(i) They add higher powers to the coefficients, so that it is no longer true that the first one is linear and the second one constant.

(ii) *At the same time they add higher derivatives.* It is therefore inconsistent to write a Fokker–Planck equation in which the full functional dependence of α_1 and α_2 is retained but the higher derivatives are cut off.

(iii) The higher orders in the jump moments, i.e., $\alpha_{1,1}, \alpha_{1,2}, \ldots, \alpha_{2,1}, \ldots$, which are due to $\Phi_1, \Phi_2, \ldots$ enter successively. In particular $\alpha_{1,1}$ adds to the first coefficient a term independent of ξ. This complication could be avoided by not expanding Φ and writing everywhere instead of our $\alpha_{v,0}$ the entire jump moments α_v. But that would amount to including higher orders of $\Omega^{-1/2}$ than is consistent with the other approximations. In particular it would also mean using (V.8.6) rather than (3.1), compare the remark in section 3.

6. HIGHER ORDERS

Remark. The original master equation that we are approximating has the properties that it conserves positivity and that all solutions tend to the stationary one. A second order Fokker–Planck equation has these properties too. However, when higher derivatives are added these properties are lost. Suppose we take the expansion (6.1) to order Ω^{-1}, that is we omit all terms represented by dots. The resulting fourth-order differential equation no longer has these two essential properties.[*] That does not mean that (6.1) is incorrect or useless: it merely means that one should not draw conclusions from it that involve higher powers of $\Omega^{-1/2}$ than (6.1) purports to describe[**]. A safe way to avoid this is by treating all terms of order $\Omega^{-1/2}$ and Ω^{-1} in (6.1) as perturbations, to be added as corrections to the solutions of (4.1). In fact, in practice this is usually the only way of obtaining a solution of (6.1). The easiest way to arrive at the desired results, however, is to write the moment equations, as will now be done.

The terms written explicitly in (6.1) are sufficient to compute $\langle \xi \rangle$ and $\langle \xi^2 \rangle$ to order Ω^{-1}. For this purpose we deduce from (6.1)

$$\partial_t \langle \xi \rangle = \alpha'_{1,0} \langle \xi \rangle + \tfrac{1}{2} \Omega^{-1/2} \alpha''_{1,0} \langle \xi^2 \rangle + \frac{1}{3!} \Omega^{-1} \alpha'''_{1,0} \langle \xi^3 \rangle$$
$$+ \Omega^{-1/2} \alpha_{1,1} + \Omega^{-1} \alpha'_{1,1} \langle \xi \rangle + \mathcal{O}(\Omega^{-3/2}), \quad (6.2\text{a})$$

$$\partial_t \langle \xi^2 \rangle = 2\alpha'_{1,0} \langle \xi^2 \rangle + \Omega^{-1/2} \alpha''_{1,0} \langle \xi^3 \rangle + \tfrac{1}{3} \Omega^{-1} \alpha'''_{1,0} \langle \xi^4 \rangle$$
$$+ 2\Omega^{-1/2} \alpha_{1,1} \langle \xi \rangle + 2\Omega^{-1} \alpha'_{1,1} \langle \xi^2 \rangle$$
$$+ \alpha_{2,0} + \Omega^{-1/2} \alpha'_{2,0} \langle \xi \rangle + \tfrac{1}{2} \Omega^{-1} \alpha''_{2,0} \langle \xi^2 \rangle$$
$$+ \Omega^{-1} \alpha_{2,1} + \mathcal{O}(\Omega^{-3/2}). \quad (6.2\text{b})$$

As the third and fourth moment appear on the right we also write the equations for them to the required order:

$$\partial_t \langle \xi^3 \rangle = 3\alpha'_{1,0} \langle \xi^3 \rangle + \tfrac{3}{2} \Omega^{-1/2} \alpha''_{1,0} \langle \xi^4 \rangle + 3\Omega^{-1/2} \alpha_{1,1} \langle \xi^2 \rangle$$
$$+ 3\alpha_{2,0} \langle \xi \rangle + \Omega^{-1/2} \alpha'_{2,0} \langle \xi^2 \rangle + \Omega^{-1/2} \alpha_{3,0} + \mathcal{O}(\Omega^{-1}), \quad (6.2\text{c})$$

$$\partial_t \langle \xi^4 \rangle = 4\alpha'_{1,0} \langle \xi^4 \rangle + 6\alpha_{2,0} \langle \xi^2 \rangle + \mathcal{O}(\Omega^{-1/2}). \quad (6.2\text{d})$$

Note that the omission of terms of order $\Omega^{-3/2}$ in $\langle \xi \rangle$ and $\langle \xi^2 \rangle$ has the effect that the hierarchy of moment equations automatically breaks off.

The strategy of solving these equations is clear. One first solves (6.2a) and (6.2b) to lowest order, i.e., omitting all terms of order $\Omega^{-1/2}$ and higher. The

---

[*] That no differential equation of higher order than Fokker–Planck can describe the evolution of a probability density rigorously has been proved by R.F. Pawula, Phys. Rev. **162**, 186 (1967).

[**] A numerical demonstration is given by H. Risken and H.D. Vollmer, Z. Phys. B **35**, 313 (1979).

resulting $\langle \xi^2 \rangle$ can be substituted in (6.2a) to find $\langle \xi \rangle$ to order $\Omega^{-1/2}$; and in (6.2d) to find $\langle \xi^4 \rangle$ in lowest approximation. These two results permit one to solve (6.2c) so as to find $\langle \xi^3 \rangle$ to order $\Omega^{-1/2}$. The information gathered so far then permits one to determine the right-hand members of (6.2a) and (6.2b) to order Ω^{-1}, and $\langle \xi \rangle$ and $\langle \xi^2 \rangle$ can be found to this order. The work is formidable for fluctuations around a non-stationary state, but becomes manageable for stationary fluctuations, in which case all coefficients α are constants.

As an application we compute the autocorrelation function of the fluctuations in the stationary state of the semiconductor defined by (2.18). The macroscopic equation is $\dot{\phi} = b - a\phi^2$ and has one stationary solution $\phi^s = \sqrt{b/a}$. The jump moments are

$$\alpha_\nu(\phi) = \alpha_{\nu,0}(\phi) = b + (-1)^\nu a\phi^2.$$

The equation (6.1) can easily be written down for $\phi = \phi^s$. To save writing we rescale the variables,

$$2t\sqrt{ab} = \theta, \qquad (a/b)^{1/4}\xi = \eta, \qquad (a/b)^{1/4}\Omega^{-1/2} = \varepsilon. \tag{6.3}$$

Then (6.1) takes the form

$$\frac{\partial \Pi}{\partial \theta} = \frac{\partial}{\partial \eta}(\eta + \tfrac{1}{2}\varepsilon\eta^2)\Pi + \frac{1}{2}\frac{\partial^2}{\partial \eta^2}(1 + \varepsilon\eta + \tfrac{1}{2}\varepsilon^2\eta^2)\Pi$$

$$+ \tfrac{1}{3}\varepsilon^2 \frac{\partial^3}{\partial \eta^3}\eta\Pi + \tfrac{1}{24}\varepsilon^2 \frac{\partial^4 \Pi}{\partial \eta^4}. \tag{6.4}$$

The equations for the moments are

$$\partial_\theta \langle \eta \rangle = -\langle \eta \rangle - \tfrac{1}{2}\varepsilon\langle \eta^2 \rangle + \mathcal{O}(\varepsilon^3), \tag{6.5a}$$

$$\partial_\theta \langle \eta^2 \rangle = -2\langle \eta^2 \rangle - \varepsilon\langle \eta^3 \rangle + 1 + \varepsilon\langle \eta \rangle + \tfrac{1}{2}\varepsilon^2\langle \eta \rangle + \mathcal{O}(\varepsilon^3), \tag{6.5b}$$

$$\partial_\theta \langle \eta^3 \rangle = -3\langle \eta^3 \rangle - \tfrac{3}{2}\varepsilon\langle \eta^4 \rangle + 3\langle \eta \rangle + 3\varepsilon\langle \eta^2 \rangle + \mathcal{O}(\varepsilon^2), \tag{6.5c}$$

$$\partial_\theta \langle \eta^4 \rangle = -4\langle \eta^4 \rangle + 6\langle \eta^2 \rangle + \mathcal{O}(\varepsilon). \tag{6.5d}$$

From this we first conclude, by setting the left-hand side equal to zero,

$$\langle \eta^4 \rangle^s = \tfrac{3}{2}\langle \eta^2 \rangle^s = \tfrac{3}{4} + \mathcal{O}(\varepsilon), \tag{6.6a}$$

$$\langle \eta^3 \rangle^s = -\tfrac{1}{8}\varepsilon + \mathcal{O}(\varepsilon^2), \tag{6.6b}$$

$$\langle \eta^2 \rangle^s = \tfrac{1}{2} - \tfrac{1}{16}\varepsilon^2 + \mathcal{O}(\varepsilon^3), \tag{6.6c}$$

$$\langle \eta \rangle^s = -\tfrac{1}{4}\varepsilon + \mathcal{O}(\varepsilon^3). \tag{6.6d}$$

Subsequently we solve the time-dependent equations, in the way indicated

6. HIGHER ORDERS

above, for given initial value η_0 so that

$$\langle\eta\rangle_0 = \eta_0, \quad \langle\eta^2\rangle_0 = \eta_0^2, \quad \langle\eta^3\rangle_0 = \eta_0^3, \quad \langle\eta^4\rangle_0 = \eta_0^4.$$

The resulting $\langle\eta(\theta)\rangle_{\eta_0}$ to order ε^2 is then utilized together with (6.6c) to construct the autocorrelation function,

$$\langle\langle\eta_0\langle\eta(\theta)\rangle_{\eta_0}\rangle\rangle^s = \tfrac{1}{2}(1 - \tfrac{1}{2}\varepsilon^2)\,\mathrm{e}^{-(1-\frac{1}{4}\varepsilon^2)\theta} + \tfrac{1}{8}\varepsilon^2\,\mathrm{e}^{-(2+\varepsilon^2)\theta}. \tag{6.7}$$

The coefficient and exponents are correct to order $\varepsilon^2 \sim \Omega^{-1}$.

Conclusions. (i) From (6.6d) it appears that in this order *the average of n no longer coincides with its macroscopic value*, not even in the stationary state:

$$\langle n\rangle^s = \Omega\phi^s + \Omega^{1/2}\langle\xi\rangle^s$$
$$= \Omega\sqrt{b/a} - \tfrac{1}{4}.$$

The difference, however, is one quarter of an electron!

(ii) From (6.7) it appears that the terms of order Ω^{-1} lower the decay rate of the leading term and also modify its coefficient. More strikingly, however, another exponential term appears, which decays roughly twice as fast. Thus *the nonlinearity gives rise to an additional Debye term in the fluctuation spectrum*, which in principle could be observed. Higher orders give rise to a sequence of such terms.

(iii) The *microscopic* treatment of nonlinearity fluctuations has led to a different conclusion, namely that nonlinearity does not affect the fluctuation spectrum at all[*]. Admittedly the present mesoscopic treatment is founded on the assumption that the master equation applies, but there are doubts about the justification of the microscopic treatment as well.[**] It would be of interest to perform an *experimentum crucis*.

Exercise. Show that $\partial_t P(x,t) = a\,\partial_x^4 P$ does not tend to a stationary solution when $a > 0$ and does not conserve positivity when $a < 0$. Extend this conclusion to general differential equations with constant coefficients.

Exercise. Let $\dot P = \mathbb{W}P$ be an equation conserving positivity. Suppose $\mathbb{W} = \mathbb{W}_0 + \varepsilon\mathbb{W}_1 + \varepsilon^2\mathbb{W}_2 + \cdots$, where $\mathbb{W}_0$ also conserves positivity, but $\mathbb{W}_0 + \varepsilon\mathbb{W}_1$ does not. Show that this can be remedied by including otherwise spurious higher orders in the following way[†]. Set $\mathbb{W} = \mathbb{U}^2$ and $\mathbb{U} = \mathbb{U}_0 + \varepsilon\mathbb{U}_1$ such that $(\mathbb{U}_0 + \varepsilon\mathbb{U}_1)^2$ coincides with $\mathbb{W}_0 + \varepsilon\mathbb{W}_1$ as far as terms of order ε are concerned.

---

[*] W. Bernard and H.B. Callen, Rev. Mod. Phys. **31**, 1017 (1959); Phys. Rev. **118**, 1466 (1960).
[**] N.G. van Kampen, Physica Norvegica **5**, 279 (1971).
[†] A. Siegel, J. Mathem. Phys. **1**, 378 (1960).

Exercise. At first sight the third and fourth derivatives in (6.1) are irrelevant for computing the first and second moments of ξ. Why is it nevertheless true that they cannot be ignored when computing an autocorrelation function such as (6.7)?

Exercise. Compute for (1.4)
$$\langle n \rangle^s = \Omega + \tfrac{1}{8} + \tfrac{3}{128}\Omega^{-1} + \mathcal{O}(\Omega^{-2}),$$
$$\langle\langle n^2 \rangle\rangle^s = \tfrac{3}{4}\Omega + \tfrac{5}{64} + \mathcal{O}(\Omega^{-1}).$$

Chapter XI

THE DIFFUSION TYPE

It has been shown that the lowest order of the Ω-expansion yields the macroscopic equation, and the next order the linear noise approximation, provided that the stability condition (X.3.4) holds. This condition is violated, albeit marginally, when $\alpha_{1,0}(\phi) = 0$. In this case the Ω-expansion takes an entirely different form: its lowest approximation is a nonlinear Fokker–Planck equation.

1. Master equations of diffusion type

An important subclass of the master equations that violate the stability condition (X.3.4) is formed by those for which

$$\alpha_{1,0}(\phi) \equiv 0. \tag{1.1}$$

According to the macroscopic equation (X.3.1) this means that $\phi(\tau) = \text{const.} = \phi(0)$. Small deviations $\delta\phi(0)$ of the initial value give rise to a $\delta\phi(\tau)$ that remains constant rather than tending to zero. Thus the macroscopic solutions are unstable[*] and one expects the fluctuations to grow. In fact, according to (X.4.2) their variance grows linearly,

$$\langle \xi^2 \rangle = \alpha_{2,0}(\phi)\tau. \tag{1.2}$$

If the initial distribution is a delta peak the fluctuations will become of the same order as the macroscopic part after a time of order

$$\tau \sim \Omega/\alpha_{2,0}(\phi). \tag{1.3}$$

After this transient period the Ω-expansion as developed so far breaks down, and the separation (X.2.9) into a macroscopic part and small fluctuations is no longer possible.

On the other hand, it may be expected that, *after* the transient period (1.3), an approximation method could be based on the idea that P is a smoothly varying function of X with a width of order Ω. Accordingly we now choose as our variable the intensive quantity $x = X/\Omega$ and write the

---

[*] In the Lyapunov classification they are called "stable but not asymptotically stable". In the theory of fluctuations it is more natural to classify this case as unstable, pursuant to the footnote in X.3.

master equation (X.2.4) in terms of this variable:

$$\frac{\partial P(x,t)}{\partial t} = f(\Omega) \int \left\{ \Phi_0\left(x - \frac{r}{\Omega}; r\right) + \frac{1}{\Omega} \Phi_1\left(x - \frac{r}{\Omega}; r\right) + \cdots \right\} P\left(x - \frac{r}{\Omega}, t\right) dr$$

$$- f(\Omega) \int \left\{ \Phi_0(x; -r) + \frac{1}{\Omega} \Phi_1(x; -r) + \cdots \right\} dr \cdot P(x,t).$$

Expansion in Ω^{-1} yields, with the aid of (X.2.13) and (1.1),

$$\frac{\partial P(x,t)}{\partial t} = \Omega^{-2} f(\Omega) \left[-\frac{\partial}{\partial x} \alpha_{1,1}(x) P + \frac{1}{2} \frac{\partial^2}{\partial x^2} \alpha_{2,0}(x) P \right]$$

$$+ \Omega^{-3} f(\Omega) \left[\frac{1}{2} \frac{\partial^2}{\partial x^2} \alpha_{2,1}(x) P - \frac{1}{3!} \frac{\partial^3}{\partial x^3} \alpha_{3,0}(x) P - \frac{\partial}{\partial x} \alpha_{2,1}(x) P \right]$$

$$+ f(\Omega) \mathcal{O}(\Omega^{-4}). \tag{1.4}$$

Comparison with (X.2.16) shows the following drastic differences. The dominant term of (X.2.16) is absent and therefore no equation for the macroscopic part of X can be extracted. In other words, on the macroscopic scale the system does not evolve in one direction rather than the other. The remaining evolution of P is merely the net outcome of the fluctuations. Accordingly the time scale of the change is a factor Ω^{-1} slower than in the preceding case, compare (X.2.14). Since P is not sharply peaked the coefficients $\alpha(x)$ cannot be expanded around some central value ϕ but they remain as nonlinear functions in the equation. The first line of (1.4) contains the main terms and is called *the diffusion approximation*

$$\frac{\partial P(x,\tau)}{\partial \tau} = -\frac{\partial}{\partial x} \alpha_{1,1}(x) P + \frac{1}{2} \frac{\partial^2}{\partial x^2} \alpha_{2,0}(x) P, \tag{1.5}$$

where our new time variable is now, in contrast with (X.2.14),

$$\tau = \Omega^{-2} f(\Omega) t. \tag{1.6}$$

The diffusion approximation (1.5) is the nonlinear Fokker–Planck equation (VIII.2.5). In fact, we have now justified the derivation in VIII.2 by demonstrating that it is actually the first term of a systematic expansion in Ω^{-1} *for those master equations that have the property* (1.1). Only under that condition is it true that the two coefficients

$$\langle \Delta x \rangle_x / \Delta t \quad \text{and} \quad \langle (\Delta x)^2 \rangle_x / \Delta t$$

are of the same order in Ω. In the usual derivation of the Fokker–Planck

1. MASTER EQUATIONS OF DIFFUSION TYPE

equation as a small jump limit this is assumed either tacitly[*] or explicitly[**]. It is also implied in the conditions on which Kolmogorov's proof is based.

Summary. The special class of master equations characterized by (1.1) will be said to be of *diffusion type*. For such master equations the Ω-expansion leads to the nonlinear Fokker–Planck equation (1.5), rather than to a macroscopic law with linear noise, as found in the previous chapter for master equations characterized by (X.3.4). The definition of both types presupposes that the transition probabilities have the canonical form (X.2.3), but does not distinguish between discrete and continuous ranges of the stochastic variable. The Ω-expansion leads uniquely to the well-defined equation (1.5) and is therefore immune from the interpretation difficulties of the Itô equation mentioned in IX.4 and IX.5.

The traditional derivation of the Fokker–Planck equation (1.5) or (VIII.1.1) is based on Kolmogorov's mathematical proof, which assumes infinitely many infinitely small jumps. In nature, however, all jumps are of some finite size. Consequently $\mathbb{W}$ is never a differential operator, but always of the type (V.1.1). Usually it also has a suitable expansion parameter and has the canonical form (X.2.3). If it then happens that (1.1) holds, the expansion leads to the nonlinear Fokker–Planck equation (1.5) as the lowest approximation. There is no justification for attributing a more fundamental meaning to Fokker–Planck and Langevin equations than in this approximate sense.

Example. Consider the continuous-time, symmetric random walk

$$\dot{p}_n = p_{n+1} + p_{n-1} - 2p_n.$$

We want to let the step length become small and therefore introduce a rescaled "macroscopic" variable $x = n/\Omega$. The probability density $P(x, t)$ obeys

$$\dot{P}(x, t) = P\left(x + \frac{1}{\Omega}, t\right) + P\left(x - \frac{1}{\Omega}, t\right) - 2P(x, t). \tag{1.7}$$

Accordingly

$$W(x; r) = \delta\left(r - \frac{1}{\Omega}\right) + \delta\left(r + \frac{1}{\Omega}\right),$$

which has the canonical form (X.2.2). Moreover

$$\alpha_1(x) = 0, \qquad \alpha_2(x) = \frac{2}{\Omega^2}.$$

Hence the master equation (1.7) is of diffusion type, and the lowest order in $1/\Omega$ is

---
[*] As in VIII.2, or in M. Planck, loc. sit.
[**] See, e.g., FELLER I, p. 323, compare VIII.2.

given by (1.5) with $\tau = \Omega^{-2} t$:

$$\frac{\partial P(x, \tau)}{\partial \tau} = \frac{\partial^2 P}{\partial x^2}. \tag{1.8}$$

It is not surprising that the diffusion equation appears, because physical diffusion is nothing but a random walk with small steps, although not necessarily all of the same size as in this simple model.

Exercise. For one-step processes whose jump probabilities have the form (X.2.6) the condition (1.1) states

$$\rho_0 \equiv \gamma_0. \tag{1.9}$$

Exercise. Verify that the first line of (1.4) is, indeed, identical with (VIII.2.5) by showing that

$$\frac{\langle \Delta x \rangle_x}{\Delta t} = \Omega^{-2} f(\Omega) \alpha_{1,1}(x), \qquad \frac{\langle (\Delta x)^2 \rangle_x}{\Delta t} = \Omega^{-2} f(\Omega) \alpha_{2,0}(x).$$

Exercise. Show that (1.8) is in agreement with (I.7.7).

Exercise. The asymmetric random walk (VIII.2.8) obeys neither (X.3.4) nor (1.1), but can be reduced to the latter type. Show that the result coincides with the treatment in VIII.2.

Exercise. The same reduction can be used whenever $\alpha_{1,0} = $ constant.

2. Diffusion in an external field

Suppose a particle moves (in one direction) in a potential field $V(X)$, which is periodic with minima separated by sharp maxima (fig. 30). The minima are the sites at which the particle resides for long periods, making occassional jumps to neighboring sites. The jump probability per unit time of this one-step process is $r_n = g_n = A$, where A involves the Arrhenius factor $\exp(-\beta E_a)$, mentioned in VII.5, see also XIII.2. This is just the example in the preceding section.[*]

Fig. 30. Periodic $V(X)$ with sharp maxima.

---

[*] For the relation with experiments, see P.A. Egelstaff, Advances in Physics **11**, 203 (1962). See also Haus and Kehr, op. cit. in VI.2.

2. DIFFUSION IN AN EXTERNAL FIELD

Now suppose that an external potential $U(x)$ is added, which varies on a macroscopic scale and is therefore written as a function of x. The number of sites per macroscopic unit of length is our parameter Ω. The presence of $U(x)$ alters the heights of potential barriers. The new jump probabilities are

$$g_n = A \exp\left[-\beta\left\{U\left(\frac{n+\frac{1}{2}}{\Omega}\right) - U\left(\frac{n}{\Omega}\right)\right\}\right], \tag{2.1a}$$

$$r_n = A \exp\left[-\beta\left\{U\left(\frac{n-\frac{1}{2}}{\Omega}\right) - U\left(\frac{n}{\Omega}\right)\right\}\right]. \tag{2.2b}$$

The matrix of transition probabilities is

$$W_{nn'} = A\delta_{n,n'+1}\left\{1 - \frac{\beta}{2\Omega}U'\left(\frac{n}{\Omega}\right) + \cdots\right\}$$
$$+ A\delta_{n,n'-1}\left\{1 + \frac{\beta}{2\Omega}U'\left(\frac{n}{\Omega}\right) + \cdots\right\}. \tag{2.2}$$

It has the canonical form (X.2.3) with

$$\Phi_0(x; r) = A(\delta_{r,1} + \delta_{r,-1}),$$
$$\Phi_1(x; r) = \tfrac{1}{2}\beta A U'(x)(-\delta_{r,1} + \delta_{r,-1}).$$

From this one obtains

$$\alpha_{1,0} = 0, \qquad \alpha_{2,0} = 2A, \qquad \alpha_{1,1} = -\beta A U'(x). \tag{2.3}$$

Thus the equation is of diffusion type; the lowest approximation is given by (1.5):

$$\frac{\partial P(x,t)}{\partial \tau} = \frac{\partial}{\partial x} U'(x) P + \theta \frac{\partial^2 P}{\partial x^2}, \tag{2.4}$$

where $\tau = \beta A t/\Omega^2$, and $\theta = kT$. This is the equation for diffusion in an external field, see (VIII.3.10).

As a modification of the model consider a potential $V(X)$ with sharp minima separated by flat maxima (fig. 31). Now suppose an external potential

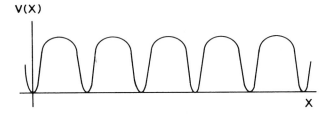

Fig. 31. Periodic $V(X)$ with flat maxima.

$U(x)$ is added, which is an increasing function of x. One sees that this leaves r_n practically equal to A but changes g_n into

$$g_n = A \exp\left[-\beta\left\{U\left(\frac{n+1}{\Omega}\right) - U\left(\frac{n}{\Omega}\right)\right\}\right] = A\{1 - \Omega^{-1}\beta U'(x)\}. \quad (2.5)$$

Consequently (**2.3**) remains valid and therefore also the final equation (**2.4**). Thus the result is insensitive to the precise form of V, provided it has pits in which the particle stays long enough to enable random interactions to wipe out all correlations between successive jumps.

Exercise. The equation (**2.4**) for the probability of one particle may be interpreted as the diffusion equation for independent particles. Show that the equilibrium distribution and the diffusion constant agree with the expressions that could have been deduced directly from the model, without the intervening calculations.

Exercise. Both models (**2.1**) and (**2.5**) satisfy detailed balance. What is in this hopping model the most general choice for r_n, g_n permitted by detailed balance?

Exercise. A pendulum obeying the equation $M\ddot{x} = -\sin x$ is suspended in air, which causes damping and fluctuations. Show that it obeys the bivariate nonlinear Fokker–Planck equation, or Kramers equation,

$$\frac{\partial P(x, v, t)}{\partial t} = -v\frac{\partial P}{\partial x} + \frac{\sin x}{M}\frac{\partial P}{\partial v} + \gamma\left(\frac{\partial}{\partial v}vP + \frac{kT}{M}\frac{\partial^2 P}{\partial v^2}\right). \quad (2.6)$$

Which quantity serves as parameter Ω?

Exercise. Solve the two-dimensional generalization of (**2.4**) with the electrostatic potential $U = -2q \log r$. What is the physical interpretation of the fact that no P^e exists?

Remark. The quasilinear diffusion equation (**2.4**) has an interesting transformation property. Write it in the form (**VIII.1.11**) taking $\theta = 1$ for convenience,

$$\frac{\partial P}{\partial \tau} = \frac{\partial}{\partial x} e^{-U(x)} \frac{\partial}{\partial x} e^{U(x)} P. \quad (2.7)$$

On multiplying with $e^{\frac{1}{2}U}$ one sees that the function $\psi(x, \tau) \equiv e^{\frac{1}{2}U} P$ obeys

$$\frac{\partial \psi}{\partial \tau} = \left\{e^{\frac{1}{2}U}\frac{\partial}{\partial x}e^{-\frac{1}{2}U}\right\}\left\{e^{-\frac{1}{2}U}\frac{\partial}{\partial x}e^{\frac{1}{2}U}\right\}\psi \quad (2.8)$$

$$= -D^+ D\psi. \quad (2.9a)$$

We have defined

$$D = e^{-\frac{1}{2}U}\frac{\partial}{\partial x}e^{\frac{1}{2}U}, \qquad D^+ = -e^{\frac{1}{2}U}\frac{\partial}{\partial x}e^{-\frac{1}{2}U}. \quad (2.9b)$$

Let ψ be a solution, then the function $\chi \equiv D\psi$ obeys

$$\frac{\partial \chi}{\partial \tau} = -DD^+ \chi. \quad (2.10)$$

Write this as an equation for $F(x, \tau) = e^{\frac{1}{2}U} \chi$,

$$\frac{\partial F}{\partial \tau} = -\frac{\partial}{\partial x} U'(x) F + \frac{\partial^2 F}{\partial x^2}. \tag{2.11}$$

Hence we have constructed a solution of the original equation with reversed potential.

Exercise. Explain the minus sign in (2.9).
Exercise. One easily finds explicitly

$$F(x, \tau) = \frac{\partial}{\partial x} e^{U(x)} P(x, \tau).$$

Hence F is not a probability density as it cannot be everywhere nonnegative.
Exercise. Each eigenfunction P_λ of (2.7) produces an eigenfunction F_λ of (2.11) with the same eigenvalue λ, with the exception of the stationary solution P_0.
Exercise. If in (2.8) one replaces τ by it it becomes a Schrödinger equation of a particle in a potential $V(x)$ given by

$$2V = \tfrac{1}{4} U'^2 - \tfrac{1}{2} U''.$$

Each eigenfunction P_λ of (2.7) produces an eigenfunction of the Schrödinger equation. This is the same correspondence as (VIII.7.15).
Exercise. Similarly (2.10) is a Schrödinger equation with the potential W,

$$2W = \tfrac{1}{4} U'^2 + \tfrac{1}{2} U''.$$

The connection between the potentials V and W is called the "Darboux transformation"[*] or "supersymmetry"[**].

3. Diffusion in an inhomogeneous medium

As an example in which the second coefficient α_2 is not constant consider diffusion in a medium whose diffusion coefficient varies in space. There are two plausible ways in which one can heuristically generalize the ordinary diffusion equation. In one dimension x, one may write either

$$\frac{\partial P(x, t)}{\partial t} = \frac{\partial^2}{\partial x^2} D(x) P \tag{3.1}$$

or

$$\frac{\partial P(x, t)}{\partial t} = \frac{\partial}{\partial x} D(x) \frac{\partial P}{\partial x}. \tag{3.2}$$

---
[*] W.M. Zheng, J. Mathem. Phys. **25**, 88 (1984).
[**] H.R. Jauslin, J. Phys. A **21**, 2337 (1988); B. Roy, P. Roy, and R. Roychoudhury, Fortschritte Phys. **39**, 211 (1991).

They differ by a term

$$\frac{\partial}{\partial x}\left[\frac{dD(x)}{dx}P\right], \tag{3.3}$$

which has the form of a flow term, again called the "spurious flow".[*] It is true that (3.2) is more attractive because it can be written

$$\frac{\partial P}{\partial t} = -\frac{\partial J}{\partial x}, \quad J = -D\frac{\partial P}{\partial x}, \tag{3.4}$$

but that is not a decisive argument. To find which one is right we shall start from a precise model and carry out the expansion in Ω. This will actually lead to (3.2), but, of course, it cannot be concluded that this is always the correct choice. In other cases, such as the diffusion of hot electrons in velocity space[**] or diffusion in a space-dependent temperature[†], one cannot make an a priori choice without examining the underlying mechanism.

For our model we suppose the following diffusion mechanism. The particle is trapped in a position x' until it gets a kick, which sends it traveling to the right or to the left. The probabilities per unit time for this to happen are $\beta(x')$ and $\alpha(x')$, respectively. It travels until it meets another trap; let the probability that there is a trap between x and $x + dx$ be $\Omega\gamma(x)\,dx$. The α, β, γ are fixed functions of the macroscopic space variable x, while Ω is a measure for the number of traps per unit length and may be scaled up. By the same considerations as in IV.6 one obtains for the probability that a particle starting at x' is trapped between x and $x + dx$, supposing $x > x'$,

$$\Omega\gamma(x)\,dx \cdot \exp\left[-\Omega\int_{x'}^{x}\gamma(x'')\,dx''\right].$$

In the scale $X = \Omega x$ of the jump size one has, if $X > X'$,

$$W(X \mid X') = \Omega\beta(x')\gamma(x)\exp\left[-\Omega\int_{x'}^{x}\gamma(x'')\,dx''\right]. \tag{3.5a}$$

Similarly for $X < X'$

$$W(X \mid X') = \Omega\alpha(x')\gamma(x)\exp\left[-\Omega\int_{x}^{x'}\gamma(x'')\,dx''\right]. \tag{3.5b}$$

---

[*] This is similar to, but not identical with the dilemma of IX.5.

[**] P.J. Price, in: *Fluctuation Phenomena in Solids* (R.E. Burgess, ed., Academic Press, New York 1965); S.V. Gantsevich, V.L. Gurevich, and R. Katilius, Rivista Nuova Cimento **2**, 1 (1979).

[†] N.G. van Kampen, J. Phys. Chem. Solids **46**, 673 (1988); J. Mathem. Phys. **29**, 1220 (1988).

3. DIFFUSION IN AN INHOMOGENEOUS MEDIUM

The values of α and β can be further specified by invoking detailed balance. Substituting (3.5) in (V.6.1) one obtains

$$\beta(x')\gamma(x)P^e(x') = \alpha(x)\gamma(x')P^e(x).$$

Consequently

$$\frac{\alpha(x)}{\gamma(x)}P^e(x) = \frac{\beta(x')}{\gamma(x')}P^e(x')$$

must be independent of x. It follows that

$$\alpha(x) = \beta(x) \quad \text{and} \quad P^e(x) = \text{const.}\,\frac{\gamma(x)}{\alpha(x)}. \tag{3.6}$$

Utilizing this identity in (3.5a) one obtains for $r > 0$

$$W(X'; r) = \Omega\alpha(x')\gamma\left(x' + \frac{r}{\Omega}\right)\exp\left[-\int_{X'}^{X'+r}\gamma\left(\frac{X''}{\Omega}\right)dX''\right]$$

$$= \Omega\alpha(x')\gamma(x')\,e^{-r\gamma(x')} + [\alpha(x')\gamma'(x')r - \tfrac{1}{2}\alpha(x')\gamma(x')\gamma'(x')r^2]\,e^{-r\gamma(x')}. \tag{3.7}$$

A similar expression obtains for $r < 0$. The conclusion is that W has the canonical form (X.2.3) with $f(\Omega) = \Omega$ and

$$\Phi_0(x; r) = \alpha(x)\gamma(x)\,e^{\mp r\gamma(x)}, \tag{3.8a}$$

$$\Phi_1(x; r) = \alpha(x)[\gamma'(x)r \mp \tfrac{1}{2}\gamma(x)\gamma'(x)r^2]\,e^{\mp r\gamma(x)}. \tag{3.8b}$$

The two signs refer to $r > 0$ and $r < 0$, respectively. It is now easy to compute

$$\alpha_{1,0} = 0, \quad \alpha_{1,1} = -2\frac{\alpha(x)\gamma'(x)}{\{\gamma(x)\}^3}, \quad \alpha_{2,0} = \frac{4\alpha(x)}{\{\gamma(x)\}^2}. \tag{3.9}$$

This shows that the M-equation is of diffusion type so that the Fokker–Planck approximation (1.5) is valid:

$$\frac{\partial P(x,t)}{\partial t} = \frac{2}{\Omega}\left[\frac{\partial}{\partial x}\frac{\alpha(x)\gamma'(x)}{\{\gamma(x)\}^3}P + \frac{\partial^2}{\partial x^2}\frac{\alpha(x)}{\{\gamma(x)\}^2}P\right]. \tag{3.10}$$

This is a nonlinear Fokker–Planck equation. If it happens that P^e is a constant and therefore $\alpha(x) = \gamma(x)$ the equation may be written

$$\frac{\partial P(x,t)}{\partial t} = \frac{\partial}{\partial x}\frac{2}{\Omega\gamma(x)}\frac{\partial P}{\partial x}, \tag{3.11}$$

which has the form (3.2). In this form of the equation one does not see a separate drift term, but, of course, there is still a drift, as is seen from the identity

$$\frac{\langle \Delta x \rangle_x}{\Delta t} = -\frac{2}{\Omega} \frac{\gamma'(x)}{\{\gamma(x)\}^2}.$$

Exercise. Argue that (3.6) could have been found without calculation.

Exercise. Write (3.1), (3.2), (3.3) for three dimensions and allow for anisotropy by taking for D a tensor.

Exercise. Show that every Fokker–Planck equation whose equilibrium solution is constant can be written in the form (3.2).

Exercise. Combine the present model with an external field as in (2.4).

Exercise. Find the next order beyond (3.10).

Exercise. The M-equation for $P(x, t)$ may be written in the form

$$\frac{\partial P(x,t)}{\partial t} = \frac{\partial}{\partial x} \int \alpha(x') \operatorname{sgn}(x'-x) P(x',t)\Omega \, \mathrm{d}x' \exp\left\{-\Omega \left| \int_{x'}^{x} \gamma(x'') \, \mathrm{d}x'' \right|\right\}.$$

Deduce from this the *exact* identity

$$\frac{\partial P}{\partial t} = \frac{2}{\Omega} \frac{\partial}{\partial x} \frac{1}{\gamma} \frac{\partial}{\partial x} \frac{\alpha}{\gamma} P + \frac{1}{\Omega^2} \frac{\partial}{\partial x} \frac{1}{\gamma} \frac{\partial}{\partial x} \frac{1}{\gamma} \frac{\partial P}{\partial t}.$$

From this equation it is again possible to find the next order beyond (3.10).

Exercise. In the Kramers equation (VIII.7.4) allow γ and T to depend on the position x. The expansion for large γ is again possible and produces the diffusion equation for inhomogeneous temperature[*]

$$\frac{\partial P(x,t)}{\partial t} = \frac{\partial}{\partial x} \frac{1}{\gamma(x)} \left[U'(x) + \frac{\partial}{\partial x} T(x) \right] P(x,t).$$

4. Multivariate diffusion equation

The most general form of a nonlinear Fokker–Planck equation in r variables x_i ($i = 1, 2, 3, \ldots, r$) is (VIII.6.1):

$$\frac{\partial P(x,t)}{\partial t} = -\frac{\partial}{\partial x_i} A_i(x) P + \frac{1}{2} \frac{\partial^2}{\partial x_i \, \partial x_j} B_{ij}(x) P. \tag{4.1}$$

Here x stands for the set $\{x_i\}$, and summation over repeated indices is implied. The matrix $B_{ij}(x)$ is symmetric and positive semi-definite. Alternatively one may write

$$\frac{\partial P}{\partial t} = -\frac{\partial J_i}{\partial x_i}, \quad J_i = A_i(x) P - \frac{1}{2} \frac{\partial}{\partial x_j} B_{ij}(x) P, \tag{4.2}$$

where $J_i(x, t)$ is the probability flux in the r-dimensional space.

---

[*] N.G. van Kampen, IBM J. Res. Developm. **32**, 107 (1988).

4. MULTIVARIATE DIFFUSION EQUATION

The stationary solution is to be found from

$$0 = -\frac{\partial}{\partial x_i} A_i P + \frac{1}{2} \frac{\partial^2}{\partial x_i \partial x_j} B_{ij} P \equiv -\frac{\partial J_i}{\partial x_i}. \tag{4.3}$$

It is physically obvious that for P to be stationary the divergence of J must vanish. One cannot conclude, however, that J itself must vanish, for it may be that the stationary state contains a circulating flow with nonzero curl. Hence, in contrast with the one-variable case, even the stationary solution of (**4.1**) cannot always be found.[*]

This difficulty does not arise in the case of closed, isolated physical systems, because there the stationary solution is known to be the thermal equilibrium distribution $P^e(x)$, as given by ordinary statistical mechanics. This knowledge implies some information about A_i and B_{ij}, but more information is available if also detailed balance (V.6.1) or extended detailed balance (V.6.14) holds. In the following we shall therefore examine the situation specified by the following stipulations.

(i) Closed, isolated, physical system with known equilibrium distribution.

(ii) Markovian description possible in r variables x_i, which therefore obey a master equation.

(iii) This equation involves a suitable expansion parameter Ω, and is of diffusion type so that (**4.1**) holds for $\Omega \to \infty$.

(iv) No external magnetic field or overall rotation, and each of the x_i is either even or odd.

Examples. A Brownian particle, *together with its surrounding fluid*, is a closed isolated system. The variables x_i are its three coordinates and Ω is its mass. $P^e(x)$ is a constant. Equation (**4.1**) for this case is the three-dimensional analog of (VIII.3.1): $A_i = 0$ and B_{ij} is constant.

The Rayleigh particle has three odd variables, viz., the components of the velocity V, and $P^e(V) = C \exp[-MV^2/kT]$. Equation (**4.1**) is the three-dimensional analog of (VIII.4.6).

The Kramers equation (VIII.7.4) has the form (**4.1**) with two variables, nonlinear $A_i(x)$, and constant B_{ij}, which is semi-definite, while

$$P^e(X, V) = C \exp[\{-\tfrac{1}{2}MV^2 - \Phi(X)\}/kT], \tag{4.4}$$

where $\Phi(X)$ is the potential of the force $F(X)$.

A particle diffusing in a three-dimensional, inhomogeneous anisotropic medium with external force with potential $U(r)$ obeys a generalization of (3.2),

$$\frac{\partial P(\mathbf{r}, t)}{\partial t} = \nabla \cdot \{\mu \cdot (\nabla U)P\} + \nabla \cdot D \cdot \nabla P, \tag{4.5}$$

where $\mu(\mathbf{r})$ is the mobility tensor and $D(\mathbf{r})$ the diffusion tensor. Onsager's relations[**] assert that D is symmetric and, of course, $P^e = C \exp[-U/kT]$.

---

[*] K. Tomita and H. Tomita, Prog. Theor. Phys. **51**, 1731 (1974); **53**, 1546 (1975).
[**] DE GROOT AND MAZUR.

In order to apply the detailed balance relation (V.6.14) to (4.1) we first write it in more suitable form.*) For each x_i we define $\varepsilon_i = \pm 1$ according as x_i is even or odd, and we write εx for the set $\{\varepsilon_i x_i\}$. Then $P^e(\varepsilon x) = P^e(x)$ and (V.6.14) reads

$$\mathbb{W}(x \mid x')P^e(x') = \mathbb{W}(\varepsilon x' \mid \varepsilon x)P^e(x).$$

In analogy with (V.7.5) this can be expressed as the identity

$$\int \frac{P_2(\varepsilon x)}{P^e(x)} \mathbb{W} P_1(x) \, dx = \int \frac{P_1(\varepsilon x)}{P^e(x)} \mathbb{W} P_2(x) \, dx \tag{4.6}$$

for all functions P_1, P_2.

Next, for the purpose of simplifying the algebra, we write (4.1) in the alternative form, suggested by (VIII.1.11),

$$\frac{\partial P}{\partial t} = -\frac{\partial}{\partial x_i} F_i P + \frac{1}{2} \frac{\partial}{\partial x_i} P^e B_{ij} \frac{\partial}{\partial x_j} \frac{P}{P^e}, \tag{4.7}$$

with

$$F_i = A_i - \frac{1}{2P^e} \frac{\partial}{\partial x_j}(B_{ij} P^e). \tag{4.8}$$

Now insert the operator $\mathbb{W}$ defined by (4.7) into the left-hand side of (4.6),

$$-\int \frac{P_2(\varepsilon x)}{P^e(x)} \frac{\partial}{\partial x_i} F_i(x) P_1(x) \, dx + \tfrac{1}{2} \int \frac{P_2(\varepsilon x)}{P^e(x)} \frac{\partial}{\partial x_i} P^e(x) B_{ij}(x) \frac{\partial}{\partial x_j} \frac{P_1(x)}{P^e(x)} \, dx$$

$$= \int F_i(\varepsilon x) P_1(\varepsilon x) \frac{\partial}{\partial \varepsilon_i x_i} \frac{P_2(x)}{P^e(x)} \, dx$$

$$+ \frac{1}{2} \int \frac{P_1(\varepsilon x)}{P^e(x)} \frac{\partial}{\partial \varepsilon_j x_j} P^e(x) B_{ij}(\varepsilon x) \frac{\partial}{\partial \varepsilon_i x_i} \frac{P_2(x)}{P^e(x)} \, dx.$$

In order that this be equal to the right-hand side of (4.6) for every P_1 one must have

$$\varepsilon_i F_i(\varepsilon x) \frac{\partial}{\partial x_i} \frac{P_2(x)}{P^e(x)} + \frac{\varepsilon_i \varepsilon_j}{2P^e(x)} \frac{\partial}{\partial x_i} P^e(x) B_{ij}(\varepsilon x) \frac{\partial}{\partial x_j} \frac{P_2(x)}{P^e(x)}$$

$$= -\frac{1}{P^e(x)} \frac{\partial}{\partial x_i} F_i(x) P_2(x) + \frac{1}{2P^e(x)} \frac{\partial}{\partial x_i} P^e(x) B_{ij}(x) \frac{\partial}{\partial x_j} \frac{P_2(x)}{P^e(x)}.$$

In order that this is true for every P_2 it is first of all necessary that the terms

---

*) S.R. de Groot and N.G. van Kampen, Physica **21**, 39 (1954). The subscript attached to ε does not affect the summation convention.

4. MULTIVARIATE DIFFUSION EQUATION

involving the second derivatives of P_2 are the same on both sides:

$$\varepsilon_i \varepsilon_j B_{ij}(\varepsilon x) = B_{ij}(x). \tag{4.9}$$

The remaining equation is

$$\varepsilon_i F_i(\varepsilon x) \frac{\partial P_2}{\partial x_i} - \frac{\varepsilon_i F_i(\varepsilon x)}{P^e} \frac{\partial P^e}{\partial x_i} P_2 = -F_i(x) \frac{\partial P_2}{\partial x_i} - P_2 \frac{\partial F_i}{\partial x_i}.$$

It follows that

$$\varepsilon_i F_i(\varepsilon x) = -F_i(x), \tag{4.10}$$

$$\varepsilon_i F_i(\varepsilon x) \frac{\partial \log P^e}{\partial x_i} = \frac{\partial F_i}{\partial x_i}. \tag{4.11}$$

From these two equations one obtains in particular

$$\frac{\partial}{\partial x_i} F_i(x) P^e(x) = 0. \tag{4.12}$$

These results lead to the following physical interpretation of the two terms in (4.7). The first term alone is the Liouville operator belonging to the deterministic equation

$$\dot{x}_i = F_i(\mathbf{x}). \tag{4.13}$$

According to (4.10) this equation is invariant for time reversal:

$$-\frac{d}{dt}(\varepsilon_i x_i) = F_i(\varepsilon \mathbf{x}).$$

According to (4.12) it conserves by itself the equilibrium distribution P^e. The second term in (4.7) of course also conserves P^e, but when time is reversed it changes sign:

$$-\frac{\partial P(\varepsilon x)}{\partial t} = -\frac{1}{2} \frac{\partial}{\partial \varepsilon_i x_i} P^e(\varepsilon x) B_{ij}(\varepsilon x) \frac{\partial}{\partial \varepsilon_j x_j} \frac{P(\varepsilon x)}{P^e(\varepsilon x)}. \tag{4.14}$$

Thus *the two terms in (4.7) are reversible and purely irreversible, respectively.*

Alternatively one may say that (4.7) separates the differential operator $\mathbb{W}$ into an anti-symmetric and a symmetric part in the sense of the scalar product (V.7.4). As a consequence the first term does not contribute to the following expression,

$$\frac{d}{dt} \frac{1}{2} \int \frac{\{P(x,t)\}^2}{P^e(x)} dx = \int \frac{P(x,t)}{P^e(x)} \mathbb{W} P(x,t) dx$$

$$= -\frac{1}{2} \int P^e B_{ij} \left(\frac{\partial}{\partial x_i} \frac{P}{P^e}\right)\left(\frac{\partial}{\partial x_j} \frac{P}{P^e}\right) dx. \tag{4.15}$$

As B_{ij} is positive definite this is negative unless $P = P^e$. The approach to equilibrium is solely due to the irreversible term. It is therefore justified to regard the terms in (4.7) as mechanical and dissipative, respectively.

Exercise. Show that (4.5) has the form (4.1) with

$$B_{ij} = 2D_{ij}, \qquad A_i = -\mu_{ij}\frac{\partial U}{\partial x_j} + \frac{\partial D_{ij}}{\partial x_j}.$$

How are these equations modified if D is not symmetric in the sense of (4.9) (e.g., in a magnetic field)?

Exercise. Show that conversely (4.9), (4.10), and (4.12) together guarantee that (4.6) is satisfied.

Exercise. The separation of $\mathbb{W}$ in a reversible and a purely irreversible term is, of course, unique. Show that the same result can be obtained by starting from (4.1) directly without anticipating the final form (4.7).

Exercise. Show that instead of the quadratic form in (4.15) one can also use $\int P \log(P/P^e)\, dx$.

Exercise. Following (4.7) the probability flux (4.2) is decomposed into a mechanical and a dissipative part. They are odd and even, respectively, with respect to time reversal. In equilibrium the dissipative part vanishes.

Exercise. Write the diffusion equation (4.5) in the form (4.7) and derive

$$D_{ij}(x) = kT\mu_{ij}(x). \tag{4.16}$$

Exercise. In all these examples $P^e(x) = C \exp[-U(x)/\theta]$ with $\theta = kT$. Two equivalent forms of (4.7) are

$$\frac{\partial P}{\partial t} = -\frac{\partial}{\partial x_i}\left\{F_i - \frac{1}{2\theta} B_{ij}\frac{\partial U}{\partial x_j}\right\}P + \frac{1}{2}\frac{\partial}{\partial x_i} B_{ij}\frac{\partial P}{\partial x_j}, \tag{4.17}$$

$$\frac{\partial P}{\partial t} = -\frac{\partial}{\partial x_i}\left\{F_i - \frac{1}{2\theta} B_{ij}\frac{\partial U}{\partial x_j} + \frac{1}{2}\frac{\partial B_{ij}}{\partial x_j}\right\}P + \frac{1}{2}\frac{\partial^2}{\partial x_i \partial x_j} B_{ij}P. \tag{4.18}$$

Exercise. Transform (4.1) to new variables $\bar{x}_k = \phi_k(x)$. The new equation has the same form with the new coefficients

$$\bar{A}_k = A_i \frac{\partial \phi_k}{\partial x_i} + \tfrac{1}{2} B_{ij}\frac{\partial^2 \phi_k}{\partial x_i \partial x_j}, \qquad \bar{B}_{kl} = B_{ij}\frac{\partial \phi_k}{\partial x_i}\frac{\partial \phi_l}{\partial x_j}. \tag{4.19}$$

[Hint: To avoid excessive calculations use the adjoint equation.] Compare (IX.4.14).

Exercise. If B_{ij} is positive definite there is a symmetric matrix h_{in} such that $(B^{-1})_{ij} = h_{ni}h_{nj}$. If h_{in} obeys the integrability condition

$$\frac{\partial h_{in}}{\partial x_m} = \frac{\partial h_{im}}{\partial x_n}$$

one can define new variables by $d\bar{x}_k = h_{kn}\, dx_n$, so that $\bar{B}_{kl} = \delta_{kl}$. Necessary and sufficient for the integrability is the existence of a $\Phi(x)$ such that $h_{in} = \partial^2 \Phi/\partial x_i \partial x_n$.[*]

---

[*] L. Garrido, D. Lurié, and M. San Miguel, Phys. Letters A **67**, 243 (1978).

Exercise. The condition that this transformation makes $\bar{A}_k$ linear is

$$(h^{-1})_{ni} \frac{\partial}{\partial x_i} \{h_{jm} A_j\} = \text{constant matrix.}$$

This is a necessary and sufficient condition for (4.1) to be transformable into a linear Fokker–Planck equation.

5. The limit of zero fluctuations

Master equations of diffusion type were characterized by the property that the lowest non-vanishing term in their Ω-expansion is not a macroscopic deterministic equation but a Fokker–Planck equation. One may ask whether it is still possible to obtain an approximation in the form of a deterministic equation, although Ω is no longer available as an expansion parameter. The naive device of omitting from the Fokker–Planck equation the term involving the second order derivatives is, of course, wrong: the result would depend on which of the various equivalent forms (4.1), (4.7), (4.17), (4.18) one chooses to mutilate in this way.

The only way to obtain a well-defined and physically meaningful approximation is by performing again an expansion in powers of a physical parameter. If the lowest order is to be deterministic the parameter has to be such that for small values of it the distribution reduces to a narrow peak. Clearly the parameter $\theta \equiv kT$ is suitable, because the low temperatures have small fluctuations. We shall show that the same method used in X for obtaining the Ω-expansion can be adapted to obtain an expansion of the Fokker–Planck equation in powers of $\theta^{1/2}$. We first demonstrate the method for the one-variable quasilinear equation (2.4).

First take $\theta = 0$; then (2.4) reduces to

$$\frac{\partial P(x,t)}{\partial t} = \frac{\partial}{\partial x} U'(x) P. \qquad (5.1)$$

That is the Liouville equation belonging to

$$\dot{x} = -U'(x). \qquad (5.2)$$

This is the deterministic equation that emerges in the limit of low temperature. Do not confuse with the "macroscopic equation", which is the deterministic equation that emerged in the limit $\Omega \to \infty$ for systems not of diffusion type.

Next, let $\boldsymbol{x}(t)$ be a solution of (5.2) and define a new variable ξ by

$$x = \boldsymbol{x}(t) + \theta^{1/2} \xi, \qquad P(x,t) = \Pi(\xi, t). \qquad (5.3)$$

Transforming (2.4) to this new variable yields

$$\frac{\partial \Pi}{\partial t} - \theta^{-1/2} \frac{d\boldsymbol{x}}{dt} \frac{\partial \Pi}{\partial \xi} = \theta^{-1/2} \frac{\partial}{\partial \xi} U'(\boldsymbol{x}(t) + \theta^{1/2}\xi)\Pi + \frac{\partial^2 \Pi}{\partial \xi^2}.$$

The terms of order $\theta^{-1/2}$ cancel and there remains

$$\frac{\partial \Pi}{\partial t} = U''(\boldsymbol{x}(t)) \frac{\partial}{\partial \xi} \xi\Pi + \frac{\partial^2 \Pi}{\partial \xi^2}, \tag{5.4}$$

apart from terms of order $\theta^{1/2}$ and higher. This is again a *linear* Fokker–Planck equation with a time-dependent coefficient for the fluctuations about the deterministic solution, provided they are small. The condition is that the temperature is low and that the deterministic solution is stable,

$$U''(\boldsymbol{x}(t)) > 0. \tag{5.5}$$

Successive higher orders can also be obtained.[*]

Remark. The properties of (2.4) for small θ are very similar to those of a general master equation for large Ω. Many methods and ideas apply to both. In fact, the distinction between both cases is often ignored. Yet, apart from their different physical meaning, there is an important mathematical difference as well.

In the case of the quasilinear Fokker–Planck equation (2.4), the free energy U defined in terms of the stationary solution by (2.6) is identical with the potential in the deterministic equation (5.2). That identity is often taken for granted when time-dependent solutions have to be constructed for systems of which only the equilibrium distribution is known. We shall now show, however, that *it holds only for systems of diffusion type whose Fokker–Planck equation is quasilinear*, i.e., of the form (2.4).

First a rough approach. Take a one-step process with generation and recombination coefficients $g(n)$, $r(n)$ as in (VI.5.1). Its macroscopic equation

$$\dot{n} = g(n) - r(n)$$

can be written as $\dot{n} = -V'(n)$ with

$$V(n) = \sum_{v=1}^{n} \{r(v) - g(v)\} + \text{const}. \tag{5.6}$$

On the other hand, its stationary distribution (VI.3.8) can be written as $p_n^s = \exp[-U(n)/\theta]$ with an arbitrary constant θ, and[**]

$$U(n)/\theta = \sum_{v=1}^{n} \{\log r(v) - \log g(v-1)\} + \text{const}. \tag{5.7}$$

Clearly *the potential V of the motion is not the same as the free energy U.*

---

[*] R.F. Rodríguez and N.G. van Kampen, Physica A **85**, 347 (1976). Another way of arriving at equivalent results is described by D. Ludwig, SIAM Review **17**, 605 (1975).

[**] This function is called ϕ by J. Ross, K.L.C. Hunt, and P.M. Hunt, J. Chem. Phys. **88**, 2719 (1988); **92**, 2372 (1990).

5. THE LIMIT OF ZERO FLUCTUATIONS

In order to make this approach more precise we substitute for g and r their Ω-expansions (X.2.6). Then (5.6) becomes

$$V(\phi) = \int^{\phi} [\rho_0(\phi') - \gamma_0(\phi') + \Omega^{-1}\{\rho_1(\phi') - \gamma_1(\phi')\} + \cdots]\Omega \, d\phi'. \tag{5.8}$$

On the other hand, (5.7) becomes

$$U(\phi)/\theta = \int^{\phi} [\log\{\rho_0(\phi') + \Omega^{-1}\rho_1(\phi')\} - \log\{\gamma_0(\phi') + \Omega^{-1}\gamma_1(\phi')\}]\Omega \, d\phi'. \tag{5.9}$$

Master equations of diffusion type are characterized by $\gamma_0 = \rho_0$. In that case (5.8) and (5.9) reduce to

$$V(\phi) = \int^{\phi} \{\rho_1(\phi') - \gamma_1(\phi')\} \, d\phi',$$

$$U(\phi)/\theta = \int^{\phi} \frac{\rho_1(\phi') - \gamma_1(\phi')}{\rho_0(\phi')} \, d\phi'.$$

If, in addition, $\rho_0 = \text{const.}$ one may choose $\theta = \rho_0 = \gamma_0$ and both potentials are identical. This additional requirement means that in (1.5)

$$\alpha_{2,0}(x) \equiv \tfrac{1}{2}\{\rho_0(x) + \gamma_0(x)\} = \theta$$

is constant, which says that the Fokker–Planck equation has the quasilinear form (2.4).

Exercise. The deterministic equation obtained by the naive device of dropping the second term from the Fokker–Planck equation is not invariant for nonlinear transformations of x.

Exercise. Any nonlinear Fokker–Planck equation (1.5) can be transformed into a quasilinear one by a suitable transformation of x.

Exercise. Rotational Brownian motion of a dipole in an external time-dependent field has been described by[*]

$$\frac{\partial f(\theta, t)}{\partial t} = \frac{1}{\sin\theta} \frac{\partial}{\partial \theta}\left[-F(\theta, t)f + \sin\theta \frac{\partial f}{\partial \theta}\right], \tag{5.10}$$

where $0 \leqslant \theta \leqslant \pi$. Transform this equation into a quasilinear Fokker–Planck equation.

Exercise. Show that the minima of U and V coincide, and also their maxima.

Exercise. Derive (to lowest order in Ω) for a one-step process

$$\exp\left[-\frac{1}{\theta}\frac{dU}{dn}\right] = 1 - \frac{1}{r(n)}\frac{dV}{dn}, \tag{5.11a}$$

$$\exp\left[\frac{1}{\theta}\frac{dU}{dn}\right] = 1 + \frac{1}{g(n)}\frac{dV}{dn}. \tag{5.11b}$$

What is the relation between U and V in the case of the diffusion type (1.5)?

---

[*] P. Debye, *Polar Molecules* (Dover Publications, New York 1945) p. 83; DE GROOT AND MAZUR. p. 234; W.T. Coffey and B.V. Paranjape, Proc. Roy. Irish Acad. A **78**, 17 (1978).

Exercise. The identification of V and U is not invariant for nonlinear transformations of x.

We now apply the temperature expansion to the multivariate Fokker–Planck equation (**4.1**) or (**4.7**). First one must display the temperature dependence explicitly. In principle this requires a further specification of the system, but in all our examples F_i is independent of θ and B_{ij} is proportional to it: $B_{ij}(x) = \theta b_{ij}(x)$. It is then convenient to start from (**4.17**),

$$\frac{\partial P}{\partial t} = -\frac{\partial}{\partial x_i}\left\{F_i - \tfrac{1}{2}b_{ij}\frac{\partial U}{\partial x_j}\right\}P + \frac{\theta}{2}\frac{\partial}{\partial x_i}b_{ij}\frac{\partial P}{\partial x_j}. \tag{5.12}$$

The limit $\theta \to 0$ yields the deterministic equation

$$\dot{x}_i = F_i(\mathbf{x}) - \tfrac{1}{2}b_{ij}(\mathbf{x})\frac{\partial U(\mathbf{x})}{\partial x_j}. \tag{5.13}$$

The first term is the reversible mechanical force, the second term is the damping. The damping force is that part of the irreversible term in (**4.7**) that survives in the limit $\theta = 0$. The other part is represented by the last term in (**5.12**) and gives rise to fluctuations. As a consequence of this common origin the damping and the fluctuations are governed by the same coefficients $b_{ij}(x)$. *This fact constitutes the generalization of the fluctuation–dissipation theorem to nonlinear systems of diffusion type. The fact that the matrix $b_{ij}(x)$ is symmetric is the generalization of Onsager's reciprocal relations for such systems.*

Warning. The idea of using a nonlinear Fokker–Planck equation as a general framework for describing fluctuating systems has attracted many authors[*]. Detailed balance, in its extended form, was a useful aid, but the link with the deterministic equation caused difficulties. It may therefore be helpful to emphasize three caveats.

(i) The nonlinear Fokker–Planck equation applies to systems of diffusion type, as defined by (**1.1**). For systems of the type treated in chapter X it does not naturally occur but can at best be manufactured as an approximation of limited validity (see X.4, comment **j**).

(ii) Even for systems of diffusion type the Fokker–Planck equation is merely the lowest term of the expression in Ω.

(iii) *It can be related to a deterministic equation only through the expansion in some other parameter* which enables one to scale down the fluctuations. Our choice of $\theta = kT$ seems to be natural, but other choices may be possible. An expansion in Boltz-

---

[*] M.S. Green, J. Chem. Phys. **20**, 1281 (1952); N.G. van Kampen, Physica **23**, 707 and 816 (1957); U. Uhlhorn, Arkiv för Fysik **17**, 361 (1960); R. Graham, in: Ergebn. exakten Naturw. **66** (Springer, Berlin 1973); H. Hasegawa, Prog. Theor. Phys. **55**, 90 (1976).

5. THE LIMIT OF ZERO FLUCTUATIONS

mann's k has been suggested[*], which should lead to the same results as our expansion in T, provided that nowhere in the coefficients a k occurs that is not combined with a T.

Exercise. Show that the damping in (5.13) has the effect of decreasing the free energy.

Exercise. Show that (5.13) yields the correct deterministic laws belonging to (VIII.7.4) and (VIII.4.11).

Exercise. Verify that the deterministic limit of (4.5) is

$$\dot{x}_i = -\mu_{ij}(\mathbf{x})\{\partial U/\partial x_j\}$$

with symmetric μ_{ij} (compare (4.16)).

Exercise. Find the next order in the expansion of (5.12) by setting $x_i = \mathbf{x}_i(t) + \theta^{1/2}\xi_i$ and obtain in analogy with (5.4)

$$\frac{\partial \Pi}{\partial t} = \left\{ -\frac{\partial F_i(\mathbf{x})}{\partial x_k} + \frac{1}{2}\frac{\partial}{\partial x_k}b_{ij}(\mathbf{x})\frac{\partial U(\mathbf{x})}{\partial x_j} \right\}\frac{\partial}{\partial \xi_i}\xi_k \Pi + \tfrac{1}{2}b_{ij}(\mathbf{x})\frac{\partial^2 \Pi}{\partial \xi_i \, \partial \xi_j}. \qquad (5.14)$$

According to VIII.6 the solution is a Gaussian.

Exercise. *Formulate the Onsager relations for nonlinear systems of diffusion type.*

Exercise. Sometimes $P^e(x)$ contains an additional phase-space factor $w(x)$, so that its dependence on θ is displayed by

$$P^e(x) = w(x)\exp[-U(x)/\theta].$$

Show that this does not affect the result (5.13).

---

[*] H. Grabert and M.S. Green, Phys. Rev. A **19**, 1747 (1979); H. Grabert, R. Graham, and M.S. Green, Phys. Rev. A **21**, 2136 (1980).

Chapter XII

FIRST-PASSAGE PROBLEMS

How long does it take a randomly varying quantity to reach a given value? The question is important in many situations, in particular for the study of unstable systems in the next chapter. We give three different approaches, suitable in different situations, but, of course, leading to the same results.

1. The absorbing boundary approach

First consider the unrestricted random walk of VI.2:

$$\dot{p}_n = p_{n+1} + p_{n-1} - 2p_n.$$

One may ask the following question. Suppose the random walker starts out at site m at $t = 0$; how long does it take him to reach a given site R for the first time? This first-passage time is, of course, different for the different realizations of his walk and is therefore a random quantity. Our purpose is to find its probability distribution, and in particular the average or *mean first-passage time*.[*]

To fix the ideas we take $m < R$. For any site $n < R$ let $q_n(t)$ denote the probability that at time t the walker who started at m is at n *without having touched site* R. This $q_n(t)$ may be visualized as the density of an ensemble of walkers, all starting at m at $t = 0$ and moving independently; whenever one of them hits R he is out, i.e., he no longer contributes to the density q_n. Thus R acts as an absorbing boundary and we have

$$\dot{q}_n = q_{n+1} + q_{n-1} - 2q_n \quad (n < R - 1),$$
$$\dot{q}_{R-1} = q_{R-2} - 2q_{R-1}.$$

These equations have to be solved with initial condition $q_n(0) = \delta_{n,m}$. Once they are solved it follows, as in VI.7, that the probability that the walker has not yet hit R at time t is

$$\sum_{-\infty}^{R-1} q_n(t) = 1 - q_*(t).$$

---

[*] J.H.B. Kemperman, *The Passage Problem for a Stationary Markov Chain* (University of Chicago Press, Chicago 1961); J. Keilson, *Markov Chain Models* (Springer, New York 1979).

1. THE ABSORBING BOUNDARY APPROACH

Let $f_{R,m}(t)\,\mathrm{d}t$ be the probability that he reaches R at a time between t and $t+\mathrm{d}t$; then

$$f_{R,m}(t) = -\frac{\mathrm{d}}{\mathrm{d}t}\sum_{-\infty}^{R-1} q_n(t) = q_{R-1}(t).$$

Next consider the general one-step process (VI.1.2):

$$\dot{p}_n = (\mathbb{E}-1)r_n p_n + (\mathbb{E}^{-1}-1)g_n p_n. \tag{1.1}$$

The first-passage problem is equivalent to the following absorbing boundary problem,

$$\dot{q}_n = (\mathbb{E}-1)r_n q_n + (\mathbb{E}^{-1}-1)g_n q_n \quad (n < R-1), \tag{1.2a}$$

$$\dot{q}_{R-1} = -r_{R-1}q_{R-1} + g_{R-2}q_{R-2} - g_{R-1}q_{R-1}. \tag{1.2b}$$

Having solved this equation with $q_n(0) = \delta_{n,m}$ one obtains

$$f_{R,m}(t) = -\sum_{-\infty}^{R-1}\dot{q}_n(t) = g_{R-1}q_{R-1}(t). \tag{1.3}$$

It follows that the total probability of ever reaching R (rather than disappearing to $-\infty$) is

$$\pi_{R,m} = \int_0^\infty f_{R,m}(t)\,\mathrm{d}t = 1 - \sum_{-\infty}^{R-1} q_n(\infty). \tag{1.4}$$

If this is equal to unity, then $f_{R,m}(t)$ is the probability density for the first-passage time at R. In that case the mean first-passage time is [compare (VI.7.5)]

$$\tau_{R,m} = \int_0^\infty t f_{R,m}(t)\,\mathrm{d}t = \sum_{-\infty}^{R-1}\int_0^\infty q_n(t)\,\mathrm{d}t. \tag{1.5}$$

Examples. (i) The diffusion-controlled reaction in VI.7 for the case $\chi = \infty$: once a molecule has reached the site $n=0$ it disappears forever.

(ii) The number n of individuals in a population: when n reaches the value 0 the population ceases to exist. The same applies to autocatalytic reactions such as

$$A + X \rightleftharpoons 2X, \quad X \to B.$$

When the number n of molecules is zero the reaction stops.

(iii) The diatomic molecule in (VII.5.7) makes a random walk over the energy levels until it reaches level $N+1$ and disappears.

(iv) A neuron receives successive random electrochemical pulses until its potential has reached a threshold value and the neuron discharges.[*]

---

[*] D.J. Amit, *Modeling brain function* (Cambridge University Press, Cambridge 1989).

Exercise. Write the corresponding equations for the case of a left exit point $L < m$.

Exercise. Solve the first-passage problem for the simple symmetric random walk. Show that any site R is reached with probability $\pi_{R,m} = 1$, but that the mean first-passage time is infinite.

Exercise. Solve the first-passage problem for the asymmetric random walk.

Exercise. Denote the operator on the right-hand side of (1.2) by $\mathbb{V}$. Then

$$f_{R,m}(t) = g_{R-1}(e^{t\mathbb{V}})_{R-1,m}. \tag{1.6}$$

Suppose one has a one-step process and one asks for the probability that a point starting at m will first reach either some given $R > m$ or a given $L < m$. This amounts to solving a problem with two absorbing boundaries:

$$\dot{q}_n = (\mathbb{E} - 1)r_n q_n + (\mathbb{E}^{-1} - 1)g_n q_n \quad (L+1 < n < R-1),$$

$$\dot{q}_{R-1} = -r_{R-1}q_{R-1} + g_{R-2}q_{R-2} - g_{R-1}q_{R-1},$$

$$\dot{q}_{L+1} = r_{L+2}q_{L+2} - r_{L+1}q_{L+1} - g_{L+1}q_{L+1}.$$

The probabilities for reaching R or L between t and $t + dt$ are

$$f_{R,m}(t) = g_{R-1}q_{R-1}(t) \quad \text{and} \quad f_{L,m} + r_{L+1}q_{L+1}(t). \tag{1.7}$$

The total probabilities to arrive at either one will be called the *splitting probabilities*

$$\pi_{R,m} = \int_0^\infty f_{R,m}(t)\,dt \quad \text{and} \quad \pi_{L,m} = \int_0^\infty f_{L,m}(t)\,dt.$$

For the subensembles of points that arrive at either R or L one has the *conditional mean first-passage times*

$$\tau_{R,m} = \frac{1}{\pi_{R,m}} \int_0^\infty t f_{R,m}(t)\,dt, \quad \tau_{L,m} = \frac{1}{\pi_{L,m}} \int_0^\infty t f_{L,m}(t)\,dt. \tag{1.8}$$

Exercise. Show that $\pi_{R,m} + \pi_{L,m} = 1$.

Exercise. If there is a reflecting boundary at some site $L < m$ one has $\pi_{R,m} = 1$.

Exercise. For the asymmetric random walk one may take $R = +\infty$, $L = -\infty$. Show that the splitting probability uphill is zero and downhill equal to unity.

Exercise. The *mean exit time* from the interval L, R is the average time the particle survives without hitting either boundary. It is given by

$$\tau_m = \int_0^\infty t\{f_{R,m}(t) + f_{L,m}(t)\}\,dt = \int_0^\infty \sum_{L+1}^{R-1} q_n(t)\,dt.$$

Exercise. A particle performs a one-step process on $-\infty < n < \infty$ starting at $n = 0$. I am interested in the probability Z to find it inside the interval $0 < N < n < M$, but also want to know whether its *last* entrance occurred through the boundary N or through M, denoted by probabilities Z^- and Z^+, respectively. Such problems are relevant for current fluctuations affecting a device with hysteresis. Show that this situation is described by the following set of equations.

$$Z^+ = \sum_{N+1}^{M-1} p_n^+, \qquad Z^- = \sum_{N+1}^{M-1} p_n^-;$$

$$\dot p_n^+ = (\mathbb{E}-1)r_n p_n^+ + (\mathbb{E}^{-1}-1)g_n p_n^+ + \delta_{nM} g_{M-1} p_{M-1}^- \quad (n > N+1),$$

$$\dot p_{N+1}^+ = r_{N+2} p_{N+2}^+ - (r_{N+1} + g_{N+1}) p_{N+1}^+,$$

$$\dot p_n^- = (\mathbb{E}-1)r_n p_n^- + (\mathbb{E}^{-1}-1)g_n p_n^- + \delta_{nN} r_{N+1} p_{N+1}^+ \quad (n < M-1),$$

$$\dot p_{M-1}^- = g_{M-2} p_{M-2}^- - (r_{M-1} + g_{M-1}) p_{M-1}^-.$$

In the limit of small steps the M-equation of the one-step process (1.1a) reduces to the Smoluchowski equation

$$\frac{\partial P(y,t)}{\partial t} = -\frac{\partial}{\partial y} A(y)P + \frac{1}{2}\frac{\partial^2}{\partial y^2} B(y)P, \tag{1.9}$$

as shown in VIII.2. Also it was shown in VIII.5 that an absorbing boundary at $y = R$ is implemented by the boundary condition $P(R,t) = 0$. Hence, in order to find the first-passage time for a particle diffusing according to (1.9), one has to solve the equation for $-\infty < y < R$ with the initial condition $P(y,t) = \delta(y - y_0)$ and the boundary condition $P(R,t) = 0$. The analog of (1.3) is

$$f_R(t|y_0) = -\frac{d}{dt}\int_{-\infty}^R P(y,t|y_0)\,dt = -\tfrac{1}{2}B(R)\left[\frac{\partial P}{\partial y}\right]_{y=R}.$$

Note that this is the diffusion flow; the convective part of the flow vanishes at an absorbing boundary. Of course, one can also have two boundaries $L < y_0 < R$ and

$$f_R(t|y_0) = -\tfrac{1}{2}B(R)\left[\frac{\partial P}{\partial y}\right]_R, \qquad f_L(t|y_0) = \tfrac{1}{2}B(L)\left[\frac{\partial P}{\partial y}\right]_L. \tag{1.10}$$

Exercise. A Brownian particle obeys the diffusion equation (VIII.3.1) in the interval $L < X < R$. Find its splitting probabilities $\pi_L(X_0)$ and $\pi_R(X_0)$ as functions of its starting point X_0. Also the conditional mean first-passage times.

Exercise. A particle diffuses vertically as in (VIII.3.5), but on reaching the bottom it gets stuck. The initial height is X_0 and for simplicity take $g/\gamma = 1$ and $D = 1$. Compute[*)]

$$f_0(t) = \frac{1}{\sqrt{4\pi}} t^{-3/2} X_0 \exp\left[-\frac{(X_0 - t)^2}{4t}\right]. \tag{1.11}$$

[*)] COX AND MILLER, p. 220.

Exercise. A molecule A diffuses in a solvent containing one fixed molecule B. When A reaches a sphere of radius b around B it reacts and is lost. Show that when A starts at a distance r_0 the probability that it will react rather than escape to infinity is b/r_0. [Hint: It is sufficient to consider the spherically symmetric case.]

We formulate the problem in more dimensions. Consider the generalized diffusion equation (VIII.6.1) for two variables. Suppose in the plane y_1, y_2 a region G is given, enclosed by a curve C, which is parametrized by its arc length l. If the particle starts diffusing at $t=0$ in a point (y_1^0, y_2^0) in G, what is the probability $f(l, t | y^0) \, dl \, dt$ that it reaches the boundary C between l and $l + dl$ at a time between t and $t + dt$? The two quantities $f_R(t|y^0)$ and $f_L(t|y^0)$ are now replaced by a quantity $f(l, t|y^0)$ for each point l on C (fig. 32).

The solution proceeds as before. Solve (VIII.6.1) inside G with initial condition $P(y, 0) = \delta(y - y^0)$ and the absorbing boundary condition $P = 0$ on C. Then the flux through a unit length of C is according to (VIII.6.6)

$$f(l, t|y^0) = \sum_i n_i J_i = -\tfrac{1}{2} \sum_{i,j} n_i B_{ij} \frac{\partial P(y, t)}{\partial y_j}\bigg|_C ,$$

where n_i is the normal on C pointing outward. Having found f one has for the total probability that the exit from G takes place between l and $l + dl$,

$$\pi(l|y^0) \, dl = dl \int_0^\infty f(l, t|y^0) \, dt. \tag{1.12}$$

The conditional mean first-passage time for those exits that take place between l and $l + dl$ is

$$\tau(l|y^0) = \frac{1}{\pi(l|y^0)} \int_0^\infty t f(l, t|y^0) \, dt.$$

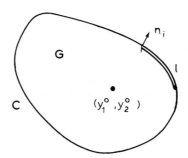

Fig. 32. The region G enclosed by the curve C with arc length l and normal n.

1. THE ABSORBING BOUNDARY APPROACH

The mean exit time from G is

$$\tau(y^0) = \oint dl \int_0^\infty tf(l,t|y^0)\,dt.$$

Remark. In each case one has to solve the M-equation with artificially added absorbing boundary conditions in order to obtain the distribution function $f(t)$. However, this is not necessary if one merely wants to know the quantities π and τ, which are the first two moments of $f(t)$. Consider the one-step problem (1.2); according to (1.3) and (1.4), one has

$$\pi_{R,m} = g_{R-1}\int_0^\infty q_{R-1}(t)\,dt.$$

Thus only the time integral of q_{R-1} is needed. According to (1.5), $\tau_{R,m}$ is also expressed in the time integral of the solution of (1.2). These time integrals (to be denoted by an overbar) obey a simpler equation, obtained from (1.2) by integrating over t:

$$(\mathbb{E}-1)r_n\bar{q}_n + (\mathbb{E}^{-1}-1)g_n\bar{q}_n = -q_n(0) = -\delta_{n,m} \quad (n < R-1),$$

$$-r_{R-1}\bar{q}_{R-1} + g_{R-2}\bar{q}_{R-2} - g_{R-1}\bar{q}_{R-1} = -\delta_{R-1,m}.$$

Similarly one sees that for the splitting probabilities and exit times in the case (1.9) one merely has to solve

$$-\frac{d}{dy}A(y)\bar{P}(y) + \tfrac{1}{2}\frac{d^2}{dy^2}B(y)\bar{P}(y) = -\delta(y-y^0).$$

In contrast to (1.9), this equation can be solved explicitly for any $A(y)$, $B(y)$. In the case of more dimensions the variable t is eliminated, but the remaining equation has still several variables and cannot be solved in general. In the next section we obtain these simpler equations for the moments in a more direct way.

Exercise. For the particle that was the subject of (1.11) find the mean survival time before it hits the bottom.

Exercise. A particle obeys the ordinary diffusion equation in three dimensions. It starts at a given point inside a given sphere. Find the probability distribution of its exit points on the sphere.

Exercise. The space outside a sphere of radius R is filled with a medium in which there are ρ independently diffusing particles per unit volume. Find for the rate at which these particles are absorbed into the sphere[*]

$$J(t) = 4\pi RD\left(1 + \frac{R}{\sqrt{\pi Dt}}\right)\rho.$$

---

[*] M. von Smoluchowski, Z. Phys. Chem. **92**, 129 (1917).

2. The approach through the adjoint equation – Discrete case

Consider again the general one-step process (1.1). We first establish an equation for the splitting probability $\pi_{R,m}$, that is, the probability that R is reached before L, after starting from a point m between both. Whenever $L+2 \leqslant m \leqslant R-2$ the particle first has to jump to $m+1$ or to $m-1$, see fig. 33. The probabilities for these jumps are $g_m/(g_m+r_m)$ and $r_m/(g_m+r_m)$, respectively. That leads to the identity

$$\pi_{R,m} = \frac{g_m}{g_m+r_m}\pi_{R,m+1} + \frac{r_m}{g_m+r_m}\pi_{R,m-1}. \tag{2.1}$$

Thus we have, for $L+2 \leqslant m \leqslant R-2$, the difference equation

$$g_m(\pi_{R,m+1}-\pi_{R,m}) + r_m(\pi_{R,m-1}-\pi_{R,m}) = 0. \tag{2.2}$$

This equation can be used to find $\pi_{R,m}$ if it is supplemented with two boundary conditions. From the physical meaning of $\pi_{R,m}$ one expects $\pi_{R,R}=1$ and $\pi_{R,L}=0$. We have to make sure that these are actually the boundary conditions to be used in (2.2). Suppose the particle starts at the site $R-1$. Then a jump to the right attains R, so that one has instead of (2.1)

$$\pi_{R,R-1} = \frac{g_{R-1}}{g_{R-1}+r_{R-1}} + \frac{r_{R-1}}{g_{R-1}+r_{R-1}}\pi_{R,R-2}.$$

This is in fact the form that (2.1) takes for $m=R-1$, provided one sets $\pi_{R,R}=1$. The same argument justifies the other boundary condition $\pi_{R,L}=0$. The conclusion is: *The splitting probability $\pi_{R,m}$, as a function of the starting site m, obeys* (2.2) *for* $L+1 \leqslant m \leqslant R-1$ *with boundary conditions* $\pi_{R,R}=1$ *and* $\pi_{R,L}=0$. The solution is given in (2.8) below.

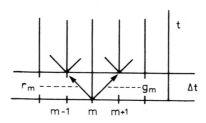

Fig. 33. The derivation of the equations for $\pi_m, \tau_m, \vartheta_m$.

Exercise. Show that the linear operator that acts in (2.2) on $\pi_{R,m}$ is the adjoint $\widetilde{\mathbb{W}}$ of the operator $\mathbb{W}$ of the underlying M-equation. (Compare V.9).

2. ADJOINT EQUATION – DISCRETE CASE

Exercise. For the random walk of VI.2 find that a walker who starts at m first reaches the left or the right wall with probabilities

$$\pi_{L,m} = \frac{R-m}{R-L} \quad \text{and} \quad \pi_{R,m} = \frac{m-L}{R-L}. \tag{2.3}$$

For $m=0$, $L<0$ this answers the question: If two gamblers with initial capitals R and $-L$ toss a coin until one is ruined, what are their chances?

Exercise. If L is a reflecting boundary one has $r_L = 0$ and (2.2) reduces to

$$g_L(\pi_{R,L+1} - \pi_{R,L}) = 0. \tag{2.4}$$

By a similar argument as above show that this is actually correct. It follows that $\pi_{R,m} = 1$ for all m.

Secondly consider the mean first-passage time $\tau_{R,m}$. We suppose that on the left there is a reflecting boundary L, so that (2.4) holds and R is reached with probability 1. At $t=0$ the particle sets out at m. In the next Δt it jumps to the right with probability $g_m \Delta t$; or to the left with probability $r_m \Delta t$; or it remains at m with probability $1 - g_m \Delta t - r_m \Delta t$. One therefore has (fig. 33):

$$\tau_{R,m} - \Delta t = g_m \Delta t \tau_{R,m+1} + r_m \Delta t \tau_{R,m-1} + (1 - g_m \Delta t - r_m \Delta t)\tau_{R,m}.$$

Hence

$$g_m(\tau_{R,m+1} - \tau_{R,m}) + r_m(\tau_{R,m-1} - \tau_{R,m}) = -1. \tag{2.5}$$

The operator on the left is again the adjoint of the $\mathbb{W}$ of the M-equation. Our derivation holds for $L+1 \leq m \leq R-2$. One sees in the same way as before that (2.5) remains valid for $m = R-1$ if one puts $\tau_{R,R} = 0$. Physically this condition expresses the obvious fact that the mean first-passage time for a particle starting at the boundary R itself is zero. Also, a similar argument applied to $m = L$ gives

$$g_L(\tau_{R,L+1} - \tau_{R,L}) = -1. \tag{2.6}$$

This is identical with (2.5) because at the reflecting boundary $r_L = 0$. The conclusion is: *When boundary L is reflecting the mean first-passage time at R is obtained by solving (2.5) for $L \leq m \leq R-1$ with the boundary condition $\tau_{R,R} = 0$.* The solution is given in (2.10).

Thirdly we take a one-step process with two exits L and R. There are two splitting probabilities $\pi_{R,m}$ and $\pi_{L,m}$ obeying (2.2) with the appropriate boundary conditions for each. There are also two conditional mean first-passage times $\tau_{R,m}$, $\tau_{L,m}$. In order to compute them we introduce the products $\vartheta_{R,m} = \pi_{R,m}\tau_{R,m}$ and $\vartheta_{L,m} = \pi_{L,m}\tau_{L,m}$. In the same way as before one argues that

$$\vartheta_{R,m} - \pi_{R,m} \Delta t = g_m \Delta t \vartheta_{R,m+1} + r_m \Delta t \vartheta_{R,m-1} + (1 - g_m \Delta t - r_m \Delta t)\vartheta_{R,m}.$$

Hence, for $L+1 \leqslant m \leqslant R-1$

$$g_m(\vartheta_{R,m+1} - \vartheta_{R,m}) + r_m(\vartheta_{R,m-1} - \vartheta_{R,m}) = -\pi_{R,m}. \tag{2.7}$$

The boundary conditions are: $\vartheta_{R,R} = 0$ (because $\tau_{R,R} = 0$) and $\vartheta_{R,L} = 0$ (because $\pi_{R,L} = 0$). Of course, $\vartheta_{L,m}$ obeys the same equation, but with $\pi_{L,m}$ on the right. Having found the splitting probabilities from (2.2) one may now compute $\vartheta_{R,m}$ and $\vartheta_{L,m}$ from (2.7), and find in this way the conditional mean first-passage times $\tau_{R,m}$ and $\tau_{L,m}$. The result is given in (2.12).

Exercise. For the radioactive decay process with M-equation (V.1.7) prove that the probability $\pi_{0,m}$ that m initial nuclei will decay completely is unity. The average time it takes is

$$\tau_{0,m} = \frac{1}{\gamma}\left(\frac{1}{m} + \frac{1}{m-1} + \frac{1}{m-2} + \cdots + 1\right).$$

This result is obvious: a step from m to $m-1$ takes on the average a time $1/\gamma m$ and averages are additive.

Exercise. For the random walk, $\pi_{R,m}$ has been computed in (2.3). Find $\vartheta_{R,m}$ and subsequently the conditional mean first-passage time*[)]

$$\tau_{R,m} = \tfrac{1}{6}(R-m)(R+m-2L) \quad (L \leqslant m \leqslant R).$$

The equations we have found for $\pi_{R,m}$, $\tau_{R,m}$, $\vartheta_{R,m}$ can be solved explicitly. First take equation (2.2) for $\pi_{R,m}$. To solve it set $\pi_{R,m+1} - \pi_{R,m} = \Delta_m$ so that the equation reduces to a one-step recursion:

$$g_m \Delta_m = r_m \Delta_{m-1} \quad (L+1 \leqslant m \leqslant R-1).$$

This can be solved to give

$$\Delta_m = \frac{r_m}{g_m} \frac{r_{m-1}}{g_{m-1}} \frac{r_{m-2}}{g_{m-2}} \cdots \frac{r_{L+1}}{g_{L+1}} \Delta_L.$$

Note that $\Delta_L = \pi_{R,L+1}$, but it is still undetermined. Subsequently

$$\pi_{R,m} = \sum_{\mu=1}^{m-1} \Delta_\mu = \pi_{R,L+1} + \sum_{\mu=L+1}^{m-1} \frac{r_\mu r_{\mu-1} r_{\mu-2} \cdots r_{L+1}}{g_\mu g_{\mu-1} g_{\mu-2} \cdots g_{L+1}} \pi_{R,L+1}.$$

The condition that this must be compatible with $\pi_{R,R} = 1$ determines $\pi_{R,L+1}$. Inserting the result one finally gets

$$\pi_{R,m} = \frac{1 + \sum_{\mu=L+1}^{m-1} \frac{r_\mu r_{\mu-1} \cdots r_{L+1}}{g_\mu g_{\mu-1} \cdots g_{L+1}}}{1 + \sum_{\mu=L+1}^{R-1} \frac{r_\mu r_{\mu-1} \cdots r_{L+1}}{g_\mu g_{\mu-1} \cdots g_{L+1}}}. \tag{2.8}$$

---

*[)] M.E. Fisher, IBM J. Res. Development **32**, 76 (1988).

Exercise. Let there be a reflecting boundary at 0, that is, $r_0 = 0$, $g_0 > 0$. Show that for $0 \leq m \leq R$ one has $\pi_{R,m} = 1$. What happens if 0 is an absorbing boundary?

Exercise. Let p^s be the stationary solution of the unlimited one-step process, not necessarily normalized. Then

$$\pi_{R,m} = \sum_{L}^{m-1} \frac{1}{g_\mu p_\mu^s} \bigg/ \sum_{L}^{R-1} \frac{1}{g_\mu p_\mu^s}$$

$$= \sum_{L+1}^{m} \frac{1}{r_\mu p_\mu^s} \bigg/ \sum_{L+1}^{R} \frac{1}{r_\mu p_\mu^s}. \tag{2.9}$$

Similar equations for $\pi_{L,m}$.

Exercise. For $L \to -\infty$ the numerator and the denominator of (2.9) either both diverge or both converge. In the former case $\pi_{R,m} \to 1$. In the latter case $\pi_{R,m} < 1$, so that there is a nonzero probability for the particle to disappear into $-\infty$.

Next consider equation (2.5) for $\tau_{R,m}$, assuming that L is reflecting as specified by (2.6). Setting $\tau_{R,m+1} - \tau_{R,m} = \Delta_m$ one gets

$$g_m \Delta_m - r_m \Delta_{m-1} = -1, \quad \Delta_L = -1/g_L.$$

This can be solved progressively to give

$$\Delta_m = -\frac{1}{g_m} - \sum_{\mu=L+1}^{m} \frac{r_m r_{m-1} \cdots r_\mu}{g_m g_{m-1} \cdots g_\mu} \frac{1}{g_{\mu-1}}$$

$$= -\sum_{\mu=L}^{m} \frac{r_m r_{m-1} \cdots r_{\mu+1}}{g_m g_{m-1} \cdots g_{\mu+1} g_\mu}.$$

In the second line it is understood that for $\mu = m$ the numerator contains no factors and is to be put equal to 1. From these Δ_m one finds directly the $\tau_{R,m}$ themselves since it is known that $\tau_{R,R} = 0$.

$$\tau_{R,m} = -\sum_{\nu=m}^{R-1} \Delta_\nu$$

$$= \sum_{\nu=m}^{R-1} \sum_{\mu=L}^{\nu} \frac{r_\nu r_{\nu-1} \cdots r_{\mu+1}}{g_\nu g_{\nu-1} \cdots g_{\mu+1} g_\mu}. \tag{2.10}$$

Exercise. Derive for a one-step process, having a stationary distribution $p_n^s > 0$,

$$\tau_{R,m} = \sum_{\nu=m+1}^{R} \frac{1}{r_\nu p_\nu^s} \sum_{-\infty}^{\nu-1} p_n^s = \sum_{\nu=m}^{R-1} \frac{1}{g_\nu p_\nu^s} \sum_{-\infty}^{\nu} p_n^s. \tag{2.11}$$

Exercise. Consider the diffusion-controlled reaction (VI.7.7). At $t = 0$ the molecule is bound. What is the mean time needed to reach some given site R?

Exercise. Find a formula for the variance of the first-passage time.[*]

---

[*] D.T. Gillespie, Physica A **101**, 535 (1980).

Exercise. Consider a one-step process with natural boundaries at $n=0$ and at $n=N$; for example, the continuous-time version of the Ehrenfest model (IV.5.4):
$$\dot p_n = (\mathbb{E}-1)(n/N)p_n + (\mathbb{E}^{-1}-1)(1-n/N)p_n.$$
Find an expression for the mean recurrence time at the site $n=0$. That is, the system starts at $n=0$ and one wants to know the average of the time at which the first transition from $n=1$ into $n=0$ occurs. Evaluate it for the above example and note that it is the reciprocal of p_0^s.

The next task is to solve (2.7) for $\vartheta_{R,m}$. The algebra is similar to the one just used, but the two boundary conditions give an additional complication. The result of the rather lengthy calculations is

$$\vartheta_{R,m} = \left[\sum_{\lambda=L}^{R-1}\frac{1}{g_\lambda p_\lambda^s}\right]^{-1}\sum_{\lambda=m}^{R-1}\frac{1}{g_\lambda p_\lambda^s}\sum_{v=L+1}^{m-1}\frac{1}{g_v p_v^s}\sum_{\mu=v+1}^{\lambda}p_\mu^s \pi_{R,m}. \quad (2.12)$$

p_n^s is the stationary solution of the underlying one-step process, $\pi_{R,m}$ is the quantity obtained in (2.9), and m runs from $L+1$ through $R-1$.

Exercise. Apply the above results to an interval of a single site (take $L=0$, $R=2$) and verify them. Also for an interval of two sites.

Exercise. Compute the mean exit time from the interval (L, R).

The quantities π, τ, ϑ considered so far are the first moments of the probability distribution $f(t)$ of the first-passage time. Specifically, for a one-step process there are two distributions $f_{R,m}(t)$ and $f_{L,m}(t)$ for the probabilities to arrive at R and L at a time t after starting out at site m. We derive an equation for them. By a similar argument as used above one obtains

$$f_{R,m}(t) = g_m\,\Delta t\, f_{R,m+1}(t-\Delta t) + r_m\,\Delta t\, f_{R,m-1}(t-\Delta t)$$
$$+ (1 - g_m\,\Delta t - r_m\,\Delta t)f_{R,m}(t).$$

This holds for $L+2 \leqslant m \leqslant R-2$ and gives the equation

$$\partial_t f_{R,m}(t) = g_m(\mathbb{E}-1)f_{R,m}(t) + r_m(\mathbb{E}^{-1}-1)f_{R,m}(t). \quad (2.13)$$

This equation requires two boundary conditions and an initial condition which determines $f_{R,m}(0)$.

For all initial sites $m \leqslant R-2$ there are at least two jumps needed to arrive at R. The probability for them to occur during Δt is of order $(\Delta t)^2$. In terms of the probability density $f_{R,m}(t)$ this means

$$\int_0^{\Delta t} f_{R,m}(t')\,dt' = \mathcal{O}(\Delta t)^2, \quad \text{so that} \quad f_{R,m}(0)=0 \quad (m \leqslant R-2).$$

For $m=R-1$, however, the integral equals $g_{R-1}\,\Delta t$ since this is the probability to reach R during the first Δt. This fact can be incorporated in (2.13) by setting $f_{R,R-1} = 0$ at $t = 0$ and

$$f_{R,R}(t) = \delta(t). \tag{2.14}$$

The physical interpretation is that, if the particle starts out at R, the time at which it reaches R is certain to be zero. It is also clear that

$$f_{R,L}(t) = 0 \quad \text{for all } t. \tag{2.15}$$

The conclusion is: *The first-passage time distribution $f_{R,m}(t)$ obeys* (2.13) *for $L + 1 \leq m \leq R - 1$ with the boundary conditions* (2.14), (2.15), *and the initial value $f_{R,m}(0) = 0$ for all m*. It is trivial to write the analogous conclusion for $f_{L,m}(t)$.

Exercise. Derive from (2.13) the equations (2.2), (2.5), and (2.7).

Exercise. Let $\mathbb{V}$ be the difference operator that belongs to the one-step process with absorbing boundaries at L and R. Its adjoint occurs in (2.13). The solution is formally[*]

$$f_{R,m}(t) = (e^{t\tilde{\mathbb{V}}})_{m,R-1} g_{R-1}.$$

Compare with (1.6).

Exercise. Obtain an equation for the Laplace transform or characteristic function

$$\hat{f}_{R,m}(\lambda) = \int_0^\infty e^{-\lambda t} f_{R,m}(t) \, dt.$$

On expanding in powers of λ one recovers the equations for π, τ, ϑ.

Exercise. Derive successive equations for higher moments of $f_{R,m}$.

Exercise. When $0 < q < 1$ the nonintegral moment of the first-passage time is[**]

$$\int_0^\infty t^q f_{R,m}(t) \, dt = \frac{-1}{\Gamma(1-q)} \int_0^\infty \lambda^{-q} \frac{d\hat{f}_{R,m}(\lambda)}{d\lambda} \, d\lambda.$$

3. The approach through the adjoint equation – Continuous case

Instead of the one-step process we now consider the generalized diffusion or Smoluchowski equation (1.9). Take a finite interval $L < y < R$ and consider the splitting probabilities $\pi_L(y)$ and $\pi_R(y)$. They obey a differential equation with the adjoint operator

$$A(y)\frac{d\pi}{dy} + \tfrac{1}{2} B(y)\frac{d^2\pi}{dy^2} = 0 \tag{3.1}$$

---

[*] This fact was applied to diffusion-controlled reactions by M. Tachiya, J. Chem. Phys. **69**, 2375 (1978). See also H.B. Rosenstock, J. Mathem. Phys. **21**, 1643 (1980).

[**] G.H. Weiss, S. Havlin and O. Matan, J. Statist. Phys. **55**, 435 (1989).

and are specified by the boundary conditions

$$\pi_R(R) = 1, \quad \pi_R(L) = 0; \quad \pi_L(R) = 0, \quad \pi_L(L) = 1.$$

If it happens that L is a reflecting boundary one has in analogy with (2.4) the boundary condition $d\pi_R/dy = 0$ at L. It then follows trivially that $\pi_R(y) = 1$ for all y, as expected. In this case one has a mean first-passage time $\tau_R(y)$. It obeys

$$A(y)\frac{d\tau_R}{dy} + \tfrac{1}{2}B(y)\frac{d^2\tau_R}{dy^2} = -1, \quad \tau_R(R) = 0. \tag{3.2}$$

The other boundary condition is

$$\frac{d\tau_R}{dy} = 0 \quad \text{at} \quad y = L. \tag{3.3}$$

To see this, start from the discrete case (2.5) and take the limit of small steps, as was done in (VIII.2.8):

$$g = \beta = \frac{1}{2}\left(\frac{B}{\varepsilon^2} + \frac{A}{\varepsilon}\right), \quad \tau_{L+1} - \tau_L = \varepsilon\frac{d\tau}{dy}.$$

Substitution in (2.6) gives in the limit the result (3.3).

When L is not a reflecting boundary there is a conditional mean first-passage time $\tau_R(y)$. It is obtained from the equation*) for the product quantity $\vartheta_R(y) = \pi_R(y)\tau_R(y)$:

$$A(y)\frac{d\vartheta_R}{dy} + \tfrac{1}{2}B(y)\frac{d^2\vartheta_R}{dy^2} = -\pi_R(y), \tag{3.4a}$$

$$\vartheta_R(R) = 0, \quad \vartheta_R(L) = 0. \tag{3.4b}$$

Exercise. To write the general solution of (3.2) we introduce the abbreviation

$$\Phi(y) = -\int_M^y \frac{2A(y')}{B(y')}\,dy', \tag{3.5}$$

where M is an arbitrary point in (L, R). Then

$$\pi_R(y) = \int_L^y e^{\Phi(y')}\,dy' \Big/ \int_L^R e^{\Phi(y')}\,dy'. \tag{3.6}$$

Exercise. The solution of (3.2) with (3.3) is

$$\tau_R(y) = \int_y^R e^{\Phi(y')}\,dy' \int_L^{y'} e^{-\Phi(y'')} \frac{2\,dy''}{B(y'')}. \tag{3.7}$$

---

*) GARDINER, section 5.2.8.

Exercise. For a particle in one dimension diffusing in a potential field according to (XI.2.4) these results reduce to

$$\pi_R(y) = \int_L^y e^{U(y')/\theta}\,dy' \bigg/ \int_L^R e^{U(y')/\theta}\,dy'. \tag{3.8}$$

$$\tau_R(y) = \frac{1}{\theta}\int_y^R e^{U(y')/\theta}\,dy' \int_L^{y'} e^{-U(y'')/\theta}\,dy''. \tag{3.9}$$

Exercise. Obtain from (3.4)

$$\vartheta_R(y) = -\int_L^y e^{\Phi(y')}\,dy' \int_L^{y'} e^{-\Phi(y'')}\frac{2\pi_R(y'')}{B(y'')}\,dy'' + C\int_L^y e^{\Phi(y')}\,dy',$$

where

$$C = \left[\int_L^R e^{\Phi(y')}\,dy'\right]^{-1}\left[\int_L^R e^{\Phi(y')}\,dy' \int_L^{y'} e^{-\Phi(y'')}\frac{2\pi_R(y'')}{B(y'')}\,dy''\right].$$

Exercise. Derive from this the analog of (2.12)

$$\vartheta_R(y) = \left[\int_L^R e^{\Phi(z)}\,dz\right]^{-1}\left[\int_y^R e^{\Phi(y')}\,dy' \int_L^{y'} e^{\Phi(y'')}\,dy'' \int_{y''}^{y'} e^{-\Phi(z)}\frac{2\pi_R(z)}{B(z)}\,dz\right]. \tag{3.10}$$

The conditional mean first-passage time has the same second factor,

$$\tau_R(y) = \left[\int_L^y e^{\Phi(z)}\,dz\right]^{-1}[\ldots\text{ditto}\ldots]. \tag{3.11}$$

Exercise. The mean exit time $\tau(y)$ obeys (3.2) with $\tau(L) = \tau(R) = 0$. Find the solution.

Exercise. Let $\mathbb{W}$ be the differential operator in the Smoluchowski equation (1.9), defined in the interval $L \leqslant y \leqslant R$ on the space of functions $P(y)$ that vanish at R and obey the reflecting boundary condition at L. Then the adjoint $\widetilde{\mathbb{W}}$ has the property that

$$\int_L^R \tau(y)\,\mathbb{W}P(y)\,dy = \int_L^R P(y)\,\widetilde{\mathbb{W}}\tau(y)\,dy$$

for any function $\tau(y)$ that vanishes at R and obeys (3.3) at L. Hence (3.3) is the adjoint of the reflecting boundary condition of P.

These considerations can be extended to more variables. For simplicity take two variables as in fig. 32. The quantity $\pi(l|y_0)$ defined in (1.11) obeys

$$\sum_i A_i(y)\frac{\partial \pi}{\partial y_i} + \tfrac{1}{2}\sum_{i,j} B_{ij}(y)\frac{\partial^2 \pi}{\partial y_i\,\partial y_j} = 0, \quad (y \text{ in } G) \tag{3.12}$$

with boundary condition

$$\pi(l|y) = 1 \quad \text{for } y \text{ on } C \text{ between } l \text{ and } l + dl;$$
$$= 0 \quad \text{for } y \text{ on } C \text{ outside } (l, dl).$$

The mean exit time $\tau(C|y)$ obeys

$$\sum_i A_i(y)\frac{\partial \tau}{\partial y_i} + \tfrac{1}{2}\sum_{i,j} B_{ij}(y)\frac{\partial^2 \tau}{\partial y_i \partial y_j} = -1 \tag{3.13}$$

with $\tau(C|y) = 0$ for y on C. The conditional mean first-passage time is obtained from

$$\sum_i A_i(y)\frac{\partial \vartheta}{\partial y_i} + \tfrac{1}{2}\sum_{i,j} B_{ij}(y)\frac{\partial^2 \vartheta}{\partial y_i \partial y_j} = -\pi(l|y) \tag{3.14}$$

with $\vartheta(l|y) = 0$ for y on C.

Exercise. A particle obeys the ordinary diffusion equation in the space between two concentric spheres. Find the splitting probability and the conditional mean first-passage times.

Exercise. For the Smoluchowski equation in one dimension show that the conditional mean first-passage time $\tau_R(y)$ obeys

$$\left\{A(y) + B(y)\frac{d \log \pi_R(y)}{dy}\right\}\frac{d\tau_R}{dy} + \tfrac{1}{2}B(y)\frac{d^2\tau_R}{dy^2} = -1. \tag{3.15}$$

Obtain a similar equation for more variables.

The observation that the operator acting on $f_{R,m}$ in (2.13) is the adjoint of $\mathbb{W}$ enables us to write down, by analogy, for the case of the Smoluchowski equation (1.9) in the interval $L < y < R$

$$\frac{\partial f_R(t|y)}{\partial y} = A(y)\frac{\partial f_R}{\partial y} + \tfrac{1}{2}B(y)\frac{\partial^2 f_R}{\partial y^2},$$

$$f_R(0|y) = 0 \quad (y < R); \qquad f_R(t|R) = \delta(t); \qquad f_R(t|L) = 0.$$

Similarly for f_L. For the multivariate case, see fig. 32,

$$\frac{\partial f(l, t|y)}{\partial y} = \sum_i A_i(y)\frac{\partial f}{\partial y_i} + \tfrac{1}{2}\sum B_{ij}(y)\frac{\partial^2 f}{\partial y_i \partial y_j},$$

$$f(l, 0|y) = 0; \qquad f(l, t|l') = \delta(l - l')\,\delta(t).$$

These equations for f contain more information than those for π, τ, ϑ but are also more difficult to solve. In fact, solving them is equivalent to solving the original Smoluchowski equation with absorbing boundaries. They lack the simplification that makes the equations for π, τ, ϑ useful.

4. The renewal approach

The idea of the third approach can again be made clear by first considering the one-step problem (1.1). We seek the distribution $f_{R,m}(t)$ of the first arrival time at the site R when starting at m. It can be found by solving the master equation (1.2) with absorbing boundary, but we now establish another connection. Let $p_{n,m}(t)$ be the solution of the original master equation, unrestricted by a boundary. Thus $p_{n,m}(t)$ is the probability that the particle starting at m is after time t at site n. All possible paths that lead from m to n can be subdivided into two classes. The first class consists of those paths that never touch R. They contribute to $p_{n,m}(t)$ an amount equal to $q_{n,m}(t)$ defined in section 1, which is the solution with absorbing boundary at R.

The second class are the paths that have touched R. The probability that that occurs for the first time between t' and $t' + dt'$ is $f_{R,m}(t')$. After that moment t' they move on according to the original master equation and contribute to the probability to be at n at time t the amount $p_{n,R}(t - t')$. Hence

$$p_{n,m}(t) = q_{n,m}(t) + \int_0^t f_{R,m}(t') p_{n,R}(t - t') \, dt'. \tag{4.1}$$

This is an identity for all n. We now substitute $n = R$ to profit from the fact that $q_{R,m} = 0$,

$$p_{R,m}(t) = \int_0^t p_{R,R}(t - t') f_{R,m}(t') \, dt'. \tag{4.2}$$

If the solution $p_{n,m}(t)$ of the unrestricted one-step process is known, this is an integral equation for $f_{R,m}(t)$, called the *renewal equation*.[*]

The integral equation is of a simple convolution type and can therefore be solved by Laplace transformation:

$$\hat{p}_{n,m}(\lambda) = \int_0^\infty e^{-\lambda t} p_{n,m}(t) \, dt, \quad \hat{f}_{R,m}(\lambda) = \int_0^\infty e^{-\lambda t} f_{R,m}(t) \, dt.$$

Notice that

$$\hat{f}_{R,m}(0) = \pi_{R,m}, \quad \hat{f}'_{R,m}(0) = -\pi_{R,m} \tau_{R,m} = -\vartheta_{R,m}. \tag{4.3}$$

In Laplace language, (4.2) reads $\hat{p}_{R,m} = \hat{p}_{R,R} \hat{f}_{R,m}$, so that it is solved by

$$\hat{f}_{R,m}(\lambda) = \frac{\hat{p}_{R,m}(\lambda)}{\hat{p}_{R,R}(\lambda)}. \tag{4.4}$$

---

[*] D.R. Cox, *Renewal Theory* (Methuen, London 1963).

Thus we have found that *it is not necessary to solve the master equation* (1.1) *with absorbing boundary*: it is sufficient to find the solution of the unbounded process and determine its Laplace transform.

Exercise. For the Poisson process (VI.1.8) one finds, when $R \geq m$,

$$\hat{p}_{R,m}(\lambda) = \left(\frac{q}{q+\lambda}\right)^{R-m}.$$

Obtain explicitly the first-passage time distribution. Compare with (II.6.1).

Exercise. For the radioactive decay (IV.6.1.) find the mean time needed for m active nuclei to dwindle to only L nuclei.

Exercise. Find $p_{n,m}$ for the random walk (VI.2.1). Show again that every site will be reached, but after a time whose average is infinite. (Thus the walker may spend an infinite time wandering in the wrong direction.)

This Exercise demonstrates that some care is needed in utilizing (4.3) for finding π and τ because $\hat{p}_{R,m}(\lambda)$ may have a singularity $\lambda = 0$. Let us suppose in particular that the M-equation has a discrete eigenvalue spectrum, as in V.7. The lowest eigenvalue is zero, with eigenfunction $p_n^s = p_{n,m}(\infty)$. Then the Laplace transform has a pole,

$$\hat{p}_{n,m}(\lambda) = \frac{1}{\lambda} p_n^s + \psi_{n,m}(\lambda), \tag{4.5}$$

where

$$\psi_{n,m}(\lambda) = \int_0^\infty [p_{n,m}(t) - p_n^s] e^{-\lambda t} \, dt \tag{4.6}$$

is regular at $\lambda = 0$. Equation (4.4) may now be written

$$\hat{f}_{R,m}(\lambda) = \frac{p_R^s + \lambda \psi_{R,m}(\lambda)}{p_R^s + \lambda \psi_{R,R}(\lambda)}. \tag{4.7}$$

From this one sees that $\pi_{R,m} = 1$. This states the obvious fact that, if there is a stationary distribution for $t \to \infty$, then every site R will be reached sooner or later. Moreover,

$$\tau_{R,m} = \frac{\psi_{R,R}(0) - \psi_{R,m}(0)}{p_R^s} \tag{4.8}$$

is finite, owing to the discrete spectrum.

4. THE RENEWAL APPROACH

Exercise. For future use show that

$$\sum_n \psi_{n,m}(\lambda) = 0. \tag{4.9}$$

Exercise. Consider a random walk with a boundary at $n = 0$, at which a fraction α is absorbed.

$$\dot p_n = p_{n+1} + p_{n-1} - 2p_n \quad (n > 1),$$
$$\dot p_1 = p_2 + (1-\alpha)p_0 - 2p_1,$$
$$\dot p_0 = p_1 - p_0.$$

Call the probability per unit time for a particle starting at m to be absorbed, $R(t|m) \equiv \alpha p_{0,m}(t)$. Verify the renewal equation

$$R(t|m) = \alpha f_{0,m}(t) + (1-\alpha)\int_0^t f_{0,m}(t')\,dt'\, R(t-t'|1).$$

[Compare J.B. Pedersen, J. Chem. Phys. **72**, 3904 (1980).]

Rather than solving the master equation and subsequently taking the Laplace transform, it is often easier to make use of the equation for $\hat p_{n,m}(\lambda)$ itself. Multiply (VI.1.2) with $e^{-\lambda t}$, integrate over t from 0 to ∞, and use partial integration in the left-hand member,

$$-\delta_{n,m} + \lambda \hat p_{n,m}(\lambda) = (\mathbb{E}_n - 1)r_n \hat p_{n,m}(\lambda) + (\mathbb{E}_n^{-1} - 1)g_n \hat p_{n,m}(\lambda). \tag{4.10}$$

This equation is no easier to solve than the master equation, but it is possible, by expanding it in λ, to obtain for the successive powers of λ equations that are easier to solve. Substitute (4.5) to get an equation for $\psi_{n,m}(\lambda)$:

$$-\delta_{n,m} + p_n^s = r_{n+1}\psi_{n+1,m}(\lambda) + g_{n-1}\psi_{n-1,m}(\lambda) - (r_n + g_n + \lambda)\psi_{n,m}(\lambda).$$

On taking $\lambda = 0$, one obtains an equation that can be written

$$-\delta_{n,m} + p_n^s = \mathbb{W}\psi_{n,m}(0). \tag{4.11}$$

It is simpler than the master equation, because it amounts to the finding of the reciprocal of $\mathbb{W}$ rather than of its entire spectrum of eigenvalues and eigenfunctions. Actually we are interested, see (4.6), in the difference

$$\psi_{n,R}(0) - \psi_{n,m}(0) \equiv \chi_n.$$

The vector χ obeys

$$-\delta_{n,m} + \delta_{n,R} = \mathbb{W}\chi_n. \tag{4.12}$$

Exercise. $\mathbb{W}$ has one eigenvalue zero. Hence (4.11) can only be solved when the inhomogeneous term on the left is orthogonal to the left null eigenvector of $\mathbb{W}$. Show that this condition is fulfilled. On the other hand, the solution is not unique; show that the requirement (4.9) takes care of that.

Exercise. For the case that $\mathbb{W}$ is a finite matrix derive

$$\hat{f}_{R,m}(\lambda) = \left(\frac{1}{\lambda - \mathbb{W}}\right)_{R,m} \bigg/ \left(\frac{1}{\lambda - \mathbb{W}}\right)_{R,R}.$$

Hence one merely needs to work out two minors of the matrix $\lambda - \mathbb{W}$. Apply this to a two-state system and show that the resulting first-passage times are the reciprocals of the corresponding transition probabilities, as they should be.

Exercise. For the random walk it is not true that $\psi_{n,m}(0)$ is finite. Accordingly the mean first-passage time is infinite (as found before).

For the case with two exit boundaries $L < R$ one easily establishes the renewal equations

$$\hat{p}_{R,m} = \hat{p}_{R,R}\hat{f}_{R,m} + \hat{p}_{R,L}\hat{f}_{L,m}, \tag{4.13a}$$

$$\hat{p}_{L,m} = \hat{p}_{L,R}\hat{f}_{R,m} + \hat{p}_{L,L}\hat{f}_{L,m}. \tag{4.13b}$$

All quantities are functions of λ. These two equations determine the first-passage time distributions $\hat{f}_{R,m}$ and $\hat{f}_{L,m}$.

For instance, if $\hat{p}_{n,m}$ has the form (4.5) the terms with $1/\lambda$ give

$$\hat{f}_{R,m}(0) + \hat{f}_{L,m}(0) = 1, \tag{4.14}$$

which states that the particle is sure to exit sooner or later through one of the boundaries. The next order gives

$$\psi_{R,m} = \psi_{R,R}\hat{f}_{R,m} + \psi_{R,L}\hat{f}_{L,m} + p_R^s(\hat{f}'_{R,m} + \hat{f}'_{L,m}),$$

$$\psi_{L,m} = \psi_{L,R}\hat{f}_{R,m} + \psi_{L,L}\hat{f}_{L,m} + p_L^s(\hat{f}'_{R,m} + \hat{f}'_{L,m}).$$

All quantities are taken at $\lambda = 0$. The terms with the derivatives can be eliminated so as to obtain an equation involving $\hat{f}_{R,m}(0)$ and $\hat{f}_{L,m}(0)$. Together with (4.14) it determines these quantities and the splitting probabilities $\pi_{R,m}$ and $\pi_{L,m}$ are now obtained. In addition one obtains the value of

$$-(\hat{f}'_{R,m}(0) + \hat{f}'_{L,m}(0)) = \pi_{R,m}\tau_{R,m} + \pi_{L,m}\tau_{L,m},$$

which is the mean exit time. To find the conditional mean first-passage times $\tau_{R,m}$ and $\tau_{L,m}$ separately one has to invoke the next order in λ.

To handle the Smoluchowski equation (1.9) define $P(y,t|y_0,0)$ as the solution in the unrestricted range, i.e., for y ranging from $-\infty$ to $+\infty$, or between natural boundaries. For $y_0 \leq R$, define $Q(y,t|y_0,0)$ as the solution with absorbing boundary at R, i.e., $Q(R,t|y_0,0) = 0$. The renewal equation (4.1) takes the form

$$P(y,t|y_0,0) = Q(y,t|y_0,0) + \int_0^t f(R,t'|y_0,0)\,dt'\, P(y,t|R,t').$$

4. THE RENEWAL APPROACH

Substitution of $y = R$ gives the analog of (4.2):

$$P(R, t|y_0, 0) = \int_0^t P(R, t|R, t')f(R, t'|y_0, 0)\,dt'.$$

Laplace transformation gives, in analogy with (4.4),

$$\hat{f}(R, \lambda|y_0) = \frac{\hat{P}(R, \lambda|y_0)}{\hat{P}(R, \lambda|R)}. \tag{4.15}$$

In the case of two boundaries one has two coupled equations for $\hat{f}(R, \lambda|y_0)$ and $\hat{f}(L, \lambda|y_0)$, which are easy to write down. In the case of two variables, as in fig. 32, the renewal equation takes the form

$$P(\mathbf{r}, t|\mathbf{r}_0, 0) = Q(\mathbf{r}, t|\mathbf{r}_0, 0) + \int_0^t dt' \oint_C dl\, f(l, t'|\mathbf{r}_0, 0)P(\mathbf{r}, t|l, t').$$

Here $\mathbf{r}$ stands for the two variables, P is the solution in the unbounded plane, Q the solution in G that vanishes on C, and l parametrizes C. Taking for $\mathbf{r}$ a point on C and applying Laplace transformation one has

$$\hat{P}(l, \lambda|\mathbf{r}_0) = \oint_C dl'\, \hat{P}(l, \lambda|l')\hat{f}(l', \lambda|\mathbf{r}_0). \tag{4.16}$$

If $\hat{P}$ is known, this is an integral equation for $\hat{f}$ in lieu of the single equation (4.2) or the two equations (4.13). The same is true for more variables. This fact severely limits the applicability of the renewal approach in more dimensions, except as a means for deriving equations for π and τ by expanding (4.16) in powers of λ.

Exercise. For the random walk on a two-dimensional square lattice, either with discrete or continuous time, show that every lattice point is reached with probability 1, but on the average after an infinite time. In three dimensions, however, the probability of reaching a given site is less than unity; there is a positive probability for disappearing into infinity.

Exercise. Consider the diffusion (1.9) on $(-\infty, \infty)$. If a particle starts from y_0 at time t_0 and arrives in y_1 at time t_1, one may ask for the point l of the intermediate path that is furthest to the left.[*] Let $Q_l(y_1, t_1|y_0, t_0)$ be the solution of (1.9) with absorbing boundary at l. Then the probability density for the left extremal point is

$$p(l) = -\frac{\partial Q_l(y_1, t_1|y_0, t_0)}{\partial l}\bigg/P(y_1, t_1|y_0, t_0).$$

Find $p(l)$ explicitly for the case of ordinary diffusion as in (VIII.3.1).

---

[*] G.H. Weiss and R.J. Rubin, op. cit. in VI.2.

5. Boundaries of the Smoluchowski equation

Consider the general one-dimensional diffusion equation (1.9) in the interval (L, R),

$$\frac{\partial P(y, t)}{\partial t} = -\frac{\partial}{\partial y} A(y)P + \frac{1}{2}\frac{\partial^2}{\partial y^2} B(y)P \quad (L < y < R). \tag{5.1}$$

To make the solution unique it is not sufficient to fix the initial distribution $P(y, 0)$ in (L, R), but one also needs suitable boundary conditions at L and R, expressing the physical nature of these boundaries. The particle has to know what to do when it arrives at a wall. We have already encountered the condition $P = 0$, which expresses absorption, and the reflecting boundary condition

$$J \equiv AP - \frac{1}{2}\frac{\partial}{\partial y} BP = 0. \tag{5.2}$$

These cases are called *regular* because of the implied assumption that $A(y)$ and $B(y)$ are smooth functions and that $B(y) > 0$, at least in some neighborhood of the boundary. It may happen, however, that at the boundary $A(y)$ becomes infinite, or $B(y)$ vanishes, in such a way that the particle is not able to reach the wall. In that case there is no room for an extraneous boundary condition; rather equation (5.1) itself determines the effect of the wall. These are called *natural boundaries*, as in VI.5.

Remark. Other boundary conditions are possible, e.g., the general or mixed boundary condition[*]

$$\alpha P + \beta \frac{\partial P}{\partial y} = 0 \quad \text{at the boundary}, \tag{5.3}$$

where α, β are constants. In a ring one has periodic conditions

$$P(L, t) = P(R, t), \quad \left[\frac{\partial P}{\partial y}\right]_L = \left[\frac{\partial P}{\partial y}\right]_R. \tag{5.4}$$

We shall mention these only occasionally and ignore more exotic possibilities, such as that the particle may get stuck temporarily on the wall or may bounce back from the wall to some point in the interior of the interval.

---

[*] In the theory of diffusion-controlled reactions it is called the "radiative boundary condition". This quaint name originated from H.S. Carslaw and J.C. Jaeger, *Conduction of Heat in Solids* (Clarendon Press, Oxford 1947) p. 13.

5. BOUNDARIES OF THE SMOLUCHOWSKI EQUATION

The various types of boundaries have been investigated by Feller[*], but we follow a slightly different scheme[**]. As the object of our study we select the right boundary R; the left boundary may be ignored. To do that we take some point M inside (L, R) at which the functions A and B are regular and restrict our attention to the interval $M < y < R$.

In order to characterize the possible types of behavior at R we recall the abbreviation introduced in (3.5):

$$\Phi(y) = -\int_M^y \frac{2A(y')}{B(y')}\,dy',$$

and define the following three integrals

$$L_1 = \int_M^{R-\varepsilon} e^{\Phi(y)}\,dy, \tag{5.5a}$$

$$L_2 = \int_M^{R-\varepsilon} e^{\Phi(y)}\,dy \int_M^y e^{-\Phi(z)} \frac{dz}{B(z)}, \tag{5.5b}$$

$$L_3 = \int_M^{R-\varepsilon} e^{-\Phi(y)} \frac{dy}{B(y)}. \tag{5.5c}$$

It will appear that there are four types of boundaries, characterized by the behavior of these integrals in the limit $\varepsilon \to 0$, namely[†]
 (i) $L_1 \to \infty$: natural repulsive;
 (ii) $L_1 < \infty$, $L_2 \to \infty$: natural attractive;
 (iii) $L_1 < \infty$, $L_2 < \infty$, $L_3 \to \infty$: adhesive;
 (iv) $L_1 < \infty$, $L_2 < \infty$, $L_3 < \infty$: regular.

Exercise. Ordinary diffusion in a field of force is described by (XI.2.4). Write the above conditions for this special case.

Exercise. In the same special case take $U(y) = \alpha \log(R - y)$. Show that (i) holds for $\alpha \leq -\vartheta$; that (iii) holds for $\alpha \geq \vartheta$; that (iv) holds for $-\vartheta < \alpha < \vartheta$; and that (ii) never holds.

Exercise. The same criteria apply when $R = \infty$. Again for the special case (XI.2.4) take $U(y) = \alpha \log y$ and find the various types of boundaries for different values of α.

Exercise. Show that the integrals L_1, L_2, L_3 are invariant for transformation of y, compare (IX.4.14).

---

[*] W. Feller, Commun. Pure Appl. Math. **8**, 203 (1955); BHARUCHA-REID, sect. 3.3.
[**] I.I. Gichman and A.W. Skorochod, *Stochastische Differentialgleichungen* (Akademie-Verlag, Berlin 1971) ch. 5; W. Horsthemke and R. Lefever, *Noise-Induced Transitions* (Springer, Berlin 1984) sect. 5.5.
[†] The names are not standardized, see S. Karlin and H.M. Taylor, *A Second Course in Stochastic Processes* (Academic Press, New York 1981).

Type (i). The particle has zero probability to reach the wall. To see this take $\varepsilon > 0$ and let $M < y < R - \varepsilon$. The probability for a particle starting at y to exit from the interval $(M, R - \varepsilon)$ through $R - \varepsilon$ rather than through M is, according to (3.6),

$$\pi_{R-\varepsilon}(y) = \int_M^y e^{\Phi(y')}\,dy' \Big/ \int_M^{R-\varepsilon} e^{\Phi(y')}\,dy'.$$

The denominator is L_1, which for this type is infinite, so that $\pi_{R-\varepsilon} \to 0$. Thus all particles return before reaching the wall and no boundary condition is needed.

Type (ii). The particle has a non-zero probability to reach the wall, but we show that it takes an infinite time to arrive. The conditional mean first-passage time at $R - \varepsilon$ is, according to (3.11),

$$\tau_{R-\varepsilon}(y) = \left[\int_M^y e^{\Phi(z)}\,dz\right]^{-1}$$

$$\times \left[\int_y^{R-\varepsilon} e^{\Phi(y')}\,dy' \int_M^y e^{\Phi(y'')}\,dy'' \int_{y''}^{y'} e^{-\Phi(z)} \pi_{R-\varepsilon}(z) \frac{2\,dz}{B(z)}\right].$$

The denominator is finite. In the numerator we may omit the factor $\pi_{R-\varepsilon}(z)$, because it is finite and non-zero and hence does not affect the convergence. The remaining triple integral can be split:

$$\int_y^{R-\varepsilon} e^{\Phi(y')}\,dy' \int_M^y e^{\Phi(y'')}\,dy'' \int_M^{y'} e^{-\Phi(z)} \frac{dz}{B(z)}$$

$$- \int_y^{R-\varepsilon} e^{\Phi(y')}\,dy' \int_M^y e^{\Phi(y'')}\,dy'' \int_M^{y''} e^{-\Phi(z)} \frac{dz}{B(z)}.$$

On the first line the integration over y'' factors out and what remains is infinite for $\varepsilon \to 0$ because L_2 diverges. The second line is finite because $L_1 < \infty$. Hence we find for type (ii) that the particles may reach the wall, but the conditional mean time for this to happen is infinite. No boundary condition is needed.

Example to show how the probability accumulates near the boundary. In the interval $0 < y < \infty$ consider

$$\frac{\partial P(y,t)}{\partial t} = \frac{1}{2}\frac{\partial^2}{\partial y^2} y^2 P. \tag{5.6}$$

Thus $A = 0$, $B = y^2$, $\Phi = 0$. The boundary at ∞ is of type (i). The boundary at 0, however, is of type (ii). There is no normalizable stationary distribution, but the time-dependent solution can be found explicitly. First set $y = e^\eta$ and subsequently

5. BOUNDARIES OF THE SMOLUCHOWSKI EQUATION

$\eta = \xi - \tfrac{1}{2}t$. In this way one finds that the solution of (5.6) with initial delta function is

$$P(y, t | y_0, 0) = \frac{1}{y\sqrt{2\pi t}} \exp\left[-\frac{1}{2t}\left\{\log\frac{y}{y_0} + \tfrac{1}{2}t\right\}^2\right].$$

This represents a peak, which moves to the left and accumulates all probability near the boundary 0. In fact, one easily finds for $\varepsilon \to 0$

$$\int_0^\varepsilon P(y, t | y_0, 0) \, dy \to 1.$$

It is possible to visualize this behavior after the transformation $y = e^{-z}$, which gives for the distribution of z

$$R(z, t | z_0, 0) = \frac{1}{\sqrt{2\pi t}} \exp\left[-\frac{1}{2t}(z - z_0 - \tfrac{1}{2}t)^2\right].$$

The center of this Gaussian peak moves to infinity at the rate $\tfrac{1}{2}t$, while its width grows at the rate $\sqrt{t}$. This picture also demonstrates for this type that it is inappropriate to describe the accumulation of $P(y, t)$ by saying that there is a stationary distribution $P^s(y) = \delta(y)$.

In the case of a boundary of one of the two remaining types the particle can arrive at the boundary so that a boundary condition is needed. One may impose the absorbing condition. The reflecting boundary condition, however, makes sense only if the particle is able to drift back away from the wall into the interval. This is the case in type (iv), but not in (iii).

Type (iv). Consider a particle starting at $y \in (M, R)$. If we impose the reflecting boundary condition at R the probability for it to leave through M is unity and the mean escape time is, according to (3.7),

$$\tau_M(y) = \int_M^y e^{\Phi(y')} \, dy' \int_{y'}^R e^{-\Phi(y'')} \frac{2 \, dy''}{B(y'')}. \tag{5.7}$$

The second integral exists since $L_3 < \infty$. Hence the time needed to pass through M is finite, even when y is close to R. Moreover, if M is close to R the time is small so that the particle does not dwell in the neighborhood of the wall. Thus for type (iv) one may either impose a reflecting or an absorbing boundary condition – which is the reason for calling it "regular".

Type (iii). The second integral in (5.7) is infinite, but we can take an interval $M < y < R - \varepsilon$, and impose a reflecting boundary condition at $R - \varepsilon$. Then the mean escape time through M is given by the same formula (5.7) with $R - \varepsilon$ in place of R. If subsequently this reflecting boundary is shifted towards R this time becomes infinite. And that remains true no matter how close M is to R. We conclude that the particle dwells an infinite time in an arbitrarily narrow boundary layer. Even though we impose the reflecting condition the effect cannot be distinguished from an absorbing wall. The boundary layer acts as a limbo state, but it is part of the interval (L, R). The probability density develops a delta peak at the boundary. In the interior of the interval the situation is the same as if we had imposed an absorbing boundary condition.

Example. In the interval $0 < y < \infty$ consider

$$\frac{\partial P(y, t)}{\partial t} = \frac{\partial^2}{\partial y^2} yP. \tag{5.8}$$

The boundary at ∞ is of type (i). The boundary at 0 is of type (iii). There is no normalizable stationary distribution, but the time-dependent solution can be found explicitly by Fourier transformation:

$$P(y, t | y_0, 0) = \frac{1}{2\pi} \int_{-\infty}^{\infty} \exp\left[iky - \frac{iky_0}{1 + ikt}\right] dk. \tag{5.9}$$

At $t = 0$ this is clearly a delta function at y_0. At first sight one might think that for $t \to \infty$ it reduces to a delta function at the boundary $y = 0$ – but that is incompatible with the fact that $\langle y \rangle$ is constant, as is seen directly from (5.8). A more careful evaluation of (5.9) is needed.

For any $y > 0$ the integration path may be shifted upwards in the complex k-plane and at $i\infty$ the integral vanishes because of the factor e^{iky}. Thus $P(y, t)$ is equal to the integral surrounding the singular point $k = i/t$. Setting $k = i/t + ir\, e^{i\vartheta}$ and choosing $r = t^{-1}\sqrt{y_0/y}$ one obtains

$$P(y, t | y_0, 0) = \frac{-1}{2\pi t}\sqrt{\frac{y_0}{y}} \exp\left[-\frac{y + y_0}{t}\right] \int_0^{2\pi} \exp\left[-\frac{2}{t}\sqrt{y_0 y}\cos\vartheta\right] e^{i\vartheta}\, d\vartheta$$

$$= \frac{1}{t}\sqrt{\frac{y_0}{y}} \exp\left[-\frac{y + y_0}{t}\right] I_1\left(\frac{2}{t}\sqrt{y_0 y}\right), \tag{5.10}$$

where I_1 is the modified Bessel function.

That this is not the final answer, however, appears on integrating (5.10) over all y. The result is

$$\int_0^\infty P(y, t | y_0, 0)\, dy = 1 - e^{-y_0/t}.$$

It shows that the total probability on the half-line $y > 0$ is drained into the sink at $y = 0$. In fact, at $y = 0$ the integral in (5.9) is infinite and it is proper to describe this behavior by adding to (5.10) the term

$$e^{-y_0/t}\delta(y) \tag{5.11}$$

5. BOUNDARIES OF THE SMOLUCHOWSKI EQUATION

Exercise. Verify explicitly for the distribution (5.10) with (5.11) that $\langle y \rangle = y_0$.

Exercise. Compute the characteristic function

$$\int_0^\infty e^{-\lambda y} P(y, t | y_0, 0) \, dy = \exp\left[-\frac{y_0 \lambda}{1 + t\lambda}\right].$$

Exercise. In the interval $0 \leqslant y < \infty$ consider

$$\frac{\partial P}{\partial t} = -\mu \frac{\partial}{\partial y} y P + \frac{1}{2} \frac{\partial^2}{\partial y^2} y^2 P. \tag{5.12}$$

Discuss the boundaries and verify the result by means of the explicit time-dependent solution.

Exercise. In the same interval consider

$$\frac{\partial P}{\partial t} = -\frac{\partial}{\partial y} y^2 P + \frac{\partial^2}{\partial y^2} y^2 P. \tag{5.13}$$

Show that the boundary zero is of type (ii) and the boundary ∞ is of type (iii).

Exercise. The one-step process (VI.4.8) has at $n = 0$ a natural boundary in the sense of VI.5. However, the corresponding Fokker–Planck equation (VIII.5.3) has a regular boundary (in the present sense).[*] Show that the reflecting condition has to be imposed to obtain agreement with the discrete M-equation. [Compare XIII.3.]

Exercise. The preceding Exercise tacitly assumed $n_A > 0$. Show that for $n_A = 0$ the boundary is adhesive. This agrees with the fact that in the discrete case the site $n = 0$ is absorbing.

Exercise. Study the equation[**]

$$\frac{\partial P}{\partial t} = \frac{\partial}{\partial y} y^3 \frac{\partial P}{\partial y} \quad (0 < y < \infty).$$

Exercise. Study for $\mu < \frac{1}{2}$, $\mu > \frac{1}{2}$, $\mu = \frac{1}{2}$ the generalization of (5,13)[†]

$$\frac{\partial P}{\partial t} = -\frac{\partial}{\partial y}(\mu y - y^2) P + \frac{1}{2} \frac{\partial^2}{\partial y^2} y^2 P \quad (0 < y < \infty).$$

We consider the stationary solutions of (5.1). On putting the left-hand side equal to zero, one is left with an ordinary second order differential equation. The general solution involves two integration constants J, K:

$$P(y) = -\frac{2J}{B(y)} e^{-\Phi(y)} \int_M^y e^{\Phi(y')} \, dy' + \frac{2K}{B(y)} e^{-\Phi(y)}. \tag{5.14}$$

J is the probability flow, M is an arbitrary point between L and R, the

---

[*] H. Grabert, P. Hänggi, and I. Oppenheim, Physica A **117**, 300 (1983).
[**] G. Casati, B.V. Chirikov, D.L. Shepelyansky, and I. Guarneri, Physics Reports **154**, 77 (1987), p. 119.
[†] Horsthemke and Lefever, op. cit., p. 125.

arbitrariness being compensated by K. This solution represents a stationary distribution when it can be normalized, i.e., when its integral over y converges. We first look at the upper bound R.

The second term diverges if $L_3 \to \infty$. In that cast the first term diverges as well and no stationary distribution exists other than zero. If L_3 is finite the second term converges and one obtains an equilibrium solution by taking $J = 0$ and using K for normalization. The first term also converges when

$$L_4 \equiv \int_M^{R-\varepsilon} e^{-\Phi(y)} \frac{dy}{B(y)} \int_M^y e^{\Phi(y')} dy' < \infty. \tag{5.15}$$

One may then take $J \neq 0$ so as to obtain a stationary distribution with an arbitrary flow, again normalizing by means of K.

Exercise. If L_3 converges, then either L_1 and L_2 are both divergent and the boundary is repulsive; or they are both convergent and the boundary is regular.

Exercise. Prove $L_2 + L_4 = L_1 L_3$. Hence if L_1 and L_3 converge, so do L_2 and L_4, and the boundary is regular: it imposes no restrictions on J and K.

Exercise. The boundaries $\pm \infty$ of the ordinary diffusion equation $\partial P/\partial t = \partial^2 P/\partial y^2$ are natural attractive, type (ii), but L_3 and L_4 diverge, so that no stationary distribution exists.

Exercise. The same is true for

$$\frac{\partial P}{\partial t} = -\mu \frac{\partial}{\partial y} yP + \frac{\partial^2 P}{\partial y^2} \quad (-\infty < y < \infty) \tag{5.16}$$

when $\mu > 0$, but not when $\mu < 0$.

The following paragraphs are added for completeness, although they are simple consequences of the preceding analysis, which the reader should be able to figure out by himself.

The existence of a normalizable stationary solution depends on *both* boundary conditions. Each one gives a homogeneous linear equation for J and K. Two regular boundaries have no other solution than $J = K = 0$, unless they obey a compatibility relation. Two reflecting boundaries are compatible because each single one gives $J = 0$. If one or both of them are absorbing no stationary solution other than zero exists. An absorbing boundary may be combined with a suitable mixed boundary condition (5.3).

If a boundary is natural the boundary condition is replaced by a convergence condition. For a natural repulsive boundary (i) one has $L_1 \to \infty$. If also $L_3 \to \infty$ there is no solution. If $L_3 < \infty$ but $L_4 \to \infty$, only a solution with $J = 0$ is possible, which is compatible with a reflecting boundary at the other end, but not with an absorbing boundary. If $L_3 < \infty$ and $L_4 < \infty$ no restriction results from the boundary, so that it is compatible with any boundary condition at the other end.

For a natural attractive boundary (ii) one has $L_1 < \infty$, $L_2 \to \infty$ and hence $L_3 \to \infty$ (see Exercise above). No normalizable stationary solution exists. For an adhesive boundary (iii) one also has $L_3 \to \infty$ and hence no stationary distribution.

Exercise. For ordinary diffusion in the interval (L, R) show that a solution exists if at each end a mixed boundary condition (5.3) is imposed, provided that they are linked by the compatibility relation

$$R + \beta_R/\alpha_R = L + \beta_L/\alpha_L.$$

Exercise. Find for the general Smoluchowski equation (5.1) between two regular boundaries L, R the condition that two mixed boundary conditions at L and R are compatible.

Exercise. Show that for $\alpha > 1$ the equation

$$\frac{\partial P}{\partial t} = \frac{\partial}{\partial y} y^\alpha P + \frac{\partial^2 P}{\partial y^2} \quad (0 < y < \infty), \quad P(0, t) = 0$$

has a stationary distribution for any $J < 0$. For $\alpha = 1$, however, one must have $J = 0$.

6. First passage of non-Markov processes

So far we studied the first passage of Markov processes such as described by the Smoluchowski equation (1.9). On a finer time scale, diffusion is described by the Kramers equation (VIII.7.4) for the joint probability of the position X and the velocity V. One may still ask for the time at which X reaches for the first time a given value R, but X by itself is not Markovian. That causes two complications, which make it necessary to specify the first-passage problem in more detail than for diffusion.

The first complication, already pointed out in IX.7, is that the process $X(t)$ is not uniquely determined by an initial value $X(t_0)$, but one needs to know $V(t_0)$ as well, either its value or its probability distribution. The second complication is that the first passage through R may occur with different values of V; and it may happen, depending on the physical context, that these different events cannot be counted indiscriminately. A simple example is a particle diffusing in the presence of a potential wall, over which it escapes when it reaches it with a sufficient velocity.

The following first-passage problem is well-defined. A Kramers particle starts out at t_0 in the point X_0 with velocity V_0. One asks for the distribution of the time t at which it first passes a point $R > X_0$, regardless of its velocity. Clearly one must have $V > 0$ and there must be total absorption at R in order to prevent return. Hence the equation to be solved is

$$\frac{\partial Q}{\partial t} + V\frac{\partial Q}{\partial X} + F(X)\frac{\partial Q}{\partial V} = \gamma\left\{\frac{\partial}{\partial V}VQ + T\frac{\partial^2 Q}{\partial V^2}\right\} \quad (6.1a)$$

for $X < R$, $-\infty < V < \infty$, $t > t_0$, with

$$Q(R, V < 0, t) = 0, \qquad Q(X, V, t_0) = \delta(X - X_0)\delta(V - V_0). \quad (6.1b)$$

The emerging probability flow at R is

$$J_R(t) = \int_0^\infty Q(R, V, t)V \, dV.$$

Its time integral equals $\pi_R(X_0, V_0)$ and its time dependence determines the first-passage time distribution.

Exercise. Formulate the problem for two boundaries L, R. Prove the physically obvious identity

$$\frac{d}{dt}\int_L^R dx \int_{-\infty}^\infty dV\, Q(X, V, t) = -J_R(t) + J_L(t).$$

Exercise. Find a similar identity for the energy and interpret the separate terms.

Exercise. For the same case derive the expressions for the two conditional mean first-passage times $\tau_L(X_0, V_0)$ and $\tau_R(X_0, V_0)$.

Exercise. What is the condition for a reflecting boundary? Formulate the first-passage problem when L is a reflecting wall.

Exercise. A particle moves in (L, R) with a velocity that at random moments jumps between two possible values c and $-c$. Show that its first-passage time distribution requires the solution of[*]

$$\frac{\partial Q_+}{\partial t} = -c\frac{\partial Q_+}{\partial x} + \gamma Q_- - \gamma Q_+,$$

$$\frac{\partial Q_-}{\partial t} = c\frac{\partial Q_-}{\partial x} + \gamma Q_+ - \gamma Q_-, \quad (6.2)$$

with

$$Q_+(L, t) = 0, \qquad Q_-(R, t) = 0, \qquad Q_+(x, 0) = \delta(x - x_0), \qquad Q_-(x, 0) = 0.$$

Unfortunately, no analytic solution of (6.1) is known, except in the case of a single boundary and constant F.[**] In analogy with (3.1) a simpler

---

[*] G.H. Weiss, J. Masoliver, K. Lindenberg, and B.J. West, Phys. Rev. A **36**, 1435 (1987).

[**] T.W. Marshall and E.J. Watson, J. Phys. A **18**, 3531 (1985); **20**, 1345 (1987). For an approximate solution, see M.E. Widder and U.M. Titulaer, J. Statist. Phys. **56**, 471 (1989). Even the case $F = 0$ is not simple, see P.S. Hagen, C.R. Doering, and C.D. Levermore, J. Statist. Phys. **54**, 1321 (1989).

6. FIRST PASSAGE OF NON-MARKOV PROCESSES

equation can be derived for the splitting probability $\pi(X, V)$:

$$-V\frac{\partial \pi}{\partial X} - F(X)\frac{\partial \pi}{\partial V} = -\gamma V\frac{\partial \pi}{\partial V} + \gamma T\frac{\partial^2 \pi}{\partial V^2} \qquad (6.3)$$

with boundary conditions

$$\pi_R(R, V>0) = 1, \qquad \pi_R(L, V<0) = 0,$$
$$\pi_L(R, V>0) = 0, \qquad \pi_L(L, V<0) = 1.$$

The differential operator in (6.3) is the adjoint of that in (6.1) and the boundary values are prescribed on the parts of the boundary on which Q was free, and vice versa. However, even this simple problem cannot be solved; we return to it in XIII.6.

Exercise. Write the analogous equations for the mean first-passage times through L and R.

Exercise. Solve (6.2) with boundary conditions for the trivial case $\gamma = 0$ and demonstrate that the result is correct. [One has three possible answers depending on X_0, V_0.]

Exercise. For the case (6.2) derive, putting $c = 1$,

$$\pi_R(x, +) = \frac{e^{2\gamma(x-L)} + 1}{e^{2\gamma(R-L)} + 1}, \qquad \pi_R(x, -) = \frac{e^{2\gamma(x-L)} - 1}{e^{2\gamma(R-L)} + 1}.$$

[See G.H. Weiss, J. Statist. Phys. **37**, 325 (1984).]

Exercise. Solve (6.2). [Hint: Determine the normal modes obeying the boundary conditions, find an orthonormality relation between them and their adjoints, and assume that they constitute a complete set in the interval (L, R) so that the initial condition can be satisfied.]

Another example of a non-Markov problem is an overdamped particle in a potential $U(y)$ and subject to a non-white Ornstein–Uhlenbeck force

$$\dot{y} = -U'(y) + \xi, \qquad (6.4a)$$
$$\dot{\xi} = -\gamma\xi + L(t). \qquad (6.4b)$$

$L(t)$ is the Langevin force of IX.1, while the parameter $\gamma = 1/\tau_c$ determines the color of the noise $\xi(t)$.[*] The joint variable (y, ξ) is Markovian and obeys the M-equation (IX.7.5):

$$\frac{\partial \mathscr{P}(y, \xi, t)}{\partial t} = \frac{\partial}{\partial y}[U'(y) - \xi]\mathscr{P} + \gamma\left[\frac{\partial}{\partial \xi}\xi\mathscr{P} + T\frac{\partial^2 \mathscr{P}}{\partial \xi^2}\right]. \qquad (6.5)$$

---

[*] P. Hänggi, F. Marchesoni, and P. Grigolini, Z. Phys. B **56**, 333 (1984); Th. Leiber and H. Risken, Phys. Rev. A **38**, 3789 (1988).

As initial condition one may choose, e.g., a fixed value y_0 and the equilibrium distribution for ξ,

$$\mathscr{P}(y, \xi, t_0) = \delta(y - y_0) e^{-\xi^2/2T}/\sqrt{2\pi T}. \tag{6.6}$$

The first time that y reaches the point $R > y_0$ it will have positive velocity: $\xi - U'(R) > 0$. Thus one has to solve (6.5) with boundary condition

$$\mathscr{P}(R, \xi, t) = 0 \quad \text{for } \xi < U'(R).$$

Exercise. Write the equations and boundary conditions for the quantities $\pi(y_0, \xi_0)$ and $\tau(y_0, \xi_0)$ belonging to (6.4). Note that they do not correspond to the initial condition (6.6); how do they relate to them?

Exercise. The equation (6.5) (without boundaries) does not have in general an equilibrium distribution of the separable form $X(\xi)Y(y)$. Hence the conditional probability $P(\xi|y)$ for given y is not the same as the marginal probability $P(\xi)$. For the case $U(y) = \tfrac{1}{2}\omega_0^2 y^2$ show explicitly that this is so.

Exercise. For the (somewhat contrived) case $U(y) = A\,e^{\gamma a y} + T a y$ with arbitrary A, a the stationary solution is separable. Although it is not normalizable one can still show that the conditional probability $P(\xi|y)$ is not the same as the unperturbed distribution of ξ in (6.6).

7. Markov processes with large jumps

Consider a Markov process $Y(t)$ in the range $(-\infty, \infty)$ governed by the M-equation

$$\dot{P}(y, t) = \int_{-\infty}^{\infty} W(y|y')P(y', t)\,dy' - P(y, t)\int_{-\infty}^{\infty} W(y'|y)\,dy'. \tag{7.1}$$

Take an interval (L, R) and an initial value y_0 in it. We want to know the splitting probabilities and the time distribution for leaving the interval through either L or R. The probability $Q(y, t)$ for being still in (L, R) without having made a jump across an end point obeys for $L < y < R$

$$\dot{Q}(y, t) = \int_{L}^{R} W(y|y')Q(y', t)\,dy' - Q(y, t)\int_{-\infty}^{\infty} W(y'|y)\,dy', \tag{7.2}$$

$$Q(y, 0) = \delta(y - y_0).$$

If one has solved this equation the probability per unit time for a jump across R is

$$J_R(t) = \int_{R}^{\infty} dy \int_{L}^{R} dy'\, W(y|y')Q(y', t).$$

A similar formula holds for $J_L(t)$ and hence everything is known. This answer

7. MARKOV PROCESSES WITH LARGE JUMPS

to the problem is the analog of the absorbing boundary approach in **1**. Unfortunately the crucial equation (7.2) can only rarely be solved.

The less ambitious approach in **2** through the adjoint equation aims only at the splitting probabilities and the mean first-passage times. We construct an equation analogous to (2.2) and (3.1). If at time t the value of Y equals x, it will have at $t + \Delta t$ another value x' with probability $\Delta t\, W(x'|x)$, or it has the same value x. The probability $\pi_R(x)$ that Y, starting at x, will exit through R obeys therefore the identity

$$\pi_R(x) = \Delta t \int_L^R W(x'|x)\pi_R(x')\,dx' + \Delta t \int_R^\infty W(x'|x)\,dx'$$
$$+ \left\{1 - \Delta t \int_{-\infty}^\infty W(x'|x)\,dx'\right\}\pi_R(x).$$

Hence we have for $L < x < R$ the integral equation

$$\int_L^R \pi_R(x')W(x'|x)\,dx' - \pi_R(x)\int_{-\infty}^\infty W(x'|x)\,dx' + \int_R^\infty W(x'|x)\,dx' = 0.$$

(7.3)

In contrast to (2.2) there is an inhomogeneous term but there are no boundary conditions. (It is not true that $\pi_R(x)$ must be equal to unity at $x = R$, because when the particle arrives at R it need not exit but may jump back into the interval.) The integral equation is sufficient to determine $\pi_R(x)$ uniquely. In fact, suppose that (7.3) had two different solutions; then their difference $\delta\pi_R(x)$ obeys

$$\int_L^R \{\delta\pi_R(x') - \delta\pi_R(x)\}W(x'|x)\,dx' = \delta\pi_R(x)\left\{\int_R^\infty + \int_{-\infty}^L W(x'|x)\,dx'\right\}.$$

In this equation substitute for x the value at which $\delta\pi_R(x)$ is maximal. It follows that $\delta\pi_R(x) = 0$ for all x.

For the quantity $\vartheta_R(x) = \pi_R(x)\tau_R(x)$ one finds by a similar argument

$$\vartheta_R(x) = \left\{1 - \Delta t \int_{-\infty}^\infty W(x'|x)\,dx'\right\}\left\{\vartheta_R(x) - \pi_R(x)\Delta t\right\}$$
$$+ \Delta t \int_L^R W(x'|x)\vartheta_R(x')\,dx' + \mathcal{O}(\Delta t)^2.$$

Thus $\vartheta_R(x)$ obeys for $L < x < R$ the integral equation

$$\int_L^R \vartheta_R(x')W(x'|x)\,dx' - \vartheta_R(x)\int_{-\infty}^\infty W(x'|x)\,dx' = -\pi_R(x). \tag{7.4}$$

Again the solution is unique without boundary conditions.

Exercise. The general M-equation (7.1) contains the one-step process as a special case. Show that in this special case, (7.3) and (7.4) reduce to (2.3) and (2.7), respectively (including their boundary conditions).

Exercise. Same question for the special case that (7.1) is the diffusion equation (1.9).

Exercise. We modify (7.1) by erecting an artificial reflecting boundary at $y = 0$; then for $y > 0$

$$\dot P(y,t) = \int_0^\infty W(y|y')P(y',t)\,dy' + \int_0^\infty W(-y|y')P(y',t)\,dy'$$

$$- P(y,t)\int_{-\infty}^\infty W(y'|y)\,dy'. \tag{7.5}$$

Show that in this case $\pi_R(x) = 1$ for any $0 < x < R$. Hence $\tau_R(x) = \vartheta_R(x)$; write the equation for $\tau_R(x)$.

Example. The equations can be solved explicitly for the case

$$W(y|y') = e^{-\gamma|y-y'|}. \tag{7.6}$$

The reason is that this function has the property

$$\left(\frac{d^2}{dy^2} - \gamma^2\right)W(y|y') = -2\gamma\,\delta(y-y').$$

Applying this differential operator to the x-dependence in (7.3) one finds

$$-2\gamma\pi_R(x) = \frac{2}{\gamma}\left[\frac{d^2\pi_R(x)}{dx^2} - \gamma^2\pi_R(x)\right]. \tag{7.7}$$

It follows that $\pi_R(x)$ is linear,

$$\pi_R(x) = ax + b. \tag{7.8}$$

The constants a, b are not fixed by boundary conditions, but by the fact that $\pi_R(x)$ does not only have to obey (7.7) but also the original integral equation (7.3).

$$\int_L^R e^{-\gamma|x-x'|}(ax'+b)\,dx' - \frac{2}{\gamma}(ax+b) + \frac{1}{\gamma}e^{\gamma(x-R)} = 0.$$

Carrying out the elementary integration one finds that many terms cancel and the remaining ones are

$$-\frac{1}{\gamma}e^{\gamma(L-x)}\left(aL+b-\frac{a}{\gamma}\right) - \frac{1}{\gamma}e^{\gamma(x-R)}\left(aR+b+\frac{a}{\gamma}-1\right).$$

7. MARKOV PROCESSES WITH LARGE JUMPS

In order that this vanishes one must have

$$a = \frac{1}{R - L + 2/\gamma}, \quad b = -\frac{L - 1/\gamma}{R - L + 2/\gamma}.$$

This fully determines (7.8). If one shifts the origin so that $L = 0$ and scales x so that $\gamma = 1$ the result is

$$\pi_R(x) = \frac{x + 1}{R + 2}.$$

Exercise. For the same special case (7.6) with $L = 0$, $\gamma = 1$ derive

$$\vartheta(x) = -\frac{x^3 + 3x^2}{12(R + 2)} + \frac{R^3 + 6R^2 + 12R + 12}{12(R + 2)^2}(x + 1).$$

Exercise. Find the mean first-passage time at R for the M-equation (7.5) with W given by (7.6).[*]

---

[*] N.G. van Kampen, in: *Instabilities and Nonequilibrium Structures* (E. Tirapegui and D. Villarroel eds., Reidel, Dordrecht 1987).

Chapter XIII

UNSTABLE SYSTEMS

The basic stability condition (X.3.4) is violated (more drastically than in chapter XI) when $\alpha_{1,0}(\phi) < 0$. Such cases are studied in this chapter. They are selected so as to demonstrate the novel features that may arise. No attempt is made to provide an exhaustive treatment of all possible types of instabilities.

1. The bistable system

To facilitate the discussion it is helpful to specify three of the numerous meanings of the word "state". We shall call a *site* any value of the stochastic variable X or n. We shall call a *"macrostate"* any value of the macroscopic variable ϕ. A time-dependent macrostate is a solution of the macroscopic equation (X.3.1), a stationary macrostate is a solution of (X.3.3). We shall call a *"mesostate"* any probability distribution P. A time-dependent mesostate is a solution of the master equation, the stationary mesostate is the time-independent solution $P^s(X)$.

First suppose that the stability condition (X.3.4) is obeyed. Then there is only one stationary macrostate ϕ^s. It is related to the stationary mesostate $P^s(X)$ in the sense that the latter consists of a sharp peak around $\Omega\phi^s$, which in the limit $\Omega \to \infty$ tends to a delta function at $\Omega\phi^s$. Moreover, thanks to (X.3.4) it is possible to relate with every time-dependent macrostate $\phi(t)$ a time-dependent mesostate $P(X, t)$ consisting of a sharp peak around $\phi(t)$ and moving along with it. This relation is neither unique nor precisely defined: to each $\phi(t)$ there are many $P(X, t)$ with these properties and there is no precise way of telling how sharp they have to be, except that the width must not be larger than order $\Omega^{1/2}$.

These mesostates are the only ones considered so far. But there are many others, namely those that do not consist of a single sharp peak and cannot therefore be related to any one macrostate ϕ. Rather they describe a probability distribution over a set of macrostates. For instance, take the mesostate

$$P(X) = \pi_1 \delta(X - X_1) + \pi_2 \delta(X - X_2), \tag{1.1}$$

with two constants $X_1 \neq X_2$, and two positive coefficients $\pi_1 + \pi_2 = 1$. This mesostate is related to *two* macrostates $\phi_1 = X_1/\Omega$ and $\phi_2 = X_2/\Omega$, in such a way that the system has a probability π_1 to be in the macrostate ϕ_1, and

1. THE BISTABLE SYSTEM

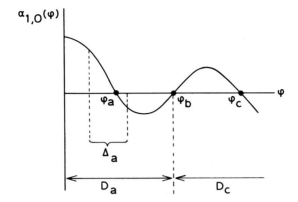

Fig. 34. The macroscopic rate equation for a bistable system and the domains of attraction.

π_2 to be in ϕ_2. In a similar way any flat distribution $P(X)$ can be decomposed as a sum of sharp peaks each of which is a mesostate related to a macrostate. Again this decomposition is not unique because the shape and width of the individual peaks can be varied within reasonable bounds.

Now consider the evolution in a situation as in fig. 34. There are three stationary macrostates ϕ_a, ϕ_b, ϕ_c, of which ϕ_a and ϕ_c are locally stable and ϕ_b is unstable. Of course, even the pure macroscopist would not regard ϕ_b as a realizable state, on the ground that a system in ϕ_b would be caused to move into either ϕ_a or ϕ_c by the smallest perturbation. Systems having a macroscopic characteristic as in fig. 34 are called "bistable". There are numerous examples; the ones that occur most often in the literature are the laser (section **9** below), the tunnel diode[*], and the Schlögl reaction (X.3.6). The macroscopic rate equation for this reaction is

$$\dot{\phi} = k_1 \phi_A \phi^2 - k_1' \phi^3 - k_2 \phi + k_2' \phi_B, \qquad (1.2)$$

where ϕ_A, ϕ_B are the constant concentrations of A and B. The right-hand side is $\alpha_{1,0}(\phi)$ and, for suitable values of the constants, has the form of fig. 34.[**]

Now examine the bistable case on the mesoscopic level. First consider the locally stable solution ϕ_a. As $\alpha'_{1,0}(\phi_a) < 0$ there will be a domain Δ_a around ϕ_a where (X.3.4) holds. Any macrostate $\phi(t)$ that starts at a $\phi(0)$ inside Δ_a

---

[*] R. Landauer, J. Appl. Phys. **33**, 2209 (1962); L. Esaki, Science **183**, 1149 (1974); A. van der Ziel, *Solid State Physical Electronics* (3rd Ed., Prentice Hall, Englewood Cliffs, NJ 1976).

[**] Other chemical instabilities are treated in A. Nitzan, P. Ortoleva, J. Deutch, and J. Ross, J. Chem. Phys. **61**, 1056 (1974); R.M. Noyes, J. Chem. Phys. **64**, 1266 (1976). See also Advances in Chemical Physics **43** (Wiley, New York 1980) and F. Schlögl, Physics Reports **62**, 267 (1980).

will move towards ϕ_a. Also the mesostate $P(X, t)$ related to it by $P(X, 0) = \delta[X - \Omega\phi(0)]$ will remain a sharp peak, provided Ω is large enough. Hence the evolution of $P(X, t)$ can still be approximated by means of the Ω-expansion and there is still a close relationship between macrostates and mesostates.

In fact, it is clear from fig. 34 that there is a larger *domain of attraction* D_a such that every solution $\phi(t)$ with $\phi(0)$ in D_a tends to ϕ_a. Two macrostates starting at two neighboring points in $D_a - \Delta_a$ will first move away from each other, but subsequently approach one another again, until they both end up in ϕ_a. This is clear from fig. 28 and also from the variational equation (X.3.5). Accordingly the fluctuations about such a $\phi(t)$ will first grow[*], but subsequently decrease again. Hence they can still be described by the Ω-expansion and there is still a relation between macrostates and suitably chosen mesostates.

Yet this description is not entirely correct, because even when the system is at a site in D_a, and even inside Δ_a, there is still a probability, however small, for a giant fluctuation to occur, which takes it across ϕ_b into D_c. It will then move to the neighborhood of ϕ_c until a similar giant fluctuation takes it back to D_a. Thus a mesostate represented by a probability peak in Δ_a does not survive forever: its probability is slowly depleted in favor of a peak near ϕ_c. Although ϕ_a is a stable solution of the macroscopic equation, the related mesostate is not strictly stable but merely long-lived, and may be called *metastable*. Indeed, a metastable state in thermodynamics, such as supersaturated vapor, also exists because it is stable with respect to small fluctuations, but an improbable giant fluctuation may carry it into a macroscopically different thermodynamic state.

In this discussion it was tacitly assumed that in fact the giant fluctuations across ϕ_b are rare. That made it possible to distinguish two time scales. The short time scale is determined by the rate at which inside Δ_a equilibrium is established, and can be seen in fig. 28. The long time scale is the time at which giant fluctuations occur. Both scales are clearly distinct provided that the metastable stationary distribution around ϕ_a is narrow compared to the distance to ϕ_b. A rough measure for the rate of occurrence of giant fluctuations is the height of the actual stationary distribution P^s at ϕ_b. A precise formula for the rate is derived in the next section. For systems of macroscopic size this "escape time" may easily be many times the age of the universe. For the dissociation of a single molecule it is determined by the Arrhenius factor. In systems in which the stabilities and instabilities are not well pronounced, as in fig. 35, the distinction between both time scales loses its meaning; this is the critical region discussed in section **5**.

---

[*] This was called the "fluctuation enhancement theorem" by M. Suzuki, Prog. Theor. Phys. **56**, 77 (1976).

1. THE BISTABLE SYSTEM

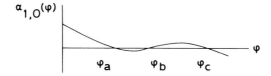

Fig. 35. Bistable system near critical point.

Remark. The state of affairs is often visualized as a random walk or diffusion in a potential $V(x)$ defined by

$$V'(\phi) = -\alpha_{1,0}(\phi).$$

The potentials corresponding to the stable, bistable, and critical situations are sketched in fig. 36. The giant fluctuations are visualized by diffusion across the potential barrier.

This suggestive picture, however, fails to take into account the difference between V and U expounded in XI.5, see in particular (XI.5.11). For instance, it suggests that $P^s = \exp(-V/\theta)$, which leads to the erroneous idea that ϕ_a and ϕ_c are equally

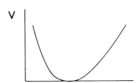

Fig. 36a. Potential illustrating global stability.

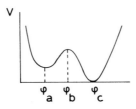

36b. Potential illustrating bistability.

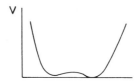

36c. Potential illustrating the situation of fig. 35.

probable if $V(\phi_a) = V(\phi_c)$, rather than $U(\phi_a) = U(\phi_c)$. The picture is valid only if one is dealing with actual diffusion in an external field and with constant diffusion coefficient θ, as described by (XI.2.4).

We return to the bistable situation with clearly separated time scales and suppose that the system starts out at a site near the boundary of D_a, i.e., close to the macroscopically unstable point ϕ_b. Then in the initial stage fluctuations across ϕ_b are not improbable. There is therefore a non-negligible probability that the system does not follow the macroscopic path towards ϕ_a but ends up in ϕ_c instead. Thus *near a point of macroscopic instability fluctuations give rise to a macroscopic effect*. It is therefore no longer possible to separate a macroscopic part from a fluctuation term as in (X.2.9) and treat the fluctuations as a small perturbation. The conclusion is that there exists no mesostate related to the stationary macrostate ϕ_b. Any probability distribution originally peaked near ϕ_b evolves in time and does not remain localized. The evolution occurs in three successive stages.*[)]

STAGE I. The distribution broadens rapidly but fluctuations across ϕ_b are still possible.

STAGE II. Fluctuations across ϕ_b have died out: the total probability is decomposed in two autonomous parts, left and right of ϕ_b. The probabilities of both parts,

$$\pi_a = \sum_{n=-\infty}^{\phi_b} p_n(t), \qquad \pi_c = \sum_{\phi_b}^{\infty} p_n(t) \tag{1.3}$$

have become practically constant in time. They are, of course, determined by the initial distribution. Each part broadens at first, but narrows into a peak when it reaches the region where the curves in fig. 28 begin to converge again.

STAGE III. The peaks have reached their local equilibrium shape around ϕ_a and ϕ_c. The mesostate does not correspond to a single macrostate but to combination of two, resembling (1.1). This mesostate is metastable, however, because on a very long time scale fluctuations across ϕ_b transfer probability between both peaks. Hence π_a and π_c vary slowly according to

$$\dot{\pi}_a = -\dot{\pi}_c = -\frac{\pi_a}{\tau_{ca}} + \frac{\pi_c}{\tau_{ac}}. \tag{1.4}$$

The constant $1/\tau_{ac}$ is the probability per unit time for a system near ϕ_c to fluctuate across ϕ_b into the domain Δ_a around ϕ_a. Ultimately π_a, π_c reach their stationary values π_a^s, π_c^s determined by

$$\pi_a^s/\tau_{ca} = \pi_c^s/\tau_{ac}. \tag{1.5}$$

---

*[)] I. Oppenheim, K.E. Shuler, and G.H. Weiss, Physica A **88**, 191 (1977).

1. THE BISTABLE SYSTEM

Example. A collection of N dipoles, each having an up and a down position. The magnetization is proportional to $2n$, being the number of up spins minus the number of down spins. The mean field theory of Weiss yields (for N even)

$$p_n^e = \text{const.} \binom{N}{\tfrac{1}{2}N + n} \exp\left[\frac{2K}{N} n^2\right], \tag{1.6}$$

where K/N is a constant measuring the strength of the ferromagnetic interaction. It is supposed[*] that n may jump by one unit, with probabilities per unit time

$$g(n) = (\tfrac{1}{2}N - n)\exp\left[\frac{2K}{N} n\right] \quad (n \to n+1), \tag{1.7a}$$

$$r(n) = (\tfrac{1}{2}N + n)\exp\left[-\frac{2K}{N} n\right] \quad (n \to n-1). \tag{1.7b}$$

N serves as parameter Ω and the macroscopic equation for $n/N = \phi$ is

$$\dot{\phi} = \alpha_1(\phi) = (\tfrac{1}{2} - \phi)e^{2K\phi} - (\tfrac{1}{2} + \phi)e^{-2K\phi}.$$

Below the Curie temperature, $K > 1$, there are two stable stationary macrostates $\pm \phi_a$ and one unstable one, $\phi_b = 0$. If the system at $t = 0$ is in $\phi_b = 0$ it will not stay there but, triggered by external perturbations or by its internal fluctuations, it will move to either ϕ_a or $-\phi_a$. This is called *symmetry breaking*: although both the equations and the initial data are symmetrical the final macrostate is not. On the other hand, the mesostate $p_n(t)$ determined by the initial condition $p_n(0) = \delta_{n,0}$ remains symmetric at all $t > 0$.

Remark. *Instability and bistability are defined as properties of the macroscopic equation.* The effect of the fluctuations is merely to make the system decide to go to one or the other macroscopically stable point. Similarly the Taylor instability and the Bénard cells are consequences of the macroscopic hydrodynamic equations.[**] Fluctuations merely make the choice between different, equally possible macrostates, and, in these examples, determine the location of the vortices or of the cells in space. (In practice they are often overruled by extraneous influences, such as the presence of a boundary.) Statements that fluctuations shift or destroy the bistability are obscure, because on the mesoscopic level there is no sharp separation between stable and unstable systems. Some authors call a mesostate (i.e., a probability distribution P) "bistable" when P has two maxima, however flat. This does not correspond to any observable fact, however, unless the maxima are well-separated peaks, which can each be related to separate macrostates, as in (1.1).

Exercise. Show that an arbitrary, smoothly varying distribution P can be written as a superposition of Gaussians with one and the same variance.

---

[*] R.J. Glauber, J. Mathem. Phys. **4**, 294 (1963); Th.W. Ruijgrok and J.A. Tjon, Physica **65**, 539 (1973).

[**] S. Chandrasekhar, *Hydrodynamic and Hydromagnetic Instability* (Clarendon, Oxford 1961; Dover, New York 1981); G. Ahlers, C.W. Meyer, and D.S. Cannell, J. Statist. Phys. **54**, 1121 (1989). H.J. Kull, Physics Reports **206**, 197 (1991).

Exercise. With the aid of (VI.3.8) verify for the one-step process (X.1.3) that P^s has maxima near ϕ_a and ϕ_c and a minimum near ϕ_b. Give a formula for the relative heights of the maxima.

Exercise. Knowledge of P^s fixes the ratio τ_{ca}/τ_{ac}. Why is this *not* an example of detailed balance? The values of τ_{ca} and τ_{ac} themselves cannot, of course, be obtained from P^s because they involve the time dimension.

Exercise. Suppose $\alpha_{1,0}(\phi)$ has A stable zeros separated by $A-1$ unstable ones. Write the equations corresponding to (1.4) and (1.5).

So far we studied instabilities in discrete one-step processes. Now consider diffusion processes as in XI. For simplicity we restrict ourselves to diffusion in an external potential $U(x)$, as described by the quasilinear Fokker–Planck equation (XI.2.4):

$$\frac{\partial P(x,t)}{\partial t} = \frac{\partial}{\partial x} U'(x) P + \theta \frac{\partial^2 P}{\partial x^2}. \tag{1.8}$$

According to XI.5 the deterministic equation is now obtained in the limit $\theta \to 0$,

$$\dot{x} = -U'(x). \tag{1.9}$$

The system is bistable when U has the shape of fig. 36b. The stationary microstates are the zeros of $U'(x)$. The same three stages as before may be distinguished. The stationary distribution is

$$P^s(x) = C\, e^{-U(x)/\theta}, \qquad C^{-1} = \int_{-\infty}^{\infty} e^{-U(x)/\theta}\, dx. \tag{1.10}$$

If θ is small one has*⁾

$$C^{-1} = e^{-U(a)/\theta} \sqrt{\frac{2\pi\theta}{U''(a)}} + e^{-U(c)/\theta} \sqrt{\frac{2\pi\theta}{U''(c)}}.$$

In the same approximation one finds as the analog of (1.3)

$$\pi_a^s = \int_{-\infty}^{b} P^s(x)\, dx = C\sqrt{\frac{2\pi\theta}{U''(a)}}, \qquad \pi_c^s = C\sqrt{\frac{2\pi\theta}{U''(c)}}.$$

The ratio between both has been called the "relative stability" of the two potential wells**⁾:

$$\frac{\pi_a^s}{\pi_c^s} = e^{-[U(a)-U(c)]/\theta} \sqrt{\frac{U''(c)}{U''(a)}}. \tag{1.11}$$

---

*⁾ Henceforth we set a, b, c for ϕ_a, ϕ_b, ϕ_c.
**⁾ R. Landauer, in: *The Maximum Entropy Formalism* (R.D. Levine and M. Tribus eds., MIT Press, Cambridge, MA 1978) p. 321.

The exponential is the expected Boltzmann factor and the second derivatives take into account the widths of the potential wells.

Exercise. Solve the diffusion equation in an inverted parabolic potential with initial delta distribution,

$$\frac{\partial P}{\partial t} = -\frac{\partial}{\partial x}xP + \frac{\partial^2 P}{\partial x^2}. \tag{1.12}$$

Unless the initial delta is located at the very top the distribution at $t>0$ consists of a single peak with almost no exchange of probability across the top.

Exercise. One-step processes have been approximated by diffusion equations in VIII.5, but the danger has been pointed out in XI.5. Show for the Schlögl reaction (X.3.6) that it is incorrect to compute the relative stability (1.11) by using the potential function V in (XI.5.6) rather than U.

2. The escape time

All information about the evolution is, of course, contained in the M-equation, but explicit solutions for bistable systems are rare[*]. Yet the one-step process and the one-dimensional Fokker–Planck equation have three features that can be found without solving the equation.

(i) The stationary solution can be obtained, and from it follow the values of π_a^s and π_c^s, as shown in the previous section. They refer to the limiting situation after a time long compared to the time needed to cross the barrier.

(ii) The average times τ_{ca}, τ_{ac} for crossing the barrier will be found in this section.

(iii) The splitting probabilities π_a, π_c in the second stage will be found in 3 as functions of the starting point.

For convenience we restrict the discussion to the quasilinear Fokker–Planck equation (1.8). The corresponding results for the one-step process will be left for Exercises. Our $U(x)$ has the shape of fig. 36b; the deterministic equation has the same form as $\alpha_{1,0}$ in fig. 34.

If the system starts from a site near a, what is the probability per unit time to arrive at a site near c? The answer will be insensitive to the precise choice of the initial site, because it runs through all sites near a in a time much shorter than the time we want to compute. It will be equally insensitive to the precise final site. The very concept of escape time is endowed with this margin of uncertainty. It is often convenient to replace this imprecise

---

[*] N.G. van Kampen, J. Statist. Phys. **17**, 71 (1977); H. Brand and A. Schenzle, Phys. Lett. A **68**, 427 (1978); M. Razavy, Phys. Lett. A **72**, 89 (1979); M.-O. Hongler and W.M. Zheng, J. Statist. Phys. **29**, 317 (1982); RISKEN, p. 114.

concept with a precise, if somewhat arbitrary convention. In the following we identify it with the mean first-passage time from a to c. It is then possible to utilize the results of the previous chapter, adapted to the present case of bistability.

The mean first-passage time is given by (XII.3.9): replace R with c; y with a; and take $L = -\infty$:

$$\tau_{ca} = \frac{1}{\theta} \int_a^c e^{U(x')/\theta}\,dx' \int_{-\infty}^{x'} e^{-U(x'')/\theta}\,dx''. \tag{2.1}$$

For bistable U, this double integral can be approximated by a simple expression. The factor $e^{U(x')/\theta}$ is large when x' is near b and otherwise exponentially smaller. It may therefore be replaced with

$$\exp\left[\frac{U(b)}{\theta} - \frac{|U''(b)|}{2\theta}(x'-b)^2\right].$$

("Parabolic approximation".) Subsequently, as $x' \approx b$, the second integral gets its main contribution from the neighborhood $x'' \approx a$ and may be replaced with

$$\int_{-\infty}^{\infty} \exp\left[-\frac{U(a)}{\theta} - \frac{U''(a)}{2\theta}(x''-a)^2\right]dx'' = \sqrt{\frac{2\pi\theta}{U''(a)}}\exp\left[-\frac{U(a)}{\theta}\right].$$

The entire expression (2.1) reduces to

$$\tau_{ca} = \frac{2\pi}{\sqrt{U''(a)|U''(b)|}} \exp\left[\frac{U(b)-U(a)}{\theta}\right]. \tag{2.2}$$

In the exponential one recognizes the Arrhenius factor mentioned in VII.5. The factor $|U''(b)|^{-1/2}$ means that a barrier with a flat top is harder to pass than a sharply peaked one of the same height. The factor $|U''(a)|^{-1/2}$ expresses the fact that in a flat-bottomed well the particle gets less often near the barrier than in a narrow one.

In the literature it is customary to put $\sqrt{U''(a)} = \omega_a$, because it is the frequency with which an undamped particle would oscillate in the bottom of the well at a. One also writes $\sqrt{|U''(b)|} = \omega_b$, although this is not a real frequency. If one finally puts W for the height of the barrier, one has for the rate of escape

$$\frac{1}{\tau_{ca}} = \frac{\omega_a \omega_b}{2\pi} \exp\left[-\frac{W}{kT}\right]. \tag{2.3}$$

On the other hand, the time needed to establish local equilibrium in each separate well does not involve the Arrhenius factor $e^{W/kT}$. It is this factor that is responsible for distinguishing the time scales between STAGE I and

STAGE II. The inherent margin of uncertainty in the definition of the escape time is therefore of relative order $e^{-W/kT}$.

Exercise. For the discrete case show that the escape time over a barrier as given by the mean first-passage time (XII.2.11) may be approximated by

$$\tau_{ca} = \left\{ \sum_{-\infty}^{b} p_n^s \right\} \left\{ \sum_{a}^{c} \frac{1}{r_v p_v^s} \right\} = \pi_a^s \sum_{a}^{c} \frac{1}{r_v p_v^s}.$$

It is manifest that (1.5) is obeyed.

Exercise. The second factor is dominated by the terms near $v = b$. They are of the order $(p_b^s)^{-1}$, which is the Arrhenius factor. In the theory of molecular dissociation this factor has suggested the idea of a "transition complex", i.e., an imagined intermediate molecule corresponding to the unstable state b. The reaction from a to c is then visualized as two successive steps: $a \to b$ and, subsequently, $b \to c$. The first step determines the overall rate. This has led some authors to identify the reaction rate with the first-passage time τ_{ba}. Show that this is only half the correct amount.

Exercise. The result (2.3) is obtained by evaluating (2.1) asymptotically for small θ. Find the next order in θ/W. [RISKEN, p. 124.]

For a symmetric bistable potential as in fig. 37 another treatment is possible, based on the eigenfunction expansion of (1.8) as given in V.7. The successive eigenfunctions $\Phi_n(x)$ and eigenvalues λ_n obey

$$\frac{d}{dx} U'(x)\Phi_n + \theta \frac{d^2 \Phi_n}{dx^2} = -\lambda_n \Phi_n. \qquad (2.4)$$

The lowest eigenvalue is $\lambda_0 = 0$ with Φ_0 equal to (1.10). This function is very small at $x = b$ and hence it only takes a minor modification to change it into $\Phi_1(x)$, which has a node at $x = b$:

$$\Phi_1(x) \approx \Phi_0(x) \quad (x > b); \qquad \Phi_1(x) \approx -\Phi_0(x) \quad (x < b).$$

The eigenvalue λ_1 is therefore very low. The next one, λ_2, is much higher

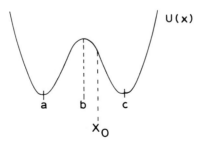

Fig. 37. Symmetric bistable potential.

because Φ_2 has nodes *inside* the two wells. Hence in the sum

$$P(x, t) = \Phi_0(x) + \sum_{n=1}^{\infty} c_n \Phi_n(x) e^{-\lambda_n t}, \tag{2.5}$$

the term $n = 1$ dies out slowly with a decay time $1/\lambda_1$. It describes the exchange of probability between both wells during STAGE III. Identification with (1.4) yields

$$\lambda_1 = \frac{1}{\tau_{ca}} + \frac{1}{\tau_{ac}}. \tag{2.6}$$

The terms in (2.5) with $n \geq 2$ decay rapidly; they describe how the local equilibrium in either well is established. The identification (2.6) is, of course, valid only with the margin of uncertainty of the definition of the escape times, i.e., disregarding terms of relative order λ_1/λ_2.

Exercise[*]. Our discussion of the eigenfunctions and eigenvalues made use of familiar facts in quantum mechanics. It is also possible to transform (2.4) into an actual Schrödinger equation by setting $\Phi_n = e^{U(x)/2\theta}\psi$, see (VIII.7.15). The potential of this Schrödinger particle is given by

$$\frac{2m}{\hbar^2} V(x) = \frac{U'^2}{4\theta^2} - \frac{U''}{2\theta}. \tag{2.7}$$

Exercise. A typical bistable potential is ("Landau–Ginzburg potential")

$$U(x) = -\tfrac{1}{2}\kappa x^2 + \tfrac{1}{4}x^4 \tag{2.8}$$

for $\kappa > 0$. Show that the associated Schrödinger potential (2.7) is also bistable as long as $\kappa^2 < 6\theta$, but for $\kappa^2 > 6\theta$ has three minima.

Exercise. Let $U(x) = \alpha x^2 + \beta x^4 + \gamma x^6 + \cdots$ be a symmetric potential and consider the corresponding equation (1.8). Rescale the variables by setting $x = \theta^{1/4} \bar{x}$ and $t = \theta^{-1/2} \bar{t}$. The result is that to lowest order in $\theta^{1/2}$ the potential is reduced to the form (2.8).

Exercise. For a symmetric potential with $x_0 > b$ one has

$$\lim_{t \to \infty} \lim_{\theta \to 0} P(x, t \mid x_0, 0) = \delta(x - c),$$

$$\lim_{\theta \to 0} \lim_{t \to \infty} P(x, t \mid x_0, 0) = \tfrac{1}{2}\delta(x - a) + \tfrac{1}{2}\delta(x - c).$$

What are the analogous formulas for asymmetric U?

---

[*] N.G. van Kampen, J. Statist. Phys. **17**, 71 (1977).

Remark. The parabolic approximation was introduced in a different way by Suzuki.[*] Consider equation (1.8) with $U(x)$ as given in fig. 37. Substitute $y = \theta^{-1/2}(x-b)$ and expand in θ,

$$\frac{\partial P(y,t)}{\partial t} = U''(b)\frac{\partial}{\partial y}yP + \frac{\partial^2 P}{\partial y^2} + \tfrac{1}{6}\theta U^{iv}(b)\frac{\partial}{\partial y}y^3 P + \cdots.$$

The zeroth order gives, putting $|U''(b)| = \omega$,

$$P(y,t|y_0,0) = \frac{1}{\sqrt{2\pi(e^{2\omega t}-1)}} \exp\left[-\frac{(y-y_0 e^{\omega t})^2}{2(e^{2\omega t}-1)}\right].$$

This is a Gaussian which rapidly broadens and moves away from b. When it reaches the nonlinear region another approximation has to be used.

This expansion in θ is *not* the same as ours, because here y_0 is kept fixed while θ tends to zero. Hence $x_0 = b + \theta^{1/2}y_0$ moves to the unstable point b, whereas we keep the starting point x_0 fixed.

3. Splitting probability

Consider again the quasilinear Fokker–Planck equation (1.8) with a bistable $U(x)$ as in fig. 34. Suppose that the particle is initially at site x between ϕ_a and ϕ_b. What are the probabilities that it will be caught in the domains of attraction D_a and D_c, respectively? This question refers to the rapid evolution during STAGE II; on the long time scale (measured by τ_{ca} and τ_{ac}) the particle cannot be said to be in one well or the other. Accordingly, the very definition of the splitting probabilities has a margin of uncertainty, but within this margin they may be indentified with the first-passage probabilities $\pi_a(x)$ and $\pi_c(x)$.

The answer is given in (XII.3.8):

$$\pi_c(x) = \int_a^x e^{U(x')/\theta}\,dx' \bigg/ \int_a^c e^{U(x')/\theta}\,dx'. \tag{3.1}$$

For the present bistable situation the denominator may be approximated by

$$\sqrt{\frac{2\pi\theta}{|U''(b)|}}\,e^{U(b)/\theta} = \frac{\sqrt{2\pi\theta}}{\omega_b}\,e^{U(b)/\theta}.$$

---

[*] M. Suzuki, J. Statist. Phys. **16**, 11 and 477 (1977). See also F. Haake, Phys. Rev. Letters **41**, 1685 (1978); B. Caroli, C. Caroli, and B. Roulet, Physica A **101**, 581 (1980).

The numerator may be written, when $x > b$,

$$\int_a^b + \int_b^x = \frac{1}{2} \frac{\sqrt{2\pi\theta}}{\omega_b} e^{U(b)/\theta} + e^{U(b)/\theta} \int_b^x e^{-(\omega_b^2/2\theta)(x'-b)^2} \, dx'.$$

Hence

$$\pi_c(x) = \tfrac{1}{2} + \frac{\omega_b}{\sqrt{2\pi\theta}} \int_0^{x-b} e^{-(\omega_b^2/2\theta)z^2} \, dz. \tag{3.2}$$

Exercise. Derive in the same way for $x < b$

$$\pi_c(x) = \tfrac{1}{2} - \frac{\omega_b}{\sqrt{2\pi\theta}} \int_0^{b-x} e^{-(\omega_b^2/2\theta)z^2} \, dz. \tag{3.3}$$

Exercise. For equation (1.10) it is possible to find the splitting probabilities exactly. The result is (3.2) and (3.3).

Now consider one-step processes. As an example we take the *Malthus–Verhulst problem* mentioned in VI.9, because it has been the subject of some discussion. The M-equation for the population growth is

$$\dot{p}_n = (\mathbb{E} - 1)\left\{\alpha n + \frac{\gamma}{\Omega} n(n-1)\right\} p_n + (\mathbb{E} - 1)\beta n p_n, \tag{3.4}$$

where the amount of available space or food Ω is displayed explicitly. Expansion in Ω according to chapter X first gives the macroscopic equation

$$\dot{\phi} = (\beta - \alpha)\phi - \gamma\phi^2. \tag{3.5}$$

It has one stable stationary solution $\phi^s = (\beta - \alpha)/\gamma$, corresponding to a population

$$n^s = \Omega\phi^s = \Omega\frac{\beta - \alpha}{\gamma}. \tag{3.6}$$

The second stationary solution is $\phi = 0$ and it is unstable. Hence ϕ^s is not globally stable although all solutions $\phi(t)$ with initial value $\phi(0) > 0$ are attracted by it. The condition (X.3.4) is not obeyed.

The mesoscopic approximation consists in solving (3.4) with given $p_n(0) = \delta_{n,m}$. If the initial population m is of order Ω one can use the Ω-expansion. The result is, of course, that $p_n(t)$ is a peak with width of order $\Omega^{1/2}$ located at a position determined by (3.5). For large t the distribution p_n becomes a stationary Gaussian around (3.6).

On the other hand, it is clear from (3.4) that $n = 0$ is an absorbing site, so that the stationary solution can be only $p_n^s = \delta_{n,0}$. All other solutions of the master equation tend towards it, i.e., with probability one the population will ultimately die out! A moment's reflection resolves this paradox. The

3. SPLITTING PROBABILITY

mesostate related to the stationary macrostate (3.6) is not stable but merely metastable. While the population is hovering around the value (3.6) there is always a small chance for a fluctuation to occur that brings it down to $n = 0$, where it remains ever after. On a very long time scale this chance builds up to unity. The Ω-expansion describes the evolution on the shorter time scale of STAGE II, while STAGE III describes how the population ultimately dies out.

First question: How long does the metastable state survive? That is, we want to calculate the probability per unit time, $1/\tau$, for a macroscopic population to die by a fluctuation. We have to compute the mean first-passage time $\tau_{0,m}$ from some m near n^s to the absorbing boundary $n = 0$. We cannot use (XII.2.11) as our p_n^s vanishes for all $n > 0$. We can use (XII.2.10) when it is adjusted to the case of a lower boundary:

$$\tau_{L,m} = \sum_{v=L+1}^{m} \sum_{\mu=v}^{R-v} \frac{g_v \, g_{v+1} \, g_{v+2} \, \cdots \, g_{v+\mu-1}}{r_v r_{v+1} r_{v+2} \cdots r_{v+\mu}}.$$

In the present problem

$$\tau = \sum_{v=1}^{n^s} \sum_{\mu=v}^{\infty} \left(\frac{\beta}{\alpha}\right)^\mu \left[1 + \frac{\gamma}{\alpha\Omega}(v-1)\right]^{-1} \left[1 + \frac{\gamma}{\alpha\Omega}v\right]^{-1}$$
$$\cdots \left[1 + \frac{\gamma}{\alpha\Omega}(v+\mu-1)\right]^{-1} \frac{1}{v+\mu}. \tag{3.7}$$

The *second question* is the survival probability of a small population. It has been mentioned that the Ω-expansion can be used when the initial population is of order Ω. In that expansion the probability for a fluctuation down to $n = 0$ is ignored, which is a very good approximation for a population of macroscopic size. But suppose that the initial population consist of a small number m of individuals. Then the Ω-expansion cannot be used, and the probability for dying out in the initial stages is not negligible.

To compute it we formulate the problem as one of splitting probability: either the population dies out during its initial stages or it survives to grow to a macroscopic size. We therefore erect a boundary at some macroscopic value $n = a$, for which one may take n^s and ask for the probabilities π_a and π_0 that n will reach a or 0 first. The answer is given by (XII.2.8), or, in our present situation

$$\frac{\pi_a}{\pi_0} = \frac{1 + \sum_{\mu=1}^{m-1} \frac{r_\mu r_{\mu-1} \cdots r_1}{g_\mu g_{\mu-1} \cdots g_1}}{1 + \sum_{\mu=m}^{a-1} \frac{r_\mu r_{\mu-1} \cdots r_1}{g_\mu g_{\mu-1} \cdots g_1}}. \tag{3.8}$$

Exercise. For the Gaussian fluctuations about (3.6) find
$$\langle (n - n^s)^2 \rangle^s = \frac{\beta(\beta - \alpha - \gamma)}{2\gamma(\beta - \alpha)} \Omega.$$

Exercise. Show directly from (3.4) that the only stationary solution is $p_n^s = \delta_{n,0}$.

Exercise. Verify that (3.7) converges. The denominator of (3.8), however, diverges; what does that mean?

Exercise. Show that the probability of a single bacterium to give rise to a macroscopic population is
$$\pi_{a,1} = 1 - \frac{\alpha}{\beta} - \frac{\gamma}{\beta - \alpha} \frac{1}{\Omega} + \mathscr{O}(\Omega^{-2}). \tag{3.9}$$

Exercise. Show that (3.8) is insensitive to the precise choice of a, provided that a is of order Ω and not much larger than $\Omega\phi^s$.

Exercise. Find a stationary solution of (3.4) in which a constant probability flow from infinity compensates the loss to p_0.

Exercise. If in a species with two sexes all births during one generation happen to be male the species ceases to exist. Estimate the probability for this to occur in homo sapiens during the next 6000 years. Compare it to the probability that the species does not survive its initial stage, e.g., if Adam and Eve had only male offspring.

Third question: Suppose again a population starts with m of order unity: what is the probability distribution of n after a time long enough to have reached a macroscopic size of order Ω (yet short compared to the ultimate dying out)? To compute this select a time t_1 such that
$$e^{(\beta - \alpha)t_1} \approx \Omega^{1/2}.$$
For $t < t_1$ the nonlinear term in (3.4) is of order $\Omega^{-1/2}$ and may be neglected. The remaining linear problem has been solved in VI.6. According to (VI.6.11)
$$p_0(t_1) = (\alpha/\beta)^m. \tag{3.10}$$
(This could also be obtained from (3.9).) The fact that this probability is independent of time means that the transitions into the absorbing site have become negligible. One also has at t_1
$$\langle n \rangle = m\, e^{(\beta - \alpha)t_1}, \qquad \langle\langle n^2 \rangle\rangle = m\, \frac{\beta + \alpha}{\beta - \alpha} (e^{2(\beta - \alpha)t_1} - e^{(\beta - \alpha)t_1}).$$

4. DIFFUSION IN MORE DIMENSIONS

These averages, however, include the site $n = 0$. The averages pertaining to the surviving population are obtained from them by renormalizing with the factor (3.10)

$$\langle n \rangle^* = \frac{m}{1 - (\alpha/\beta)^m} e^{(\beta - \alpha)t_1},$$

$$\langle\langle n^2 \rangle\rangle^* = \frac{\langle\langle n^2 \rangle\rangle}{1 - (\alpha/\beta)^m} - \frac{(\alpha/\beta)^m}{[1 - (\alpha/\beta)^m]^2} \langle n \rangle^2.$$

The subsequent population can now be computed by means of the Ω-expansion from these initial data at $t = t_1$.

Incidentally, suppose one replaces the M-equation (3.4) by the naive Fokker–Planck approximation (VIII.5.3), obtained by breaking off the Kramers–Moyal expansion after the second term rather than by the systematic expansion of chapter X. This cannot be correct for small n and cannot therefore reproduce the evolution starting from small initial m. It is therefore not paradoxical that the absorbing site $n = 0$ does *not* translate into an absorbing boundary condition of the Fokker–Planck equation – as remarked in an Exercise of XII.5.

4. Diffusion in more dimensions

Two-dimensional diffusion in a potential $U(x, y)$ is sufficient to demonstrate some of the essential features:

$$\frac{\partial P(x, y, t)}{\partial t} = \frac{\partial}{\partial x}\left(\frac{\partial U}{\partial x} P\right) + \frac{\partial}{\partial y}\left(\frac{\partial U}{\partial y} P\right) + \theta\left(\frac{\partial^2 P}{\partial x^2} + \frac{\partial^2 P}{\partial y^2}\right). \quad (4.1)$$

Let U have a minimum at the origin, surrounded by a ridge, beyond which U decreases again, see fig. 38. We set $U(0, 0) = 0$, call the lowest point of the

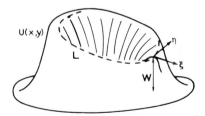

Fig. 38. A two-dimensional potential crater.

ridge its "col", with height W, and assume $W \gg \theta$. If the particle starts somewhere near the bottom the mean escape time will involve the Arrhenius factor, $\tau = A\,e^{W/\theta}$, because escape through any higher point on the ridge is exponentially less likely. Our aim is to compute the prefactor A.

Again τ is defined apart from terms of order unity, i.e., terms without the Arrhenius factor. Within this margin it may be identified with the first-passage time from any point inside the valley to some curve surrounding the entire crater. The mean first-passage time $\tau(x, y)$ starting from the point x, y obeys

$$-\frac{\partial U}{\partial x}\frac{\partial \tau}{\partial x} - \frac{\partial U}{\partial y}\frac{\partial \tau}{\partial y} + \theta\left(\frac{\partial^2 \tau}{\partial x^2} + \frac{\partial^2 \tau}{\partial y^2}\right) = -1. \tag{4.2}$$

It will be of order $e^{W/\theta}$ when x, y lies inside, and of order 1 when (x, y) lies outside.

Set $\tau(x, y) = e^{W/\theta} w(x, y)$ so that the function w obeys

$$-\frac{\partial U}{\partial x}\frac{\partial w}{\partial x} - \frac{\partial U}{\partial y}\frac{\partial w}{\partial y} + \theta\left(\frac{\partial^2 w}{\partial x^2} + \frac{\partial^2 w}{\partial y^2}\right) \approx 0, \tag{4.3}$$

and $w(x, y) \approx 0$ outside, while inside the valley w is practically constant, $w(x, y) \approx A$. It drops very abruptly from A to 0 on passing the ridge from the inside to the outside, except near the col, where it changes more smoothly. By computing the behavior in the vicinity of the col we shall be able to determine A.[*]

Near the col one may expand U to second order. Introducing suitable local coordinates ξ, η one has

$$U(x, y) = W - \tfrac{1}{2}\omega_1^2\xi^2 + \tfrac{1}{2}\omega_2^2\eta^2. \tag{4.4}$$

Accordingly (4.3) reduces to a linear equation:

$$\omega_1^2\xi\frac{\partial w}{\partial \xi} - \omega_2^2\eta\frac{\partial w}{\partial \eta} + \theta\left(\frac{\partial^2 w}{\partial \xi^2} + \frac{\partial^2 w}{\partial \eta^2}\right) = 0. \tag{4.5}$$

When ξ points to the outside our solution must obey

$$w(\xi \to \infty) = 0, \qquad w(\xi \to -\infty) = A. \tag{4.6}$$

The equation can be solved by separating variables; let $w = X(\xi)Y(\eta)$, then

$$\theta X'' + \omega_1^2\xi X' + \lambda X = 0, \qquad \theta Y'' - \omega_2^2\eta Y' - \lambda Y = 0.$$

---

[*] Z. Schuss and B.J. Matkowski, SIAM J. Appl. Math. **35**, 604 (1970); Z. Schuss, *Theory and Applications of Stochastic Differential Equations* (Wiley, New York 1980).

4. DIFFUSION IN MORE DIMENSIONS

The separation constant λ must vanish in order that X can tend to a constant A, as in (4.6). The solution is

$$X(\xi) = C_1 \int_0^{\xi} e^{-(\omega_1^2/2\theta)\xi'^2} \, d\xi' + C_2,$$

$$Y(\eta) = C_3 \int_0^{\eta} e^{-(\omega_2^2/2\theta)\eta'^2} \, d\eta' + C_4.$$

The integration constant C_3 must vanish because w is symmetric in η and cannot become negative. C_4 may be taken unity and C_1 and C_2 are determined by (4.6). The result is

$$X(\xi) = A \sqrt{\frac{\omega_1^2}{2\pi\theta}} \int_{\xi}^{\infty} e^{-(\omega_1^2/2\theta)\xi'^2} \, d\xi', \qquad Y(\eta) = 1.$$

Hence we have found that in the vicinity of the col

$$\tau(x, y) = A\, e^{W/\theta} \frac{\omega_1}{\sqrt{2\pi\theta}} \int_{\xi}^{\infty} e^{-(\omega_1^2/2\theta)\xi'^2} \, d\xi'. \tag{4.7}$$

However, the crucial constant A still appears as an undetermined factor. The reason is that we have reduced the problem to a homogeneous equation by putting the right-hand side of (4.3) equal to zero. We therefore return to (4.2), which may be written

$$\theta \nabla \cdot e^{-U/\theta} \nabla \tau = -e^{-U/\theta}.$$

Integrate this identity over the region Ω of the valley enclosed by the curve L along the top of the ridge, using Gauss' theorem in two dimensions,

$$\theta \oint e^{-U/\theta} \frac{\partial \tau}{\partial n} \, dl = - \iint_{\Omega} e^{-U/\theta} \, dx \, dy. \tag{4.8}$$

The line integral on the left is dominated by the contribution of the vicinity of the col owing to the factor $e^{-U/\theta}$. In that vicinity the expression (4.4) may be used for U, and $\partial \tau/\partial n$ is obtained from the previously obtained result (4.7),

$$\frac{\partial \tau}{\partial n} = \frac{\partial \tau}{\partial \xi} = -A\, e^{W/\theta} \frac{\omega_1}{\sqrt{2\pi\theta}}.$$

Thus the left side of (4.8) is

$$-\theta A\, e^{W/\theta} \frac{\omega_1}{\sqrt{2\pi\theta}} \int_{-\infty}^{\infty} e^{-(\omega_2^2/2\theta)\eta^2} \, d\eta = -\theta A\, e^{W/\theta} \frac{\omega_1}{\omega_2}.$$

The right-hand side of (4.8) is dominated by the minimum of U at the origin and becomes

$$-\frac{2\pi\theta}{\sqrt{\Delta}}, \quad \Delta = \left[\frac{\partial^2 U}{\partial x^2}\frac{\partial^2 U}{\partial y^2} - \left(\frac{\partial^2 U}{\partial x \partial y}\right)^2\right]_{x=0, y=0}. \quad (4.9)$$

Collecting results one finally obtains

$$\tau = \frac{2\pi}{\sqrt{\Delta}}\frac{\omega_2}{\omega_1},$$

where ω_1, ω_2 are defined by (4.4) and Δ by (4.9). When ω_2 is large, the col is narrow and therefore τ is long; when ω_1 is large the col is steep and short, and τ is short. When Δ is small the valley is spacious so that the particle does not often approach the col.

Exercise. Suppose the crater ridge has several minima of the same height W. Compute the probability of escape through each of them.

Exercise. Suppose the potential ridge is circularly symmetric. Compute τ.

Exercise. Suppose the ridge has constant height but varying width. Study the problem of obtaining τ in this case.

5. Critical fluctuations

Let $\phi_c \equiv c$ be a stationary macrostate, $\alpha_{1,0}(c) = 0$. We have seen that c is at least locally stable if $\alpha'_{1,0}(c) < 0$ and unstable if $\alpha'_{1,0}(c) > 0$. If $\alpha'_{1,0}(c) = 0$ it is in general unstable, but it may still be stable, for instance, if at c

$$\alpha_{1,0} = \alpha'_{1,0} = \alpha''_{1,0} = 0, \quad \alpha'''_{1,0} < 0. \quad (5.1)$$

The fact that in this case c is stable is immediately clear from fig. 39a. From comparison with fig. 35 one sees that this situation may be regarded as a limiting case in which the three stationary states a, b, c all coincide. The name "critical point" is therefore appropriate for a macrostate with the properties (5.1). It has many features in common with the critical points in equilibrium statistical mechanics.

The stability of the critical point c is much weaker than that of the normal stable points ϕ_a and ϕ_c in fig. 34. It cannot be found from the linear

5. CRITICAL FLUCTUATIONS

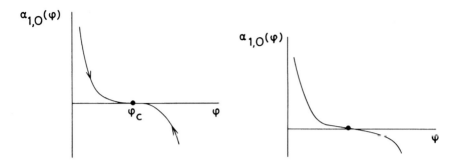

Fig. 39a. The critical point.   39b. Slightly above the critical point.

variational equation (X.3.5), but higher powers of $\delta\phi$ have to be included. In fact, according to (5.1) one has for $\phi - c \equiv \delta\phi$

$$\frac{d}{dt}\delta\phi = -\tfrac{1}{6}|\alpha'''_{1,0}(c)|(\delta\phi)^3 + \mathcal{O}(\delta\phi^4).$$

When $\delta\phi$ starts out with a sufficiently small value $\delta\phi_0$ at $t = t_0$ for the last term to be negligible, one has the solution

$$\delta\phi = \frac{\delta\phi_0}{\sqrt{1 + \tfrac{1}{2}|\alpha'''_{1,0}(c)|(t - t_0)}}. \tag{5.2}$$

It follows that $\delta\phi$ does approach zero, but only as $t^{-1/2}$ rather than exponentially. This is the *critical slowing down of the macroscopic approach to equilibrium*.

We want to compute the fluctuations around c, i.e., we want to compute the mesostate related to the macrostate c. In the situation in fig. 39a that mesostate is identical with the stationary solution P^s of the M-equation because c is the only stationary macrostate. Expansion (X.2.16) is no longer the correct starting point. For, the terms written on the first line of the right-hand member are all zero according to (5.1), and only the (unwritten) term with $\alpha'''_{1,0}$, which is of order Ω^{-1}, does not vanish. But that is the term that is responsible for restraining the fluctuations, which are caused by the terms on the second line. The conclusion is that the fluctuations will be proportional to a higher power of Ω than anticipated (X.2.9). This is the *enhancement of fluctuations near the critical point*, as in critical opalescence.

Accordingly we now try instead of (X.2.9)

$$x = \Omega c + \Omega^\mu \xi. \tag{5.3}$$

The constant μ is still adjustable subject to the condition $\mu < 1$. The transformed master equation now becomes

$$\frac{\partial \Pi}{\partial \tau} = -\Omega^{1-\mu} \frac{\partial}{\partial \xi} \alpha_{1,0}(c + \Omega^{\mu-1}\xi)\Pi$$

$$+ \tfrac{1}{2} \Omega^{1-2\mu} \frac{\partial^2}{\partial \xi^2} \alpha_{2,0}(c + \Omega^{\mu-1}\xi)\Pi + \cdots$$

$$= -\frac{1}{3!} \Omega^{2\mu-2} \alpha'''_{1,0}(c) \frac{\partial}{\partial \xi} \xi^3 \Pi$$

$$+ \tfrac{1}{2} \Omega^{1-2\mu} \alpha_{2,0}(c) \frac{\partial^2}{\partial \xi^2} \Pi + \cdots. \tag{5.4}$$

It appears that one must take $\mu = \tfrac{3}{4}$ in order that the fluctuation producing term on the second line and the restraining term on the first line are of the same order.[*] The result is again a Fokker–Planck equation, but no longer a linear one:

$$\frac{\partial \Pi}{\partial \tau} = \Omega^{-1/2} \left[\tfrac{1}{6} |\alpha'''_{1,0}(c)| \frac{\partial}{\partial \xi} \xi^3 \Pi + \tfrac{1}{2} \alpha_{2,0}(c) \frac{\partial^2 \Pi}{\partial \xi^2} \right]. \tag{5.5}$$

Thus one finds that at the critical point the fluctuations are of order $\Omega^{3/4}$ rather than $\Omega^{1/2}$. In addition, the factor $\Omega^{-1/2}$ shows that their relaxation time is magnified by an order $\Omega^{1/2}$. Moreover their distribution Π is not Gaussian.

This expansion in powers of $\Omega^{-3/4}$ is valid only *at the exact critical point*. If one alters a parameter so as to be in the situation of fig. 39b, i.e., slightly "above the critical point", the original expansion in $\Omega^{-1/2}$ must be used, no matter how close to the critical point. It is true that on approaching the critical point the expansion in $\Omega^{-1/2}$ becomes less useful, as larger and larger values for Ω are needed to ensure that the higher order terms are small, but formally it is the correct asymptotic expansion. The fact that *at* the critical point the expansion changes abruptly is called Stokes' phenomenon.[**] For an expansion that covers the critical point together with its neighborhood another method is needed.[†]

Exercise. Show that a stationary solution with $\alpha_{1,0}(\phi) = 0$, $\alpha'_{1,0}(\phi) = 0$, $\alpha''_{1,0}(\phi) \neq 0$ is unstable.

---

[*] R. Kubo, K. Matsuo, and K. Kitahara, J. Statist. Phys. **9**, 51 (1973).

[**] P.M. Morse and H. Feshbach, *Methods of Theoretical Physics* I (McGraw-Hill, New York 1953) p. 609.

[†] H. Dekker, Physica A **103**, 55 and 80 (1980); T. Nakanishi and K. Yamamoto, Phys. Letters A **147**, 257 (1990).

Exercise. Verify that all terms omitted in (5.4) are of higher order than the ones kept.

Exercise. Find the stationary solution of (5.5) and observe that it is not Gaussian.

Exercise. Repeat the discussion for the case that all derivatives of $\alpha_{1,0}$ vanish up to $\alpha_{1,0}^{(q)}(\phi_c)$.

Exercise. Find the equilibrium fluctuations at the critical point of the Schlögl reaction (X.3.6).

6. Kramers' escape problem

In VII.5 the dissociation of a molecule into two fragments was described by the M-equation (5.4), in which the coefficients $W_{\nu\mu}$ and Γ_ν are yet to be obtained from a physical picture of the actual mechanism. Christiansen[*] replaced the various states ν of the molecule with a "reaction coordinate" x representing the distance between both fragments. Kramers[**] assumed that x undergoes Brownian motion as described by equation (VIII.7.4); in appropriate units,

$$\frac{\partial P(x,v,t)}{\partial t} = -v\frac{\partial P}{\partial x} + U'(x)\frac{\partial P}{\partial v} + \gamma\left(\frac{\partial}{\partial v}vP + T\frac{\partial^2 P}{\partial v^2}\right). \tag{6.1}$$

The interaction potential U is supposed to have a shape as sketched in fig. 40. The problem is again to find the average time for x to escape across the potential barrier W.[†]

First the naive approach. The escape rate will involve the Arrhenius factor and is therefore very small provided that $W \gg T$. Hence the left-hand side

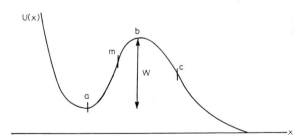

Fig. 40. Kramers' potential for a chemical reaction.

---

[*] J.A. Christiansen, Z. Phys. Chemie B **33**, 145 (1936).

[**] H.A. Kramers, Physica **7**, 284 (1940).

[†] The nucleation problem is of the same type, see R. Becker and W. Döring, Ann. Physik **24**, 719 (1935); J.L. Katz and M.D. Donohue, in: Advances in Chemical Physics **40** (Interscience, New York 1979).

may be taken equal to zero, as in (**4**.3), and the remaining time-independent equation has the stationary solution

$$P^s(x, v) = C \exp\left[-\frac{U(x) + \tfrac{1}{2}v^2}{T}\right]. \tag{6.2}$$

This cannot, of course, be true for all x but, since P^s is very small at $x = b$, no great sin is committed if one cuts it off by setting $P^s = 0$ for $x > b$. The normalization is then given by

$$C^{-1} = \sqrt{2\pi T} \int_{-\infty}^{b} e^{-U(x)/T}\, dx \approx \frac{2\pi T}{\omega_a} e^{-U(a)/T},$$

where $\omega_a^2 = U''(a)$ as in **2**. Subsequently the escape rate is obtained by computing the outward flow across the top:

$$\frac{1}{\tau} = \int_0^\infty vP(b, v)\, dv = C\, e^{-U(b)/T} \int_0^\infty v\, e^{-v^2/2T}\, dv = \frac{\omega_a}{2\pi} e^{-W/T}. \tag{6.3}$$

This assumes that whenever the Brownian particle is at the top with positive velocity it will escape, as if there were an absorbing wall or an infinitely deep abyss. A rough interpretation of (**6**.3) is that the particle oscillates in the potential $\tfrac{1}{2}\omega_a^2 x^2$ and therefore hits the wall $\omega_a/2\pi$ times per second, and each time has the probability $e^{-W/T}$ to get across it. At the end of this section we discuss the validity of (**6**.3).

For a more sophisticated treatment one has to solve (**6**.1), with an initial distribution located inside the well. In view of the inherent margin of uncertainty the precise form is immaterial and we may take the initial distribution

$$P(x, v, 0) = \frac{\omega_a}{2\pi T} \exp\left[-\frac{\omega_a^2 (x-a)^2 + v^2}{2T}\right]. \tag{6.4}$$

The escape is determined by the flow past a point c sufficiently far from the top to neglect the probability of return.

This is Kramers' escape problem. Since no analytic solution is known for any metastable potential of the shape in fig. 40 the quest is for suitable approximation methods. This problem has received an extraordinary amount of attention from physicists, chemists and mathematicians.[*] We describe the main features – all present already in the seminal paper by Kramers.

First the problem is again simplified by neglecting the left-hand side of (**6**.1) and looking for a stationary solution. It is then necessary to specify

---

[*] For an extensive review, see P. Hänggi, P. Talkner, and M. Borkovich, Rev. Mod. Phys. **62**, 251 (1990).

6. KRAMERS' ESCAPE PROBLEM

this solution by requiring that no flow enters from infinity:

$$P(c, v) = 0 \quad \text{for} \quad v < 0. \tag{6.5}$$

Of course, strictly speaking no stationary solution exists obeying (6.5) except $P = 0$. To overcome this objection one may imagine a source of particles near the bottom of the well, but it makes no difference as it does not enter the equation anyway.

For the purpose of solving the stationary problem different approximation methods are available depending on whether γ is large, intermediate, or small. For *large* γ it was shown in VIII.7 that it is possible to eliminate the velocity v so as to be left with a diffusion equation (VIII.7.14), for which the mean first-passage time has been computed in (2.2) to be

$$\tau_{ca} = \frac{2\pi\gamma}{\omega_a \omega_b} e^{W/T}, \quad \omega_b^2 = -U''(b). \tag{6.6}$$

Exercise. Carry out the derivation of (6.6) from (6.1), including the factor γ (which in (2.2) had been taken equal to unity).

Exercise. Compute for large γ the escape rate in the case of a sharply cut off potential as in fig. 41.

Fig. 41. A sharply cut off variety of Kramers' potential.

For *intermediate* γ subdivide the interval a, b by choosing a point m on the slope such that

(i) for $x < m$ the stationary solution P is virtually the local equilibrium distribution (6.2);

(ii) for $x > m$ one may approximate $U(x)$ by the parabola

$$U(x) = U(b) - \tfrac{1}{2}\omega_b^2(x - b)^2, \tag{6.7}$$

previously used in **2**.

We shall see that such an m exists unless γ is too small. There is no need for a third interval on the right because one may locate c inside the parabolic interval.

In the parabolic interval it is possible to find a suitable solution in the following way. Take b as the origin of x and set

$$P(x, v) = Q(x, v) \exp\left[\frac{\omega_b^2 x^2 - v^2}{2T}\right] \tag{6.8}$$

so that Q obeys

$$-v\frac{\partial Q}{\partial x} - \omega_b^2 x\frac{\partial Q}{\partial v} - \gamma v\frac{\partial Q}{\partial v} + \gamma T\frac{\partial^2 Q}{\partial v^2} = 0.$$

Try as a special solution a function of a single variable, viz.,

$$Q(x, v) = f(v - \alpha x) \equiv f(z).$$

Substitution gives

$$\alpha v f'(z) - \omega_b^2 x f'(z) - \gamma v f'(z) + \gamma T f''(z) = 0.$$

This equation can be satisfied if the coefficient of $f'(z)$ is a function of z alone:

$$\alpha v - \omega_b^2 x - \gamma v = \lambda(v - \alpha x).$$

This determines the parameters λ and α,

$$\omega_b^2 = \lambda\alpha, \qquad \alpha - \gamma = \lambda, \qquad \alpha^2 - \gamma\alpha - \omega_b^2 = 0.$$

Hence α must have one of the two values

$$\alpha_\pm = \tfrac{1}{2}\gamma \pm \tfrac{1}{2}\sqrt{\gamma^2 + 4\omega_b^2}, \qquad \lambda_\pm = -\alpha_\mp.$$

For either choice one has

$$\lambda z f'(z) + \gamma T f''(z) = 0.$$

The solution involves two integration constants:

$$f(z) = B + C\int_0^z dz' \exp\left[-\frac{\lambda z'^2}{2\gamma T}\right].$$

Thus we have found for P in the parabolic interval

$$P(x, v) = \exp\left[\frac{\omega_b^2 x^2 - v^2}{2T}\right]\left\{B + C\int_0^{v-\alpha x} dz' \exp\left[-\frac{\omega_b^2 z'^2}{2\gamma\alpha T}\right]\right\}. \quad (6.9)$$

In order to obey (6.5) this should vanish as $x \to \infty$ while $v < 0$. That is achieved by taking for α the positive root α_+, replacing the lower limit of integration in (6.9) by $-\infty$, and putting $B = 0$,

$$P(x, v) = \exp\left[-\frac{U(x) - U(b) + \tfrac{1}{2}v^2}{T}\right]C\int_{-\infty}^{v-\alpha_+ x} dz \exp\left[-\frac{\omega_b^2 z^2}{2\gamma\alpha_+ T}\right], \quad (6.10)$$

In this expression let x become negative so as to move *into* the well:

$$P(x, v) \simeq \exp\left[-\frac{U(x) - U(b) + \tfrac{1}{2}v^2}{T}\right]C\sqrt{\frac{2\pi\gamma\alpha_+ T}{\omega_b^2}}.$$

6. KRAMERS' ESCAPE PROBLEM

This expression merges smoothly with the distribution (6.2) inside the well provided that

$$C = e^{-W/T}(2\pi T)^{-3/2} \frac{\omega_a \omega_b}{\sqrt{\gamma \alpha_+}}. \tag{6.11}$$

Collecting the results we have as an approximation for all x

$$P(x, v) = (2\pi T)^{-3/2} \exp\left[-\frac{U(x) - U(a) + \frac{1}{2}v^2}{T} \right] \frac{\omega_a \omega_b}{\sqrt{\gamma \alpha_+}}$$

$$\times \int_{-\infty}^{v - \alpha_+ x} dz \exp\left[-\frac{\omega_b^2 z^2}{2\gamma \alpha_+ T} \right]. \tag{6.12}$$

To find the escape rate we compute the net probability flow across the top and find, using partial integration,

$$\frac{1}{\tau} = \int_{-\infty}^{\infty} v P(b, v) \, dv = e^{-W/T} \frac{\omega_a \omega_b}{2\pi} (\gamma \alpha_+ + \omega_b^2)^{-1/2}$$

$$= e^{-W/T} \frac{\omega_a \omega_b}{\pi} (\gamma + \sqrt{\gamma^2 + 4\omega_b^2})^{-1}. \tag{6.13}$$

Exercise. For large γ this result coincides with (6.6). Thus Kramers' treatment for intermediate γ covers the case of large γ as well. The large γ limit is valid for $\gamma \gg \omega_b$. Verify that this condition agrees with (VIII.7.16).

Exercise. For $\gamma \ll \omega_b$ the result (6.13) coincides with the naive approximation (6.3). (It will appear presently, however, that there is a restriction on the smallness of γ.)

Exercise. Show that (6.12) does indeed vanish asymptotically as $x \to \infty$ with $v < 0$, supposing that (6.7) holds all the way.

Exercise. It cannot be true that (6.12) also vanishes for $x \to \infty$ with $v > 0$, because the net current (6.13) cannot vanish, as it is the same for all x. How does this show up in the asymptotic evaluation of (6.12)?

Exercise. Formulate the escape problem for the bistable potential in fig. 37. Show that the result is again (6.13).[*]

The general picture is that inside the potential well the distribution has practically the stationary form (6.2), but in the vicinity of the top b it deviates, owing to the draining of probability across the barrier. The rate at which the equilibrium is established is seen from (6.1) to be γ. If for a given potential one decreases γ, the deviation extends deeper into the well. When it extends beyond the interval where U may be replaced with a parabola the preceding

---

[*] H.C. Brinkman, Physica **22**, 149 (1956).

treatment breaks down. For this case Kramers developed another approximation method geared to *small* γ.[*]

For $\gamma = 0$ the equation (6.1) reduces to the Liouville equation for particles moving acccording to the deterministic equations

$$\dot{x} = v, \qquad \dot{v} = -U'(x). \tag{6.14}$$

The energy E is conserved and the orbits inside the potential well are closed curves given by

$$\tfrac{1}{2}v^2 + U(x) = E. \tag{6.15}$$

Define action-angle variables I, w by

$$I(E) = \frac{1}{2\pi} \oint \sqrt{2E - 2U(x)}\, \mathrm{d}x, \tag{6.16a}$$

$$w(x, v) = \frac{\mathrm{d}E}{\mathrm{d}I} \int_0^x \frac{1}{\sqrt{2E - 2U(x')}}\, \mathrm{d}x'. \tag{6.16b}$$

In these new variables the equations (6.14) and the corresponding Liouville equation take the form

$$\dot{I} = 0, \qquad \dot{w} = 1, \qquad \frac{\partial P}{\partial t} = -\frac{\partial P}{\partial w}.$$

Since the orbit is periodic in w (with period 1) one may integrate over w and thereby verify that the integrated distribution $\bar{P}(I)$ is independent of t.

Now reinstate the terms with γ and transform the entire equation (6.1) to the same variables I, w, using (XI.4.19). If one now integrates again over w the terms with $\partial/\partial w$ disappear. Yet one does not obtain an equation for $\bar{P}(I, t)$ because the remaining terms have coefficients involving w. One cannot integrate these coefficients separately, because they are multiplied with $P(I, w, t)$ or its derivatives. That is the reason for assuming that γ is small, namely so small that the term with γ in (6.1) has negligible effect during one period. Then $P(I, w, t)$ is practically independent of w, and equal to $\bar{P}(I, t)$.

After all these steps one winds up with the equation

$$\frac{\partial \bar{P}}{\partial t} = \gamma \frac{\partial}{\partial I} \overline{v \frac{\partial I}{\partial v}} \bar{P} - \gamma T \frac{\partial}{\partial I} \overline{\frac{\partial^2 I}{\partial v^2}} \bar{P} + \gamma T \frac{\partial^2}{\partial I^2} \overline{\left(\frac{\partial I}{\partial v}\right)^2} \bar{P}. \tag{6.17}$$

The overbars indicate averages over one period of w. From (6.16)

$$\frac{\partial I}{\partial v} = \frac{\mathrm{d}I}{\mathrm{d}E} v, \qquad \frac{\partial^2 I}{\partial v^2} = \frac{\mathrm{d}I}{\mathrm{d}E} + \frac{\mathrm{d}^2 I}{\mathrm{d}E^2} v^2,$$

---

[*] See also H. Risken and H.D. Vollmer, Phys. Letters A **69**, 387 (1979).

6. KRAMERS' ESCAPE PROBLEM

and from (6.15)

$$\overline{v^2} = \int_0^1 \{2E - 2U(x)\} \, dw = \frac{dE}{dI} \oint \sqrt{2E - 2U(x)} \, dx = I \frac{dE}{dI}.$$

Substituting in (6.17) and rearranging terms one gets

$$\frac{\partial \overline{P}(I, t)}{\partial t} = \gamma \frac{\partial}{\partial I} I \overline{P} + \gamma T \frac{\partial}{\partial I} I \frac{dI}{dE} \frac{\partial \overline{P}}{\partial I}.$$

The problem has been reduced to a diffusion equation in the single variable I, which is a known function of the energy E. Hence it is possible to utilize equation (XII.3.2) for the mean time $\tau(E)$ for a particle with initial energy E to arrive at the top W for the first time:

$$-\gamma I \frac{d\tau}{dI} + \gamma T \frac{d}{dI} I \frac{dI}{dE} \frac{d\tau}{dI} = -1.$$

The solution with boundary condition $\tau(W) = 0$ is

$$\tau(E) = \frac{1}{\gamma T} \int_E^W \frac{dE'}{I'} \exp\left[\frac{E'}{T}\right] \int_0^{I'} \exp\left[-\frac{E''}{T}\right] dI''. \tag{6.18}$$

Exercise. What happened to the second integration constant?

Exercise. When the potential U is harmonic one has $E = (\omega_a/2\pi)I$. Then (6.18) reduces to

$$\tau(E) = \frac{1}{\gamma} \int_E^W \frac{dE'}{E'} (e^{E'/T} - 1).$$

[Remember the quantum mechanical version in (VII.5.7).]

Exercise. Show that no serious error is made when (6.16) is replaced with

$$\tau(E) = e^{W/T} \frac{2\pi T}{\gamma \omega_a} \frac{1}{I(W)}. \tag{6.19}$$

This demonstrates that the mean escape time is independent of the starting energy E as long as it is not too close to W; more precisely $W - E \gg T$. For a harmonic U this reduces to

$$\tau = e^{W/T} \frac{T}{\gamma W}. \tag{6.20}$$

Exercise. The assumption that the damping has negligible effect during one period means $\gamma \ll \omega_a/2\pi$.

Actually the result is restricted to potentials that are sharply cut off as in fig. 41. When the barrier top is parabolic the motion along the orbits near the top is very slow, the period very long, and the averaging is no longer permitted. Nevertheless the result will be approximately correct if it is true that the escape is mainly determined by the supply of particles to the barrier region rather than by the delay in getting across the top. I shall make these considerations more precise.

For definiteness I take for U two parabolas as in fig. 42.

$$U(x) = \tfrac{1}{2}\omega_a^2(x-a)^2 \qquad \text{when } x < m,$$
$$U(x) = W - \tfrac{1}{2}\omega_b^2(b-a)^2 \qquad \text{when } x > m,$$
$$U(m) = \tfrac{1}{2}\omega_a^2(m-a)^2 = W - \tfrac{1}{2}\omega_b^2(b-m)^2.$$

Moreover let us take ω_a and ω_b of the same order and γ much smaller. Then the rate at which the potential well supplies particles to the density at m can be obtained from (6.20) on replacing W with $U(m)$:

$$\frac{1}{\tau_m} = \gamma \frac{U(m)}{T} e^{-U(m)/T}. \tag{6.21}$$

We compare this with the rate at which the density at m is depleted by the escape across the top.

First case: The supply rate is much smaller. Then the supply rate may be identified with the overall escape rate, $1/\tau_m = 1/\tau$, and the calculation leading to (6.19) and (6.20) is valid.

Second case: The supply rate is larger than the depletion rate, so that the overall rate is determined by the bottleneck at the top. Then the assumptions on which (6.13) was based are fulfilled; and also (6.3) is valid because we restricted ourselves to $\gamma \ll \omega_b$. Thus the condition for the second case to apply is that (6.21) is large compared to (6.3), or

$$\gamma \gg \frac{\omega_a}{2\pi} \frac{T}{W} \exp\left[-\frac{W - U(m)}{T}\right].$$

This is compatible with $\gamma \ll \omega_a/2\pi$ if $W - U(m) > T$, i.e., if the parabola at the top extends far enough.

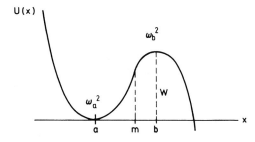

Fig. 42. Potential consisting of two parabolas.

Apart from these two extreme cases there is an intermediate region, where the supply to the parabolic top and the depletion are of comparable size. Many efforts have been made to cover this region[*], but I feel that this cannot be done without solving Kramers' equation exactly. On the other hand, modifications of the original problem have been studied, e.g., replacing the Brownian statistics with a memory function (non-white noise)[**] or including a bath of harmonic oscillators to represent the cause of damping and noise[†]. The problem has been generalized to potentials with many minima ("washboard potential")[††] and to more dimensions[†]. Finally there is the problem of quantum mechanical tunneling through a potential barrier in the presence of damping and noise, see XVII.6.

7. Limit cycles and fluctuations

It was demonstrated in X.5 that the Ω-expansion applies to multivariate master equations, provided that the macroscopic equations possess a single stationary solution, and that it is globally stable. The difference with the one-variable case was that no general method exists to solve the macroscopic equations. With respect to unstable situations, however, there is the added difference that the variety of possible instabilities is much larger than for a single variable.

One novel phenomenon is that a set of two nonlinear macroscopic equations may have a *limit cycle* (fig. 43). When a solution curve in the phase

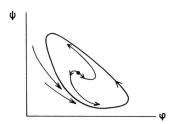

Fig. 43. Limit cycle surrounding an unstable point.

---

[*] For example, V.I. Mel'nikov and S.V. Meshkov, J. Chem. Phys. **85**, 1018 (1986). See also the review quoted; and V.I. Mel'nikov, Physics Reports **209**, 1 (1991).
[**] R.F. Grote and J.T. Hynes, J. Chem. Phys. **73**, 2715 (1980). Review by G.H. Weiss, J. Statist. Phys. **42**, 3 (1986).
[†] P. Hänggi, H. Grabert, P. Talkner, and H. Thomas, Phys. Rev. A **29**, 371 (1984).
[††] RISKEN, ch. 11.
[‡] R. Landauer and J.A. Swanson, Phys. Rev. **121**, 1668 (1961); J.S. Langer, Annals Phys. **54**, 258 (1969).

space is a closed cycle it represents a periodic solution – or rather a family of periodic solutions differing only in phase. Such a closed solution curve is a limit cycle when all other solution curves, at least within some domain of attraction, tend towards it. That means that all solutions of the differential equations (in this domain) are ultimately periodic, with one and the same period and amplitude *determined by the equation*, while only the phase depends on the initial data of the particular solution. In other words, the limit cycle is asymptotically stable *as an orbit* in phase space, and the periodic solutions corresponding with it are called "orbitally stable". They are not asymptotically stable in the usual sense since the phase difference between them never dies out.

The oscillations produced by a frequency generator are of this type and have inspired Van der Pol to construct his classic example of a differential equation having a limit cycle. The best known example in chemistry is the Zhabotinskii reaction.[*] In biology many periodic phenomena are known that can presumably be described in this way.[**]

An imagined chemical reaction with two compounds X, Y exhibiting a limit cycle is the "Brusselator"[†]

$$A \to X, \quad B + X \to Y + D, \quad 2X + Y \to 3X, \quad X \to E. \tag{7.1}$$

It describes an open system with a net transport from A to E and from B to D. The model is somewhat unrealistic inasmuch as it involves a reaction for which three molecules must collide. Another model, not having this drawback, is the "Oregonator"[††], which, however, has the added complication that three compounds X, Y, Z are involved. This is inevitable, for it can be proved that no limit cycles exist in systems involving just two reactants with only bimolecular reactions.[‡]

The bivariate M-equation for the reaction (7.1) is, in suitable units,

$$\dot{p}_{nm} = \Omega\alpha(\mathbb{E}_n^{-1} - 1)p_{nm} + \beta(\mathbb{E}_n \mathbb{E}_m^{-1} - 1)np_{nm}$$
$$+ \Omega^{-2}(\mathbb{E}_n^{-1}\mathbb{E}_m - 1)n^2 m p_{nm} + (\mathbb{E}_n - 1)np_{nm}. \tag{7.2}$$

Setting, as in X.5,

$$n = \Omega\phi(t) + \Omega^{1/2}\xi, \quad m = \Omega\psi(t) + \Omega^{1/2}\eta$$

---

[*] J.J. Tyson, *The Belousov–Zhabotinskii Reaction* (Lecture Notes in Biomathematics **10**; Springer, Berlin 1976).

[**] See, e.g., G. Nicolis and I. Prigogine, *Self-Organization in Non-Equilibrium Systems* (Wiley–Interscience, New York 1977).

[†] P. Glansdorff and I. Prigogine, *Structure, Stability and Fluctuations* (Wiley–Interscience, London 1971).

[††] R.J. Field and R.M. Noyes, J. Chem. Phys. **60**, 1877 (1974).

[‡] P. Hanusse, Comptes Rendus (Paris) C **274**, 1245 (1972); J. Tyson and J. Light, J. Chem. Phys. **59**, 4164 (1973).

7. LIMIT CYCLES AND FLUCTUATIONS

one obtains the macroscopic rate equations for the concentrations ϕ, ψ

$$\dot\phi = \alpha + \phi^2\psi - \beta\phi - \phi, \tag{7.3a}$$

$$\dot\psi = \beta\phi - \phi^2\psi. \tag{7.3b}$$

There are, of course, two integration constants, but one of them is merely a shift in t and determines the phase. The solution *curves* in the (ϕ, ψ)-plane are therefore a one-parameter family. A careful study*[)] reveals that for a certain range of values of α, β there is one closed curve and that it is a limit cycle.

These properties of the macroscopic equations have their effect on the fluctuations around the periodic solutions. The effect can be computed in some detail**[)], hampered only by the lack of explicit solutions of (7.3). We shall here merely outline a qualitative description.

Consider the solution of the M-equation that at $t = 0$ consists of a delta peak located at some point (ϕ_0, ψ_0) on the macroscopic limit cycle. The shape of the probability distribution is (in linear noise approximation) governed by the equation

$$\frac{\partial \Pi(\xi, \eta, t)}{\partial t} = (-2\phi\psi + \beta + 1)\frac{\partial}{\partial\xi}\xi\Pi - \phi^2\frac{\partial}{\partial\xi}\eta\Pi$$

$$+ (-\beta + 2\phi\psi)\frac{\partial}{\partial\eta}\xi\Pi + \phi^2\frac{\partial}{\partial\eta}\eta\Pi$$

$$+ \frac{\alpha + \phi}{2}\frac{\partial^2\Pi}{\partial\xi^2} + \frac{\beta\phi + \phi^2\psi}{2}\left(\frac{\partial}{\partial\xi} - \frac{\partial}{\partial\eta}\right)^2\Pi, \tag{7.4}$$

where (ϕ, ψ) is the time-dependent periodic solution with initial values ϕ_0, ψ_0. The terms on the last line tend to spread out Π. This tendency would be counteracted by the terms with first derivatives if the equations (7.3) were asymptotically stable, because that means that all deviations tend to zero. Hence the fluctuations would remain small as in X.5. In our case of a limit cycle this counteraction is still operative for fluctuations perpendicular to it, since all macroscopic solutions lead back to it.

However, a fluctuation in the direction along the limit cycle has no tendency to return. It only changes the phase and the system has no mechanism by which it is forced back to the original phase. The next fluctuation *may* bring it back closer to the original phase, but may just as well increase the difference. The result is that Π spreads out in the direction of the limit

---

*[)] R. Lefever and G. Nicolis, J. Theor. Biol. **30**, 267 (1971); G. Nicolis, Advances in Chemical Physics **19**, 209 (Wiley-Interscience, New York, 1971).
[)] K. Tomita, T. Ohta, and H. Tomita, Prog. Theor. Phys. **52, 1744 (1974).

cycle, just as the probability density of a Brownian particle in one dimension spreads out. The width along the limit cycle will therefore grow proportionally with $\sqrt{t}$.

A consequence is that the linear noise approximation, and the very Ω-expansion itself, become invalid when $t \sim \Omega$. Yet it is intuitively clear what happens after that. The probability density continues to spread until the whole limit cycle is covered, although in the perpendicular direction it remains narrow. The ultimate distribution $P(n, m, \infty) = P^s(n, m)$ has the form of a crater ridge covering the macroscopic limit cycle. This means that all information about the original phase has been lost. In frequency generators this "phase slip" gives rise to a broadening of the generated frequency band. All this is a consequence of the absence of asymptotic stability for variations parallel to the limit cycle.

Exercise. Verify the above mentioned estimate $t \sim \Omega$ for the time after which the fluctuations are no longer small.

Exercise. Find the stationary solution of (7.3) and show that it is unstable when $\beta > \alpha^2 + 1$.

Exercise. Prove that inside a limit cycle there must be an unstable stationary point.

Exercise. Take in (7.3) the special values $\alpha = 1$, $\beta = 3$. Prove in the following way that there is a limit cycle[*]. There is one unstable stationary solution. On the other hand, there exists a closed curve surrounding it with the property that all solutions it intersects are directed towards its interior; viz., the curve formed by the axes and the lines $y = 5.84 + x$, $y = 7.84 - x$.

Exercise. Find from (7.4) the moments $\langle \xi^2 \rangle, \langle \xi\eta \rangle, \langle \eta^2 \rangle$. They obey (VIII.6.9). How does the fact that the moments are unbounded show up in that equation?

The limit cycle is an attractor. A slightly different kind occurs in the theory of the *laser*.[**] Consider the electric field in the laser cavity interacting with the atoms, and select a single mode near resonance, having a complex amplitude E. One then derives from a macroscopic description laced with approximations the evolution equation

$$\dot{E} = aE - b|E|^2 E. \tag{7.5}$$

a and b are positive constants, a represents the pumping and b is due to the interaction with the nonlinear medium formed by the atoms. One then realizes that there are losses due to the escape of radiation and to spontaneous emission by the atoms into other modes. Moreover there is noise due

---

[*] J.J. Tyson, op. cit.

[**] M. Sargent, M.O. Scully, and W.E. Lamb, *Laser Physics* (Addison-Wesley, Reading, MA 1974) ch. 8; H. Haken, *Laser Theory* (Springer, Berlin 1984).

7. LIMIT CYCLES AND FLUCTUATIONS

to the randomness of that emission. These effects are taken into account by adding *ad hoc* a damping term and a random force,

$$\dot{E} = (a - c)E - b|E|^2 E + L(t). \tag{7.6}$$

Here $L(t)$ is a complex Langevin process,

$$\langle L(t) \rangle = 0, \quad \langle L(t)L(t') \rangle = 0, \quad \langle L(t)L(t')^* \rangle = \Gamma \, \delta(t - t').$$

It should be emphasized that this way of including fluctuations has no other justification than that it is convenient and bypasses a description of the noise sources, compare IX.4. It may provide some qualitative insight into the effect of noise, but does not describe its actual mechanism. For instance, fluctuations in the pumping should give rise to randomness in the coefficient a, rather than to an additive term. Yet the equation (7.6) has been the subject of extensive study and it is famous in statistical mechanics under the name of generalized Ginzburg–Landau equation. It may well serve us as an illustration for a stochastic process.[*]

According to IX.3 one may cast (7.6) into the form of an equivalent Fokker–Planck equation. Since (7.6) is actually a set of two equations for the real and imaginary parts E', E'' of E one will obtain an equation for the bivariate distribution $P(E', E'', t) \, dE' \, dE''$. It is convenient to write $P(E', E'', t)$ as $P(E, E^*, t)$ (which is still the same probability density in the E', E'' plane, though one often writes $dE \, dE^*$ for $dE' \, dE''$). Then the Fokker–Planck equation is[**]

$$\frac{\partial P(E, E^*, t)}{\partial t} = -\frac{\partial}{\partial E}(a - c - b|E|^2)EP$$

$$- \frac{\partial}{\partial E^*}(a - c - b|E|^2)E^*P + \Gamma \frac{\partial^2 P}{\partial E \, \partial E^*}. \tag{7.7}$$

Exercise. Derive (7.7) by showing that the first and second moments agree with those obtained from (7.6).

Exercise. Find the stationary solution of (7.7). Why is it not possible to determine Γ by some kind of fluctuation–dissipation theorem?

Exercise. Prove that all solutions of (7.7) approach the stationary solution.

Exercise. Introduce polar coordinates in the (E', E'')-plane by setting $E = \sqrt{s} \, e^{i\phi}$ and transform (7.7) into

---

[*] A more sophisticated treatment was given by K. Hepp and E.H. Lieb, Annals Phys. **76**, 360 (1973); see also P.A. Martin, *Modèles en mécanique statistique des processus irréversibles* (Lecture Notes in Physics **103**; Springer, Berlin 1979) ch. IV.

[**] H. Risken, Z. Phys. **186**, 85 (1965) and **191**, 302 (1966).

$$\frac{\partial P(s, \phi, t)}{\partial t} = -2\frac{\partial}{\partial s}(a - c - bs)sP + \Gamma\left\{\frac{\partial}{\partial s}s\frac{\partial P}{\partial s} + \frac{1}{4s}\frac{\partial^2 P}{\partial \phi^2}\right\}. \tag{7.8}$$

In the absence of an explicit mesoscopic description of the fluctuations it is not possible to extract a deterministic macroscopic equation in a systematic way. Indeed one simply omits in (7.6) the term $L(t)$, or equivalently one sets $\Gamma = 0$ in (7.7) and (7.8). The resulting deterministic equation can be written as an equation for $|E|^2 \equiv s$,

$$\dot{s} = 2(a - c)s - 2bs^2 \equiv -V'(s), \tag{7.9a}$$

$$V(s) = -(a - c)s^2 + \tfrac{2}{3}bs^2. \tag{7.9b}$$

This describes the motion of an underdamped particle in the potential V, see fig. 44. Below pumping threshold, $a < c$, the damping dominates and the only stationary state is $s = 0$. At the threshold this state becomes critical and for $a > c$ it is unstable, while a new stable state has appeared at

$$s = |E|^2 = \frac{a - c}{b}. \tag{7.10}$$

In the plane E', E'' this is a circular attractor, but there is no motion along it. Each point is a stationary solution, marginally stable in the tangential direction.

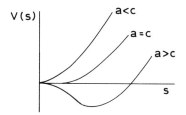

Fig. 44. Sketch of the potential given by (7.9b).

Now take $\Gamma > 0$. The stationary solution of (7.8) can be found explicitly:

$$P^s(s) = \text{const. } \exp\left[\frac{2}{\Gamma}\{(a - c)s - \tfrac{1}{2}bs^2\}\right]. \tag{7.11}$$

For $a < c$ its maximum lies at $s = 0$, which is the stable stationary solution of (7.5) and (7.9). For $a > c$ its maximum lies at $d = (a - c)/b$.[*] Then P^s is a

---

[*] This is the same d as used by Risken, loc. cit. His W, r, β, q, a are our $P, \sqrt{s}, b, \tfrac{1}{4}\Gamma$, $2(c - a)/\sqrt{b\Gamma}$, respectively. See also H. Risken and H.D. Vollmer, Z. Phys. B **39**, 89 (1980).

distribution in the (E', E'')-plane, which is independent of the angle ϕ. Hence the maximum is a circular crater ridge and the phase is entirely undetermined.

The rate at which s approaches d can be found from (7.9) to be roughly $\tau_s^{-1} \approx 2(a-c)$. The time needed for ϕ to spread out over the full circle is $\tau_\phi \sim \pi^2(2d/\Gamma)$. For small Γ this is much longer than τ_s, so that the field E first attains its stationary amplitude and subsequently the phase spreads out along the ridge. In the emitted radiation this uncertainty of the phase creates a line width [compare (VIII.3.12)]

$$\Delta\omega \sim \frac{1}{\tau_\phi} \sim \frac{\Gamma}{|E|^2}. \tag{7.12}$$

Although there is no limit cycle the spreading of the phase occurs in much the same way as in the previous section.

Exercise. Why is the free energy in (7.11) not the same function as the potential V? [Compare the Remark in XI.5.]

Exercise. In the stationary distribution (7.11) the average $\langle s \rangle$ differs from the most probable value d by a very small amount, of order $\Gamma^{1/2} \exp[-bd^2/\Gamma]$. But the average *amplitude* is less than $\sqrt{d}$ by an amount $(\Gamma/16b)d^{-3/2}$. [Compare the discussion of the Brillouin paradox in IX.5.]

Exercise. Far below threshold the nonlinear terms in (7.6) and (7.7) may be neglected. In this case E is an Ornstein–Uhlenbeck process and

$$\langle E(t_1)E^*(t_2)\rangle^s = \frac{\Gamma}{2(c-a)} e^{-(c-a)|t_1-t_2|}.$$

The condition for the validity is $c - a \gg \sqrt{b\Gamma}$. The autocorrelation of the intensity is the square of this,

$$\langle\langle |E(t_1)|^2 |E(t_2)|^2 \rangle\rangle^s = \frac{\Gamma^2}{4(c-a)^2} e^{-2(c-a)|t_1-t_2|}.$$

Exercise. Far above threshold, $a - c \gg \sqrt{b\Gamma}$, equation (7.8) can be linearized by setting $s = d + \sigma$,

$$\frac{\partial P(\sigma, \phi, t)}{\partial t} = 2bd \frac{\partial}{\partial \sigma}\sigma P + \Gamma d \frac{\partial^2 P}{\partial \sigma^2} + \frac{\Gamma}{4d}\frac{\partial^2 P}{\partial \phi^2} \tag{7.13}$$

in order to approximate the fluctuations around the steady lasering solution. Show that now

$$\langle E(t_1)E^*(t_2)\rangle^s = e^{-(\Gamma/4d)|t_1-t_2|}\left\{d - \frac{\Gamma}{8bd}(1 - e^{-2bd|t_1-t_2|})\right\},$$

$$\langle\langle |E(t_1)|^2 |E(t_2)|^2 \rangle\rangle^s = \frac{\Gamma}{2b} e^{-2bd|t_1-t_2|}.$$

Verify (7.12).

Exercise. For the number of photons n in the single-mode laser the following master equation has been derived[*]

$$\dot{p}_n = (\mathbb{E}^{-1} - 1)\frac{A(n+1)}{1 + B(n+1)} p_n + C(\mathbb{E} - 1)np_n. \tag{7.14}$$

Oblivious of the message of chapter X and using some simplifications we replace this with the naive Fokker–Planck approximation of VIII.5:

$$\dot{p}_n = \frac{\partial}{\partial n}\{An(1 - Bn) - Cn\}p_n + \frac{1}{2}\frac{\partial^2}{\partial n^2}\{An(1 - Bn) + Cn\}p_n.$$

Show that it leads to similar results below and above threshold as found above, but with two different values for Γ.

Exercise. If Ω is the size of the laser, the coefficient B in (7.14) turns out to be proportional to $1/\Omega$. Find the same results as in the preceding Exercise in a systematic way by means of the Ω-expansion.

---

[*] M.O. Scully and W.E. Lamb, Phys. Rev. **159**, 208 (1967).

Chapter XIV

FLUCTUATIONS IN CONTINUOUS SYSTEMS

So far our random variables represented the *total* number of molecules, electrons, individuals, etc., in a system. In order to describe *local* fluctuations one has to introduce the spatial density as random object. Rather than having one or a few random variables one now has to deal with an infinite set of them.

1. Introduction

In studying chemical reactions we have so far always described the state of the mixture by means of the *total* numbers of the molecules of various kinds. This assumes that the mixture is homogeneous, and remains homogeneous even during the fluctuations of these total numbers. It means that a fluctuation in the total number must have time to spread throughout the volume Ω before it disappears. Any molecule created somewhere must on the average have time to move through the volume before it is annihilated again by another reaction. Unless the reactions are very fast, this can be achieved by stirring, or in favorable circumstances diffusion might suffice. A similar comment applies to our applications to semiconductors, populations, etc.

Suppose, however, that there is no stirring and that diffusion does not suffice. Then a fluctuation originating at one point will disappear before it has been able to make itself felt in remote regions of the volume: it is a local phenomenon. Such a situation requires a description in more variables, viz., the density of each compound at the different points in space. This can be done as follows. For simplicity we suppose that there is just one species of particles, as in the reaction (X.1.1):

$$A \to X, \quad 2X \to B.$$

Subdivide the total volume Ω into cells Δ and call n_λ the number of particles in cell λ. The cells must be so small that inside each of them the above mentioned condition of homogeneity prevails. Let $P(\{n_\lambda\}, t)$ be the joint probability distribution of all n_λ. At $t + dt$ it will have changed because of two kinds of possible processes. Firstly, the n_λ inside each separate cell λ may change by an event that creates or annihilates a particle. In the master equation for $P(\{n_\lambda\}, t)$ this gives a corresponding term for each separate cell.

For example, in this reaction one obtains according to (X.1.4)

$$\dot{P}(\{n_\lambda\}, t) = \Delta \sum_\lambda (\mathbb{E}_\lambda - 1)P + (2\Delta)^{-1} \sum_\lambda (\mathbb{E}_\lambda^2 - 1)n_\lambda(n_\lambda - 1)P. \qquad (1.1)$$

Secondly, P changes because during the time dt a particle may move from a cell λ into a cell μ (which need not be an adjacent one). The probability for a particle in cell λ to arrive in μ is proportional to Δ and dt and will therefore be denoted by $w_{\mu\lambda} \Delta\, dt$. The corresponding contribution to the M-equation is

$$\dot{P}(\{n_\lambda\}, t) = \Delta \sum_{\lambda,\mu} w_{\mu\lambda}(\mathbb{E}_\mu^{-1}\mathbb{E}_\lambda - 1)n_\lambda P. \qquad (1.2)$$

The total change of P is the sum of both contributions. The resulting M-equation contains all information needed for computing the local fluctuations in the chemically reacting system and has actually been used for that purpose.[*]

The artificial discretization of space, however, is awkward. In particular, it is responsible for introducing a large set of constants $w_{\mu\lambda}$ as a substitute for the diffusion coefficient D. This difficulty is even more serious if one has to take into account the free flight in space of the particles, as in the case of neutrons in a reactor, see 4. Hence it is desirable to cast (1.1) and (1.2) in a form in which the spatial coordinates occur as continuously varying parameters instead of the discrete subscripts λ.

For this purpose we replace the set of numbers n_λ by a particle density function $u(r)$ such that

$$u(r) = n_\lambda / \Delta, \qquad (1.3)$$

where r is the position of cell λ in space, e.g., of its center.[**] The joint distribution $P(\{n_\lambda\}, t)$ then turns into a probability in function space, $P([u(r)], t)$, which is a functional of $u(r)$. Summation over all n_λ turns into an integration over the space of functions $u(r)$. This can be done formally and leads to correct results for the moments and correlation functions. A more direct, and mathematically safer way, however, consists in first deriving equations for the moments from the discrete formulation of the master equation, and subsequently expressing them as continuous functions of space. This *method of compounding moments* is described in the next two sections.

---

[*] G. Nicolis and I. Prigogine, Proc. Nat. Acad. Sci. USA **68**, 2102 (1971); C. van den Broeck, W. Horsthemke, and M. Malek-Mansour, Physica A **89**, 339 (1977); M. DelleDonne and P. Ortoleva, J. Statist. Phys. **18**, 319 (1978); M.A. Burschka, J. Statist. Phys. **45**, 715 (1986); GARDINER, ch. 8.

[**] I am forced to the unfamiliar use of the symbol u for the density by the poverty of the alphabet.

Remark. Mathematically the probability density in function space and the integration over all functions is not defined. The reason is that this inadvertently introduces an overwhelming amount of very rapidly varying functions $u(r)$. Physically they are meaningless, because (1.3) defines $u(r)$ as an interpolation of numbers on a grid. The problem is therefore to find a mathematically consistent and physically satisfactory method for restricting function space to sufficiently smooth functions. This problem need not be solved, however, because the resulting equations for the moments lead to correct results.

Exercise. Write the M-equation for the local distribution of charge carriers in the semiconductor of VI.9, assuming that they are transported by diffusion.

Exercise. The criterion that diffusion suffices to maintain homogeneity is

$$D\tau \gg l^2, \tag{1.4}$$

where τ is the average life time of a particle and l the diameter of the vessel.

2. Diffusion noise

We first consider the diffusion part (1.2) of the M-equation for $P(\{n_\lambda\}, t)$. Multiply (1.2) with n_α and sum over all $\{n_\lambda\}$,

$$\partial_t \langle n_\alpha \rangle = \Delta \sum_\lambda \{w_{\alpha\lambda} \langle n_\lambda \rangle - w_{\lambda\alpha} \langle n_\alpha \rangle\} = \Delta \sum_\lambda W_{\alpha\lambda} \langle n_\lambda \rangle. \tag{2.1}$$

In the same way one finds by laborious algebra an equation for the second moments $\langle n_\alpha n_\beta \rangle$, which simplifies when expressed in the factorial cumulants

$$\partial_t [n_\alpha n_\beta] = \Delta \sum_\lambda W_{\alpha\lambda}[n_\lambda n_\beta] + \Delta \sum_\lambda W_{\beta\lambda}[n_\alpha n_\lambda]. \tag{2.2}$$

Remember that according to the definition (I.3.14)

$$[n_\alpha n_\beta] = \langle n_\alpha n_\beta \rangle - \langle n_\alpha \rangle \langle n_\beta \rangle - \delta_{\alpha\beta} \langle n_\alpha \rangle. \tag{2.3}$$

Actually the result (2.2) could have been established without calculation, see VII.6, because we are dealing with a collection of independent particles, each of which moves as prescribed by the jump probabilities $w_{\mu\lambda}$.

We rewrite these equations in continuous notation. Use (1.3) and replace $w_{\alpha\lambda}$ with $w(r|r')$. The summation together with the factor Δ becomes an integration, so that (2.1) takes the form

$$\partial_t \langle u(r) \rangle = \int w(r|r') \langle u(r') \rangle \, dr'. \tag{2.4}$$

To rewrite (2.2) in the same way we first have to define the factorial cumulant in the continuous description. Divide (2.3) by Δ^2 and replace α, β with r_1, r_2:

$$[u(r_1)u(r_2)] = \langle u(r_1)u(r_2) \rangle - \langle u r_1 \rangle \langle u r_2 \rangle - \delta r_1 - r_2) \langle u(r_1) \rangle. \tag{2.5}$$

Here we have used the identification

$$\frac{\delta_{\alpha\beta}}{\Delta} \to \delta(r_1 - r_2). \tag{2.6}$$

For the factorial cumulant (2.5) of the density one now obtains from (2.2) the equation

$$\partial_t[u(r_1)u(r_2)] = \int \mathrm{w}(r_1|r')[u(r')u(r_2)]\,dr' + \int \mathrm{w}(r_2|r')[u(r_1)u(r')]\,dr'. \tag{2.7}$$

All this is true regardless of the specific form of the transition probabilities $\mathrm{w}_{\mu\lambda}$ or $\mathrm{w}(r|r')$. If they are isotropic and the jumps are small compared with the distance over which $u(r)$ varies one may replace the operator by its diffusion approximation:

$$\mathrm{w} \to D\nabla^2.$$

As a result (2.4) becomes the diffusion equation

$$\partial_t \langle u(r) \rangle = D\nabla^2 \langle u(r) \rangle. \tag{2.8}$$

This holds for the average density; for the fluctuations one obtains from (2.7)

$$\partial_t[u(r_1)u(r_2)] = D(\nabla_1^2 + \nabla_2^2)[u(r_1)u(r_2)]. \tag{2.9}$$

From this one readily finds for the covariance

$$\{\partial_t - D\nabla_1^2 - D\nabla_2^2\}\langle\langle u(r_1)u(r_2)\rangle\rangle = 2D\nabla_1 \cdot \nabla_2\{\delta(r_1 - r_2)\langle u(r_1)\rangle\}. \tag{2.10}$$

Exercise. Derive (2.1) and (2.2) directly from (1.2).

Exercise. Justify (2.6) by showing that for arbitrary, smoothly varying sets $\{n_\alpha\}$, $\{m_\alpha\}$ one has

$$\sum_{\alpha\beta} n_\alpha (\delta_{\alpha\beta}/\Delta) m_\beta = \int\int u(r)\,\delta(r - r')v(r')\,dr\,dr',$$

where u and v are the densities corresponding to n and m.

Exercise. For electrons in a semiconductor subject to a constant field $\boldsymbol{F}$ one has, in proper units,

$$\partial_t \langle u(r) \rangle = -\boldsymbol{F} \cdot \nabla \langle u(r) \rangle + D\nabla^2 \langle u(r) \rangle, \tag{2.11a}$$

$$\{\partial_t + \boldsymbol{F} \cdot (\nabla_1 + \nabla_2) - D(\nabla_1^2 + \nabla_2^2)\}\langle\langle u(r_1)u(r_2)\rangle\rangle = 2D\,\nabla_1 \cdot \nabla_2\{\delta(r_1 - r_2)\langle u(r_1)\rangle\}. \tag{2.11b}$$

Exercise. The density fluctuations in diffusion were first calculated by van Vliet[*] using the following Langevin approach. The density u and current $\boldsymbol{J}$ are assumed to obey

$$\partial_t u = -\nabla \cdot \boldsymbol{J}, \quad \boldsymbol{J} = F u - D \nabla u + \boldsymbol{L}(\boldsymbol{r}, t);$$

$$\langle \boldsymbol{L}(\boldsymbol{r}, t) \rangle = 0,$$

$$\langle L_i(\boldsymbol{r}, t) L_j(\boldsymbol{r}', t') \rangle = 2D\, \delta_{ij}\, \delta(\boldsymbol{r} - \boldsymbol{r}')\, \delta(t - t') \langle u(\boldsymbol{r}, t) \rangle. \tag{2.12}$$

Show that this leads formally to the same result (2.11). Notice, however, that L is not a true Langevin force, because its stochastic properties depend on the particular solution of (2.11a).

Exercise. Show for the one-dimensional diffusion in an inhomogeneous medium described by (XI.3.2):

$$\{\partial_t - \nabla_1 D(x_1)\nabla_1 - \nabla_2 D(x_2)\nabla_2\} \langle\langle u(x_1) u(x_2) \rangle\rangle = 2\nabla_1 \nabla_2 D(x_1) \langle u(x_1) \rangle\, \delta(x_1 - x_2).$$

The corresponding Langevin term is obtained by replacing (2.12) with

$$\langle L(x, t) L(x', t') \rangle = 2D(x)\, \delta(x - x')\, \delta(t - t') \langle u(x, t) \rangle.$$

3. The method of compounding moments

Now consider the fluctuations arising from the chemical reaction. We take a simpler example than (1.1), namely the reaction (VI.4.7):

$$A \rightleftharpoons X.$$

Its M-equation (VI.4.8), applied to the separate cells, is (in slightly different notation)

$$\dot{P}(\{n_\lambda\}, t) = a \sum_\lambda (\mathbb{E}_\lambda - 1) n_\lambda P + b\varDelta \sum_\lambda (\mathbb{E}_\lambda^{-1} - 1) P. \tag{3.1}$$

The reader who has done the Excercises will easily derive

$$\partial_t \langle n_\alpha \rangle = b\varDelta - a\langle n_\alpha \rangle, \tag{3.2a}$$

$$\partial_t [n_\alpha n_\beta] = -2a[n_\alpha n_\beta]. \tag{3.2b}$$

Divide by $\varDelta$ and use the continuous notation:

$$\partial_t \langle u(\boldsymbol{r}) \rangle = b - a\langle u(\boldsymbol{r}) \rangle, \tag{3.3a}$$

$$\partial_t [u(\boldsymbol{r}_1) u(\boldsymbol{r}_2)] = -2a[u(\boldsymbol{r}_1) u(\boldsymbol{r}_2)]. \tag{3.3b}$$

The effects of the reaction and the diffusion can now be combined by noting that the moments vary in time due to both agencies. Hence for the

---

[*] K.M. van Vliet, J. Mathem. Phys. **12**, 1981 and 1998 (1971).

reaction in the presence of diffusion one obtains from (3.3a) and (2.8)

$$\partial_t \langle u(r) \rangle = b - a\langle u(r) \rangle + D\nabla^2 \langle u(r) \rangle \qquad (3.4a)$$

and from (3.3b) and (2.9)

$$\partial_t [u(r_1)u(r_2)] = \{D(V_1^2 + V_2^2) - 2a\}[u(r_1)u(r_2)]. \qquad (3.4b)$$

This is the *method of compounding moments*, which avoids the explicit use of a probability in function space. It is, of course, possible to write similar equations for higher moments.

As an application we study the fluctuations in equilibrium. From (3.4a) one obtains the equilibrium solution

$$\langle u \rangle^e = b/a. \qquad (3.5)$$

From (3.4b) one obtains for $[u(r_1)u(r_2)]^e \equiv g(|r_1 - r_2|)$

$$\{2D\nabla^2 - 2a\}g(r) = 0.$$

As the factorial cumulant must remain finite both at large and at small distance the only admissible solution is $g = 0$. The covariance of the fluctuations is therefore

$$\langle\langle u(r_1)u(r_2)\rangle\rangle^e = \delta(r_1 - r_2)\langle u(r_1)\rangle^e = \frac{b}{a}\delta(r_1 - r_2). \qquad (3.6)$$

Next consider the correlation of the equilibrium fluctuations at two different points and different times

$$\langle\langle u(r_1, t_1)u(r_2, t_2)\rangle\rangle^e = G(r_2 - r_1, t_2 - t_1). \qquad (3.7)$$

Take $t_2 > t_1$ and compute from (3.4a) the average density $\langle u(r_2, t_2)\rangle$, conditional on a given $u(r_1, t_1)$:

$$\langle u(r_2, t_2)\rangle_{\text{cond}} - \langle u \rangle^e = e^{-a(t_2 - t_1)}[4\pi D(t_2 - t_1)]^{-3/2}$$

$$\times \int \exp\left[-\frac{(r_2 - r')^2}{4D(t_2 - t_1)}\right]\{u(r', t_1) - \langle u \rangle^e\}\,dr'.$$

Multiply with $u(r_1, t_1) - \langle u \rangle^e$ and average the product over the values of $u(r_1, t_1)$ as they occur in equilibrium with the aid of (3.6); the result is that for $t > 0$

$$G(r, t) = \frac{b}{a}e^{-at}(4\pi Dt)^{-3/2}e^{-(r^2/4Dt)}. \qquad (3.8)$$

Exercise. Verify (3.2b).

Exercise. Let V_1 and V_2 be two regions in space, which may overlap. Let $N(V_1), N(V_2)$ be the number of particles in either one. Compute $\langle\langle N(V_1)N(V_2)\rangle\rangle^e$.

3. THE METHOD OF COMPOUNDING MOMENTS

Exercise. The Fourier transform of (3.8) (needed for scattering data) is

$$S(k, \omega) = \frac{2\langle u \rangle^e}{(2\pi)^2} \frac{a + Dk^2}{(a + Dk^2)^2 + \omega^2}. \tag{3.9}$$

Exercise. Study in the same way the reaction

$$B \to 2X, \quad X \to A.$$

Exercise. Solve (3.4) for the time-dependent case specified by the initial condition $u(r, 0) = 0$.

Exercise. Consider the density fluctuations of a radioactive substance dissolved in a fluid and subject to diffusion. If at $t = 0$ the density is precisely u_0 compute $\langle\langle u(r_1) u(r_2) \rangle\rangle$ at $t > 0$.

The treatment of our example was facilitated by the fact that the master equation (3.1) for the reaction part was linear. Now consider again the reaction (X.1.1), studied in X.1 for the case of perfect stirring. In order to obtain equations for the first and second moments one has to apply the Ω-expansion to each separate cell, i.e., one has to expand the multivariate master equation (1.1) in powers of $\Delta^{-1/2}$. This imposes an additional condition on the cell size that Δ must be large enough to contain many particles. We invite the reader to do the calculation and give the result.

$$\partial_t \langle n_\alpha \rangle = \Delta - \Delta^{-1} \langle n_\alpha \rangle^2, \tag{3.10a}$$

$$\partial_t \langle\langle n_\alpha n_\beta \rangle\rangle = -2\Delta^{-1}(\langle n_\alpha \rangle + \langle n_\beta \rangle)\langle\langle n_\alpha n_\beta \rangle\rangle$$
$$+ \delta_{\alpha\beta}\{\Delta + 2\Delta^{-1} \langle n_\alpha \rangle^2\}. \tag{3.10b}$$

After this preliminary work we go over to the continuous description by dividing these equations with Δ and Δ^2, respectively, and writing them in terms of the density $u(r)$:

$$\partial_t \langle u(r) \rangle = 1 - \langle u(r) \rangle^2, \tag{3.11a}$$

$$\partial_t \langle\langle u(r_1) u(r_2) \rangle\rangle = -2\{\langle u(r_1) \rangle + \langle u(r_2) \rangle\}\langle\langle u(r_1) u(r_2) \rangle\rangle$$
$$+ \delta(r_1 - r_2)\{1 + 2\langle u(r_1) \rangle^2\}. \tag{3.11b}$$

For the factorial cumulant this last equation gives

$$\partial_t [u(r_1) u(r_2)] = -2\{\langle u(r_1) \rangle + \langle u(r_2) \rangle\}[u(r_1) u(r_2)]$$
$$- \delta(r_1 - r_2)\langle u(r_1) \rangle^2. \tag{3.11c}$$

It is now possible to add the effect of diffusion:

$$\partial_t \langle u(r) \rangle = 1 - \langle u(r) \rangle^2 + D \nabla^2 \langle u(r) \rangle, \tag{3.12a}$$

$$\partial_t [u(r_1) u(r_2)] = -2\{\langle u(r_1) \rangle + \langle u(r_2) \rangle\}[u(r_1) u(r_2)]$$
$$- \delta(r_1 - r_2)\langle u(r_1) \rangle^2 + D(\nabla_1^2 + \nabla_2^2)[u(r_1) u(r_2)]. \tag{3.12b}$$

These equations describe the local fluctuations in the chemical reaction.

Let us employ these equations to find the fluctuations in equilibrium, taking the total volume infinite. According to (3.12a) the macrostate corresponds (in our units) to $\langle u(r) \rangle^e = 1$. Hence (3.12b) reduces to

$$D(\nabla_1^2 + \nabla_2^2)[u(r_1)u(r_2)]^e = 4[u(r_1)u(r_2)]^e + \delta(r_1 - r_2). \tag{3.13}$$

Since $[u(r_1)u(r_2)]^e$ must be a function of $|r_1 - r_2|$ alone and cannot grow exponentially the only admissible solution is

$$[u(r_1)u(r_2)]^e = -\frac{1}{8\pi D} \frac{e^{-(2/D)^{1/2}|r_1 - r_2|}}{|r_1 - r_2|}.$$

For the covariance this gives in the original units of (X.1.3)

$$\langle\langle u(r_1)u(r_2)\rangle\rangle^e = \langle u\rangle^e \left\{ \delta(r_1 - r_2) - \frac{\kappa^2}{16\pi} \frac{e^{-\kappa|r_1 - r_2|}}{|r_1 - r_2|} \right\}, \tag{3.14}$$

where $\langle u \rangle^e = \sqrt{k\phi_A/2k'}$ and $\kappa^2 = 2\sqrt{2kk'\phi_A}/D$.

According to (X.1.3) the probability per unit time for a molecule X to disappear is (in equilibrium) $2k'\langle u\rangle^e/\Omega = \sqrt{2kk'\phi_A}$ and its average lifetime is the reciprocal of this. It follows that κ^{-1} is a measure for the distance over which a molecule diffuses during its lifetime. Take in our infinite volume a region V whose linear dimensions are large compared to κ^{-1}. The variance of the number N_V of molecules X in V is obtained by integrating (3.14), with the result[*]

$$\langle\langle N_V \rangle\rangle^e = \langle N_V \rangle^e (1 - \tfrac{1}{4}) = \tfrac{3}{4}\langle N_V \rangle^e. \tag{3.15}$$

This is the same as the result (X.1.13) for the thoroughly stirred case; the reason is that V is too large for the diffusion to establish a material exchange with the surroundings.

On the other hand, when V is small compared with κ^{-1} the diffusion is important, but its effect depends of course on the shape of V. Yet one may make an estimate of the integral of (3.14):

$$\langle\langle N_V^2 \rangle\rangle^e = \langle u\rangle^e \left\{ V - \frac{\kappa^2}{16\pi} \frac{V^2}{l} \right\}, \tag{3.16}$$

where l is of the order of the diameter of V. In the limit of small V this tends to $\langle N_V \rangle^e$ as in a Poisson distribution; the reason is that the fluctuations caused by diffusion dominate the effect of the chemical reaction.[**]

---

[*] Y. Kuramoto, Prog. Theor. Phys. **49**, 1782 (1973); A. Nitzan and J. Ross, J. Statist. Phys. **10**, 379 (1974).

[**] This difference between large and small V was pointed out by Y. Kuramoto, Prog. Theor. Phys. **52**, 711 (1974). I propose to call κ^{-1} the Kuramoto length. See also L. Brenig and C. van den Broeck, Phys. Rev. A **21**, 1039 (1980).

Finally, if the reaction is enclosed in a finite volume Ω one must solve (3.13) in this volume with the boundary condition that the normal derivative of u on the wall vanishes. If the diameter of Ω is small compared with κ^{-1} the solution is

$$[u(\mathbf{r}_1)u(\mathbf{r}_2)]^e = -\tfrac{1}{4}\delta(\mathbf{r}_1 - \mathbf{r}_2), \tag{3.17}$$

and from this one obtains again (X.1.13).

Exercise. Derive (3.10).
Exercise. Find for the Fourier transform of (3.14)

$$\int \langle\langle u(\mathbf{r}_1)u(\mathbf{r}_2)\rangle\rangle^e \, e^{i\mathbf{k}\cdot(\mathbf{r}_1-\mathbf{r}_2)} \, d(\mathbf{r}_1 - \mathbf{r}_2) = \langle u \rangle^e \frac{3 + 4k^2/\kappa^2}{4 + 4k^2/\kappa^2},$$

which was the form given by Kuramoto. The two limiting cases can easily be obtained.

Exercise. Solve (3.13) in a finite cube and obtain (3.17) in the limit $D \to 0$.

4. Fluctuations in phase space density

So far we only considered transport of particles by diffusion. As mentioned in **1** the continuous description was not strictly necessary, because diffusion can be described as jumps between cells and therefore incorporated in the multivariate master equation. Now consider particles that move freely and should therefore be described by their velocity v as well as by their position r. The cells Δ are six-dimensional cells in the one-particle phase space. As long as no reaction occurs v is constant but r changes continuously. As a result the probability distribution varies in a way which cannot be described as a succession of jumps but only in terms of a differential operator. Hence the continuous description is indispensable, but the method of compounding moments can again be used.

We demonstrate the method on the following concrete – if somewhat trivial – example. A swarm of particles is moving freely in space, but each particle has a probability a per unit time to disappear, through spontaneous decay or through a reactive collision. To cover the latter possibility we allow a to depend on v. The (r, v)-space is decomposed in cells Δ and n_λ is the number of particles in cell λ. The joint probability distribution $P(\{n_\lambda\}, t)$ varies through decay and through the motion of the particles. The decay is described by

$$\dot{P} = \sum_\lambda a_\lambda (\mathbb{E}_\lambda - 1) n_\lambda P. \tag{4.1}$$

From it follows for the moments

$$\partial_t \langle n_\alpha \rangle = -a_\alpha \langle n_\alpha \rangle, \qquad (4.2\text{a})$$

$$\partial_t \langle\langle n_\alpha n_\beta \rangle\rangle = -(a_\alpha + a_\beta)\langle\langle n_\alpha n_\beta \rangle\rangle + \delta_{\alpha\beta} a_\alpha \langle n_\alpha \rangle, \qquad (4.2\text{b})$$

$$\partial_t [n_\alpha n_\beta] = -(a_\alpha + a_\beta)[n_\alpha n_\beta]. \qquad (4.2\text{c})$$

In the continuous description

$$\partial_t \langle u(r, v) \rangle = -a(v)\langle u(r, v) \rangle, \qquad (4.3\text{a})$$

$$\partial_t [u(r_1, v_1)u(r_2, v_2)] = -\{a(v_1) + a(v_2)\}[u(r_1, v_1)u(r_2, v_2)]. \qquad (4.3\text{b})$$

Note that in this case $\delta_{\alpha\beta}/\Delta \to \delta(r_1 - r_2)\delta(v_1 - v_2)$.

It is now easy to add the terms due to the flow:

$$\partial_t \langle u(r, v) \rangle = -a(v)\langle u(r, v) \rangle - v \cdot \nabla \langle u(r, v) \rangle, \qquad (4.4\text{a})$$

$$\partial_t [u(r_1, v_1)u(r_2, v_2)] = -\{a(v_1) + a(v_2) + v_1 \cdot \nabla_1 + v_2 \cdot \nabla_2\}$$
$$\times [u(r_1, v_1)u(r_2, v_2)]. \qquad (4.4\text{b})$$

In the diffusion case it was essential to add the transfer terms to the factorial cumulant rather than to the variance in order to do justice to the noisy character of diffusion, but in the present case it makes no difference. We merely use the factorial cumulant because it makes the equations simpler.

These equations are easy to solve. Suppose the initial density is given to be $u_0(r, v)$. Then (4.4) yields

$$\langle u(r, v) \rangle_t = e^{-a(v)t} u_0(r - vt, v), \qquad (4.5\text{a})$$

$$[u(r_1, v_1)u(r_2, v_2)]_t = e^{-\{a(v_1)+a(v_2)\}t} [u(r_1 - v_1 t, v_1)u(r_2 - v_2 t, v_2)]_0$$
$$= -e^{-2a(v_1)t} \delta(r_1 - r_2)\delta(v_1 - v_2) u_0(r_1 - v_1 t, v_1). \qquad (4.5\text{b})$$

Consequently

$$\langle\langle u(r_1, v_1)u(r_2, v_2) \rangle\rangle_t = \delta(r_1 - r_2)\delta(v_1 - v_2)\{1 - e^{-a(v_1)t}\}\langle u(r_1, v_1) \rangle_t, \qquad (4.6)$$

and one finds, in analogy with (3.8), for $t_2 \geqslant t_1$

$$\langle\langle u(r_1, v_1, t_1)u(r_2, v_2, t_2) \rangle\rangle = \delta\{r_1 - r_2 + v_1(t_2 - t_1)\}\delta(v_1 - v_2)$$
$$\times e^{-a(v_1)t_2}\{1 - e^{-a(v_1)t_1}\} u_0(r_1 - v_1 t_1, v_1). \qquad (4.7)$$

Of course, no equilibrium exists other than everything zero.

The following example is less trivial, but unfortunately the equations for the fluctuations are too complicated to work out in detail. Consider a semiconductor in which the charge carriers occur in states with pseudo-momentum k, which is also the velocity in that state. They are generated at

4. FLUCTUATIONS IN PHASE SPACE DENSITY

a rate b_k per unit volume and recombine at the rate $a_k w$, where w is the spatial density of empty donors. Let the coordinate space be subdivided in cells Δ, labelled λ; momentum space still consists of discrete levels k. The joint probability distribution of the numbers m_λ of empty donors and of the numbers $n_{\lambda k}$ of charge carriers obeys

$$\dot{P}(\{n_{\lambda k}\}, \{m_\lambda\}, t) = \Delta \sum_{\lambda k} b_k(\mathbb{E}_{m_\lambda}^{-1}\mathbb{E}_{n_{\lambda k}}^{-1} - 1)P + \Delta^{-1} \sum_{\lambda k} a_k(\mathbb{E}_{m_\lambda}\mathbb{E}_{n_{\lambda k}} - 1)m_\lambda n_{\lambda k} P. \tag{4.8}$$

One derives from it for the first moments

$$\partial_t \langle m_\alpha \rangle = \Delta \sum_k b_k - \Delta^{-1} \langle m_\alpha \rangle \sum_k a_k \langle n_{\alpha k} \rangle + \mathcal{O}(\Delta^0),$$

$$\partial_t \langle n_{\alpha k} \rangle = \Delta b_k - \Delta^{-1} a_k \langle m_\alpha \rangle \langle n_{\alpha k} \rangle + \mathcal{O}(\Delta^0).$$

Now make coordinate space continuous by introducing the densities $w(r)$ and $u_k(r)$,

$$\partial_t \langle w(r) \rangle = \sum_k b_k - \langle w(r) \rangle \sum_k a_k \langle u_k(r) \rangle, \tag{4.9a}$$

$$\partial_t \langle u_k(r) \rangle = b_k - a_k \langle w(r) \rangle \langle u_k(r) \rangle. \tag{4.9b}$$

In order to make the description also continuous in k-space we suppose that there are $\rho(k) \, d^3k$ levels in a momentum space cell d^3k and write for the number of charge carriers in it $u(r, k) \, d^3k$. Then (4.9) takes the form

$$\partial_t \langle w(r) \rangle = \int b_k \rho(k) \, d^3k - \langle w(r) \rangle \int a_k \langle u(r, k) \rangle \, d^3k, \tag{4.10a}$$

$$\partial_t \langle u(r, k) \rangle = b_k \rho(k) - a_k \langle w(r) \rangle \langle u(r, k) \rangle. \tag{4.10b}$$

It is now easy to add the terms describing the spatial motion under influence of an applied force F. Of course, (4.10a) is not affected, because the donors are fixed, but (4.10b) becomes

$$\partial_t \langle u(r, k) \rangle = b_k \rho(k) - a_k \langle w(r) \rangle \langle u(r, k) \rangle$$
$$- (k \cdot \nabla) \langle u(r, k) \rangle - (F \cdot \nabla_k) \langle u(r, k) \rangle. \tag{4.11}$$

These are the macroscopic equations, which could have been concluded directly from the macroscopic picture. In order to find the fluctuations as well one has to expand (4.8) in $\Delta^{-1/2}$, find the equations for the second moments in the linear approximation, and supplement them with the flow terms. However, as there are two random fields, $u(r, k)$ and $w(r)$, the equations become complicated and will not be worked out here.

Exercise. It has tacitly been assumed that the density w of empty donors is much smaller than the total density of donors. How do the equations look if one does not make this simplifying assumption?

Exercise. Simplify the semiconductor model by assuming that the recombination probability $a_k w = c_k$ does not depend on the density of available donors. Determine for this linearized model the function $G(r, t)$ defined in (3.7).

Exercise. Find for the stationary solution of (4.10a) with (4.11) in one dimension

$$\langle w(x) \rangle^s = w = \text{const.},$$

$$\langle u(x, k) \rangle^s = \frac{1}{F} \int_{-\infty}^{k} b_{k'} \rho(k') \, dk' \exp\left[-\frac{w}{F} \int_{k'}^{k} a_{k''} \, dk'' \right]. \tag{4.12}$$

(It is assumed that $b_k \rho(k)$ tends to zero for large $|k|$, and that a_k does not.) How is w to be determined?

Exercise. Also find the stationary solution of (4.11) in three dimensions.

Exercise. Derive from (4.12) for small F

$$\langle u(x, k) \rangle^s = \frac{b_k \rho(k)}{w a_k} - \frac{F}{w^2 a_k} \frac{d}{dk} \frac{b_k \rho(k)}{a_k} + \mathcal{O}(F^2).$$

Deduce from it for the linear part of the conductivity

$$\sigma = e^2 \int_{-\infty}^{\infty} \frac{b_k \rho(k)}{a_k} \left(\frac{d}{dk} \frac{k}{a(k)} \right) dk \bigg/ \int_{-\infty}^{\infty} \frac{b_k \rho(k)}{a_k} \, dk.$$

Exercise. The neutrons in a nuclear reactor behave as free particles until they are absorbed, scatter, or cause fission and thereby produce more neutrons. The master equation for the joint probability distribution of the occupation numbers n_λ of the phase space cells λ is

$$\dot{P}(\{n\}, t) = \sum_{\lambda} \sum_{m=0}^{\infty} \sum_{\{\mu\}} q_m(\mu_1, \mu_2, \ldots, \mu_m | \lambda) \{ \mathbb{E}_{\mu_1}^{-1} \mathbb{E}_{\mu_2}^{-1} \cdots \mathbb{E}_{\mu_m}^{-1} \mathbb{E}_\lambda - 1 \} n_\lambda P.$$

$q_0(\lambda)$ describes absorption, $q_1(\mu | \lambda)$ scattering, and $q_m(m > 1)$ fission. Derive the equations for first and second moments, translate them into continuous notation, and add the transfer terms. The result is an equation for neutron transfer including fluctuations.[*]

5. Fluctuations and the Boltzmann equation

This section contains a rather elaborate application of the method of compounding moments. The subject is the celebrated Boltzmann equation[**]

---

[*] For a different approach and for references see J. Lewins, Proc. Roy. Soc. A **362**, 537 (1978).

[**] L. Boltzmann, *Vorlesungen über Gastheorie* I (J.A. Barth, Leipzig 1896); S. Chapman and T.G. Cowling, *The Mathematical Theory of Non-Uniform Gases* (University Press, Cambridge 1939); P. Résibois and M. de Leener, *Classical Kinetic Theory of Fluids* (Wiley, New York 1977); C. Cercignani, *Mathematical Methods in Kinetic Theory* (Plenum, New York 1990).

5. FLUCTUATIONS AND THE BOLTZMANN EQUATION

$$\frac{\partial f(r_1, p_1)}{\partial t} + \frac{p_1}{m} \cdot \frac{\partial f}{\partial r_1} + F(r_1) \cdot \frac{\partial f}{\partial p_1}$$

$$= \int w(p_1, p_2 | p_3, p_4) f(r_1, p_3) f(r_1, p_4) \, d^3p_3 \, d^3p_4 \, d^3p_2$$

$$- f(r_1, p_1) \int w(p_3, p_4 | p_1, p_2) f(r_1, p_2) \, d^3p_2 \, d^3p_3 \, d^3p_4. \quad (5.1)$$

Here $f(r, p) \, d^3r \, d^3p$ is the number of molecules in a cell $d^3r \, d^3p$ of the one-particle phase space (μ-space), provided that the cell is large enough to contain many molecules. F is an external space-dependent force on all particles. The equation is based on the "Stosszahlansatz", which is the following assumption: The number of collisions (per unit volume and unit time) during which two molecules with momenta p_1, p_2 (within margins d^3p_1, d^3p_2) collide and emerge with momenta p_3, p_4 (margins d^3p_3, d^3p_4), is proportional to the product

$$f(r_1, p_1) \, d^3p_1 \cdot f(r_2, p_2) \, d^3p_2$$

of the number of available molecules. The proportionality factor w conserves momentum and energy,

$$w(p_3, p_4 | p_1, p_2) = \sigma \, \delta(p_1 + p_2 - p_3 - p_4) \, \delta\left(\frac{p_1^2 + p_2^2 - p_3^2 - p_4^2}{2m}\right).$$

Here σ is the differential cross-section, and depends only on $|p_1 - p_2| \equiv |p_3 - p_4|$ and on $(p_1 - p_2) \cdot (p_3 - p_4)$.

The precise number of molecules in the cell fluctuates around the value given by the Boltzmann equation, because the collisions occur at random, and only their probability is given by the Stosszahlansatz. Our aim is to compute these fluctuations. If f differs little from the equilibrium distribution one may replace the Boltzmann equation by its linearized version. It is then possible to include the fluctuations by adding a Langevin term, whose strength is determined by means of the fluctuation–dissipation theorem.[*] As demonstrated in IX.4, however, the Langevin approach is unreliable outside the linear domain. We shall therefore start from the master equation and use the Ω-expansion. The whole procedure consists of four steps.

The first step is the construction of an M-equation that describes the effect of the random collisions. Subdivide coordinate space in cells Δ', labelled λ', μ', etc., and momentum space in cells Δ'', labelled $\lambda'', \mu'', \ldots$. The one-

---

[*] A.A. Abrikosov and I.M. Khalatnikov, Sov. Phys. JETP **34**, 135 (1958); M. Bixon and R. Zwanzig, Phys. Rev. **187**, 267 (1969); R.F. Fox and G.E. Uhlenbeck, Phys. Fluids **13**, 1893 and 2881 (1970).

particle phase space is thereby subdivided in cells $\Delta = \Delta'\Delta''$ with labels $\lambda = (\lambda', \lambda'')$. Let n_λ denote the number of molecules in cell λ, and $P(\{n\}, t)$ the joint probability distribution of all occupation numbers. A collision takes two molecules from cells ρ, σ into cells λ, μ, where, of course, $\rho' = \sigma' = \lambda' = \mu'$. The Stosszahlansatz tells us that the probability per unit time for such a collision to occur is $w_{\lambda\mu\rho\sigma} n_\rho n_\sigma$. Hence the M-equation is*[)]

$$\dot{P}(\{n\}, t) = \sum_{\lambda\mu\rho\sigma} w_{\lambda\mu\rho\sigma}(\mathbb{E}_\lambda^{-1}\mathbb{E}_\mu^{-1}\mathbb{E}_\rho\mathbb{E}_\sigma - 1) n_\rho n_\sigma P. \tag{5.2}$$

To find the relation of $w_{\lambda\mu\rho\sigma}$ with $w(\mathbf{p}_1, \mathbf{p}_2 | \mathbf{p}_3, \mathbf{p}_4)$ we extract from (5.2) the macroscopic equation

$$\dot{n}_\alpha = \sum_{\lambda\mu\rho\sigma} w_{\lambda\mu\rho\sigma}(\delta_{\alpha\lambda} + \delta_{\alpha\mu} - \delta_{\alpha\rho} - \delta_{\alpha\sigma}) n_\rho n_\sigma$$

$$= 2 \sum_{\mu\rho\sigma} w_{\alpha\mu\rho\sigma} n_\rho n_\sigma - 2 n_\alpha \sum_{\lambda\mu\sigma} w_{\lambda\mu\alpha\sigma} n_\sigma.$$

On the other hand, the first term in the collision part of the Boltzmann equation may be written

$$\int w(\mathbf{p}_1, \mathbf{p}_2 | \mathbf{p}_3, \mathbf{p}_4) \, \delta(\mathbf{r}_1 - \mathbf{r}_2) \, \delta(\mathbf{r}_1 - \mathbf{r}_3) \, \delta(\mathbf{r}_1 - \mathbf{r}_4)$$

$$\times f(\mathbf{r}_3, \mathbf{p}_3) f(\mathbf{r}_4, \mathbf{p}_4) \, d\mathbf{r}_2 \, d\mathbf{p}_2 \, d\mathbf{r}_3 \, d\mathbf{p}_3 \, d\mathbf{r}_4 \, d\mathbf{p}_4.$$

As $n = f\Delta'\Delta''$, it is possible to compare both first terms; one finds with the aid of (2.6)

$$2 w_{\lambda\mu\rho\sigma} = \Delta^2 w(\mathbf{p}_{\lambda''}, \mathbf{p}_{\mu''} | \mathbf{p}_{\rho''}, \mathbf{p}_{\sigma''}) \frac{\delta_{\lambda'\mu'}}{\Delta'} \frac{\delta_{\lambda'\rho'}}{\Delta'} \frac{\delta_{\lambda'\sigma'}}{\Delta'}.$$

The second step is the expansion of the M-equation in $\Delta^{-1/2}$.**[)] Setting $n_\lambda = \Delta\phi_\lambda + \Delta^{1/2}\xi_\lambda$ one obtains from the largest terms the macroscopic equation

$$\dot{\phi}_\alpha = \Delta \sum_{\lambda\mu\rho\sigma} w_{\lambda\mu\rho\sigma}(\delta_{\alpha\lambda} + \delta_{\alpha\mu} - \delta_{\alpha\rho} - \delta_{\alpha\sigma}) \phi_\rho \phi_\sigma, \tag{5.3}$$

which is the nonlinear Boltzmann equation. It will be convenient to introduce the abbreviation

$$w_{\lambda\mu\rho\sigma}(\delta_{\alpha\lambda} + \delta_{\alpha\mu} - \delta_{\alpha\rho} - \delta_{\alpha\sigma}) = W(\alpha | \lambda\mu\rho\sigma).$$

The next order in $\Delta^{-1/2}$ yields for the distribution $\Pi(\{\xi_\lambda\}, t)$ the multivari-

---

*[)] A.J.F. Siegert, Phys. Rev. **76**, 1708 (1949); H.K. Janssen, Z. Phys. **258**, 243 (1973).
[)] N.G. van Kampen, Phys. Letters A **50, 237 (1974); M. Kac and J. Logan, in: *Studies in Statistical Mechanics* VII. *Fluctuation Phenomena* (E.W. Montroll and J.L. Lebowitz eds.; North-Holland, Amsterdam 1979).

5. FLUCTUATIONS AND THE BOLTZMANN EQUATION

ate Fokker–Planck equation

$$\frac{\partial \Pi}{\partial t} = 2\varDelta \sum_{\lambda\mu\rho\sigma} w_{\lambda\mu\rho\sigma}\left(-\frac{\partial}{\partial \xi_\lambda} - \frac{\partial}{\partial \xi_\mu} + \frac{\partial}{\partial \xi_\rho} + \frac{\partial}{\partial \xi_\sigma}\right)\phi_\rho \xi_\sigma \Pi$$

$$+ \varDelta \sum_{\lambda\mu\rho\sigma} w_{\lambda\mu\rho\sigma}\phi_\rho\phi_\sigma\left(-\frac{\partial}{\partial \xi_\lambda} - \frac{\partial}{\partial \xi_\mu} + \frac{\partial}{\partial \xi_\rho} + \frac{\partial}{\partial \xi_\sigma}\right)^2 \Pi. \tag{5.4}$$

From it one obtains for the first moments

$$\partial_t \langle \xi_\alpha \rangle = 2\varDelta \sum_{\lambda\mu\rho\sigma} W(\alpha|\lambda\mu\rho\sigma)\phi_\rho \langle \xi_\sigma \rangle. \tag{5.5}$$

This is the Boltzmann equation (5.3) linearized around the macroscopic solution ϕ. It shows that the average $\langle n_\alpha \rangle = \varDelta \phi_\alpha + \varDelta^{1/2}\langle \xi_\alpha \rangle$ obeys the nonlinear Boltzmann equation not only in order $\varDelta$, but even including the next order $\varDelta^{1/2}$.

For the second moments one obtains from (5.4)

$$\partial_t \langle \xi_\alpha \xi_\beta \rangle = 2\varDelta \sum W(\alpha|\lambda\mu\rho\sigma)\phi_\rho \langle \xi_\sigma \xi_\beta \rangle$$

$$+ 2\varDelta \sum W(\beta|\lambda\mu\rho\sigma)\phi_\rho \langle \xi_\alpha \xi_\sigma \rangle$$

$$+ 2\varDelta \sum W(\alpha|\lambda\mu\rho\sigma)\phi_\rho \phi_\sigma (\delta_{\beta\lambda} - \delta_{\beta\rho})$$

$$+ 2\varDelta \sum W(\beta|\lambda\mu\rho\sigma)\phi_\rho \phi_\sigma (\delta_{\alpha\lambda} - \delta_{\alpha\rho}).$$

The last two lines consist of a mass of terms, which can be reduced by introducing the factorial cumulants

$$[n_\alpha n_\beta] = \varDelta \langle\langle \xi_\alpha \xi_\beta \rangle\rangle + \delta_{\alpha\beta}\varDelta \phi_\alpha.$$

After rather lengthy calculations one ends up with

$$\partial_t [n_\alpha n_\beta] = 2\varDelta \sum_{\lambda\mu\rho\sigma} W(\alpha|\lambda\mu\rho\sigma)\phi_\rho [n_\sigma n_\beta]$$

$$+ 2\varDelta \sum_{\lambda\mu\rho\sigma} W(\beta|\lambda\mu\rho\sigma)\phi_\rho [n_\alpha n_\sigma]$$

$$+ 2\varDelta^2 \sum_{\rho\sigma} (w_{\alpha\beta\rho\sigma}\phi_\rho\phi_\sigma - w_{\rho\sigma\alpha\beta}\phi_\alpha\phi_\beta). \tag{5.6}$$

The third step consists in writing these equations in continuous notation. Each index is replaced with six parameters r, p and

$$w_{\lambda\mu\rho\sigma} \to \varDelta^2 w(p_1, p_2|p_3, p_4)\,\delta(r_1 - r_2)\,\delta(r_1 - r_3)\,\delta(r_1 - r_4),$$

$$W(\alpha|\lambda\mu\rho\sigma) \to \varDelta^3 w(p_1, p_2|p_3, p_4)\,\delta(r_0 - r_1)\,\delta(r_0 - r_2)\,\delta(r_0 - r_3)\,\delta(r_0 - r_4)$$

$$\times \{\delta(p_0 - p_1) + \delta(p_0 - p_2) - \delta(p_0 - p_3) - \delta(p_0 - p_4)\}.$$

Then (5.3) together with (5.5) yields

$$\partial_t \langle u(r_1, p_1) \rangle = 2 \int w(p_1, p_2 | p_3, p_4) \langle u(r_1, p_3) \rangle \langle u(r_1, p_4) \rangle \, dp_2 \, dp_3 \, dp_4$$

$$- 2 \int w(p_3, p_4 | p_1, p_2) \langle u(r_1, p_1) \rangle \langle u(r_1, p_2) \rangle \, dp_2 \, dp_3 \, dp_4. \tag{5.7}$$

Finally as *the fourth step* one supplements the right-hand side with the flow terms, including an external force $F(r)$:

$$-\frac{p}{m} \cdot \frac{\partial}{\partial r} \langle u(r, p) \rangle - F \cdot \frac{\partial}{\partial p} \langle u(r, p) \rangle. \tag{5.8}$$

One has thus obtained the full Boltzmann equation for $\langle u(r, p) \rangle$. Similarly one obtains from (5.6)

$$\partial_t [u(r_1, p_1) u(r_2, p_2)]$$

$$= 4 \int w(p_1, p_4 | p_5, p_6) \langle u(r_1, p_5) \rangle [u(r_1, p_6) u(r_2, p_2)] \, dp_4 \, dp_5 \, dp_6$$

$$- 2 \int w(p_3, p_4 | p_1, p_6) \langle u(r_1, p_1) \rangle [u(r_1, p_6) u(r_2, p_2)] \, dp_3 \, dp_4 \, dp_6$$

$$- 2 \int w(p_3, p_4 | p_5, p_1) \langle u(r_1, p_5) \rangle \, dp_3 \, dp_4 \, dp_5 \cdot [u(r_1, p_1) u(r_2, p_2)]$$

$$+ 4 \int w(p_2, p_4 | p_5, p_6) \langle u(r_2, p_5) \rangle [u(r_1, p_1) u(r_2, p_6)] \, dp_4 \, dp_5 \, dp_6$$

$$- 2 \int w(p_3, p_4 | p_2, p_6) \langle u(r_2, p_2) \rangle [u(r_1, p_1) u(r_2, p_6)] \, dp_3 \, dp_4 \, dp_6$$

$$- 2 \int w(p_3, p_4 | p_5, p_1) \langle u(r_1, p_5) \rangle \, dp_3 \, dp_4 \, dp_5 \cdot [u(r_1, p_1) u(r_2, p_2)]$$

$$+ 2\delta(r_1 - r_2) \int w(p_1, p_2 | p_3, p_4) \langle u(r_1, p_3) \rangle \langle u(r_1, p_4) \rangle \, dp_3 \, dp_4$$

$$- 2\delta(r_1 - r_2) \int w(p_3, p_4 | p_1, p_2) \, dp_3 \, dp_4 \cdot \langle u(r_1, p_1) \rangle \langle u(r_1, p_2) \rangle$$

$$+ \left\{ -\frac{p_1}{m} \cdot \frac{\partial}{\partial r_1} - \frac{p_2}{m} \cdot \frac{\partial}{\partial r_2} - F(r_1) \cdot \frac{\partial}{\partial p_1} - F(r_2) \cdot \frac{\partial}{\partial p_2} \right\} [u(r_1, p_1) u(r_2, p_2)]. \tag{5.9}$$

5. FLUCTUATIONS AND THE BOLTZMANN EQUATION

This equation determines (in linear noise approximation) the fluctuations about the solution $\langle u(r, p)\rangle$ of the Boltzmann equation.

It is less forbidding than it looks. The first three lines are the linearized Boltzmann operator acting on the factor $u(r_1, p_1)$ in the factorial cumulant; the next three lines are the same operator acting on the factor $u(r_2, p_2)$; lines seven and eight represent the sources of the fluctuations; and on the last line the flow terms have been added for both factors. The equation has therefore the general form (VIII.6.8), when A is identified with the linearized Boltzmann operator including the streaming term.

We now utilize equation (5.9) to compute the fluctuations in equilibrium.[*] Accordingly we take $F = 0$ and

$$\langle u(r, p)\rangle^e = \frac{N}{\Omega}(2\pi mkT)^{-3/2} \exp\left[-\frac{p^2}{2mkT}\right] \equiv \frac{N}{\Omega}\psi_0(p).$$

In that case the seventh and eighth lines vanish, see Exercise below. Also the factorial cumulant will be independent of time, and we write it as a matrix Θ,

$$[u(r_1, p_2)u(r_2, p_2)] = (r_1, p_1|\Theta|r_2, p_2).$$

Then (5.9) reduces to the matrix equation

$$0 = A\Theta + \Theta\tilde{A}, \tag{5.10}$$

where A is the linearized Boltzmann operator Λ plus streaming term:

$$(r_1, p_1|A|r_2, p_2) = \delta(r_1 - r_2)(p_1|\Lambda|p_2) + \delta(r_1 - r_2)\delta(p_1 - p_2)\frac{p_1}{m}\cdot\frac{\partial}{\partial r_1},$$

$$(p_1|\Lambda|p_2) = 4\int w(p_1, p'|p'', p_2)\langle u(p'')\rangle^e\, dp'\, dp''$$

$$- 2\int w(p', p''|p_1, p_2)\, dp'\, dp''\, \langle u(p_1)\rangle^e.$$

Λ, as an operator on the p-dependence alone, has one zero eigenvalue with five eigenfunctions,

$$\psi_0(p), \qquad p\psi_0(p), \qquad p^2\psi_0(p), \tag{5.11}$$

---

[*] Fluctuations in a non-equilibrium electron gas were studied by S.V. Gantsevich, V.L. Gurevich, and R. Katilius, Rivista Nuovo Cimento **2**, 1 (1979) and in non-equilibrium stationary fluids by D. Ronis, I. Proccacia and I. Oppenheim, Phys. Rev. A **19**, 1324 (1979); T. Kirkpatrick, E.G.D. Cohen, and J.R. Dorfman, Phys. Rev. Letters **44**, 472 (1980).

which will be denoted respectively by $\psi_j(p) = Q_j(p)\psi_0(p)$ ($j = 0, \ldots, 4$). All other eigenvalues are negative. It can be shown that these ψ_j are also the only eigenfunctions of A with zero eigenvalue; all other eigenvalues of A have negative real part (see Exercise). It follows from (5.10) that Θ must have the form

$$(r_1, p_1 | \Theta | r_2, p_2) = \sum_{i,j} \theta_{ij} \psi_i(p_1) \psi_j(p_2). \tag{5.12}$$

To determine the coefficients θ_{ij} we invoke the conservation of total particle number, total momentum, and total energy. They give the five identities ($k = 0, \ldots, 4$)

$$\int [u(r_1, p_1) u(r_2, p_2)]^e Q_k(p_2) \, dr_2 \, dp_2 = -Q_k(p_1) \langle u(r_1, p_1) \rangle^e.$$

Substituting (5.12) and performing the integration one finds

$$\sum_{i,j} \theta_{ij} Q_i(p_1) B_{jk} = -(N/\Omega^2) Q_k(p_1),$$

where $B_{jk} = \int Q_j(p_2) Q_k(p_2) \psi_0(p_2) \, dp_2$ is the matrix

$$\begin{bmatrix} 1 & 0 & 0 & 0 & 3mkT \\ 0 & mkT & 0 & 0 & 0 \\ 0 & 0 & mkT & 0 & 0 \\ 0 & 0 & 0 & mkT & 0 \\ 3mkT & 0 & 0 & 0 & 15(mkT)^2 \end{bmatrix}$$

Hence $\theta_{ij} = -(N/\Omega^2)(B^{-1})_{ij}$ and after some algebra one obtains for the covariance matrix of the density

$$\langle\langle u(r_1, p_1) u(r_2, p_2) \rangle\rangle^e = \frac{N}{\Omega} \bigg[\delta(r_1 - r_2) \delta(p_1 - p_2) \psi_0(p_1)$$
$$- \Omega^{-1} \psi_0(p_1) \psi_0(p_2) \bigg\{ \frac{5}{2} - \frac{(p_1 - p_2)^2}{2mkT} + \frac{p_1^2 p_2^2}{6(mkT)^2} \bigg\} \bigg]. \tag{5.13}$$

Conclusions. (i) In the literature the f in the Boltzmann equation is variously defined as the number, the average number, the probable number or the most probable number (of molecules in a unit phase cell). In our treatment, based on the master equation, it appears as the macroscopic value of that number obtained in the limit of zero fluctuations. Moreover, to order Δ^{-1} the fluctuations are symmetric, so that f is also the average. It cannot be called "the most probable number" because it is not an integer.

5. FLUCTUATIONS AND THE BOLTZMANN EQUATION

(ii) Although the master equation is more informative than the Boltzmann equation its physical input is precisely the same. The main ingredient is the Stosszahlansatz, which is also responsible for the Markov character.

(iii) The first term in (5.13) is Poissonian. The second term merely tells us that if any fluctuation occurs at one point in space, the excess in molecules, momentum, or energy has to come from the remaining volume Ω. For large Ω this correction vanishes and the correlation is purely Poissonian, in spite of the fact that the molecules interact.

Exercise. Show that for hard spheres with mass m, diameter d,

$$w(p_1, p_2 | p_3, p_4) = \frac{d^2}{m^2} \delta(p_1 + p_2 - p_3 - p_4) \, \delta\left(\frac{p_1^2 + p_2^2 - p_3^2 - p_4^2}{2m}\right). \tag{5.14}$$

[Hint: Start from the expression

$$\frac{|p_3 - p_4|}{m} d^2 \int \delta[p_3 + n\{n \cdot (p_4 - p_3)\} - p_1] \delta[p_4 - n\{n \cdot (p_4 - p_3)\} - p_2] \, \mathrm{d}^2 n,$$

where n is the unit vector pointing from the center of sphere 2 to the point of contact; it is integrated over the hemisphere $(p_4 - p_3) \cdot n > 0$.]

Exercise. For more general interaction between molecules the constant d^2/m^2 in (5.14) becomes a function of p_1, p_2, p_3, p_4, viz., the differential cross-section.[*]

Exercise. Show that in equilibrium the last three lines of (5.9) vanish, provided that $w(p_1, p_2 | p_3, p_4) = w(p_3, p_4 | p_1, p_2)$, which means that to each collision corresponds an inverse or restituting collision with equal cross-section.[**]

Exercise. Prove the following *lemma*: If H is a positive semi-definite Hermitian matrix, and F anti-Hermitian then the eigenvalues of $A = H + F$ have nonnegative real parts. Moreover, if the real part is zero the corresponding eigenvector is an eigenvector of H and F separately. Use this lemma to show that (5.12) is the solution of (5.10).

Exercise. How does one compute for the Boltzmann case the function G defined in (3.7)?

Exercise. Apply the same method to the Lorentz gas (although in that case it is *a priori* clear that the fluctuations are Poissonian according to VII.6).

Exercise. Consider an ideal Fermi gas whose particles can jump to another level because of the interaction with some heat bath. Subdivide the levels in cells labelled $\mu, \nu, \ldots$, each containing N levels, and call their occupation numbers $n_\mu \, (= 0, 1, \ldots, N)$. A particle in cell μ has a probability $w_{\nu\mu}$ per unit time to jump to ν when cell ν is

---

[*] P. Résibois and M. de Leener, op. cit. p. 89.

[**] See K. Huang, *Statistical Mechanics* (Wiley, New York 1963). An ample but unsatisfactory discussion of the case where this symmetry does not hold is given by R.C. Tolman, *Statistical Mechanics* (Clarendon, Oxford 1938).

empty. The probability distribution obeys therefore

$$\dot{P}(\{n\}) = \sum_{\mu,\nu} w_{\nu\mu}(\mathbb{E}_\mu \mathbb{E}_\nu^{-1} - 1)n_\mu \frac{N - n_\nu}{N} P.$$

Set $n_\mu = N\phi_\mu + N^{1/2}\xi_\mu$ and expand in $N^{-1/2}$. The resulting macroscopic equation is

$$\dot{\phi}_\nu = \sum_\mu \{w_{\nu\mu}\phi_\mu(1 - \phi_\nu) - w_{\mu\nu}\phi_\nu(1 - \phi_\mu)\}.$$

Since $\phi_\mu^e = \{\exp[\varepsilon_\mu/kT] + 1\}^{-1}$ the detailed balance relation takes the form

$$w_{\nu\mu} \exp[-\varepsilon_\mu/kT] = w_{\mu\nu} \exp[-\varepsilon_\nu/kT].$$

Find the equation for the correlations $\langle \xi_\mu \xi_\nu \rangle$ in equilibrium and verify that it is satisfied by

$$\langle\langle n_\mu n_\nu \rangle\rangle^e = N\delta_{\mu\nu} \frac{e^{\varepsilon_\nu/kT}}{(e^{\varepsilon_\nu/kT} + 1)^2} = -kT \frac{\partial}{\partial \varepsilon_\mu} \langle n_\nu \rangle^e.$$

Remark. A great deal of attention has been paid in recent years to non-equilibrium stationary processes that are unstable and also extended in space. They can give rise to different phases that exist side by side, so that translation symmetry is broken. The name "dissipative structures" has been coined for them, and the prime examples are the Bénard cells and the Zhabotinski reactions, but they also occur in biology and meteorology. However, *these are features of the macroscopic equations.* They are only relevant for fluctuation theory inasmuch as the fluctuation becomes very large at the point where the instability sets in. The critical fluctuations in XIII.5 are an example. There are many more varieties, in particular in the case of more variables.

Chapter XV

THE STATISTICS OF JUMP EVENTS

Many of the examples so far were concerned with the statistics of quantities that change through randomly occurring jumps. This chapter investigates the statistics of the time points at which those jumps occur. It also develops some formalism (section 3) needed in the next chapter.

1. Basic formulae and a simple example

Consider a one-step process in which particles are generated and recombine with probabilities $g(n)$ and $r(n)$ per unit time. Each generation or recombination is an event in time and we want to know the statistics of these events. An example is a photoconductor in which the recombination takes place under emission of a photon, which can be registered. We shall concentrate on the recombination events, but the generation events can be treated in the same way. To describe the statistics of these events we employ the functions f introduced in II.3.

We consider the one-step process, governed by the M-equation (VI.1.2)*[)]

$$\dot{p}(n, t) = r(n+1)p(n+1) + g(n-1)p(n-1) - [r(n) + g(n)]p(n, t). \quad (1.1)$$

What is the probability $f_1(t_1) \, dt_1$ for a recombination event to take place between t_1 and $t_1 + dt_1$? If there are n_1 particles available at t_1 the probability is $r(n_1) \, dt_1$. The total probability for an event is obtained by multiplying this with the probability $p(n_1, t_1)$ for having n_1 particles at t_1 and summing over all n_1,

$$f_1(t_1) = \sum_{n_1=1}^{\infty} r(n_1) p(n_1, t_1). \quad (1.2)$$

Next consider the probability $f_2(t_1, t_2) \, dt_1 \, dt_2$ for having one event in $t_1, t_1 + dt_1$ and another in $t_2, t_2 + dt_2$. Without loss of generality one may take $t_2 > t_1$. The probability for the first event is (1.2). After this event one is left with $n_1 - 1$ available particles; hence the probability for having n_2 particles at t_2 is given by the transition probability $p(n_2, t_2 | n_1 - 1, t_1)$, which is obtained from (1.1) by taking its solution for $t > t_1$ with the initial

---

*[)] In this chapter we write n in parentheses rather than as a subscript for typographical convenience.

condition*⁾

$$p(n_2, t_1 | n_1 - 1, t_1) = \delta(n_2, n_1 - 1).$$

The probability for having both events is therefore

$$f_2(t_1, t_2) = \sum_{n_1, n_2} r(n_2) p(n_2, t_2 | n_1 - 1, t_1) r(n_1) p(n_1, t_1). \tag{1.3}$$

The simplest example is the decay process treated in IV.6, but there the result is trivial since the decay events are independent by definition. The same remark applies to all linear one-step processes, see VII.6. In order to avoid the complications of nonlinear processes we here choose an example which is linear but not a one-step process. The recombinations, however, take place in one step, so that the formulas (1.2) and (1.3) remain valid.

Consider the chemical reaction

$$B \to 2X, \quad X \to A, \tag{1.4}$$

whose M-equation is

$$\dot{p}(n, t) = \alpha(\mathbb{E} - 1)np + \beta\Omega(\mathbb{E}^{-2} - 1)p. \tag{1.5}$$

The recombination probability is $r(n) = \alpha n$ and hence according to (1.1)

$$f_1(t_1) = \alpha \langle n(t_1) \rangle. \tag{1.6}$$

Moreover (1.3) reduces to, in abbreviated notation,

$$f_2(t_1, t_2) = \alpha^2 \langle \langle n(t_2) | n_1(t_1) - 1 \rangle n_1(t_1) \rangle, \tag{1.7}$$

where $\langle n | m \rangle$ denotes a conditional average of n. We compute these quantities in the stationary state.

The simplifying feature of our linear example is that f_1 and f_2 involve the low moments of n alone, which can be obtained without solving (1.5) explicitly. One easily finds with the initial condition $p(n, 0) = \delta(n, n_0)$

$$\langle n(t) \rangle = n_0 e^{-\alpha t} + 2(\beta\Omega/\alpha)(1 - e^{-\alpha t}), \tag{1.8a}$$

$$\langle n \rangle^s = 2(\beta\Omega/\alpha), \tag{1.8b}$$

$$\langle n^2 \rangle^s = 3(\beta\Omega/\alpha) + 4(\beta\Omega/\alpha)^2. \tag{1.8c}$$

Hence one has in the stationary state

$$f_1(t_1) = 2\beta\Omega, \tag{1.9}$$

---

*⁾ The remark that one should take the initial value $n_1 - 1$ rather than n_1 is due to J.T. Ubbink, Physica **52**, 253 (1971). For a similar remark in connection with photon counting, see E. Jakeman and R. Loudon, J. Physics A **24**, 5339 (1991).

which merely asserts that the density of recombination events is constant and equal to the rate at which the molecules are generated.

Next one obtains from (1.8a)

$$\langle n(t_2) | n(t_1) - 1 \rangle = \{n(t_1) - 1\} e^{-\alpha(t_2 - t_1)} + 2(\beta\Omega/\alpha)\{1 - e^{-\alpha(t_2 - t_1)}\}.$$

Multiplication with $n(t_1)$ and averaging over the stationary distribution yields

$$f_2(t_1, t_2) = 4(\beta\Omega/\alpha)^2 + (\beta\Omega/\alpha) e^{-\alpha(t_2 - t_1)} \quad (t_2 \geqslant t_1)$$

The significance of this result is better displayed by expressing it in the correlation function:

$$g_2(t_1, t_2) = \tfrac{1}{2} \langle n \rangle^s e^{-\alpha|t_2 - t_1|}. \tag{1.10}$$

Remark. The reaction scheme (1.4) corresponds to an overall reaction

$$B \to 2A \tag{1.11}$$

with one intermediate product. The overall rate is according to (1.9) equal to $2\beta\Omega$, as is immediately clear. Are the fluctuations in this overall reaction affected by the fact that it proceeds via an intermediate step? To answer this question one must distinguish between the short time scale of the order of the lifetime α^{-1} of the X, and a long time scale on which the reaction is adequately described by (1.11). On the *short* time scale one cannot talk about fluctuations in the reaction (1.11), because the disappearance of one molecule B does not coincide with the appearance of two molecules A. In fact, the disappearance of B is a Poisson process, while the appearance of two A's is not, according to (1.10). On the *long* time scale the correlation (1.10) is not visible and (1.11) is practically a Poisson process, as it would also be if no intermediate X occurred.[*]

Exercise. Compute (1.6) and (1.7) for the M-equation of the radioactive decay process.

Exercise. Find the statistics of recombination events for the chemical reaction

$$B \to qX, \quad X \to A \quad (q = 2, 3, 4, \ldots).$$

How can one understand that the correlation becomes more positive when q increases?

Exercise. Let $\mathbb{R}$ denote the diagonal matrix whose diagonal elements are $r(n)$. Then (1.3) can be written as a scalar product of the type (V.7.4):

$$f_2(t_1, t_2) = (p^s, \mathbb{R} e^{(t_2 - t_1)\mathbb{W}} \mathbb{E}\mathbb{R} p(t_1)).$$

Exercise. Write the general formula for the functions f_n, both for the recombination and for the generation processes in a one-step process.

Exercise. Compute $f_3(t_1, t_2, t_3)$ for the recombination events of (1.4).

---

[*] J.A.M. Janssen, J. Statist. Phys. **57**, 171 and 187 (1989).

386 XV. THE STATISTICS OF JUMP EVENTS

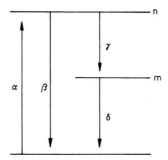

Fig. 45. The transitions in the three-level atom.

Exercise. The two-state Markov process has jumps up and down. Compute the functions (in obvious notation) $f_2(t_1\downarrow, t_2\downarrow)$ and $f_2(t_1\downarrow, t_2\uparrow)$.

Exercise. Consider an ensemble of N atoms with three levels as in fig. 45. Write the M-equation for the probability $p(n, m, t)$ of having n atoms in the upper level and m in the intermediate level. Compute $f_2(t_1\gamma, t_2\gamma), f_2(t_1\gamma, t_2\delta)$, etc. These functions describe the statistics of the quantum jumps of a single atom.[*]

2. Jump events in nonlinear systems

Although the example in the previous section had a linear master equation the identities (1.2) and (1.3) are general. They can be evaluated when the equation (1.1) can be solved. For nonlinear systems this can be done by means of the Ω-expansion. It turns out, however, that one has to go beyond the linear noise approximation in order to find a correlation between the jump events. Unfortunately this makes the calculations rather formidable.[**] We shall here treat an example which has been constructed to be as simple as possible.

Consider the chemical reaction scheme

$$B \to X, \quad X \to A, \quad 2X \to C,$$

whose master equation is nonlinear,

$$\dot{p}(n, t) = \alpha(\mathbb{E} - 1)np + \beta\Omega(\mathbb{E}^{-1} - 1)p + (\tfrac{1}{2}\gamma/\Omega)(\mathbb{E}^2 - 1)n(n-1)p. \quad (2.1)$$

As before we are interested in the time statistics of the events $X \to A$, given by (1.6) and (1.7). To save writing we choose units in which $\alpha = \gamma = 1$, and

---

[*] G. Nienhuis, Phys. Rev. A **35**, 4639 (1987); T. Erber, P. Hammerling, G. Hockney, M. Porrati, and S. Putterman, Annals Phys. **190**, 254 (1989).

[**] For an example, see N.G. van Kampen, in: Advances in Chemical Physics **34** (Wiley, New York 1976) sect. 12.

2. JUMP EVENTS IN NONLINEAR SYSTEMS

again we consider only the stationary state. The Ω-expanded form of (1.6) is

$$f_1(t_1) = \Omega\phi + \Omega^{1/2}\langle\xi\rangle^s,$$

where ϕ stands for the stationary value. Equation (1.7) takes the form

$$f_2(t_1, t_2) = \Omega^2\phi^2 + \Omega^{3/2}\phi\langle\xi\rangle^s + \Omega^{3/2}\phi\langle\{\langle\xi(t_2)|\xi(t_1)\rangle - \Omega^{-1/2}\rangle\}\rangle^s$$
$$+ \Omega\langle\langle\xi(t_2)|\xi(t_1)\rangle - \Omega^{-1/2}\rangle\xi(t_1)\rangle^s.$$

The correlation is expressed by (we simplify the notation)

$$g_2(t_1, t_2) = \Omega^{3/2}\phi\{\langle\langle\xi_2|\xi_1 - \Omega^{-1/2}\rangle\rangle^s - \langle\xi\rangle^s\}$$
$$+ \Omega\{\langle\langle\xi_2|\xi_1 - \Omega^{-1/2}\rangle\xi_1\rangle^s - (\langle\xi\rangle^s)^2\}. \tag{2.2}$$

Evidently there is no contribution in order $\Omega^{3/2}$ and it is therefore necessary to compute the factor { } on the first line to order $\Omega^{-1/2}$. We outline the calculation.

The standard expansion method first yields the macroscopic equation, from which one finds for the stationary ϕ

$$\beta = \phi + \phi^2. \tag{2.3}$$

Subsequently one has to write the equation for Π including terms of order $\Omega^{-1/2}$. That gives

$$\partial_t\langle\xi\rangle = -(1 + 2\phi)\langle\xi\rangle + \Omega^{-1/2}\phi - \Omega^{-1/2}\langle\xi^2\rangle + \mathcal{O}(\Omega^{-1}), \tag{2.4a}$$

$$\partial_t\langle\xi^2\rangle = -2(1 + 2\phi)\langle\xi^2\rangle + \beta + \phi + 2\phi^2 + \mathcal{O}(\Omega^{-1/2}). \tag{2.4b}$$

From the latter one obtains for initial value ξ_0

$$\langle\xi^2\rangle_t = \xi_0^2 e^{-2\tau} + \sigma^2(1 - e^{-2\tau}) + \mathcal{O}(\Omega^{-1/2}),$$

where the following two abbreviations have been used

$$\tau = (1 + 2\phi)t, \qquad \sigma^2 = \frac{\beta + \phi + 2\phi^2}{2(1 + 2\phi)} = \frac{2\phi + 3\phi^2}{2 + 4\phi}.$$

Having found $\langle\xi^2\rangle_t$ to the required order one may insert it into (2.4a) to get

$$\langle\xi\rangle_t = \xi_0 e^{-\tau} + \frac{\Omega^{-1/2}}{1 + 2\phi}(1 - e^{-\tau})\{\phi - \sigma^2 + (\sigma^2 - \xi_0^2)e^{-\tau}\}. \tag{2.5}$$

It follows in particular that

$$\langle\xi\rangle^s = \Omega^{-1/2}\frac{\phi - \sigma^2}{1 + 2\phi} + \mathcal{O}(\Omega^{-1}).$$

Now in (2.5) take $\xi_0 = \xi_1 - \Omega^{-1/2}$ and average over ξ_1,

$$\langle\langle\xi_2|\xi_1 - \Omega^{-1/2}\rangle\rangle^s = \Omega^{-1/2}\left(\frac{\phi - \sigma^2}{1 + 2\phi} - e^{-\tau}\right) + \mathcal{O}(\Omega^{-1}),$$

where $\tau = (1 + 2\phi)(t_2 - t_1)$. In the same way

$$\langle\langle\xi_2|\xi_1 - \Omega^{-1/2}\rangle\xi_1\rangle^s = \sigma^2 e^{-\tau} + \mathcal{O}(\Omega^{-1/2}).$$

Substitution in (2.2) yields

$$g_2(t_1, t_2) = \Omega(\sigma^2 - \phi)e^{-\tau} = \Omega\frac{-\phi^2}{2 + 4\phi}e^{-(1+2\phi)|t_1-t_2|}.$$

Exercise. In a one-step process with constant generation rate g, but arbitrary $r(n)$, equation (1.3) in the stationary regime reduces to

$$f(t_1, t_2) = \sum_{n_1,n_2} r(n_2)p(n_2, t_2|n_1 - 1, t_1)gp^s(n_1 - 1) = g^2.$$

Show that the same reduction applies to the higher f_n, so that in these circumstances the recombination events are Poissonian, as well as the generation events.

Exercise. Show for the reaction

$$B \to qX, \quad qX \to A$$

that the recombination events are Poissonian.

3. Effect of incident photon statistics

The following problem is in a certain sense the inverse of the one treated in the two preceding sections. Consider a photoconductor in which the electrons are excited into the conduction band by a beam of incoming photons. The arrival times of the incident photons constitute a set of random events, described by distribution functions f_n or correlation functions g_m. If they are independent (Poisson process or shot noise) they merely give rise to a constant probability per unit time for an electron to be excited, and (VI.9.1) applies. For any other stochastic distribution of the arrival events, however, successive excitations are no longer independent and therefore the number of excited electrons is not a Markov process and does not obey an M-equation. The problem is then to find how the statistics of the number of charge carriers is affected by the statistics of the incident photon beam. Their statistical properties are supposed to be known; and furthermore it is supposed that they have the cluster property, i.e., their correlation functions g_m obey (II.5.8). The problem was solved by Ubbink[*] in the form of a

---

[*] J.T. Ubbink, loc. cit.

3. EFFECT OF INCIDENT PHOTON STATISTICS

systematic expansion. True to our general strategy we shall choose a simple example to demonstrate his method.

Consider a collection of N independent atoms, each of which can be in its ground level or in one excited level. An excited atom has a probability α per unit time to fall back, but excitation can occur only through an incident photon. Each photon has a probability β to excite an atom. We suppose that the number n of excited atoms is always much less than N, so that this probability may be taken independent of n.

Let the photons arrive at times $\tau_1, \tau_2, \ldots$. Between two arrivals the probability distribution $p(t)$ obeys

$$\dot{p} = \alpha(\mathbb{E} - 1)np \equiv \mathbb{W} p. \tag{3.1}$$

At each arrival time τ the probability changes discontinuously:

$$p(\tau + 0) = p(\tau - 0) + \beta(\mathbb{E}^{-1} - 1)p(\tau - 0)$$
$$\equiv (1 + \mathbb{B})p(\tau - 0). \tag{3.2}$$

Suppose a shutter is opened at $t = 0$ admitting the photons and suppose $p(0)$ is known; one wants to know $p(t)$. If s photons arrived between 0 and t at times

$$0 < \tau_1 < \tau_2 < \cdots < \tau_s < t, \tag{3.3}$$

the distribution is, as in (VII.7.6),

$$[p(t)]_{\text{cond}} = e^{(t - \tau_s)\mathbb{W}}(1 + \mathbb{B}) e^{(\tau_s - \tau_{s-1})\mathbb{W}}(1 + \mathbb{B}) \cdots (1 + \mathbb{B}) e^{\tau_1 \mathbb{W}} p(0). \tag{3.4}$$

We introduce a kind of interaction representation by setting

$$e^{-\tau \mathbb{W}} \mathbb{B} e^{\tau \mathbb{W}} = \bar{\mathbb{B}}(\tau). \tag{3.5}$$

Then the probability conditional on the photon history (3.3) is

$$[p(t)]_{\text{cond}} = e^{t\mathbb{W}}\{1 + \bar{\mathbb{B}}(\tau_s)\}\{1 + \bar{\mathbb{B}}(\tau_{s-1})\} \cdots \{1 + \bar{\mathbb{B}}(\tau_1)\}p(0).$$

The final distribution is obtained from this by averaging over all photon histories,

$$p(t) = e^{t\mathbb{W}} \langle \{1 + \bar{\mathbb{B}}(\tau_s)\}\{1 + \bar{\mathbb{B}}(\tau_{s-1})\} \cdots \{1 + \bar{\mathbb{B}}(\tau_1)\}\rangle p(0). \tag{3.6}$$

The average that occurs here has the same form as the average in (II.3.6), but the $\bar{\mathbb{B}}$'s are operators and do not commute. To remedy this we introduce the time ordering symbol $\lceil \cdots \rceil$, which stipulates that all operators between these hooks have to be taken in chronological order, i.e., according to decreasing time arguments. Once this stipulation is made one can freely manipulate the operators as if they commute, provided in the final formula they are put back into chronological order. Then (3.6) may be transformed according to (II.5.7):

$$p(t) = e^{tW} \left\langle \left[\prod_{\sigma=1}^{s} \{1 + \bar{\mathbb{B}}(\tau_\sigma)\} \right] \right\rangle p(0)$$

$$= e^{tW} \left[\left\langle \prod_{\sigma=1}^{s} \{1 + \bar{\mathbb{B}}(\tau_\sigma)\} \right\rangle \right] p(0)$$

$$= e^{tW} \left[\exp \left\{ \int_0^t g_1(t_1) \bar{\mathbb{B}}(t_1) \, dt_1 \right. \right.$$

$$\left. \left. + \tfrac{1}{2} \int_0^t \int_0^t g_2(t_1, t_2) \bar{\mathbb{B}}(t_1) \bar{\mathbb{B}}(t_2) \, dt_1 \, dt_2 + \cdots \right\} \right] p(0). \quad (3.7)$$

This formula is exact, but less simple than it looks. The time ordering requires that the exponential be expanded in a series and that in each term of that series the operators $\bar{\mathbb{B}}$ are written in chronological order. That means that the multiple integrals have to be broken up in a number of terms for different parts of the integration domain. Before proceeding, however, we collect a number of properties of the time ordering in the form of Exercises.[*]

Exercise. Let $A(t)$ be a time-dependent matrix. Write explicitly the first five terms of

$$\left[\exp \int_0^t A(t') \, dt' \right]. \quad (3.8)$$

Exercise. Let $y(t)$ be a vector obeying the equation

$$\dot{y}(t) = A(t)y(t). \quad (3.9a)$$

Show that the solution of this equation can be written

$$y(t) = \left[\exp \int_0^t A(t') \, dt' \right] y(0). \quad (3.9b)$$

[Hint: Solve (3.9a) by iteration.] It follows that (3.9b) is a formal expression for the evolution matrix defined in (VIII.6.15).

Exercise. Prove for $0 < t_1 < t$ the identity

$$\left[\exp \int_0^t A(t') \, dt' \right] = \left[\exp \int_{t_1}^t A(t') \, dt' \right] \left[\exp \int_0^{t_1} A(t') \, dt' \right].$$

[Hint: Either use the result (3.9) or expand.]

---

[*] For a general treatment of such identities, see R. Kubo, J. Phys. Soc. Japan **17**, 1100 (1962).

3. EFFECT OF INCIDENT PHOTON STATISTICS

Exercise. Prove

$$\left[\exp \int_0^t A(t') \, dt' \right]^{-1} = \left\lfloor \exp - \int_0^t A(t') \, dt' \right\rfloor .$$

where $\lfloor \cdots \rfloor$ indicates anti-chronological ordering, i.e., with later times on the right.

Exercise. Prove*)

$$\left[\exp \int_0^t \{A(t') + B(t')\} \, dt' \right] = \left[\exp \int_0^t A(t') \, dt' \right] \left[\exp \int_0^t \bar{B}(t') \, dt' \right], \tag{3.10}$$

where $\bar{B}(t)$ is the interaction representation of $B(t)$:

$$\bar{B}(t) = \left[\exp \int_0^t A(t') \, dt' \right]^{-1} B(t) \left[\exp \int_0^t A(t') \, dt' \right].$$

We begin by taking in the exponent of (3.7) only the first term. That amounts to omitting *in the exponent* all terms containing β^2 and higher powers of β. At the same time all correlation functions $g_2, g_3, \ldots$ are omitted, so that the incident photons are treated as if they were uncorrelated. Expression (3.7) mutilated in this fashion becomes

$$p(t) = e^{t\mathbb{W}} \left[\exp \int_0^t g_1(t_1) \bar{\mathbb{B}}(t_1) \, dt_1 \right] p(0).$$

According to (3.9) we conclude

$$\frac{d}{dt} e^{-t\mathbb{W}} p(t) = g_1(t) \bar{\mathbb{B}}(t) \, e^{-t\mathbb{W}} p(t).$$

Hence, remembering (3.5),

$$\dot{p}(t) = \{\mathbb{W} + g_1(t) \mathbb{B}\} p(t)$$
$$= \alpha(\mathbb{E} - 1) n p + g_1(t) \beta (\mathbb{E}^{-1} - 1) p. \tag{3.11}$$

As expected, p obeys an M-equation since the photons are treated as uncorrelated and merely create a generation probability $g_1(t)\beta$ per unit time.

Exercise. Find $\langle n \rangle_t$ and $\langle n^2 \rangle_t$ from (3.11) with fixed initial $n(0)$. In particular, for $\alpha t \gg 1$

$$\langle\langle n^2 \rangle\rangle_t = \beta \int_0^\infty e^{-\alpha \tau} g_1(t - \tau) \, d\tau.$$

---
*) R.P. Feynman, Phys. Rev. **84**, 108 (1951).

Exercise. Find the autocorrelation function $\langle\langle n(t_1)n(t_2)\rangle\rangle$ from (3.11) with initial $n(0)$.

Exercise. Solve (3.11) explicitly by means of the generating function of p.

4. Effect of incident photon statistics – continued

The next approximation includes the effect of the correlation function g_2. For simplicity we suppose that the photon beam is stationary, so that g_1 does not depend on t. The term $g_1 \mathbb{B}$ in (3.11) cannot be regarded as a small perturbation, because this term is responsible for maintaining the stationary state. As unperturbed equation one must therefore take

$$\dot{p}(t) = \{\mathbb{W} + g_1 \mathbb{B}\}p(t) \equiv \mathbb{W}^*p(t). \tag{4.1}$$

The question is how to rearrange the terms in (3.7) in such a way that the term with g_2 appears as a correction to (4.1). This can actually be done using (3.10), but a more direct method is the following.

The starting point is equation (3.4), in which the interaction representation (3.5) has not yet been introduced. This equation can also be rewritten in a condensed form using time ordering by means of the following additional trick. Although $\mathbb{W}$ does not depend on time we write $\mathbb{W}(t)$, where the argument t merely serves to indicate its position in the chronological order. Similarly we write $\mathbb{B}(t)$ for the constant operator $\mathbb{B}$. Then (3.4) is the same as

$$[p(t)]_{\text{cond}} = \left\lceil \exp\left\{\int_0^t \mathbb{W}(t_1)\,dt_1\right\} \prod_{\sigma=1}^s \{1 + \mathbb{B}(\tau_\sigma)\} \right\rceil p(0).$$

The time ordering refers jointly to t_1 and the τ.

The averaging over histories (3.3) can again be performed as if the operators commute,

$$p(t) = \left\lceil \exp\left\{\int_0^t \mathbb{W}(t_1)\,dt_1\right\} \left\langle \prod_{\sigma=1}^s \{1 + \mathbb{B}(\tau_\sigma)\} \right\rangle \right\rceil p(0)$$

$$= \left\lceil \exp\left\{\int_0^t [\mathbb{W}(t_1) + g_1 \mathbb{B}(t_1)]\,dt_1 \right.\right.$$

$$\left.\left. + \tfrac{1}{2} \int_0^t\!\int_0^t g_2(t_1,t_2)\mathbb{B}(t_1)\mathbb{B}(t_2)\,dt_1\,dt_2 + \cdots \right\} \right\rceil p(0). \tag{4.2}$$

The operator [ ] is equal to $\mathbb{W}^*(t_1)$ defined in (4.1) and may be extracted by means of the identity (3.10),

$$p(t) = e^{t\mathbb{W}^*}\left\lceil \exp\left\{\tfrac{1}{2}\int_0^t\!\int_0^t g_2(t_1,t_2)\mathbb{B}^*(t_1)\mathbb{B}^*(t_2)\,dt_1\,dt_2\right\} \right\rceil p(0). \tag{4.3}$$

4. EFFECT OF INCIDENT PHOTON STATISTICS – CONTINUED

Here $\mathbb{B}^*(t)$ is another interaction representation, based on $\mathbb{W}^*$ rather than $\mathbb{W}$:

$$\mathbb{B}^*(t) = e^{-t\mathbb{W}^*} \mathbb{B} e^{t\mathbb{W}^*}. \tag{4.4}$$

Terms with higher powers of β than β^2 have been omitted in the exponent of (4.3). No other approximations have been made, but again the formula (4.3) is less simple than it looks, owing to the time ordering.

The time ordering is partly carried out if one replaces (4.3) with

$$p(t) = e^{t\mathbb{W}^*} \exp\left\{ \int_0^t dt_1 \int_0^{t_1} dt_2 \, g_2(t_1, t_2) \mathbb{B}^*(t_1) \mathbb{B}^*(t_2) \right\} p(0). \tag{4.5}$$

It will be shown presently that the remaining error is of higher order, but let us first examine (4.5). It constitutes an expression for $p(t)$, which in this case can actually be explicitly evaluated, because (4.1) can be solved so that $e^{t\mathbb{W}^*}$ can be found.

Alternatively the result (4.5) may be expressed in the form of a differential equation. On differentiating $e^{-t\mathbb{W}^*} p(t)$ one obtains

$$\dot{p}(t) = \left\{ \mathbb{W}^* + \int_0^t dt' \, g_2(t, t') \mathbb{B} e^{(t-t')\mathbb{W}^*} \mathbb{B} e^{(t'-t)\mathbb{W}^*} \right\} p(t)$$

$$= \left\{ \mathbb{W}^* + \int_0^t g_2(\tau) \mathbb{B} e^{\tau\mathbb{W}^*} \mathbb{B} e^{-\tau\mathbb{W}^*} d\tau \right\} p(t). \tag{4.6}$$

The operator { } still involves the time t elapsed since the initial time. This reflects the special role of the initial time, due to the fact that the probability at that time is prescribed. However, as the cluster property has been assumed, $g_2(\tau)$ vanishes for large τ, and we assume more specifically that there is a τ_c such that $g_2(\tau) \approx 0$ for $\tau > \tau_c$. It is then clear that as soon as $t > \tau_c$ in (4.6) one makes no error by replacing the upper limit of integration with ∞. The consequence is that $p(t)$ obeys a differential equation with constant coefficients,

$$\dot{p}(t) = \mathbb{K} p(t), \tag{4.7a}$$

where $\mathbb{K}$ is a renormalized transition matrix,

$$\mathbb{K} = \mathbb{W}^* + \int_0^\infty g_2(\tau) \mathbb{B} e^{\tau\mathbb{W}^*} \mathbb{B} e^{-\tau\mathbb{W}^*} d\tau. \tag{4.7b}$$

This is the final result. Higher approximations involving $g_3, g_4, \ldots$ do not change (4.7a) but merely add more terms to (4.7b).

Exercise. Derive (4.6) from (4.5).
Exercise. Prove that (4.6) conserves total probability.

Exercise. For the present example the integral in (4.7b) turns out to be very simple. First prove the commutator identity $[\mathbb{B}, \mathbb{W}^*] = \alpha \mathbb{B}$ and then derive from it $\mathbb{B}^*(t) = e^{\alpha t} \mathbb{B}$. The result is

$$\mathbb{K} = \mathbb{W}^* + \left\{ \int_0^\infty e^{-\alpha \tau} g_2(\tau) \, d\tau \right\} \mathbb{B}^2.$$

Exercise. Solve the equation (4.7a) with this $\mathbb{K}$.

Exercise. Since in this model the operators $\mathbb{B}^*(t)$ at different t commute, the time ordering in (4.3) may be omitted. It is therefore also possible to supply the higher terms and to construct the corresponding equation (4.7a) to all orders. In fact, it can even be solved. Show in this way that if $p(0) = \delta(n, n_0)$ the k-th moment $\langle n^k \rangle_t$ does not depend on the correlation functions g_m with $m > k$.

We still have the duty to examine the approximations made in the course of deriving (4.7).

a. A typical integral represented by the dots in (4.2) and omitted in (4.3) is

$$\int_0^t \cdots \int_0^t g_m(t_1, t_2, \ldots, t_m) \mathbb{B}(t_1) \mathbb{B}(t_2) \cdots \mathbb{B}(t_m) \, dt_1 \, dt_2 \cdots dt_m. \tag{4.8}$$

The integration extends over an m-dimensional cube of volume t^m; but owing to the cluster property the integrand vanishes unless all variables t lie within a distance of order τ_c from each other. The region where that is so is the neighborhood of the diagonal of the cube and its volume is $t\tau_c^{m-1}$. Hence (4.8) is of order

$$\beta^m t \tau_c^{m-1} = (\beta t)(\beta \tau_c)^{m-1}. \tag{4.9}$$

It is true that in the time-ordered exponential (4.2) the integral (4.8) does not occur as such, but is broken up in pieces and infiltrated with factors $\mathbb{B}$ coming from elsewhere, but that cannot raise its contribution to a higher order than (4.9). Hence the terms omitted in (4.3) are corrections to the coefficient of t in the exponent which are of higher order in $\beta \tau_c$.

b. On replacing (4.3) with (4.5) errors in ordering have been committed, of which the simplest one consists in ignoring the time ordering in the term

$$\frac{1}{2!} \int_0^t dt_1 \int_0^{t_1} dt_2 \int_0^t dt_1' \int_0^{t_1'} dt_2' \, g_2(t_1, t_2) g_2(t_1', t_2')$$

$$\times \lceil \mathbb{B}^*(t_1) \mathbb{B}^*(t_2) \mathbb{B}^*(t_1') \mathbb{B}^*(t_2') \rceil.$$

The difference arises from that part of the four-dimensional integration region where the pair t_1, t_2 interlaces with the pair t_1', t_2', and where moreover neither of the factors g_2 vanishes. The volume of that part is $t\tau_c^3$ and the order of the error therefore $(\beta t)(\beta \tau_c)^3$. As this is linear in t it must be

4. EFFECT OF INCIDENT PHOTON STATISTICS – CONTINUED

compared with the contribution from

$$\int_0^t dt_1 \int_0^{t_1} dt_2\, g_2(t_1, t_2)\mathbb{B}^*(t_1)\mathbb{B}^*(t_2),$$

which is of order $(\beta t)(\beta\tau_c)$. Hence the error is of relative order $(\beta\tau_c)^2$. This conclusion remains true for higher terms in the expansion of the exponential in (4.3).

c. Equation (4.7) was derived for arbitrary $p(0)$, but it was supposed that the shutter has been opened at $t = 0$. The initial time has a special role because it is the only time at which the variable n and the incident photons are statistically independent. At any $t_0 > 0$ this is not so: a large value of n, say, indicates that many photons have arrived in the recent past, and via the photon correlation that influences the probability for a photon to arrive at t_0. It also influences the photon arrival probability in the immediate future after t_0, which is the reason why n is not Markovian. It is also the reason why we *had* to take as our initial time the moment at which the shutter opened.

In equation (4.7) the initial time is no longer mentioned, and it therefore applies to all distributions, regardless of the time the shutter opened, provided that occurred more than τ_c earlier than t. This proviso means that (4.7) is *not* a master equation (see the Warning in V.1). Yet it may be approximately treated as such *if p does not change much during τ_c*.

This restriction can be alleviated by the following consideration. Write the operator $\mathbb{K}$ in (4.7b) as $\mathbb{K} = \mathbb{W}^* + \mathbb{K}_1$. Then (4.7a) is, in interaction representation,

$$\dot{p}^* = \mathbb{K}_1^* p^*, \tag{4.10}$$

where $p^* = e^{-t\mathbb{W}^*} p$ and $\mathbb{K}_1^*$ is the interaction representation (4.4) of $\mathbb{K}_1$. This is approximately a master equation *if p^* does not change much during τ_c*.

This weakened restriction can be quantified. The magnitude of $\mathbb{K}_1$, given by the integral in (4.7b), is determined by the factor β^2, multiplied with the integral over g_2, which is τ_c. Hence $\mathbb{K}_1 \sim \beta^2 \tau_c$ and the change of p^* during τ_c is of order $(\beta\tau_c)^2$. Thus (4.10), and consequently also (4.7a), are approximate master equations provided that

$$\beta\tau_c \ll 1. \tag{4.11}$$

But approximations **a** and **b** above were also based on this, and therefore *this is the overall condition for the validity of* (4.7) *as a master equation*. The parameter $\beta\tau_c$ is a measure for the effect of the external influence during one correlation time, and has been called the *Kubo number*.

Chapter XVI

STOCHASTIC DIFFERENTIAL EQUATIONS

When a system is subject to fluctuating external influences its equations of motion are differential equations with random coefficients. In a number of cases these equations can be solved exactly or approximately. In particular, when the fluctuations are weak and rapid a systematic expansion leads to explicit approximate equations.

1. Definitions

A stochastic differential equation is a differential equation whose coefficients are random numbers or random functions of the independent variable (or variables). Just as in normal differential equations, the coefficients are supposed to be given, independently of the solution that has to be found. Hence stochastic differential equations are the appropriate tool for describing systems with external noise (see IX.5).

The general form of a stochastic differential equation is

$$\dot{u} = F(u, t; Y(t)), \qquad (1.1)$$

where u and F may be vectors, and $Y(t)$ stands for one or more random functions whose stochastic properties are given. This equation, together with an initial condition $u(t_0) = a$, determines for each particular realization $y(t)$ a function $U(t; [y], a)$, which is a functional of $y(t)$, i.e., it depends on all values $y(t')$ for $0 \leqslant t' \leqslant t$. The ensemble of solutions $U(t; [y], a)$ for all possible $y(t')$ constitutes a stochastic process. Equation (1.1) is solved when the stochastic properties of this process have been found.

Sometimes the initial value a is also a random quantity (or vector). Then the resulting stochastic process $U(t; [y], a)$ is a function of the random variable a, as well as a functional of y. As this is only a trivial generalization of the problem with fixed initial a there is no need to treat random initial values separately.

One example of a stochastic differential equation is the Langevin equation (IX.1.1). In fact, in most of the mathematical literature the name "stochastic differential equation" is restricted to this equation, which they call Itô equation.[*] The emphasis on this special case is misdirected. If the equation

---

[*] Gikhman and Skorohod, op. cit. in XII.5; L. Arnold, *Stochastische Differentialgleichungen* (Oldenbourg, Munich 1973); A. Friedman, *Stochastic Differential Equations and Applications*, two volumes (Academic Press, New York 1975 and 1976).

1. DEFINITIONS

represents external noise it is an approximation for a more general equation with non-white noise, such as considered here.*) If the equation represents internal noise it is an approximation for a master equation, and should be treated as such, see chapter X. The misconception that the Langevin equation is the universal stochastic differential equation for all kinds of noisy systems is responsible for the difficulties mentioned in IX.5. We therefore start from the wider class of stochastic differential equations (1.1) and subsequently study as a special case the approximations that are valid when the autocorrelation time of $Y(t)$ is short. This will be done in sections 2–5, but the case of long autocorrelation time is also of interest and is the subject of section 6.**)

Another example of a stochastic differential equation (1.1) is

$$\dot{u} = -i\omega(t)u, \quad (1.2)$$

where u is complex and $\omega(t)$ a random function of time. This example can be solved explicitly and has been used to illustrate paramagnetic resonance[†] and line broadening[††]. Note that (1.2) is distinguished from the Langevin equation in that the random coefficient *multiplies* u. An important example of an equation that cannot be solved explicitly is

$$\ddot{x} + \omega^2(t)x = 0, \quad (1.3)$$

describing a harmonic oscillator whose frequency is a random function of time.

One is led to the following subdivision in three categories.

I. *Linear* differential equations in which only the inhomogeneous term is a random function, such as the Langevin equation. Such equations have been called *additive*[‡] and can be solved in principle.

II. *Linear* differential equations in which one or more of the coefficients multiplying u are random functions. They have been called *multiplicative*[‡] and can be solved only in special cases, but a rather general approximation method will be given in sections 2 and 3.[‡‡]

---

*) The more general definition is also used by H. Bunke, in: *Gewöhnliche Differentialgleichungen mit zufälligen Parametern* (Akademie-Verlag, Berlin 1972) and T.T. Soong, *Random Differential Equations in Science and Engineering* (Academic Press, New York 1973).

) We follow the review in Physics Reports **24, 171 (1976).

†) P.W. Anderson and P.R. Weiss, Rev. Mod. Phys. **25**, 269 (1953).

††) R. Kubo, in: *Fluctuation, Relaxation and Resonance in Magnetic Systems* (D. ter Haar ed., Oliver and Boyd, Edinburgh 1961).

‡) R.F. Fox, J. Mathem. Phys. **13**, 1196 (1972).

‡‡) For other surveys and literature, see A. Brissaud and U. Frisch, J. Mathem. Phys. **15**, 524 (1974); V.I. Klyatskin and V.I. Tatarskii, Sov. Phys. Usp. **16**, 494 (1974).

III. *Nonlinear* equations, i.e., equations that are nonlinear in the unknown function u. Here the distinction between additive and multiplicative is moot. In section 4 they will be transformed into linear equations with multiplicative noise.

Note that the subdivision refers to the form of the equation, not to the process described by it; the term "multiplicative noise" is a misnomer. There are other categories, such as stochastic partial differential equations, eigenvalue problems[*], and random boundaries[**], but they will not be treated here.

Remark. The following difference with ordinary, non-stochastic differential equations needs to be emphasized. All solutions of a *non-stochastic* equation are obtained by imposing at an arbitrary t_0 an initial condition $u(t_0) = a$, and then considering all possible values of a. For a *stochastic* differential equation, however, one gets in this way only a subclass of solutions, namely those that happen to have no dispersion at this particular t_0.

To be more precise, consider the set S of all stochastic functions $U(t)$ that satisfy (1.1). A subset $S_0(a) \subset S$ is formed by those that also obey $U(t_0) = a$. Let S_0 be the union of all $S_0(a)$ obtained for the various possible values of a. Then $S_0 \subset S$ is the set of all stochastic functions obeying (1.1) that have zero variance at t_0. Now take $t_1 \neq t_0$ and form the corresponding S_1. For non-stochastic equations $S_1 = S_0 = S$ but in the stochastic case $S_1 \neq S_0$. In general S_0 and S_1 do not even overlap, $S_0 \cap S_1 = 0$.

Note that we are talking about the *stochastic processes* U, not about their individual realizations. Any individual realization u of a $U \in S_0$ is a solution of $\dot u = F(u, t; y(t))$ for some $y(t)$ and is therefore also a realization of one of the U's in S_1.

Exercise. Solve (IX.1.1) when $L(t)$ is an arbitrary random function, whose stochastic properties are given through its generating functional.

Exercise. Consider the additive equation

$$\ddot u + u = F(t),$$

where $F(t)$ is a random force, determined through its generating functional. Find the generating functional of the solution $u(t)$ determined by the initial values $u(0) = a$, $\dot u(0) = b$.

Exercise. Solve the same equation with initial conditions $u(t_0) = \alpha$, $\dot u(t_0) = \beta$, where $t_0 \neq 0$. Show that none of these solutions coincides with any solution found in the preceding exercise, unless F is non-stochastic.

Exercise. Solve the multiplicative equation (1.2).

---

[*] W.E. Boyce, in: *Probabilistic Methods in Applied Mathematics* 1 (A.T. Bharucha-Reid ed., Academic Press, New York 1968); W. Puckert and J. vom Scheidt, Reports Mathem. Phys. **15**, 206 (1979).

[**] F.G. Bass and I.M. Fuks, *Wave Scattering from Statistically Rough Surfaces* (Pergamon, Oxford 1979).

Exercise. Derive for a magnetic dipole moment S in a magnetic field B the equation $\dot{S} = -g B \wedge S$. If B has a fixed direction but random strength this equation can be written in the form (1.2).

Exercise. Show that a monochromatic electromagnetic plane wave, propagating in the x-direction through a medium whose dielectric constant varies with x, obeys an equation of the form (1.3).

Exercise. All solutions in S are obtained from the initial value a at t_0 if one allows a to be random.

2. Heuristic treatment of multiplicative equations

Consider a linear equation of the type

$$\dot{u} = A(t)u = \{A_0 + \alpha A_1(t)\}u, \tag{2.1}$$

where u is a vector, A_0 a constant matrix, $A_1(t)$ a *random matrix*, and α a parameter measuring the magnitude of the fluctuations in the coefficients. Moreover $A_1(t)$ is assumed to have a finite autocorrelation time τ_c, in the sense that for any two time points t_1, t_2 such that $|t_1 - t_2| \gtrsim \tau_c$ one may treat all matrix elements of $A_1(t_1)$ as statistically independent of those of $A_1(t_2)$. The quantity $\alpha \tau_c$ is the Kubo number introduced in (XV.4.11) and is again supposed to be small. Our result will be an approximate solution of (2.1) in the form of a series expansion in powers of $\alpha \tau_c$. A proper derivation will be given in the next section. In the present section a more unsophisticated method is used, which, however, is easier to handle and can therefore also readily be applied to other situations, as in XVII.3.

Although not strictly necessary, it is convenient to assume that $A_1(t)$ is a *stationary* matrix process. Then $\langle A_1(t) \rangle$ is independent of time and can be incorporated in A_0 by setting

$$A'_0 = A_0 + \alpha \langle A_1(t) \rangle, \qquad A'_1(t) = A_1(t) - \langle A_1(t) \rangle, \tag{2.2}$$

so that $\langle A'_1(t) \rangle = 0$. We shall suppose that this has been done and omit the primes.[*] Thus we are dealing with (2.1) with $\langle A_1(t) \rangle = 0$.

First eliminate A_0 by transforming to the interaction representation,

$$u(t) = e^{tA_0} v(t), \tag{2.3a}$$

$$\dot{v} = \alpha e^{-tA_0} A_1(t) e^{tA_0} v \equiv \alpha V(t)v. \tag{2.3b}$$

The solution to second order in α with $v(0) = u(0) = a$ is

$$v(t) = a + \alpha \int_0^t dt_1 \, V(t_1)a + \alpha^2 \int_0^t dt_1 \int_0^{t_1} dt_2 \, V(t_1)V(t_2)a + \cdots. \tag{2.4}$$

---

[*] The coming expansion in α will refer to the α in front of A'_1 alone, not to the α embodied in the definition of A'_0.

Now take the average with fixed a,

$$\langle v(t)\rangle = a + \alpha^2 \int_0^t dt_1 \int_0^{t_1} dt_2 \, \langle V(t_1)V(t_2)\rangle a. \tag{2.5}$$

This second order approximation can be used only as long as the higher orders are small. As each successive term involves one more integration over time this restriction amounts to $\alpha t \ll 1$. Hence for $t \ll \alpha^{-1}$

$$\langle v(t)\rangle = a + \alpha^2 \int_0^t dt_1 \int_0^{t_1} d\tau \, \langle V(t_1)V(t_1-\tau)\rangle a.$$

We want to use this for $t > \tau_c$ and therefore have to suppose $\alpha\tau_c \ll 1$. Then for $t_1 > \tau_c$ the upper limit in the integral may be replaced with ∞ as the integrand vanishes anyway. Although t_1 runs from 0 to t it is large compared to τ_c for the major part of the integration interval. Hence we have approximately for $\tau_c \ll t \ll \alpha^{-1}$

$$\langle v(t)\rangle = a + \alpha^2 \int_0^t dt_1 \int_0^\infty d\tau \, \langle V(t_1)V(t_1-\tau)\rangle a.$$

This, however, is also the solution to order α^2 of the linear differential equation

$$\partial_t \langle v(t)\rangle = \alpha^2 \left[\int_0^\infty \langle V(t)V(t-\tau)\rangle \, d\tau \right] \langle v(t)\rangle. \tag{2.6}$$

The conclusion is that the evolution of $\langle v(t)\rangle$ may be described by this equation. In the original representation it reads

$$\partial_t \langle u(t)\rangle = \left[A_0 + \alpha^2 \int_0^\infty \langle A_1(t) e^{tA_0} A_1(t-\tau)\rangle e^{-\tau A_0} \, d\tau \right] \langle u(t)\rangle. \tag{2.7}$$

It must be remembered, however, that the derivation applied to an interval $\Delta t \ll \alpha^{-1}$ after the initial time $t_0 = 0$. This initial time is special because at that time the value of u was given to be equal to a non-stochastic vector a. It is easily seen that the result is equally valid when the initial value is stochastic, provided it is statistically independent of A_1. That fact enables us to apply the same equation (2.7) again to the next interval Δt. Since the values of A_1 in the next interval are uncorrelated with those in the previous one owing to the shortness of τ_c. Admittedly there is an overlap at the boundary between both intervals, but this can only give a small error since it extends over a range of order τ_c out of the total interval Δt. Hence (2.7) is approximately valid for all times, provided $\alpha\tau_c \ll 1$. Thus *the average of $u(t)$ by itself obeys approximately the non-stochastic differential equation* (2.7).

2. HEURISTIC TREATMENT OF MULTIPLICATIVE EQUATIONS

Now suppose that not only $\alpha\tau_c \ll 1$ but also

$$\tau_c |A_0| \ll 1. \tag{2.8}$$

This additional condition[*] states that the free motion of u is slow compared to the fluctuations in A_1. Then (2.7) reduces to

$$\partial_t \langle u(t) \rangle = \left[A_0 + \alpha^2 \int_0^\infty \langle A_1(t) A_1(t-\tau) \rangle \, d\tau \right] \langle u(t) \rangle. \tag{2.9}$$

As A_1 was supposed stationary the integral is independent of time. The effect of the fluctuations is therefore to "renormalize" A_0 by adding a constant term of order α^2 to it. The added term is the integrated autocorrelation function of A_1. In particular, if one has a non-dissipative system described by A_0, this additional term due to the fluctuations is usually dissipative. This relation between dissipation and the autocorrelation function of fluctuations is analogous to the Green–Kubo relation in many-body systems[**]; but not identical to it, because there the fluctuations are internal, rather than added as a separate term as in (2.1).

Exercise. Apply the result to the harmonic oscillator (1.3) with frequency $\omega^2(t) = \omega_0^2 \{1 + \alpha \xi(t)\}$, where $\xi(t)$ is a stationary random process with zero mean and autocorrelation time τ_c. The answer is

$$\frac{d^2 \langle x \rangle}{dt^2} + \tfrac{1}{2} \alpha^2 \omega_0^2 c_2 \frac{d \langle x \rangle}{dt} + \omega_0^2 (1 - \tfrac{1}{2} \alpha^2 \omega_0 c_1) \langle x \rangle = 0, \tag{2.10}$$

$$c_1 = \int_0^\infty \langle \xi(t) \xi(t-\tau) \rangle \sin 2\omega_0 \tau \, d\tau,$$

$$c_2 = \int_0^\infty \langle \xi(t) \xi(t-\tau) \rangle (1 - \cos 2\omega_0 \tau) \, d\tau.$$

Exercise. The constants c_1 and c_2 in (2.10) depend on the random term at twice the frequency of the free oscillator. Express them in the spectral density of ξ.

Exercise. Apply the result (2.10) to a plane electromagnetic wave in a medium whose dielectric constant varies randomly with x. How is it possible for the wave to be damped although the medium gains no energy?

Exercise. When A_0 is also a function of t (although not stochastic) the same method can be used provided that the definition of the interaction representation is modified [as in (VIII.6.15)].

---

[*] The imprecise formulation "τ_c must be short" ignores the distinction between both approximating assumptions and is responsible for much confusion, compare IX.7.

[**] M.S. Green, J. Chem. Phys. **20**, 1281 (1952); R. Kubo, Can. J. Phys. **34**, 1274 (1956); R. Kubo, M. Toda, and N. Hashitsume, *Statistical Physics* II (Springer, Berlin 1985) ch. 4.

Exercise. In deriving (2.7) we supposed $\langle A_1(t)\rangle = 0$. If that is not true one has instead of (2.7)

$$\partial_t \langle u(t)\rangle = \left[A_0 + \alpha\langle A_1(t)\rangle + \alpha^2 \int_0^\infty \langle\langle A_1(t) \, \mathrm{e}^{\tau A_0} A_1(t-\tau)\rangle\rangle \, \mathrm{e}^{-\tau A_0} \, \mathrm{d}\tau \right] \langle u(t)\rangle. \quad (2.11)$$

Exercise. Keeping the same order in $\alpha\tau_c$ one can improve on (2.9) by including the next order in $\tau_c|A_0|$. The additional term involves a commutator:

$$\alpha^2 \int_0^\infty \langle A_1(t)[A_0, A_1(t-\tau)]\rangle \tau \, \mathrm{d}\tau. \quad (2.12)$$

An extension of the equation considered hitherto is the inhomogeneous equation

$$\dot u = A(t) + f(t). \quad (2.13)$$

$A(t)$ is a given random $n \times n$ matrix and $f(t)$ a given random n-vector. This case can be reduced to the previous case by the following device.[*] Extend the vector u by adding an $(n+1)$-th component $u_{n+1} = 1$. Then (2.13) may be written

$$\frac{\mathrm{d}}{\mathrm{d}t}\begin{pmatrix} u \\ 1 \end{pmatrix} = \begin{pmatrix} A & f \\ 0 & 0 \end{pmatrix}\begin{pmatrix} u \\ 1 \end{pmatrix}. \quad (2.14)$$

This enables us to use the previously developed methods.

To find an approximate solution suppose as in (2.1) that

$$A(t) = A_0 + \alpha A_1(t), \qquad f(t) = f_0 + \alpha f_1(t)$$

with constant non-random A_0, f_0 and vanishing $\langle A_1(t)\rangle$ and $\langle f_1(t)\rangle$. In order to apply the result (2.7) one has to compute

$$\exp\left[\tau \begin{pmatrix} A_0 & f_0 \\ 0 & 0 \end{pmatrix}\right] = \begin{pmatrix} \mathrm{e}^{\tau A_0} & \dfrac{\mathrm{e}^{\tau A_0} - 1}{A_0} f_0 \\ 0 & 1 \end{pmatrix}. \quad (2.15)$$

Substitution in (2.7) gives an equation for the average of the $(n+1)$-vector, of which the last component is trivial. The other n components are given by

$$\partial_t \langle u(t)\rangle = \left[A_0 + \alpha^2 \int_0^\infty \langle A_1(t) \, \mathrm{e}^{\tau A_0} A_1(t-\tau)\rangle \, \mathrm{e}^{-\tau A_0} \, \mathrm{d}\tau \right]\langle u(t)\rangle$$

$$+ f_0 + \alpha^2 \int_0^\infty \langle A_1(t) \, \mathrm{e}^{\tau A_0} A_1(t-\tau)\rangle \frac{\mathrm{e}^{-\tau A_0} - 1}{A_0} \, \mathrm{d}\tau \cdot f_0$$

---

[*] J.B.T.M. Roerdink, Physica A **109**, 23 (1981); A **112**, 557 (1982). He considers higher orders as well.

$$+ \alpha^2 \int_0^\infty \langle A_1(t) \, e^{\tau A_0} f_1(t-\tau) \rangle \, d\tau. \tag{2.16}$$

Exercise. Verify (2.15).

Exercise. When $A_1(t)$ and $f_1(t)$ are statistically independent of one another the last term of (2.16) is absent. The remaining terms can be found directly, without the aid of the device (2.14).

Exercise. Let u be a complex scalar function of t obeying

$$\dot{u} = -i\omega_0 \{1 + \alpha \xi(t)\} u + \alpha \eta(t),$$

where $\xi(t)$, $\eta(t)$ are stochastic functions with zero mean. Find an exact expression for $\partial_t \langle u \rangle$ and verify that to order α^2 it coincides with (2.16).

Exercise. If one drops the simplifying assumptions $\langle A_1 \rangle = \langle f_1 \rangle = 0$ the result is

$$\partial_t \langle u(t) \rangle = \left[A_0 + \alpha \langle A_1 \rangle + \alpha^2 \int_0^\infty \langle\langle A_1(t) \, e^{\tau A_0} A_1(t-\tau) \rangle\rangle \, e^{-\tau A_0} \, d\tau \right] \langle u(t) \rangle$$

$$+ f_0 + \alpha \langle f_1 \rangle + \alpha^2 \int_0^\infty \langle\langle A_1(t) \, e^{\tau A_0} A_1(t-\tau) \rangle\rangle \frac{e^{-\tau A_0} - 1}{A_0} \, d\tau \cdot f_0$$

$$+ \alpha^2 \int_0^\infty \langle\langle A_1(t) \, e^{\tau A_0} f_1(t-\tau) \rangle\rangle \, d\tau. \tag{2.17}$$

Another extension is the application to the higher moments of the component of u. Let $u = \{u_\nu\}$ be a real n-component vector obeying (2.1); more explicitly

$$\partial_t u_\nu = \sum_{\lambda=1}^n A_{\nu\lambda}(t) u_\lambda \quad (\nu = 1, 2, \ldots, n).$$

Then the products $u_\nu u_\mu$ obey again a linear stochastic differential equation

$$\partial_t (u_\nu u_\mu) = \sum_\lambda A_{\nu\lambda}(u_\lambda u_\mu) + \sum_\lambda A_{\mu\lambda}(u_\nu u_\lambda)$$

$$\equiv \sum_{\lambda\rho} \mathscr{A}_{\nu\mu;\lambda\rho}(u_\lambda u_\rho), \tag{2.18}$$

$$\mathscr{A}_{\nu\mu;\lambda\rho}(t) = A_{\nu\lambda}(t) \delta_{\mu\rho} + A_{\mu\rho}(t) \delta_{\nu\lambda}.$$

Hence their averages obey an equation similar to (2.11). To write it explicitly we consider the $\tfrac{1}{2}n(n+1) = N$ products $u_\nu u_\mu$ as components of a single vector $\mathscr{u}_a$ so that (2.18) takes the form

$$\partial_t \mathscr{u}_a = \sum_{b=1}^N \mathscr{A}_{ab}(t) \mathscr{u}_b \quad (a = 1, 2, \ldots, N). \tag{2.19}$$

Then, with $\mathscr{A}(t) = \mathscr{A}_0 + \alpha \mathscr{A}_1(t)$,

$$\partial_t \langle u_a \rangle = \left[\mathscr{A}_0 + \alpha \langle \mathscr{A}_1(t) \rangle \right.$$
$$\left. + \alpha^2 \int_0^\infty \langle\langle \mathscr{A}_1(t) \, \mathrm{e}^{\tau \mathscr{A}_0} \mathscr{A}_1(t-\tau) \rangle\rangle \, \mathrm{e}^{-\tau \mathscr{A}_0} \, \mathrm{d}\tau \right] \langle u_a \rangle.$$

If u is complex one can, of course, first reduce it to the preceding case by writing the $2n$ equations for its real and imaginary parts. Usually, however, one is only interested in the quantities $u_\nu u_\mu^*$; they obey a set of n^2 linear equations, from which one can again find n^2 approximate equations for $\langle u_\nu u_\mu^* \rangle$.

Exercise. Harmonic oscillator (1.3) with frequency $\omega^2(t) = \omega_0^2 \{1 + \alpha \xi(t)\}$. Find the equations for $\langle x^2 \rangle$, $\langle \dot{x}^2 \rangle$, $\langle x\dot{x} \rangle$. Show that the average energy grows exponentially. Hence *the random frequency oscillator is unstable energy-wise, although its average amplitude in general goes to zero* according to (2.10).

Exercise. A two-level atom is subject to a random field acting on its dipole moment, so that the Hamiltonian is

$$\mathscr{H} = \begin{pmatrix} E_1 & 0 \\ 0 & E_2 \end{pmatrix} - \alpha \xi(t) \begin{pmatrix} 0 & 1 \\ 1 & 0 \end{pmatrix}. \tag{2.20}$$

Find the equation for its average density matrix and show that for $t \to \infty$ both levels are equally occupied. Thus the temperature tends to infinity, owing to the fact that no damping by spontaneous emission is included in (2.20).

Exercise. For the same two-level atom show that the autocorrelation function of the dipole moment p (the second of the two matrices in (2.20)) is

$$\langle \mathrm{Tr}\, p(t) p(0) \rho^s \rangle = \tfrac{1}{2} \mathrm{e}^{-2at} \left(\cos wt - 3 \frac{a}{w} \sin wt \right).$$

where

$$a + ib = \alpha^2 \int_0^\infty \langle \xi(t) \xi(t-\tau) \rangle \, \mathrm{e}^{-i(E_2 - E_1)\tau} \, \mathrm{d}\tau,$$

$$w = \sqrt{(E_2 - E_1 - 2b)^2 - 4(a^2 + b^2)}.$$

Exercise. The criterion for the energy of the damped random frequency oscillator

$$\ddot{x} + 2\gamma \dot{x} + \omega_0^2 \{1 + \alpha \xi(t)\} x = 0$$

to tend to zero is, to order α^2,

$$\gamma > \tfrac{1}{2} \alpha^2 \omega_0^2 \int_0^\infty \langle \xi(t) \xi(t-\tau) \rangle \cos 2\omega_0 \tau \, \mathrm{d}\tau.$$

3. The cumulant expansion introduced

For the reader who felt uneasy with the various ad hoc approximations in the preceding section we now present a more systematic derivation of the same result. That also opens the door to higher approximations, although we shall not go through it.

Consider again the linear stochastic differential equation (2.1). There is no need now to assume $A_1(t)$ stationary, nor to eliminate its average as was done in (2.2). Transform (2.1) to the interaction representation (2.3). According to (XV.3.9) a formal solution can be written by means of the time-ordered exponential

$$v(t) = \left\lceil \exp\left\{\alpha \int_0^t V(t')\,dt'\right\} \right\rceil a. \tag{3.1}$$

This is merely an abbreviated notation of the same expansion of which (2.4) gives the first two terms. Now take again the average with fixed a. The averaging commutes, of course, with the time ordering, so that

$$\langle v(t) \rangle = \left\lceil \left\langle \exp\left\{\alpha \int_0^t V(t')\,dt'\right\}\right\rangle \right\rceil a.$$

Since the operators inside $\lceil\ \rceil$ may be treated as if they commute the identity (III.4.6) for the cumulants can be utilized,

$$\langle v(t)\rangle = \left\lceil \exp\left\{\alpha \int_0^t dt_1 \langle V(t_1)\rangle + \tfrac{1}{2}\alpha^2 \int_0^t dt_1\,dt_2 \langle\langle V(t_1)V(t_2)\rangle\rangle + \cdots \right.\right.$$
$$\left.\left. + \frac{\alpha^m}{m!}\int_0^t dt_1\,dt_2\cdots dt_m \langle\langle V(t_1)V(t_2)\cdots V(t_m)\rangle\rangle + \cdots\right\}\right\rceil a. \tag{3.2}$$

The difference with the series (2.4) is that now the expansion in α appears in the exponent. Of course, one must bear in mind that this is not an actual exponent because the time ordering can only be carried out after expanding the exponential – which brings us back to (2.4). Yet the following estimates remain true.

The assumption that $A_1(t)$ and therefore $V(t)$ have a correlation time τ_c implies that each cumulant vanishes unless the time points in it are close together within a domain of order τ_c. Hence the entire contribution to the m-fold integral in (3.2) arises from a domain of order $t\tau_c^{m-1}$. Accordingly the m-th term in the exponent is of order

$$(\alpha t)(\alpha\tau_c)^{m-1}.$$

Thus (3.2) is an expansion in powers of $(\alpha\tau_c)$, each term being roughly linear in t. This is the advantage of the cumulant expansion as compared with the series (2.4), which is an expansion in successive powers of αt and is therefore limited to small times.

The time ordering of (3.2) is a complicated procedure, which in practice can be carried out only in the first two orders. Breaking off after $m = 1$ one obtains

$$\langle v(t) \rangle = \left[\exp \left\{ \alpha \int_0^t dt_1 \, \langle V(t_1) \rangle \right\} \right] a. \tag{3.3}$$

In this approximation the exponent is of order αt and terms of order $(\alpha t)(\alpha \tau_c)$ are omitted. Equation (3.3) is precisely the solution of the differential equation

$$\partial_t \langle v(t) \rangle = \alpha \langle V(t) \rangle \langle v(t) \rangle. \tag{3.4}$$

This is the naive approximation in which the randomness is simply averaged.

Next we break off after $m = 2$:

$$\langle v(t) \rangle = \left[\exp \left\{ \alpha \int_0^t \langle V(t_1) \rangle \, dt_1 + \alpha^2 \int_0^t dt_1 \int_0^{t_1} dt_2 \, \langle\langle V(t_1) V(t_2) \rangle\rangle \right\} \right] a. \tag{3.5}$$

This includes terms of order $(\alpha t)(\alpha \tau_c)$ in the exponent and omits terms $(\alpha t)(\alpha \tau_c)^2$. If we put

$$\int_0^{t_1} dt_2 \, \langle\langle V(t_1) V(t_2) \rangle\rangle = K(t_1),$$

the equation has the form

$$\langle v(t) \rangle = \left[\exp \int_0^t \{ \alpha \langle V(t') \rangle + \alpha^2 K(t') \} \, dt' \right] a. \tag{3.6}$$

This is the solution of the differential equation

$$\partial_t \langle v(t) \rangle = \{ \alpha \langle V(t) \rangle + \alpha^2 K(t) \} \langle v(t) \rangle. \tag{3.7}$$

In the original representation it reads

$$\partial_t \langle u(t) \rangle = \left[A_0 + \alpha \langle A_1(t) \rangle + \alpha^2 \int_0^t \langle\langle A_1(t) \, e^{\tau A_0} A_1(t - \tau) \rangle\rangle \, e^{-\tau A_0} \, d\tau + \mathcal{O}(\alpha^3 \tau_c^2) \right] \langle u(t) \rangle. \tag{3.8}$$

The upper limit of the integral is still a reminder of the special role of the initial time but may be replaced with infinity as soon as $t > \tau_c$. Hence we have confirmed the results (2.7) and (2.11).

Remark. We have made a mistake, because (3.6) is not quite the same as (3.5). The time ordering of (3.5) implies that the exponential must be expanded and that in each term the factors V have to be ordered, also those occurring in $\langle\langle V(t_1)V(t_2)\rangle\rangle$. In (3.2), however, they have been combined into the single symbol $K(t')$, which in the subsequent time ordering is treated as one and indivisible. The factor $V(t_2)$ has been erroneously placed according to the value of t_1 rather than of t_2. On the other hand, if K were not treated in the ordering as a single operator it would no longer be true that (3.6) is the solution of (3.7). In short, (3.5) is not really equivalent to (3.7). Fortunately the error is of higher order in $\alpha\tau_c$, as will appear in the next section.

Exercise. Suppose the matrices $A(t)$ for different values of t commute with each other and also with A_0. Then the time ordering may be omitted and (3.3) can be evaluated. Show that the result is the same as can be found more directly by diagonalizing the matrix $A(t)$ in (2.1).

Exercise. A special but not uncommon form of $A_1(t)$ is $\xi(t)B$ with constant matrix B and scalar random function $\xi(t)$. If, moreover (2.8) holds it is again possible to write (3.3) without time ordering. Give the resulting differential equation for $u(t)$ to all orders in $\alpha\tau_c$.

Exercise. When A and B are two statistically dependent random matrices one has the identity

$$\langle B\,e^A\rangle = \left\{\langle B\rangle + \langle\langle BA\rangle\rangle + \text{----}\langle\langle\langle BA^2\rangle\rangle\rangle + \text{----}\langle\langle\langle BA^3\rangle\rangle\rangle + \cdots\right\}\langle e^A\rangle, \quad (3.9)$$

where the triple brackets are derived in the same way as the "ordered cumulants" in Physica **74**, 239 (1974).

4. The general cumulant expansion

Having established the existence of a formal expansion in powers of $\alpha\tau_c$ we may now employ a more streamlined method for obtaining the successive terms. It will be sufficient to count the powers of α.

Consider the linear stochastic differential equation (2.1). For fixed initial values $u(t_0) = a$ the solution has the form

$$u(t) = Y(t|t_0)\,a \quad (t > t_0). \tag{4.1}$$

The stochastic evolution operator $Y(t|t_0)$ might be written as a time-ordered product as in (3.1) but there is no need for an explicit form. From (4.1) one finds, on the one hand,

$$\langle u(t)\rangle = \langle Y(t|t_0)\rangle a \tag{4.2}$$

and consequently
$$\partial_t \langle u(t) \rangle = \langle \dot{Y}(t|t_0) \rangle a, \tag{4.3}$$
where the dot denotes differentiation with respect to t. On the other hand, one obtains from (4.2)
$$a = \langle Y(t|t_0) \rangle^{-1} \langle u(t) \rangle. \tag{4.4}$$
Substitution in (4.4) yields
$$\partial_t \langle u(t) \rangle = \langle \dot{Y}(t|t_0) \rangle \langle Y(t|t_0) \rangle^{-1} \langle u(t) \rangle$$
$$= K(t|t_0) \langle u(t) \rangle. \tag{4.5}$$
By construction the operator K is not stochastic, so that (4.5) is the desired equation for the average of $u(t)$.

There is no need to worry about the existence of the reciprocal in (4.4) as we are merely concerned with a bookkeeping method for obtaining the successive terms of the cumulant expansion. Nor do we have to worry about the dependence of K on t_0, because we have seen that this dependence disappears as soon as $t - t_0 > \tau_c$. After this transient the fluctuations around $\langle u(t) \rangle$ are determined by $A(t)$ alone and hence also their influence on the evolution of $\langle u(t) \rangle$ itself. In this regime, u moves surrounded by a cloud of fluctuations, not unlike an electron surrounded by its cloud of virtual photons.

To solve (4.1) in successive powers of α we transform to the interaction representation as in (2.3), and set
$$Y(t|t_0) = e^{tA_0} Z(t|t_0) e^{-t_0 A_0},$$
so that
$$\partial_t \langle v(t) \rangle = \langle \dot{Z}(t|t_0) \rangle \langle Z(t|t_0) \rangle^{-1} \langle v(t) \rangle. \tag{4.6}$$
For Z one has the same expansion as in (2.4)
$$Z(t|t_0) = 1 + \alpha \int_{t_0}^{t} dt_1 \langle V(t_1) \rangle + \alpha^2 \int_{t_0}^{t} dt_1 \int_{t_0}^{t_1} dt_2 \langle V(t_1) V(t_2) \rangle + \cdots.$$
Consequently, to second order in α,
$$\langle \dot{Z}(t|t_0) \rangle = \alpha \langle V(t) \rangle + \alpha^2 \int_{t_0}^{t} dt' \langle V(t) V(t') \rangle.$$
Since this is already small of order α is suffices to write for the other factor of (4.6)
$$\langle Z(t|t_0) \rangle^{-1} = 1 - \alpha \int_{t_0}^{t} dt_1 \langle V(t_1) \rangle + \mathcal{O}(\alpha^2).$$

4. THE GENERAL CUMULANT EXPANSION

The result is to second order in α

$$\partial_t \langle v(t) \rangle = \left[\alpha \langle V(t) \rangle + \alpha^2 \int_{t_0}^t dt' \{ \langle V(t)V(t') \rangle - \langle V(t) \rangle \langle V(t') \rangle \} \right] \langle v(t) \rangle.$$

This is the same as (3.7) and (3.8). The cumulant entered automatically on collecting the terms of order α^2.

This method is simple enough to make a general formulation in all orders of α possible.[*] The reason why I called it a bookkeeping device is that the expansion is formally correct for any value of $\alpha \tau_c$. Yet the condition $\alpha \tau_c \ll 1$ is crucial, because the result has no real meaning as long as it depends on t_0.

Remark. The following alternative treatment of stochastic differential equations needs to be mentioned. From (2.3b) one finds

$$v(t) = a + \alpha \int_0^t dt_1\, V(t_1) a + \alpha^2 \int_0^t dt_1 \int_0^{t_1} dt_2\, V(t_1) V(t_2) v(t_2).$$

In contrast to (2.4) this equation is exact, but it does not solve (2.3b), as it contains the unknown $v(t)$ on the right. If one takes the average, one obtains $\langle v(t) \rangle$ expressed in the equally unknown quantity $\langle V(t_1)V(t_2)v(t_2) \rangle$. However, if one now *assumes* that this average may be broken up into $\langle V(t_1)V(t_2) \rangle \langle v(t_2) \rangle$ the result is an integral equation for $\langle v(t) \rangle$ alone[**]

$$\langle v(t) \rangle = a + \alpha \int_0^t dt_1\, \langle V(t_1) \rangle a + \alpha^2 \int_0^t dt_1 \int_0^{t_1} dt_2\, \langle V(t_1)V(t_2) \rangle \langle v(t_2) \rangle.$$

Again we take for simplicity $\langle V(t) \rangle = 0$. Then by differentiation one obtains an integro-differential equation in which the initial value a no longer appears:

$$\partial_t \langle v(t) \rangle = \alpha^2 \int_0^t dt'\, \langle V(t)V(t') \rangle \langle v(t') \rangle. \tag{4.7}$$

This result differs from (2.6) in that the rate of change of $\langle v(t) \rangle$ is expressed not in the instantaneous value $\langle v(t) \rangle$ alone, but as an integral over the previous values $\langle v(t') \rangle$. Only when one makes the additional approximation that $\langle v(t) \rangle$ varies so slowly that in the integral $\langle v(t') \rangle$ may be replaced with $\langle v(t) \rangle$ does (4.7) reduce to (2.6).

Yet this difference is only apparent. First it can be shown that breaking up $\langle VVv \rangle$ into $\langle VV \rangle \langle v \rangle$ is correct only if terms of relative order $\alpha \tau_c$ are neglected. On the other hand, as the integral over t' in (4.7) virtually extends only over the interval

---

[*] Higher orders are given by R.C. Bourret, Can. J. Phys. **40**, 782 (1962); J.B. Keller, Proc. Symp. Appl. Math. **16** (Amer. Mathem. Soc., Providence 1964) p. 84; N.G. van Kampen, Physica **74**, 215 and 239 (1974); R.H. Terwiel, Physica **74**, 248 (1974).

[**] R.C. Bourret, Nuovo Cimento **26**, 1 (1962).

$(t-\tau_c, t)$, the equation tells us that $\partial_t \langle v \rangle \sim \alpha^2 \tau_c \langle v \rangle$. Hence replacing $\langle v(t') \rangle$ with $\langle v(t) \rangle$ changes the integral by an amount of order

$$\tfrac{1}{2}\tau_c^2 |\partial_t \langle v \rangle| \sim \alpha^2 \tau_c^2 \langle v \rangle;$$

i.e., of relative order $(\alpha \tau_c)^2$. This difference is negligible compared to the error already made. Hence *the integral equation* (4.7) *is not more accurate than the differential equation* (2.6).

Exercise. For the case that $\langle A_1(t) \rangle$ does not vanish derive

$$\partial_t \langle u(t) \rangle = \{A_0 + \alpha \langle A_1(t) \rangle\} \langle u(t) \rangle$$
$$+ \alpha^2 \int_0^t d\tau \, \langle\langle A_1(t) \, \mathrm{e}^{\tau A_0} A_1(t-\tau) \rangle\rangle \langle u(t-\tau) \rangle.$$

Show that it is not more accurate than (2.11).

Exercise. The equation for a monochromatic wave in a three-dimensional medium with random refractive index[*] is

$$\nabla^2 \phi(\mathbf{r}) + \{k^2 + \xi(\mathbf{r})\}\phi(\mathbf{r}) = 0.$$

Derive for $\langle \phi(\mathbf{r}) \rangle$ the integral equation

$$(\nabla^2 + k^2)\langle \phi(\mathbf{r}) \rangle = - \int \frac{\mathrm{e}^{\mathrm{i}k|\mathbf{r}-\mathbf{r}'|}}{4\pi|\mathbf{r}-\mathbf{r}'|} \langle \xi(\mathbf{r})\xi(\mathbf{r}') \rangle \langle \phi(\mathbf{r}') \rangle \, \mathrm{d}\mathbf{r}'.$$

What are the conditions for its validity?

Exercise. Show that under certain conditions this integral equation reduces to a wave equation with renormalized wave number

$$\nabla^2 \langle \phi \rangle + (k^2 + \beta + \mathrm{i}\gamma) \langle \phi \rangle = 0.$$

Here $\gamma > 0$ and causes damping; where does the dissipated energy go?

5. Nonlinear stochastic differential equations

Having treated in the previous two sections linear stochastic differential equations we now return to the general case (1.1). Just as normal differential equations, it can be translated into a linear equation by going to the associated Liouville equation. To do this we temporarily take a single realization $y(t)$ of $Y(t)$ and consider the non-stochastic equation

$$\dot{u}_\nu = F_\nu(u, t; y(t)). \tag{5.1}$$

---

[*] S.M. Rytov, Yu.A. Kravtsov, and V.I. Tatarskii, *Principles of Statistical Radiophysics*. Vol. 4: *Wave Propagation Through Random Media* (Springer, Berlin 1989); Yu.A. Kravtsov, Reports Prog. Phys. **55**, 39 (1992).

5. NONLINEAR STOCHASTIC DIFFERENTIAL EQUATIONS

It describes a flow in u-space; the density of this flow varies according to

$$\frac{\partial \rho(u, t)}{\partial t} = -\sum_v \frac{\partial}{\partial u_v} \{F_v(u, t; y)(t))\rho\}. \tag{5.2}$$

This is merely the hydrodynamic continuity equation $\dot\rho = -\operatorname{div} \rho v$ written for an arbitrary number of dimensions.[*] If one now lets $y(t)$ run over all realizations of $Y(t)$, with their appropriate probabilities, equation (5.2) becomes a linear stochastic differential equation for $\rho(u, t)$.

Equation (5.2) has the same linear form as (2.1), if one regards ρ as the analog of u in (2.1) and the present vector u as the analog of the unwritten label v in (2.1). The linear operator $\sum (\partial/\partial u_v) F_v \cdots$ is the analog of the matrix A. Formally one can therefore apply the same method to obtain an approximate equation for the average $\langle \rho(u, t) \rangle$ with given $\rho(u, 0)$. Suppose this has been done: what does the result tell us about the solution of the original equation (5.1)? The answer is provided by the following lemma.

Suppose (5.1) has been solved with initial condition $u_v(0) = a_v$, where a is a fixed vector. Let $P(u, t)$ be the resulting probability distribution of u at time $t \geq 0$. On the other hand, let (5.2) be solved with initial condition

$$\rho(u, 0) = \delta(u - a) = \prod_v \delta(u_v - a_v).$$

Then the lemma states

$$\langle \rho(u, t) \rangle = P(u, t). \tag{5.3}$$

Proof. Observe that (5.1) for each $y(t)$ determines a transformation of the initial value a into the value u at time t – which we denote by a^t. If now all $y(t)$ are admitted one has according to (I.5.2)

$$P(u, t) = \langle \delta(u - a^t) \rangle. \tag{5.4}$$

On the other hand, for each $y(t)$ the flow density in u-space obeys

$$\rho(a^t, t)\, \mathrm{d}a^t = \rho(a, 0)\, \mathrm{d}a. \tag{5.5}$$

It follows that

$$\rho(u, t) = \rho(u^{-t}, 0) \frac{\mathrm{d}(u^{-t})}{\mathrm{d}(u)}$$

$$= \delta(u^{-t} - a) \frac{\mathrm{d}(u^{-t})}{\mathrm{d}(u)}, \tag{5.6}$$

---

[*] The familiar Liouville equation of statistical mechanics is a special case, in which the flow is incompressible, i.e., the divergence $\sum \partial F_v / \partial u_v$ vanishes, so that the factor F_v in (5.2) may be written in front of $\partial/\partial u_v$. This restriction was not made, however, by J. Liouville, J. Mathém. Pures Appl. **3**, 342 (1838).

where $d(u^{-t})/d(u)$ is the Jacobian of the transformation. According to the transformation property of the delta function, (5.6) is equal to $\delta(u - a^t)$. Now take the average over all $y(t)$: the result is (5.3).   Q.E.D.

Thanks to this lemma the problem reduces to solving the linear equation (5.2). Of course, this can be done explicitly only in rare cases, but we are interested in an expansion for small fluctuations. Suppose

$$F_v(u, t; y(t)) = F_v^{(0)}(u) + \alpha F_v^{(1)}(u; t). \tag{5.7}$$

$F_v^{(0)}$ is not stochastic and we take $\langle F_v^{(1)}(u; t)\rangle = 0$ (for fixed values of u and t). For simplicity we consider only situations that are stationary in the following sense. $F_v^{(0)}(u)$ does not depend on time (other than through u) and $F_v^{(1)}(u; t)$ depends on time only through $y(t)$ (and u) and moreover $y(t)$ is a stationary process. Substituting (5.7) in (5.2) gives

$$\dot\rho(u, t) = \{A_0(u) + \alpha A_1(u; t)\}\rho, \tag{5.8}$$

$$A_0\rho = -\frac{\partial}{\partial u_v} F_v^{(0)}(u)\rho, \qquad A_1\rho = -\frac{\partial}{\partial u_v} F_v^{(1)}(u; t)\rho.$$

Summation over v is implied.

Equation (5.8) has the form (2.1). Hence the expansion in $\alpha\tau_c$ gives (2.7), which in the present case is

$$\partial_t\langle\rho(u, t)\rangle = -\frac{\partial}{\partial u_v} F_v^{(0)}(u)\langle\rho(u, t)\rangle$$

$$+ \alpha^2 \frac{\partial}{\partial u_v}\int_0^\infty d\tau \left\langle F_v^{(1)}(u, \tau)\, e^{\tau A_0} \frac{\partial}{\partial u_\mu} F_\mu^{(1)}(u, t-\tau)\right\rangle e^{-\tau A_0}\langle\rho(u, t)\rangle.$$
$$\tag{5.9}$$

It remains to evaluate e^{tA_0}.

Now e^{tA_0} acting on any function $f(u)$ is by definition the solution of the unperturbed part of the Liouville equation (5.8). Hence $e^{tA_0} f(u)$ is the solution $\phi(u, t)$ of

$$\frac{\partial\phi(u, t)}{\partial t} = -\frac{\partial}{\partial u_v} F_v^{(0)}(u)\phi(u, t), \qquad \phi(u, 0) = f(u).$$

This solution is given in (5.6):

$$\phi(u, t) = \phi(u^{-t}, 0)\frac{d(u^{-t})}{d(u)},$$

where the trajectory $u \to u^t$ is determined from

$$\dot u^t = F_v^{(0)}(u^t), \qquad u^0 = u.$$

Thus we have obtained

$$e^{\tau A_0} f(u) = f(u^{-\tau}) \frac{d(u^{-\tau})}{d(u)}. \tag{5.10}$$

To use this in (5.9) first note, with the aid of the lemma,

$$e^{-\tau A_0} \langle \rho(u, t) \rangle = P(u^\tau, t) \frac{d(u^\tau)}{d(u)}.$$

Subsequently

$$e^{\tau A_0} \frac{\partial}{\partial u_\mu} F_\mu^{(1)}(u, t-\tau) P(u^\tau, t) \frac{d(u^\tau)}{d(u)} = \frac{d(u^{-\tau})}{d(u)} \frac{\partial}{\partial u_\mu^{-\tau}} F_\mu^{(1)}(u^{-\tau}, t-\tau) P(u, t) \frac{d(u)}{d(u^{-\tau})}.$$

As a result, (5.9) becomes

$$\dot{P}(u, t) = -\frac{\partial}{\partial u_\nu} F_\nu^{(0)}(u) P(u, t) + \alpha^2 \frac{\partial}{\partial u_\nu} \int_0^\infty d\tau$$

$$\times \left\langle F_\nu^{(1)}(u; t) \frac{d(u^{-\tau})}{d(u)} \frac{\partial}{\partial u_\mu^{-\tau}} F_\mu^{(1)}(u^{-\tau}; t-\tau) \right\rangle \frac{d(u)}{d(u^{-\tau})} \cdot P(u, t). \tag{5.11}$$

Note that the differentiation $\partial/\partial u_\mu^{-\tau}$ acts on everything behind it; hence the right-hand side involves first and second derivatives of P and has the form of a Fokker–Planck equation.

Exercise. In case of a single variable u the result (5.11) may be written

$$\frac{\partial P}{\partial t} = -\frac{\partial}{\partial u} G(u) P + \frac{\partial^2}{\partial u^2} B(u) P,$$

$$B(u) = \alpha^2 \int_0^\infty d\tau \, \langle F^{(1)}(u; t) F^{(1)}(u; t-\tau) \rangle \frac{du}{du^{-\tau}},$$

$$G(u) = F^{(0)}(u) + \alpha^2 \int_0^\infty d\tau \left\langle \frac{\partial F^{(1)}(u, t)}{\partial u} F^{(1)}(u^{-\tau}, t-\tau) \right\rangle \frac{du}{du^{-\tau}}. \tag{5.12}$$

Exercise. If A_0 is a function of t a similar result can be obtained using the same interaction representation as in (VIII.6.15).

Application. Consider the following problem connected with the heating of a plasma.[*] A charged particle moves in one dimension in an electric field E,

$$\dot{x} = v, \qquad \dot{v} = \alpha E(x, t).$$

---

[*] P.A. Sturrock, Phys. Rev. **114**, 186 (1966); D.E. Hall and P.A. Sturrock, Phys. Fluids **10**, 2620 (1967); M.B. Silevitch and K.I. Golden, J. Statist. Phys. **7**, 65 (1973).

$E(x, t)$ is random with zero mean, stationary in time and space, and has an autocorrelation time τ_c. The density of an ensemble of such particles obeys (5.2),

$$\frac{\partial \rho(x, v, t)}{\partial t} = -\frac{\partial}{\partial x} v\rho - \alpha \frac{\partial}{\partial v} E(x, t)\rho.$$

Thus $u_1 = x$, $u_2 = v$ and

$$F_1^{(0)} = v, \qquad F_2^{(0)} = 0, \qquad F_1^{(1)} = 0, \qquad F_2^{(1)} = E(x, t).$$

One has $v^t = v$, $x^t = x + vt$ so that

$$\frac{d(x, v)}{d(x^{-\tau}, v^{-\tau})} = 1 \quad \text{and} \quad \frac{\partial}{\partial v^{-\tau}} = \frac{\partial}{\partial v} + \tau \frac{\partial}{\partial x}.$$

Substitution in (5.9) yields

$$\frac{\partial P(x, v, t)}{\partial t} = -v \frac{\partial P}{\partial x} + \alpha^2 \frac{\partial}{\partial v} \int_0^\infty \langle E(x, t) E(x - \tau v, t - \tau) \rangle \left(\frac{\partial P}{\partial v} + \tau \frac{\partial P}{\partial x} \right) d\tau,$$

or

$$\frac{\partial P}{\partial t} + v \frac{\partial P}{\partial x} = \alpha^2 \frac{\partial}{\partial v} c_0(v) \frac{\partial P}{\partial v} + \alpha^2 \frac{\partial}{\partial v} c_1(v) \frac{\partial P}{\partial x}, \tag{5.13}$$

$$c_0(v) = \int_0^\infty \langle E(x, t) E(x - \tau v, t - \tau) \rangle \, d\tau,$$

$$c_1(v) = \int_0^\infty \tau \langle E(x, t) E(x - \tau v, t - \tau) \rangle \, d\tau.$$

It is not surprising that the autocorrelation function involves the field *as felt by the particle* while it moves with the unperturbed velocity v.

Exercise. Derive from (5.8) an equation for the velocity distribution (which is the relevant quantity for plasma heating).

Exercise. A population grows with a fluctuating rate coefficient

$$\dot{u} = \{1 + \alpha \xi(t)\} u. \tag{5.14}$$

Apply the same method to obtain the probability distribution at time t to second order in α. Compare the result with the exact solution.

Exercise. If not only $\alpha \tau_c \ll 1$ but also $\tau_c |F^{(0)}| \ll 1$, as in (2.8), the general result (5.11) reduces to

$$\frac{\partial P(u, t)}{\partial t} = -\frac{\partial}{\partial u_\nu} F_\nu^{(0)}(u) P + \frac{\partial^2}{\partial u_\nu \partial u_\mu} B_{\nu\mu}(u) P,$$

$$B_{\nu\mu}(u) = \alpha^2 \int_0^\infty \langle F_\nu^{(1)}(u; t) F_\mu^{(1)}(u^{-\tau}; t - \tau) \rangle \, d\tau. \tag{5.15}$$

Application. A particle moves in one dimension x, subject to a driving force $K(x)$, a friction force $-\gamma \dot{x}$, and a random force $\alpha \xi(t)$,

$$\ddot{x} + \gamma \dot{x} = K(x) + \alpha \xi(t).$$

5. NONLINEAR STOCHASTIC DIFFERENTIAL EQUATIONS

If ξ were white noise this would be the Kramers model of VIII.7, but we now assume only that ξ has a short τ_c. Equation (5.15) takes the form

$$\frac{\partial P(x, v, t)}{\partial t} = -v\frac{\partial P}{\partial x} - K(x)\frac{\partial P}{\partial v} + \gamma\frac{\partial}{\partial v}vP + \alpha^2\left[\int_0^\infty \langle\xi(t)\xi(t-\tau)\rangle\,d\tau\right]\frac{\partial^2 P}{\partial v^2}.$$

Thus we have derived the Kramers equation (VIII.7.4) as an approximation for short τ_c. It becomes exact in the white noise limit (3.12). The coefficient of the fluctuation term is the integrated autocorrelation function of the fluctuating force, in agreement with (IX.3.5) and (IX.3.6).[*]

Exercise. Show that for this case the correction (2.12) has the form

$$-\alpha^2\gamma c_1\frac{\partial^2 P}{\partial v^2} + \alpha^2 c_1\frac{\partial^2 P}{\partial x\,\partial v}, \quad c_1 = \int_0^\infty \langle\xi(t)\xi(t-\tau)\rangle\tau\,d\tau.$$

Exercise. For the following Malthus–Verhulst equation with random coefficient,

$$\dot{u} = u - \{1 + \alpha\xi(t)\}u^2,$$

obtain the master equation for $P(u, t)$ and show[**]

$$P^s(u) = \frac{1}{u^2}\exp\left[\frac{1}{\alpha^2 c_0}\left(\frac{1}{u} - \frac{1}{2u^2}\right)\right].$$

Obtain the same result by transforming the equation into a linear one for u^{-1}.

Exercise. Examine in the same way

$$\dot{u} = \{1 + \alpha\xi(t)\}u - \{1 + \beta\eta(t)\}u^2,$$

with two random terms, not necessarily uncorrelated.

Exercise. Consider the one-component equation

$$\dot{y} = A(y) + C(y)\xi(t).$$

It has the same form as the nonlinear Langevin or Itô equation (IX.4.5) but ξ is supposed to have a small but non-zero τ_c. Show that the corresponding master equation (5.7) in the limit $\tau_c \to 0$ takes the Stratonovich form (IX.4.8) rather than the Itô form (IX.4.12). This confirms by explicit calculation what has been argued in IX.5.[†]

Exercise. In the circuit of fig. 46 the flux in the core is a nonlinear function $\Phi(I)$ of the current I. The generator produces an e.m.f.

$$V(t) = V_0(t) + \alpha V_1(t),$$

where $V_1(t)$ is random with zero mean. In the limit (3.12) the probability density of

---

[*] For a more general application, combined with the theory of composite processes, see C.D. Levermore, G.C. Pomraning, D.L. Sanzo, and J. Wong, J. Mathem. Phys. **27**, 2526 (1986); **29**, 995 (1988).

[**] O.J. Heilmann and N.G. van Kampen, Physica A **93**, 476 (1978).

[†] A different derivation is given in STRATONOVICH, sect. IV.8.

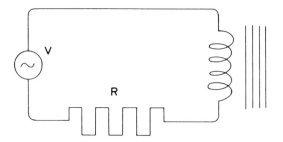

Fig. 46. Nonlinear circuit with random potential generator.

the current obeys the Fokker–Planck equation

$$\frac{\partial P(I, t)}{\partial t} = \frac{\partial}{\partial I} \left\{ \frac{RI - V(t)}{\Phi'(I)} + \alpha^2 c_0 \frac{\Phi''(I)}{\Phi'(I)^2} \right\} P + \alpha^2 c_0 \frac{\partial^2}{\partial I^2} \frac{P}{\Phi'(I)^2},$$

where

$$c_0 = \int_0^\infty \langle V(t) V(t-\tau) \rangle \, d\tau.$$

Conclusion. The expansion in $\alpha\tau_c$, which was developed in 3 and 4 for linear equations, can be adapted to nonlinear equations by the simple device of introducing the corresponding Liouville equation. Moreover, this results in an equation for the entire probability density $P(u, t)$ rather than just the average $\langle u(t) \rangle$. In fact, the same device can be applied to linear equations as well, if one wants to find the entire distribution, as in the case of (5.14); or to inhomogeneous equations.

In all these cases the result is an equation of the form of a master equation

$$\dot{P}(u, t) = \mathbb{K} P(u, t), \tag{5.16}$$

where $\mathbb{K}$ is a differential operator acting on the u-dependence. Higher orders of $\alpha\tau_c$ merely add corrections to $\mathbb{K}$ (including higher derivatives). Of course, for any $\tau_c > 0$ the process cannot be truly Markovian and (5.16) is at best approximately true. Thus the net effect of our expansion in the Kubo number $\alpha\tau_c$ is to approximate a non-Markovian process by a Markov process whose M-equation is obtained in successive orders of $\alpha\tau_c$. In the white noise limit (3.12) only the first term survives.

6. Long correlation times

The preceding sections showed that for short τ_c an expansion is possible that gives definite and physically useful results. The basic reason why such a general expansion exists appeared in section 2: the effect of the fluctuating term during τ_c is small and can be computed using perturbation theory; subsequently the long-time behavior is obtained by adding up successively all these independent contributions. No such method is available when the autocorrelation time of the random coefficients $Y(t)$ in F is long, i.e., $\alpha\tau_c \gtrsim 1$. However, special methods apply to certain classes of equations. They help to gain insight in the effect of long autocorrelation times.

A *first class* has an infinite τ_c: the coefficients in (1.1) are random constants X, rather than random functions $Y(t)$. It may then happen that the equation can be solved for every particular value x in the range of X. Let that solution with initial value a be $U(t; x, a)$. The probability density of u at time t is then

$$P_U(u, t) = \int \delta[u - U(t; x, a)] P_X(x)\, dx.$$

A famous and only partly solved problem of this type is the linear chain of harmonically bound particles, in which the masses and spring constants are random.[*] A related problem is the determination of the distribution of eigenvalues of a random matrix.[**]

As will be seen in the Exercise, the initial time $t = 0$ does not disappear from the result. The reason is that the system has an infinite memory, and never forgets that at that particular time the value of u was fixed, independently of the values of the coefficients. One cannot therefore expect that u is even approximately Markovian, let alone that $\langle u \rangle$ obeys a time-independent differential equation as in (2.7).

A *second class* can be illustrated by the single-variable equation

$$\dot{u} = -A(t)u = -\{A_0 + \alpha\xi(t)\}u, \qquad u(0) = a.$$

A_0 is a positive constant, $\xi(t)$ random with long τ_c, and α a small parameter. Set $u = u_0 + \alpha u_1 + \alpha^2 u_2 + \cdots$ and apply straightforward perturbation theory. To second order

$$u(t) = \mathrm{e}^{-A_0 t}\left[1 - \alpha \int_0^t \xi(t_1)\, dt_1 + \alpha^2 \int_0^t dt_1 \int_0^{t_1} dt_2\, \xi(t_1)\xi(t_2)\right] a.$$

Take the average and suppose for convenience $\langle \xi(t) \rangle = 0$:

$$\langle u(t) \rangle = \mathrm{e}^{-A_0 t}\left[1 + \alpha^2 \int_0^t dt_1 \int_0^{t_1} dt_2\, \langle \xi(t_1)\xi(t_2)\rangle\right] a. \tag{6.1}$$

This is the same formula as (2.4), but we cannot proceed in the same way owing to our long τ_c. If, however, A_0 is large, then *the whole solution goes*

---

[*] E.H. Lieb and D.C. Mattis, *Mathematical Physics in One Dimension* (Academic Press, New York 1966); J.L. van Hemmen and R.G. Palmer, J. Physics (London) A **12**, 563 (1979).

[**] References in II.1.

to zero rapidly, and it is sufficient that the perturbation solution applies during a time long compared with A_0^{-1}. The condition for validity of (6.1) is therefore

$$\alpha/A_0 \ll 1.$$

Exercise. Consider the equation $\dot u = -i\omega u$ in which the random constant ω is distributed around its average ω_0 according to
  (i) the Lorentz distribution;
  (ii) the Gauss distribution;
  (iii) the exponential distribution.
Find the distribution $P(u, t\,|\,a, 0)$ of u at t for fixed initial value a.

Exercise. For the equation $\ddot x + \omega^2 x = 0$ with ω as above find $P(x, \dot x, t\,|\,a, b, 0)$.

Exercise. Compute (6.1) when $\xi(t)$ is the Ornstein–Uhlenbeck process.

Exercise. Write the generalization of (6.1) for the multivariate equation (2.1). What is the condition for its validity?

Exercise. Formulate the analogous method and the validity condition for non-linear equations.

A *third class* of equations, which permit to study the effect of autocorrelation times of arbitrary length, has been encountered in IX.7. This class consists of equations (1.1) in which $Y(t)$ is a Markov process. We write Π for its transition probability density $\Pi(y, t\,|\,y_0, t_0)$ and

$$\dot\Pi = \mathbb{W}\Pi \tag{6.2}$$

for its M-equation. Then the process described by the joint variables (u, y) is again Markovian. Its joint probability density $\mathscr{P}(u, y, t\,|\,u_0, y_0, t_0)$ obeys the M-equation[*]

$$\frac{\partial \mathscr{P}}{\partial t} = -\sum_\nu \frac{\partial}{\partial u_\nu} F_\nu(u, t; y)\mathscr{P} + \mathbb{W}\mathscr{P}. \tag{6.3}$$

Integration of $\mathscr{P}$ over y gives the probability of u alone, conditional on given values u_0, y_0 at t_0. In most cases one is not interested in one y_0, but in the average over all y_0,

$$P(u, t\,|\,u_0, 0) = \int \mathrm{d}y \int \Pi_1(y_0, t_0)\,\mathrm{d}y_0\,\mathscr{P}(u, y, t\,|\,u_0, y_0, t_0).$$

Accordingly one needs to solve (6.3) with initial condition

$$\mathscr{P}(u, y, t_0) = \delta(u - u_0)\,\Pi_1(y_0, t_0).$$

Usually $Y(t)$ is a stationary process, so that Π_1 is $\Pi^s(y)$.

---

[*] Introduced by R. Kubo, J. Mathem. Phys. **4**, 174 (1963) under the title "Stochastic Liouville Equation". The adjective "stochastic" is here used in the sense of "pertaining to stochastic phenomena"; in contrast to our use as a synonym of "random" – as in the title of this chapter.

6. LONG CORRELATION TIMES

Unfortunately (6.3) can be solved only in rare cases. The situation is more fortunate, however, if F is linear in u and does not explicitly depend on time, so that (1.1) takes the form

$$\dot{u}_\nu = \sum_\mu A_{\nu\mu}(Y(t)) u_\mu. \tag{6.4}$$

Then (6.3) becomes

$$\frac{\partial \mathcal{P}(u, y, t)}{\partial t} = -\sum_{\nu,\mu} A_{\nu\mu}(y) \frac{\partial}{\partial u_\nu} u_\mu \mathcal{P} + \mathbb{W}\mathcal{P}. \tag{6.5}$$

Now define the marginal averages

$$m_\nu(y, t) = \int u_\nu \mathcal{P}(u, y, t) \, du. \tag{6.6}$$

Multiplying (6.5) with u_ν and integrating yields

$$\frac{\partial m_\nu(y, t)}{\partial t} = \sum_\mu A_{\nu\mu}(y) m_\mu + \mathbb{W} m_\nu. \tag{6.7}$$

These coupled equations have to be solved with initial values

$$m_\nu(y, 0) = u_{0\nu} \Pi_1(y, 0). \tag{6.8}$$

This problem is simpler than (6.5) itself, but, of course, its solution can only furnish the average

$$\langle u_\nu(t) \rangle = \int m_\nu(y, t) \, dy.$$

Example. Kubo constructed the following model to illustrate line broadening and narrowing due to random perturbations.[*] In equation (1.2) suppose that $\omega = \omega_0 + \alpha \xi(t)$, where ω_0 and α are constants and $\xi(t)$ is the dichotomic Markov process (IV.2.3). As ξ only takes two values we may abbreviate

$$\mathcal{P}(u, \pm 1, t) = \mathcal{P}_\pm(u, t)$$

and write (6.6) as two coupled equations,

$$\frac{\partial \mathcal{P}_+}{\partial t} = i \frac{\partial}{\partial u} (\omega_0 + \alpha) u \mathcal{P}_+ - \gamma \mathcal{P}_+ + \gamma \mathcal{P}_-,$$

$$\frac{\partial \mathcal{P}_-}{\partial t} = i \frac{\partial}{\partial u} (\omega_0 - \alpha) u \mathcal{P}_- - \gamma \mathcal{P}_- + \gamma \mathcal{P}_+.$$

---

[*] R. Kubo, J. Phys. Soc. Japan **9**, 935 (1954); P.W. Anderson, J. Phys. Soc. Japan **9**, 316 (1954); R. Kubo, in: *Fluctuation, Relaxation and Resonance in Magnetic Systems* (D. ter Haar ed., Oliver and Boyd, Edinburgh 1962).

Equation (6.7) for the marginal averages is obtained by multiplying either equation by u and integrating:

$$\dot{m}_+ = -i(\omega_0 + \alpha)m_+ - \gamma m_+ + \gamma m_-,$$
$$\dot{m}_- = -i(\omega_0 - \alpha)m_- - \gamma m_- + \gamma m_+. \tag{6.9}$$

The initial condition (6.8) reduces to $m_\pm(0) = \tfrac{1}{2}a$ and the corresponding solution for $\langle u(t) \rangle = m_+(t) + m_-(t)$ is

$$\langle u(t) \rangle = a\, e^{-(i\omega_0 + \gamma)t} \left\{ \cos(t\sqrt{\alpha^2 - \gamma^2}) + \frac{\gamma}{\sqrt{\alpha^2 - \gamma^2}} \sin(t\sqrt{\alpha^2 - \gamma^2}) \right\}. \tag{6.10}$$

This result shows that for *slow* perturbations, i.e., $\gamma \ll \alpha$, the average behaves as a superposition of two harmonic oscillators, with frequencies close to $\omega_0 \pm \alpha$, each being slightly damped. In Fourier language that amounts to two separate Lorentzians. When γ grows, both peaks *broaden* and merge into a single broad line. On the other hand, for rapid perturbations, $\gamma \gg \alpha$, one obtains from (6.9) to first order in γ^{-1},

$$\langle u(t) \rangle \sim \exp\left[-i\omega_0 t - \frac{\alpha^2}{2\gamma} t\right].$$

This represents a single Lorentzian peak near ω_0. It shows how it is possible that more rapid perturbations lead to a *narrower* line.[*]

Exercise. Express the hierarchy distributions P_n of the process u in terms of the solution $\mathscr{P}(u, y, t \,|\, u_0, y_0, t_0)$ of (6.3), thereby proving that the process is completely defined by (6.3).

Exercise. In Kubo's example the line shape is expressed by the normalized "line shape function"[**]

$$I(\omega - \omega_0) = \frac{1}{2\pi} \int_{-\infty}^{\infty} \frac{\langle u^*(0)u(t) \rangle^s}{\langle u^*(0)u(0) \rangle^s}\, e^{-i\omega t}\, dt.$$

Compute it and confirm the above statements.

Exercise. Again in (1.2) take $\omega = \omega_0 + \alpha\xi$ but let ξ be the Ornstein–Uhlenbeck process. Show that in proper units

$$\langle u^*(0)u(t) \rangle = e^{-i\omega_0 t} \exp[1 - t - e^{-t}].$$

Exercise. In the random oscillator (1.3) take $\omega(t)^2 = \omega_0^2(1 + \alpha\xi)$, where $\xi(t)$ is again

---

[*] Further literature on the influence of stochastic perturbations on the line shape is given by G.S. Agarwal, Z. Phys. B **33**, 111 (1979).

[**] P.C. Martin, *Measurements and Correlation Functions* (Gordon and Breach, New York 1968).

a dichotomic Markov process.[*] There are now four equations (**6.7**) for $\langle x \rangle_+, \langle x \rangle_-, \langle \dot{x} \rangle_+, \langle \dot{x} \rangle_-$. The four eigenvalues give the frequencies in the averaged process $\langle x(t) \rangle$. Derive again for small γ two Lorentzians near $\omega_0 \sqrt{1 \pm \alpha}$, and for large γ a single one near ω_0.

Exercise. The equations for the survival of a diseased gene[**] can be transformed into

$$\dot{u} = (a + b\xi)u + c,$$

with constant a, b, c and ξ a dichotomic Markov process. As the actual number of surviving genes is $u/(1 + u)$, it is of no use to know just $\langle u \rangle$. Find the probability distributions as a function of t in the special case $c = 0$.

Exercise. For a dichotomic Markov process $Y(t)$ the generating function of its integral is

$$G(\alpha, t) = \left\langle \exp\left[\alpha \int_0^t Y(t') \, dt' \right] \right\rangle.$$

Show that it obeys (**6.9**) with $\omega_0 = 0$. Its explicit value can therefore be read off from (**6.10**).[†]

Exercise. A flagellated bacterium in a chemical gradient moves with constant speed along the x-axis. At random moments it stops and departs with equal probability in the $+x$ or $-x$ direction. However, the probability per unit time to stop depends on the direction in which it is moving. The net displacement $X(t)$ does not average out. Find the characteristic function of $X(t)$ and hence its average and variance.[††]

Exercise. Take in (**3.16**) for $\xi(t)$ the dichotomic Markov process and find the exact condition for the energy to tend to zero.

---

[*] R.C. Bourret, U. Frisch, and A. Pouquet, Physica **65**, 303 (1973); N.G. van Kampen, Physica **70**, 222 (1973).

[**] H. Falk and W.J. Ventevogel, Physica A **95**, 191 (1979).

[†] This was implicitly used by P. Hu and S.R. Hartmann, Phys. Rev. B **9**, 1 (1974), for the effect on a central spin of the magnetic field due to another spin, which they assumed to flip at random.

[††] H.C. Berg. Scientific American **233**, nr. 2, p. 36 (1975); R.J. Nossal and G.H. Weiss, J. Statist. Phys. **10**, 245 (1974).

Chapter XVII

STOCHASTIC BEHAVIOR OF QUANTUM SYSTEMS

So far quantum mechanics entered only incidentally, as in the applications involving discrete energy states. In this chapter it serves as the starting point from which fluctuations and damping are derived. Detailed applications to lasers and quantum optics are beyond the scope of this book.

1. Quantum probability

Probability, as introduced in chapter I, is a mathematical concept, which is defined prior to applications. It is a tool, which may be utilized in such diverse fields as classical physics, populations and gambling. It may also be applied to quantum systems. Suppose I have a system that has been prepared in a quantum state ψ; suppose this ψ is one of a set of states $\psi^{(\nu)}$ but it is not known which one. Then it may be possible to assign a probability w_ν to each of them in such a way that w_ν is the probability that our system is in $\psi^{(\nu)}$. Alternatively one may visualize this situation as an ensemble of identical systems, each in one of the states $\psi^{(\nu)}$, the number of them in $\psi^{(\nu)}$ being proportional to w_ν. For example, one hydrogen molecule in a hot gas is in any one eigenstate with a probability given by the Boltzmann factor; alternatively one may treat all hydrogen molecules of the gas as an ensemble in which the various eigenstates are occupied by a number of molecules proportional to that probability.

Quantum mechanics, however, is different from the other applications in that the states themselves require a concept of probability for their physical interpretation. I shall call this "intrinsic probability" or "quantum probability" to distinguish it from the above "classical" or "statistical" probability. Intrinsic probability is not covered by the definition in I.1 and cannot be regarded as an ensemble.*[)]

We summarize the relevant elements of quantum mechanics. A particle or any other system, whose state is classically described by coordinates q and momenta p, is described in quantum mechanics by a wave function $\psi(q)$.

---

*[)] In spite of stubborn efforts to reduce it to a statistical probability distribution over states of hidden variables: D. Bohm, Phys. Rev. **85**, 166 and 180 (1952); F.J. Belinfante, *A Survey of Hidden-Variables Theories* (Pergamon, Oxford 1973); E. Nelson, *Quantum Fluctuations* (Princeton University Press, Princeton, NY 1985); J.S. Bell, *Speakable and Unspeakable in Quantum Mechanics* (Cambridge University Press, Cambridge 1987).

1. QUANTUM PROBABILITY

The functions $\psi(q)$ obeying the requirement that

$$(\psi|\psi) \equiv \int \psi^*(q)\psi(q)\,dq < \infty \tag{1.1}$$

constitute the Hilbert space **H** of all possible states of the system. (I use round brackets for the scalar product.) We shall always normalize them so that $(\psi|\psi) = 1$. Then the quantity $|\psi(q)|^2\,dq$ is the (intrinsic) probability to find the system in a cell dq of its configuration space. Any observable physical quantity A is represented by a Hermitian matrix $(q|A|q')$; in the state $\psi(q)$ it does not have one definite value, but it has an *expectation value*

$$\langle A \rangle = \int\int \psi^*(q)(q|A|q')\psi(q')\,dq\,dq' = (\psi|A|\psi). \tag{1.2}$$

The fact that this is *not* an integral over the probability density $|\psi(q)|^2$, but requires the "probability amplitude" ψ itself, is the essential difference with classical statistical probability. As a reminder of that difference I use the name "expectation value" rather than "average".

The statement (1.2) implies that the characteristic function of A is

$$G(k) = \langle e^{ikA} \rangle = \int\int \psi^*(q)(q|e^{ikA}|q')\psi(q')\,dq\,dq'. \tag{1.3}$$

This can be written more explicitly in terms of the eigenvalues a_n and eigenfunctions $\chi_n(q)$ of A,

$$G(k) = \langle e^{ikA} \rangle = \sum_n e^{ika_n} |(\chi_n|\psi)|^2.$$

One sees that the only possible values that A can take are the a_n, each occurring with a probability equal to the absolute square of the projection of ψ on the corresponding eigenvector χ_n. The case of a continuous spectrum requires some adjustment.

The probability distribution of A is

$$P_A(x) = (\psi|\delta(A - x)|\psi).$$

A second observable B has a similar distribution $P_B(y)$. However, there is no joint probability for A and B together, unless their matrices commute. Obviously, a joint distribution $P_{AB}(x, y)$ would not be able to reproduce the difference between the expectation values $\langle AB \rangle$ and $\langle BA \rangle$.

Exercise. It is tempting to regard as a joint characteristic function of A and B

$$G(k, l) = \langle e^{ikA + ilB} \rangle. \tag{1.4}$$

Show, however, that its expansion does not generate the moments $\langle AB \rangle$ and $\langle BA \rangle$ separately but only their sum.

Exercise. Take $A = q$, $B = p = -i(\partial/\partial q)$ and compute for the Fourier transform of (1.4)

$$P_{AB}(x, y) = \frac{1}{2\pi} \int_{-\infty}^{\infty} \psi^*(x + \tfrac{1}{2}s)\psi(x - \tfrac{1}{2}s)\, e^{isy}\, ds. \tag{1.5}$$

("Wigner distribution function"[*]). Show that this is *not* a probability density since it may become negative – e.g., when $\chi(x) = (\sin x)/x\sqrt{\pi}$.

Next take an ensemble of replicas of a system, distributed over wave functions $\psi^{(v)}$ with probabilities w_v. The $\psi^{(v)}$ may be any set of normalized functions, not necessarily orthogonal to one another. An observable A has in each $\psi^{(v)}$ a quantum-mechanical expectation (1.2), and the statistical average over the ensemble is

$$\overline{(A)} = \sum_v w_v (\psi^{(v)} | A | \psi^{(v)}). \tag{1.6}$$

The round brackets refer to the quantum expectation and the overbar to the ensemble average. In the future I shall simply write $\overline{(A)} = \langle A \rangle$, as is customary.

It is not possible to write (1.6) as a matrix sandwiched between a single wave function and its conjugate as in (1.2). To write it nonetheless in a compact fashion one defines a density matrix ρ by

$$(q|\rho|q') = \sum_v w_v \psi^{(v)}(q) \psi^{(v)*}(q'). \tag{1.7}$$

It is then possible to write (1.6) in the form

$$\langle A \rangle = \int\!\!\int (q|A|q')(q'|\rho|q)\, dq\, dq'$$
$$= \mathrm{Tr}\, A\rho. \tag{1.8}$$

The density matrix provides a complete description in the sense that all averages can be expressed in it; but it does not uniquely identify the $\psi^{(v)}$ and w_v used for the construction of the ensemble.

Exercise. Write the probability distribution and the characteristic function of A in terms of ρ.

Exercise. Show that ρ has the following three essential properties (the dagger indicates Hermitian conjugation):

$$\rho^\dagger = \rho, \quad \mathrm{Tr}\, \rho = 1, \quad (\phi|\rho|\phi) \geq 0 \quad \text{for all } \phi. \tag{1.9}$$

---

[*] See, e.g., L.E. Reichl, *A Modern Course in Statistical Mechanics* (University of Texas Press, Austin, 1980) p. 205.

1. QUANTUM PROBABILITY

Exercise. Give the explicit 2×2 density matrix for a beam of electrons with spin up. Also for a beam with spins in the x-direction. Also for a mixture of both. Find the average of the y-component in this mixed ensemble.

Exercise. Every 2×2 density matrix can be written as

$$\rho = \tfrac{1}{2}(1 + x\sigma^x + y\sigma^y + z\sigma^z), \tag{1.10}$$

where $\sigma^x, \sigma^y, \sigma^z$ are the Pauli matrices and x, y, z are real parameters restricted by $x^2 + y^2 + z^2 \leqslant 1$.

Exercise. In any representation (i.e., in any choice of orthogonal basis in **H**) the diagonal elements of ρ form a probability distribution. In particular $(q|\rho|q)$ is the probability density in configuration space.

Exercise. Give the explicit expression for $(q|\rho|q')$ of a harmonic oscillator in thermal equilibrium.

Exercise. For a certain system let ρ_1 and ρ_2 be two different density matrices (describing two different ensembles). "Convex addition" of them (involving two positive numbers λ_1, λ_2) is defined by

$$\rho = \lambda_1 \rho_1 + \lambda_2 \rho_2, \quad \lambda_1, \lambda_2 \geqslant 0, \quad \lambda_1 + \lambda_2 = 1. \tag{1.11}$$

Show that the sum ρ is a density matrix with the properties (1.9). What does it represent?

Exercise. Construct an example of two different sets $\psi^{(\nu)}$, each with a set of w_ν, giving rise to the same ρ.

Remark. The density matrix is a convenient tool but it conceals the distinction between quantum probability and classical probability. The former is intrinsic to the system, the latter describes our knowledge about it. Certain authors call ρ "the state of the system" rather than the state of the ensemble. What we called a state ψ is then represented by a density matrix of the special form

$$(q|\rho|q') = \psi(q)\psi^*(q'), \tag{1.12}$$

which they call a "pure state". Any other density matrix is a "mixed state". The drawback of this parlance is that their "state" is affected by the observer's knowledge, just as the "state" of a die before I look at it.

Exercise. Show that the criterion for a density matrix to be a pure state is

$$\rho^2 = \rho. \tag{1.13}$$

Exercise. When is (1.10) pure?

Exercise. Show that (1.11) cannot be pure unless ρ_1 is pure and $\lambda_2 = 0$, or *vice versa*.

Exercise. By construction, (1.7) is the convex sum of pure states. Show that any matrix obeying (1.11) can be so written, i.e., any mixed state is a mixture of pure states. The pure states may be chosen orthonormal.

The density matrix also enters in another, more subtle way. Suppose one has two systems S_1, S_2 with coordinates q_1, q_2 and a joint wave function $\Psi(q_1, q_2)$. Suppose one observes only S_1; i.e., one is concerned with operators

A_1 that are unit matrices with respect to q_2:

$$(q_1, q_2 | A_1 | q_1', q_2') = (q_1 | A_1 | q_1') \delta(q_2 - q_2'). \tag{1.14}$$

The quantum-mechanical expectation is

$$(A_1) = \int\int \Psi^*(q_1, q_2)(q_1 | A | q_1')\Psi(q_1', q_2)\, dq_1\, dq_1'\, dq_2. \tag{1.15}$$

Again it is not possible to write this as a quantum expectation value for the single system S_1 as in (1.2).*) Rather one must introduce the density matrix

$$(q_1 | \rho_1 | q_1') = \int \Psi(q_1, q_2)\Psi^*(q_1', q_2)\, dq_2. \tag{1.16}$$

With its aid it is possible to write for (1.15)

$$(A_1) = \operatorname{Tr} A_1 \rho_1 = \langle A_1 \rangle.$$

Here the round brackets indicate the quantum expectation in the joint system, while the angular brackets denote an average over an ensemble of systems S_1.

To show that (1.16) has the form (1.7) identify v with q_2 and the sum over v with the integral over q_2. Moreover, adjust the normalization by setting

$$\int |\Psi(q_1, q_2)|^2\, dq_1 = w(q_2), \qquad \Psi(q_1, q_2) = \sqrt{w(q_2)}\, \psi^{(q_2)}(q_1).$$

Then (1.16) has the form (1.7). One sees that S_1 is described by a mixed state, a mixture of pure states belonging each to one value of q_2.

Remark. This fact can be expressed in a more pregnant way. Suppose that S_1 and S_2 are two molecules, which initially are independent, so that

$$\Psi(q_1, q_2, t) = \psi_1(q_1, t)\psi_2(q_2, t) \quad \text{for } t \to -\infty. \tag{1.17}$$

Suppose they approach each other, interact, then move away. This scattering process is described by a joint Schrödinger equation for $\Psi(q_1, q_2, t)$ with initial condition (1.17). The point is that after the interaction has taken place the wave function $\Psi(q_1, q_2, t)$ no longer factorizes as in (1.17). The separated molecules cannot be described by separate wave functions, but only by the joint wave function, or else by their density matrices ρ_1 and ρ_2. This is considered a paradox by those who have not learned to live with quantum mechanics.**)

---

*) This fact is called the "quantum non-separability" by B. d'Espagnat, *Conceptual Foundations of Quantum Mechanics* (Benjamin, Menlo Park, CA 1971).

) A. Einstein, B. Podolski, and N. Rosen, Phys. Rev. **47, 777 (1935); F. Selleri ed., *Quantum Mechanics Versus Local Realism – The Einstein–Rosen–Podolski Paradox* (Plenum, New York 1988).

Exercise. Show that ρ_1 has the properties (1.9).

Exercise. Two particles with spin $\frac{1}{2}$ are combined in a state with total spin 0. Find the ρ_1 of one of them.

Exercise. Two harmonic oscillators are linearly coupled with total Hamiltonian

$$H = \tfrac{1}{2}(p_1^2 + \omega_1^2 q_1^2) + \tfrac{1}{2}(p_2^2 + \omega_2^2 q_2^2) + \varepsilon q_1 q_2. \tag{1.18}$$

They are in the ground state. What is ρ_1?

So far we considered only instantaneous quantum states. The way a wave function evolves in time is given (for a closed isolated system) by the Schrödinger equation

$$\dot\psi(q, t) = -\mathrm{i}H\psi(q, t). \tag{1.19}$$

H is a Hermitian operator independent of t. A formal solution is

$$\psi(q, t) = \mathrm{e}^{-\mathrm{i}(t-t_0)H}\psi(q, t_0).$$

The evolution of the density matrix is given by a commutator

$$\dot\rho(t) = -\mathrm{i}[H, \rho(t)] \tag{1.20}$$

("Neumann equation") with the formal solution

$$\rho(t) = \mathrm{e}^{-\mathrm{i}(t-t_0)H}\rho(t_0)\mathrm{e}^{\mathrm{i}(t-t_0)H}. \tag{1.21}$$

Exercise. It was said that (1.17) describes two independent systems. Show that any measurement of a property A_1 of S_1 gives a result that does not depend on the state of S_2.

Exercise. The same is true when the density matrix ρ of the combined system factorizes

$$(q_1, q_2|\rho|q_1', q_2') = (q_1|\rho_1|q_1')(q_2|\rho_2|q_2'). \tag{1.22}$$

Note that ρ_1 and ρ_2 operate in two different Hilbert spaces $\mathbf{H}_1$ and $\mathbf{H}_2$, and ρ in the total Hilbert space $\mathbf{H}$. This is indicated by the "tensor product"

$$\mathbf{H} = \mathbf{H}_1 \otimes \mathbf{H}_2, \qquad \rho = \rho_1 \otimes \rho_2. \tag{1.23}$$

Exercise. When the combined Hamilton operator of two systems is additive, $H = H_1 + H_2$, the Neumann equation (1.20) has solutions of the form (1.22). The two systems are independent at all times.

Exercise. For any observable A one defines the *time-dependent* or *Heisenberg operator* $A(t)$ by

$$A(t) = \mathrm{e}^{\mathrm{i}tH} A \mathrm{e}^{-\mathrm{i}tH}. \tag{1.24}$$

Show that $\dot A(t) = \mathrm{i}[H, A(t)]$, which differs from (1.20)! The average of A at time t is

$$\langle A \rangle_t = \mathrm{Tr}\, A\rho(t) = \mathrm{Tr}\, A(t)\rho(0).$$

One more notational tool will be needed. Consider the Hilbert space $\mathbf{H}$ of wave vectors ψ obeying (1.1). Let A, B be linear operators acting on ψ.

One may define a *scalar product of operators* by

$$(A, B) = \operatorname{Tr} A^\dagger B. \tag{1.25}$$

The norm $\|A\|$ of an operator is $(A, A)^{1/2}$. The operators whose norm is finite constitute a super Hilbert space. A linear mapping $\mathscr{L}$ of operators $A \to B$,

$$B = \mathscr{L} A \tag{1.26}$$

is a *superoperator* in this super Hilbert space. An example of a (time-dependent) superoperator is (1.24).

Exercise. Prove $(A, A) \geq 0$ with equality only if $A = 0$.
Exercise. Prove the Schwartz inequality

$$|(A, B)|^2 \leq \|A\| \, \|B\|.$$

2. The damped harmonic oscillator

In classical theory the prototype of a fluctuating system was the Brownian particle in VIII.3. Its equation of motion is the Langevin equation (IX.1.1), obtained by supplementing Newton's mechanics with a damping term and a fluctuating force. These two terms serve as an easy way to take the surrounding medium into account. In quantum mechanics, however, a damping term added to the Schrödinger equation would dissipate probability rather than energy. In order to describe damping one therefore includes explicitly the surrounding medium (or bath) as part of the total system. In the present section a special example is treated, which has the unique feature that it can be solved analytically. It has an important role for the proper understanding of the way in which interaction with the surroundings results in damping and fluctuations[*], and for the testing of future approximations.

The model consists of a system S in interaction with a bath B. The system S is a harmonic oscillator with frequency Ω. Its canonical variables are q_0, p_0 and its Hamiltonian is

$$H_S = \tfrac{1}{2}(p_0^2 + \Omega^2 q_0^2) = \Omega(a_0^\dagger a_0 + \tfrac{1}{2}). \tag{2.1}$$

Here $a_0, a_0^\dagger$ are annihilation and creation operators defined by

$$p_0 + i\Omega q_0 = \sqrt{2\Omega}\, a_0^\dagger, \qquad p_0 - i\Omega q_0 = \sqrt{2\Omega}\, a_0.$$

The zero-point energy $\tfrac{1}{2}\Omega$ is immaterial for the time evolution and will be omitted.

---

[*] P. Ullersma, Physica **23**, 27, 56, 74, 90 (1966); H. Dekker, Physics Reports **80**, 1 (1981).

2. THE DAMPED HARMONIC OSCILLATOR

The bath B is an assembly of small oscillators with Hamiltonian

$$H_B = \tfrac{1}{2} \sum_n (p_n^2 + k_n^2 q_n^2) = \sum_n k_n a_n^\dagger a_n, \qquad (2.2)$$

where the k_n are a sequence of frequencies to be specified later. Apart from being explicitly solvable this bath is a realistic description of the electromagnetic field in vacuum and of the lattice vibrations in a crystal.

As interaction we choose

$$H_I = \sum_n v_n(a_n a_0^\dagger + a_n^\dagger a_0),$$

where the v_n are a set of coupling constants. The total Hamiltonian is

$$H_T = \Omega a_0^\dagger a_0 + \sum_n k_n a_n^\dagger a_n + \sum_n v_n(a_n a_0^\dagger + a_n^\dagger a_0). \qquad (2.3)$$

It is a quadratic form in the operators a, $a^\dagger$ and may therefore be diagonalized by means of a linear transformation among them. The special form of the interaction has the advantage that the transformation does not mix the a with the $a^\dagger$. It will later appear that this special form is less contrived than it looks.*[)]

Exercise. The annihilation and creation operators obey the commutation rules

$$[a_n, a_m^\dagger] = \delta_{nm}, \qquad [a_n, a_m] = 0, \qquad [a_n^\dagger, a_m^\dagger] = 0.$$

(I use temporarily roman n, m to include zero.) Show that the Heisenberg operators $a_n(t)$, $a_n^\dagger(t)$ obey the same rules provided they are taken at the same t. The commutation relations of two of these operators taken at different times are not simple: they involve the solution of the equations of motion.

Exercise. Show that $H(t) = H$, where $H(t)$ is the same expression (2.3) with a_n, $a_n^\dagger$ replaced with $a_n(t)$, $a_n^\dagger(t)$.

Exercise. For a single oscillator the eigenstates of the Hamiltonian are the states $|N\rangle$ having $N = 0, 1, 2, \ldots$ excitations. One has

$$a|N\rangle = \sqrt{N}\,|N-1\rangle, \qquad a^\dagger|N\rangle = \sqrt{N+1}\,|N+1\rangle. \qquad (2.4)$$

Exercise. The eigenstates of the unperturbed part of (2.3) are the states with given occupation numbers of all oscillators $n = 0, 1, 2, \ldots$:

$$|N_0, N_1, N_2, \ldots\rangle = \frac{(a_0^\dagger)^{N_0}}{\sqrt{N_0!}} \frac{(a_1^\dagger)^{N_1}}{\sqrt{N_1!}} \frac{(a_2^\dagger)^{N_2}}{\sqrt{N_2!}} \cdots |0, 0, 0, \ldots\rangle.$$

Show that they are normalized provided that the ground state $|0, 0, 0, \ldots\rangle$ is normalized.

Exercise. How do the present a_n, $a_n^\dagger$ relate to the a_q, a_q^* in (III.5.11)?

---

*[)] The case of several oscillators in a common bath has been considered by P. Ullersma and J.A. Tjon, Physica **71**, 294 (1974); N.G. van Kampen, Physica A **147**, 165 (1987).

The equations of motion of the Heisenberg operators $a_0(t)$, $a_n(t)$ are

$$\dot{a}_0(t) = \mathrm{i}[H, a_0(t)] = -\mathrm{i}\Omega a_0(t) - \mathrm{i}\sum_m v_m a_m(t), \tag{2.5a}$$

$$\dot{a}_n(t) = \mathrm{i}[H, a_n(t)] = -\mathrm{i}k_n a_n(t) - \mathrm{i}v_n a_0(t), \tag{2.5b}$$

This is a set of linear equations. Normal modes, proportional to $\mathrm{e}^{-\mathrm{i}\omega t}$, are found from

$$\omega a_0 = \Omega a_0 + \sum_m v_m a_m, \tag{2.6a}$$

$$\omega a_n = k_n a_n + v_n a_0. \tag{2.6b}$$

Solve the a_n from the second line and substitute them into the first line:

$$a_n = \frac{v_n}{\omega - k_n} a_0, \quad (\omega - \Omega) a_0 = \sum_n \frac{v_n^2}{\omega - k_n} a_0. \tag{2.7}$$

In order that there is a non-zero solution, ω must satisfy the eigenvalue equation

$$\omega - \Omega = \sum_n \frac{v_n^2}{\omega - k_n}. \tag{2.8}$$

In fig. 47 the left- and right-hand members of this equation are sketched. One sees that there is a sequence of eigenvalues ω_ν. For each of them the linear set (2.6) has one solution, which we denote by $X_{0\nu}$, $X_{n\nu}$. According to (2.7)

$$X_{n\nu} = \frac{v_n}{\omega_\nu - k_n} X_{0\nu}, \tag{2.9}$$

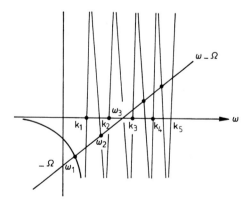

Fig. 47. Graphical solution of (2.8).

2. THE DAMPED HARMONIC OSCILLATOR

and $X_{0\nu}$ is a common factor, to be determined from the normalization condition

$$1 = X_{0\nu}^2 + \sum_n X_{n\nu}^2 = X_{0\nu}^2 \left\{ 1 + \sum_n \frac{v_n^2}{(\omega_\nu - k_n)^2} \right\}. \tag{2.10}$$

That makes the sets of eigenvectors orthonormal,

$$X_{0\nu} X_{0\mu} + \sum_n X_{n\nu} X_{n\mu} = \delta_{\nu\mu}. \tag{2.11}$$

The general solution of the time dependent problem (2.5) is a superposition of the normal modes

$$a_0(t) = \sum_\nu c_\nu X_{0\nu} e^{-i\omega_\nu t}, \qquad a_n(t) = \sum_\nu c_\nu X_{n\nu} e^{-i\omega_\nu t}.$$

The superposition constants c_ν can be adjusted so as to reproduce the initial values. One then gets the solution in the form

$$a_0(t) = U(t) a_0(0) + \sum_m V_m(t) a_m(0), \tag{2.12a}$$

$$a_n(t) = W_n(t) a_0(0) + \sum_m S_{nm}(t) a_m(0). \tag{2.12b}$$

The coefficients can easily be found using the orthogonality (2.11). For instance,

$$U(t) = \sum_\nu X_{0\nu}^2 e^{-i\omega_\nu t}. \tag{2.13}$$

This solves the problem. The fact that a_0, a_n are operators rather than classical variables made no difference thanks to the linearity.

Exercise. Show that (2.8) has no complex solutions.
Exercise. Show

$$\sum_\nu X_{n\nu} X_{m\nu} = \delta_{nm}, \qquad \sum_\nu X_{n\nu} X_{0\nu} = 0, \qquad \sum_\nu X_{0\nu}^2 = 1. \tag{2.14}$$

Exercise. Check (2.14) using the explicit expressions for the $X_{n\nu}$.
Exercise. Find $V_m(t), W_n(t), S_{nm}(t)$ and verify that $V_n = W_n$ and $S_{nm} = S_{mn}$.

We write (2.13) in a more pliable form. Define the analytic function

$$G(z) = z - \Omega - \sum_n \frac{v_n^2}{z - k_n}. \tag{2.15}$$

G is meromorphic; its poles k_n are the frequencies of the bath oscillators and the residues v_n^2 are the interaction strengths. The zeros are according to (3.8) the eigenfrequencies ω_ν of the total system, and the derivative of $G(z)$ at a zero is

$$G'(\omega_\nu) = 1 + \sum_n \frac{v_n^2}{(\omega_\nu - k_n)^2} = \frac{1}{X_{0\nu}^2}. \tag{2.16}$$

Hence, If C denotes a contour surrounding all zeros,

$$\frac{1}{2\pi i} \int_C \frac{e^{-izt}}{G(z)} dz = \sum_v \frac{e^{-i\omega_v t}}{G'(\omega_v)} = \sum_v X_{0v}^2 e^{-i\omega_v t} = U(t). \tag{2.17}$$

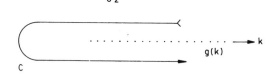

Fig. 48. The poles k_n and the contour C.

The contour C may be stretched into two horizontal lines at some distance ε above and below the real axis (see Fig. 48). For $t > 0$ the one below may be shifted to $-i\infty$ and its contribution vanishes. One is left with

$$U(t) = -\frac{1}{2\pi i} \int_{-\infty}^{\infty} \frac{e^{-ixt+\varepsilon t}}{G(x+i\varepsilon)} dx \quad (t > 0). \tag{2.18}$$

So far everything is exact. Of course, the bath variables are still present in the equations of motion (2.12). We shall eliminate them so as to obtain an equation for S itself, at the expense of two approximations.

First one has to assume that the poles k_n lie very densely on the real axis and that the v_n vary smoothly from one k_n to the next. Then it makes sense to define a "strength function" $g(x)$ by setting, as in (III.3.3),

$$\sum_{x < k_n < x + \Delta x} v_n^2 = g(x) \Delta x. \tag{2.19}$$

With its aid one may write for (2.15), see fig. 49,

$$G(z) = z - \Omega - \int_0^\infty \frac{g(k) \, dk}{z - k}. \tag{2.20}$$

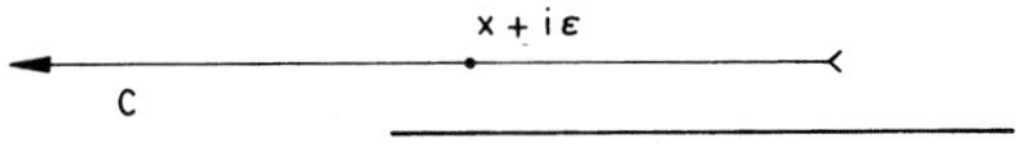

Fig. 49. The poles have merged to form a cut.

provided that z does not lie on the real axis – in fact it must lie at a distance large compared to the distance between successive k_n. On the other hand, on substituting $z = x + i\varepsilon$ in (2.20) one has for small ε

$$G(x+i\varepsilon) = x - \Omega - \int \frac{g(k) \, dk}{x - k} + i\pi g(x), \tag{2.21}$$

2. THE DAMPED HARMONIC OSCILLATOR

where the principal-value integral is defined by

$$\int \frac{g(k)\,dk}{x-k} = \lim_{\delta \to 0} \left\{ \int_0^{x-\delta} + \int_{x+\delta}^{\infty} \right\}.$$

One may use (2.21) as an approximation in (2.18), provided that ε is *small* compared to the distance in the x-scale over which $g(x)$ varies appreciably; as well as *large* compared to the distance between successive k_n. This can always be achieved by a suitable choice of the set of oscillators that serve as our bath.*)

A more serious approximation is needed, namely that the interaction is weak, g small. Then in expression (2.21) the contribution $x - \Omega$ dominates over the other two terms unless $|x - \Omega| \sim g$. Hence in both these terms one may replace the variable quantity x with the constant Ω:

$$G(x + i\varepsilon) = x - \Omega - \int \frac{g(k)\,dk}{\Omega - k} + i\pi g(\Omega)$$

$$= x - \Omega' + i\Gamma. \tag{2.22}$$

The consequence is that (2.18) simplifies to

$$U(t) = -\frac{1}{2\pi i} \int_{-\infty}^{\infty} \frac{e^{-ixt}}{x - \Omega' + i\Gamma}\,dx = e^{-i\Omega' t - \Gamma t}. \tag{2.23}$$

The exponential is a damped oscillation with shifted frequency

$$\Omega' = \Omega + \int_0^{\infty} \frac{g(k)\,dk}{\Omega - k}. \tag{2.24}$$

We still have to link it up with actual observations.

Suppose the initial state at $t = 0$ is $|N_0; \{0\}\rangle$: all bath oscillators in their ground state and the oscillator S in level N_0. According to (2.4)

$$a_0(0)|N_0; \{0\}\rangle = \sqrt{N_0}\,|N_0 - 1; \{0\}\rangle, \qquad a_n(0)|N_0; \{0\}\rangle = 0.$$

According to (2.12a), (2.17) and (2.23)

$$a_0(t)|N_0; \{0\}\rangle = e^{-i\Omega' t - \Gamma t}\sqrt{N_0}\,|N_0 - 1; \{0\}\rangle.$$

The occupation number of this oscillator is the operator $a_0^\dagger a_0 \equiv \mathcal{N}$; its expectation value at time t is

$$\langle \mathcal{N}(t) \rangle = \langle N_0; \{0\}|a_0^\dagger(t)a_0(t)|N_0; \{0\}\rangle = N_0\,e^{-2\Gamma t}. \tag{2.25}$$

---

*) Mathematicians like to meet this requirement by making the bath infinitely large, but that creates more difficulties than it solves – and is unrealistic.

XVII. STOCHASTIC BEHAVIOR OF QUANTUM SYSTEMS

This shows that the energy is emitted into the empty bath at the rate

$$2\Gamma = 2\pi g(\Omega). \tag{2.26}$$

Exercise. Show that (2.26) is the Golden Rule mentioned in V.1.

Exercise. Many authors write instead of (2.19)

$$g(x) = \sum_n v_n^2 \, \delta(x - k_n)$$

and then treat g as a smooth function (although it is of delta peaks). Convince yourself that the result is the same. [Compare the discussion in connection with (I.7.7).]

Exercise. Using the same approximations as in (2.22) obtain

$$V_n = W_n = \frac{v_n}{\Omega' - k_n - i\Gamma} \left\{ e^{-i\Omega' t - \Gamma t} - e^{-ik_n t} \right\}. \tag{2.27}$$

Exercise. With the aid of (2.27) one finds, for the same initial state as in (2.26), in the limit $t \to \infty$:

$$(N_0; \{0\} | a_n^\dagger(\infty) a_n(\infty) | N_0; \{0\}) = \frac{v_n^2}{(\Omega' - k_n)^2 + \Gamma^2}. \tag{2.28}$$

This is the Wigner–Weisskopf formula: the emitted energy has a Lorentz distribution about the shifted frequency.[*]

Next suppose at $t = 0$ the bath is in thermal equilibrium with the temperature β^{-1}, while S is in level N_0. The average occupation number of the n-th bath oscillator is given by

$$\langle a_n^\dagger(0) a_m(0) \rangle = \delta_{nm}(e^{\beta k} - 1)^{-1}.$$

The expectation value of $\mathcal{N}$ at time t is found with the aid of (2.12a), (2.27), and (2.26):

$$\langle \mathcal{N}(t) \rangle = \langle a_0^\dagger(t) a_0(t) \rangle = |U(t)|^2 N_0 + \sum_m |V_m(t)|^2 (e^{\beta k_m} - 1)^{-1}$$

$$= e^{-2\Gamma t} N_0 + \int \frac{g(k) \, dk}{e^{\beta k} - 1} \frac{e^{-2\Gamma t} + 1 - 2e^{-\Gamma t} \cos(\Omega - k)t}{(\Omega - k)^2 + \Gamma^2}$$

$$\approx e^{-2\Gamma t} N_0 + \frac{g(\Omega)}{e^{\beta \Omega} - 1} \frac{\pi}{\Gamma} (1 - e^{-2\Gamma t})$$

$$= e^{-2\Gamma t} N_0 + (e^{\beta \Omega} - 1)^{-1} (1 - e^{-2\Gamma t}). \tag{2.29}$$

The equilibrium value is $\langle \mathcal{N}(\infty) \rangle = (e^{\beta \Omega} - 1)^{-1}$, as it should.

---

[*] H.M. Nussenzveig, *Introduction to Quantum Optics* (Gordon and Breach, New York 1973); R. Loudon, *The Quantum Theory of Light* (2nd Ed., Clarendon, Oxford 1983).

2. THE DAMPED HARMONIC OSCILLATOR

Conclusions. (i) For irreversible emission to occur it is necessary that the frequencies k_n of the bath oscillators are dense.[*] If there had been only a few oscillators the energy would shuttle back and forth between them and the main oscillator. The return time (or "Poincaré period") would be of the order of the reciprocal of the distance between the k_n.

(ii) The damping is achieved by the bath oscillators whose frequencies k_n lie in the vicinity of Ω – within a range of the order of the line width Γ. The Lorentzian form (2.28) of the emission line requires that the strength function $g(k)$ may be taken constant in that range. $g'(\Omega) \ll 1$.

(iii) The higher and lower frequencies merely contribute to the frequency shift (2.24). This is the same phenomenon as the Lamb shift.[**] If, however, Ω is itself small (of order Γ) a different treatment is needed.

(iv) The bath has been eliminated simply by postulating its initial state. The point is, however, that the resulting average occupation (2.29) of S obeys a differential equation by itself

$$\frac{d}{dt}\langle N_0(t)\rangle = -2\Gamma\langle N_0(t)\rangle + \frac{2\Gamma}{e^{\beta\Omega}-1}. \tag{2.30}$$

The special initial time is no longer visible and the state of the bath enters through β only.

Remark. To show that the model (2.3) is less contrived than it looks consider a charged particle with the electromagnetic field. In obvious notation

$$H = \frac{1}{2M}\{\boldsymbol{P} - e\boldsymbol{A}(\boldsymbol{R})\}^2 + V(\boldsymbol{R}) + \tfrac{1}{2}\sum_{\boldsymbol{k},\lambda}(p_{\boldsymbol{k}\lambda}^2 + k^2 q_{\boldsymbol{k}\lambda}^2). \tag{2.31}$$

k is the wave vector of the electromagnetic modes, λ their polarization. The vector potential is

$$\boldsymbol{A}(\boldsymbol{R}) = L^{-3/2}\sum_{\boldsymbol{k},\lambda}(\boldsymbol{e}_{\boldsymbol{k}\lambda}/k)\{p_{\boldsymbol{k}\lambda}\cos\boldsymbol{k}\cdot\boldsymbol{R} + q_{\boldsymbol{k}\lambda}\sin\boldsymbol{k}\cdot\boldsymbol{R}\}.$$

Now suppose: (a) the potential V is quadratic; and (b) the dipole approximation $|\boldsymbol{k}\cdot\boldsymbol{R}| \ll 1$ may be used. Then (2.31) decomposes into three terms of the form

$$H = P^2/2M + \tfrac{1}{2}M\Omega^2 R^2 + \tfrac{1}{2}\sum_n(p_n^2 + k_n^2 q_n^2)$$

$$-\frac{e}{M\sqrt{L}}\sum\frac{p_n}{k_n}P + \frac{e^2}{2ML}\sum\frac{p_n p_m}{k_n k_m}.$$

---

[*] This is an essential assumption, appearing already in the original derivations of the Golden Rule by P.A.M. Dirac, Proc. Roy. Soc. (London) **114**, 243 (1927) and by W. Pauli, op. cit. in V.1.

[**] Loudon, op. cit.; W.R. Louisell, *Quantum Statistical Properties of Radiation* (Wiley, New York 1973).

This is again a quadratic form, which can be diagonalized.[*] In terms of the operators a_n, $a_n^\dagger$ it becomes

$$H = \Omega a_0^\dagger a_0 + \sum_n \omega_n a_n^\dagger a_n - \sum_n v_n(a_n - a_n^\dagger)(a_0 - a_0^\dagger).$$

This reduces to (2.3) if one supposes: (c) *the rotating wave approximation*: terms with $a_n a_0$ and $a_n^\dagger a_0^\dagger$ may be omitted. They are associated with the high frequencies $\Omega + \omega_n$ and therefore negligible in most applications. For our present case this approximation was not indispensable, but simplified the algebra.

3. The elimination of the bath

Consider the general case of a system S interacting with a bath B and described by the total Hamiltonian

$$H_T = H_S + H_B + \alpha H_I. \tag{3.1}$$

The factor α is a coupling constant and serves to monitor the strength of the interaction. The problem is to obtain a self-contained equation for S alone, which will inevitably be approximate. According to (1.16) it is necessary to use density matrices. The density matrix of the total system obeys

$$\dot{\rho}_T = -i[H_T, \rho_T], \tag{3.2a}$$

or, in terms of superoperators,

$$\dot{\rho}_T = \mathscr{L}_T \rho_T = (\mathscr{L}_S + \mathscr{L}_B + \alpha \mathscr{L}_I)\rho_T. \tag{3.2b}$$

This equation determines $\rho_T(t)$ when $\rho_T(0)$ is given:

$$\rho_T(t) = e^{-itH_T} \rho_T(0) e^{itH_T} = e^{t\mathscr{L}_T} \rho_T(0). \tag{3.3}$$

Suppose S and B were uncoupled before $t = 0$, so that the total density matrix factorizes as in (1.23),

$$\rho_T(0) = \rho_B(0) \otimes \rho_S(0). \tag{3.4}$$

The density matrix $\rho_S(0)$ is an operator in the Hilbert space $\mathbf{H}_S$ of S and represents an arbitrary initial condition. The matrix $\rho_B(0)$ operates in $\mathbf{H}_B$; as in the preceding section we take for it an equilibrium distribution

$$\rho_B(0) = \rho_B^e, \qquad \mathscr{L}_B \rho_B^e = 0.$$

Their product $\rho_T(0)$ operates in the total Hilbert space $\mathbf{H}_T = \mathbf{H}_B \otimes \mathbf{H}_S$, see

---

[*] N.G. van Kampen, Kong. Danske Vid. Selsk. Mat.-fys. Medd. **26**, Nr. 15 (1951).

3. THE ELIMINATION OF THE BATH

(1.23). We have

$$\rho_T(t) = e^{t\mathcal{L}_T}[\rho_B^e \otimes \rho_S(0)] = e^{t\mathcal{L}_T} \rho_B^e \rho_S(0). \tag{3.5}$$

I use the latter, simplified way of writing, which is customary in physics, and trust that the reader will remember that the superoperator $\mathcal{L}_T$ is *not* a matrix multiplication but transforms the entire operator following it.

We are interested in S alone, i.e., we are concerned with operators that are diagonal in the bath variables, compare (1.14). For the average value of such operators one does not need all the information comprised in ρ_T but the density matrix of S alone suffices,

$$\rho_S(t) = \operatorname*{Tr}_B \rho_T(t) = \operatorname*{Tr}_B e^{t\mathcal{L}_T} \rho_B^e \rho_S(0).$$

The trace is taken over the bath only, as in (1.15), thereby reducing the operator in $\mathbf{H}_T$ to an operator in $\mathbf{H}_S$. This equation defines a mapping of $\rho_S(0)$ onto $\rho_S(t)$. However, this mapping is contingent on the special choice for $\rho_B(0)$ and cannot, therefore be utilized again to get, e.g., $\rho_S(2t)$ from $\rho_S(t)$. There is no semigroup property and no differential equation of the type

$$\dot{\rho}_S = \mathcal{L} \rho_S. \tag{3.6}$$

This problem is the subject of section **6**.

The best one can hope is that there is an approximate equation of type (3.6), called "Redfield equation"[*] – or in the present context the "quantum master equation". The approximation requires an expansion parameter; the obvious choice is the parameter α. To prepare for this expansion we transform ρ_T to its interaction representation σ_T,

$$\rho_T(t) = e^{t(\mathcal{L}_S + \mathcal{L}_B)} \sigma_T(t) \equiv e^{t\mathcal{L}_0} \sigma_T(t), \tag{3.7}$$

so that,

$$\dot{\sigma}_T(t) = \alpha\, e^{-t\mathcal{L}_0} \mathcal{L}_1 e^{t\mathcal{L}_0} \sigma_T(t) = \alpha \mathcal{L}_1(t) \sigma_T(t). \tag{3.8}$$

Also note that

$$\rho_S(t) = e^{t\mathcal{L}_S} \operatorname*{Tr}_B e^{t\mathcal{L}_B} \sigma_T(t) = e^{t\mathcal{L}_S} \operatorname*{Tr}_B \sigma_T(t). \tag{3.9}$$

because Tr_B commutes with $\mathcal{L}_S$, and $\mathcal{L}_B$ does not affect the trace.

Exercise. When $\alpha = 0$ one has

$$\rho_T(t) = e^{-itH_S} \rho_S(0)\, e^{itH_S} \otimes \rho_B^e.$$

The system S evolves independently of the bath.

---

[*] A.G. Redfield, IBM J. Research Devel. **1**, 19 (1957), and in: *Advances in Magnetic Resonance* **1** (Academic Press, New York 1965); C.P. Slichter, *Principles of Magnetic Resonance* (Harper and Row, New York 1963).

Exercise. What are the eigenvalues and eigenvectors of the superoperator $\mathscr{L}_0$ defined in (3.7)?

We aim for the same cumulant expansion as in XVI.4 so as to obtain an expansion of $\mathscr{L}$ in powers of a quantity $\alpha\tau_c$. The strategy is again: (i) compute explicitly $\rho_T(t)$ as a function of $\rho_T(0)$ to second order in αt; (ii) take the trace over B to obtain $\rho_S(t)$; (iii) differentiate so as to get $\dot\rho_S(t)$ as a function of $\rho_T(0) = \rho_B^e \rho_S(0)$; (iv) finally solve the equation obtained in (ii) so as to express $\rho_S(0)$ in terms of $\rho_S(t)$. These successive operations are facilitated by the interaction representation (3.7), (3.9).

(i) Solve (3.8) to second order in α

$$\sigma_T(t) = \sigma_T(0) + \alpha \int_0^t dt_1\ \mathscr{L}_1(t_1)\sigma_T(0) + \alpha^2 \int_0^t dt_1 \int_0^{t_1} dt_2\ \mathscr{L}_1(t_1)\mathscr{L}_1(t_2)\sigma_T(0). \tag{3.10}$$

(ii) The trace over B of the second term involves the integrand

$$\operatorname*{Tr}_B e^{-t_1(\mathscr{L}_S + \mathscr{L}_B)} \mathscr{L}_1 e^{t_1(\mathscr{L}_S + \mathscr{L}_B)} \rho_B^e \rho_S(0) = e^{-t_1 \mathscr{L}_S} \operatorname*{Tr}_B \mathscr{L}_1 \rho_B^e e^{t_1 \mathscr{L}_S} \rho_S(0).$$

We abbreviate temporarily $e^{t_1 \mathscr{L}_S} \rho_S(0) \equiv \tilde\rho$. Remembering the definition of $\mathscr{L}_1$ one finds for this integrand

$$-i e^{-t_1 \mathscr{L}_S} \operatorname*{Tr}_B [H_1 \rho_B^e \tilde\rho - \rho_B^e \tilde\rho H_1] = -i e^{-t_1 \mathscr{L}_S}\left[\left\{\operatorname*{Tr}_B H_1 \rho_B^e\right\}\tilde\rho - \tilde\rho\left\{\operatorname*{Tr}_B \rho_B^e H_1\right\}\right].$$

Here $\operatorname{Tr}_B H_1 \rho_B^e$ is the average of H_1 over the equilibrium state of the bath but it is still an operator in H_S. For simplicity we assume that this average vanishes; otherwise it could be absorbed into H_S, as in (XVI.2.2). Then the trace of (3.10) is

$$\operatorname*{Tr}_B \sigma_T(t) = \rho_S(0) + \alpha^2 \int_0^t dt_1 \int_0^{t_1} dt_2 \left\{\operatorname*{Tr}_B \mathscr{L}_1(t_1)\mathscr{L}_1(t_2)\rho_B^e\right\}\rho_S(0). \tag{3.11}$$

(iii) As we are looking for an equation of the type (3.6) we differentiate:

$$\frac{d}{dt}\operatorname*{Tr}_B \sigma_T(t) = \alpha^2 \int_0^t dt_2 \left\{\operatorname*{Tr}_B \mathscr{L}_1(t)\mathscr{L}_1(t_2)\rho_B^e\right\}\rho_S(0)$$

$$= \alpha^2 \int_0^t dt' \left\{\operatorname*{Tr}_B e^{-t\mathscr{L}_0} \mathscr{L}_1 e^{(t-t')\mathscr{L}_0} \mathscr{L}_1 e^{t'\mathscr{L}_0} \rho_B^e\right\}\rho_S(0). \tag{3.12}$$

(iv) Finally solve (3.11) for $\rho_S(0)$. Only the lowest order is needed,

$$\rho_S(0) = \operatorname*{Tr}_B \sigma_T(t) = e^{-t\mathscr{L}_S}\rho_S(t).$$

3. THE ELIMINATION OF THE BATH

Substitution in (3.12) yields

$$\frac{d}{dt}\operatorname*{Tr}_{B}\sigma_T(t) = \alpha^2 e^{-t\mathscr{L}_S} \int_0^t dt' \left\{ \operatorname*{Tr}_{B} \mathscr{L}_1 e^{(t-t')\mathscr{L}_0} \mathscr{L}_1 e^{t'\mathscr{L}_0} \rho_B^e \right\} \operatorname*{Tr}_{B}\sigma_T(t).$$

Returning to the original representation we get

$$\dot{\rho}_S(t) = \left[\mathscr{L}_S + \alpha^2 \int_0^t d\tau \left\{ \operatorname*{Tr}_{B} \mathscr{L}_1 e^{\tau\mathscr{L}_0} \mathscr{L}_1 e^{-\tau\mathscr{L}_0} \rho_B^e \right\} \right] \rho_S(t). \tag{3.13}$$

This is an equation for the system S by itself. It is not yet quite of the type (3.6) inasmuch as the superoperator [···] still depends on the time elapsed since that special time at which (3.4) had been postulated. Suppose, however, that there exists a finite time interval τ_c such that the integrand practically vanishes for $t > \tau_c$. Then the integral may be extended to infinity as soon as $t > \tau_c$ and one finds the quantum master equation:

$$\dot{\rho}_S(t) = (\mathscr{L}_S + \alpha^2 \mathscr{K})\rho_S(t) \quad (t > \tau_c), \tag{3.14a}$$

$$\mathscr{K} = \int_0^\infty d\tau \left\{ \operatorname*{Tr}_{B} \mathscr{L}_1 e^{\tau\mathscr{L}_0} \mathscr{L}_1 e^{-\tau\mathscr{L}_0} \rho_B^e \right\} \tag{3.14b}$$

$$= \int_0^\infty d\tau \operatorname*{Tr}_{B} \mathscr{L}_1 \mathscr{L}_1(-\tau)\rho_B^e. \tag{3.14c}$$

Whether or not such a τ_c exists depends on the nature of the bath. It is certainly necessary that the bath has a continuous, or at least very dense, spectrum with eigenvalues that contribute equally to the interaction in order to have a smooth strength function $g(k)$. Also this function must be virtually constant over the range over which the interaction is felt, i.e., the line width, as seen in **2**. It would be desirable to have more concrete criteria, but it is hard to formulate them.

Exercise. Where has it been used that the bath must have many degrees of freedom?

Exercise. Argue that the second term in (3.14a) is of order $\alpha^2 \tau_c$, and that higher terms involve additional powers of $\alpha\tau_c$.

Exercise. Suppose not only $\alpha\tau_c \ll 1$ but in addition $\tau_c \ll \tau_m$, where $1/\tau_m$ is the rate of change caused by $\mathscr{L}_S$. Then (3.14b) reduces to

$$\mathscr{K} = \int_0^\infty d\tau \left\{ \operatorname*{Tr}_{B} \mathscr{L}_1 e^{\tau\mathscr{L}_B} \mathscr{L}_1 \rho_B^e \right\}. \tag{3.15}$$

This describes the case that the force exerted by the bath fluctuates rapidly compared to the free motion of S.

Exercise. If all eigenvalues of $\mathscr{L}_B$ were negative, one might write for (3.15)

$$\mathscr{K} = -\operatorname*{Tr}_B \mathscr{L}_1 \mathscr{L}_B^{-1} \mathscr{L}_1 \rho_B^e. \tag{3.16}$$

Actually all eigenvalues are imaginary, but we also know that in a suitable macroscopic approximation the bath tends to the equilbrium ρ_B^e. That amounts to replacing $\mathscr{L}_B$ with an operator having only negative eigenvalues. The highest one determines τ_c.

Exercise. Suppose that H_T consists of a product $S \cdot B$ of an operator S of the system and an operator B of the bath. Then (3.14) may be written more explicitly

$$\dot{\rho}_S = -\mathrm{i}[H_S, \rho_S] - \alpha^2 \int_0^\infty \mathrm{d}\tau \left\{ \langle BB(-\tau)\rangle^e [S, S(-\tau)\rho_S] - \langle B(-\tau)B\rangle^e [S, \rho_S S(-\tau)] \right\}, \tag{3.17}$$

where $\langle \cdots \rangle^e = \operatorname{Tr}_B(\ldots \rho_B^e)$.

Exercise. Alternatively (3.17) may be written

$$\dot{\rho}_S = -\mathrm{i}[H_S, \rho_S] - \tfrac{1}{2}\alpha^2 \int_0^\infty \mathrm{d}\tau \, \langle [B, B(-\tau)]_+ \rangle^e [S, [S(-\tau), \rho_S]]$$

$$- \tfrac{1}{2}\alpha^2 \int_0^\infty \mathrm{d}\tau \, \langle [B, B(-\tau)]\rangle^e [S, [S(-\tau), \rho_S]_+], \tag{3.18}$$

where the subscript $+$ denotes the anticommutator.

Exercise. The additional restriction $\tau_c \ll \tau_m$ reduces the equation (3.17) to

$$\dot{\rho}_S = -\mathrm{i}[H_S, \rho_S] - \alpha^2 \int_0^\infty \mathrm{d}\tau \left\{ \langle BB(-\tau)\rangle^e S[S, \rho] - \langle B(-\tau)B\rangle^e [S, \rho]S \right\}. \tag{3.19}$$

The importance of this restriction will be seen in section **6**.

4. The elimination of the bath – continued

In this section we work out an example and make two comments. The reader should be warned that (3.14) is not yet the final result but needs a modification, which will be given in **6**. In the following example, however, this modification is not yet needed.

Take the harmonic oscillator of **2** in its bath of small oscillators:

$$H_S = \Omega a_0^\dagger a_0, \qquad H_B = \sum_n k_n a_n^\dagger a_n, \qquad H_1 = \sum_n v_n(a_n a_0^\dagger + a_n^\dagger a_0). \tag{4.1}$$

The unperturbed motion is given by

$$\dot{a}_0 = \mathrm{i}[H_S, a_0] = -\mathrm{i}\Omega a_0, \qquad a_0(-\tau) = \mathrm{e}^{\mathrm{i}\Omega\tau} a_0.$$

Similarly

$$a_n(-\tau) = \mathrm{e}^{\mathrm{i}k_n \tau} a_n, \qquad a_n^\dagger(-\tau) = \mathrm{e}^{-\mathrm{i}k_n \tau} a_n^\dagger.$$

4. THE ELIMINATION OF THE BATH – CONTINUED

We also need the familiar formula

$$\rho_B^e = Z^{-1} \exp\left[-\sum_n k_n a_n^\dagger a_n\right], \quad Z = \prod_n (1 - e^{-\beta k_n})^{-1}. \tag{4.2}$$

Hence

$$\operatorname*{Tr}_B a_m^\dagger a_n \rho_B^e = \langle a_m^\dagger a_n \rangle^e = \delta_{mn}(e^{\beta k_n} - 1)^{-1},$$

$$\operatorname*{Tr}_B a_m^\dagger \rho_B^e a_n = \langle a_n a_m^\dagger \rangle^e = \delta_{mn}(1 - e^{-\beta k_n})^{-1}.$$

In order to find $\mathcal{K}\rho_S$ we first have to compute the double commutator

$$\mathcal{L}_1 \mathcal{L}_1(-\tau) \rho_B^e \rho_S$$

$$= -\sum_{m,n} v_m v_n \left[a_m a_0^\dagger + a_m^\dagger a_0, [a_n a_0^\dagger e^{-i(\Omega - k_n)\tau} + a_n^\dagger a_0 e^{i(\Omega - k_n)\tau}, \rho_B^e \rho_S]\right]. \tag{4.3}$$

On taking the trace only the terms with $a_n a_n^\dagger$ or $a_n^\dagger a_n$ survive,

$$\operatorname*{Tr}_B \mathcal{L}_1 \mathcal{L}_1(-\tau) \rho_B^e \rho_S$$

$$= -\sum_n v_n^2 \left\{ e^{i(\Omega - k_n)\tau} \left(\frac{a_0^\dagger a_0 \rho_S - a_0 \rho_S a_0^\dagger}{1 - e^{-\beta k_n}} + \frac{\rho_S a_0 a_0^\dagger - a_0^\dagger \rho_S a_0}{e^{\beta k_n} - 1} \right) \right.$$

$$\left. + e^{-i(\Omega - k_n)\tau} \left(\frac{a_0 a_0^\dagger \rho_S - a_0^\dagger \rho_S a_0}{e^{\beta k_n} - 1} + \frac{\rho_S a_0^\dagger a_0 - a_0 \rho_S a_0^\dagger}{1 - e^{-\beta k_n}} \right) \right\}. \tag{4.4}$$

The subsequent integration over τ does not converge, but one can give a meaning to it. First turn the summation over n into an integration over k with the aid of (2.19). Then use the formal identity

$$\int_0^\infty e^{i(\Omega - k)\tau} \, d\tau = \pi \, \delta(\Omega - k) + \frac{i}{\Omega - k}. \tag{4.5}$$

The last term is to be read as a principal value. The justification of this procedure is provided by the more exact treatment in **2**. Performing the integration in (4.4) one finds, apart from the frequency shift (2.24),

$$-\frac{\pi g(\Omega)}{e^{\beta \Omega} - 1}(a_0 a_0^\dagger \rho_S - 2a_0^\dagger \rho_S a_0 + \rho_S a_0 a_0^\dagger)$$

$$-\frac{\pi g(\Omega)}{1 - e^{-\beta \Omega}}(a_0^\dagger a_0 \rho_S - 2a_0 \rho_S a_0^\dagger + \rho_S a_0^\dagger a_0).$$

The term with the principal value renormalizes Ω into Ω', cf. (2.24). The result is

$$\dot\rho_S = -i\Omega[a_0^\dagger a_0, \rho_S]$$
$$+ \frac{2\pi g(\Omega)}{e^{\beta\Omega}-1}(a_0^\dagger \rho_S a_0 - \tfrac{1}{2} a_0 a_0^\dagger \rho_S - \tfrac{1}{2}\rho_S a_0 a_0^\dagger)$$
$$+ \frac{2\pi g(\Omega)}{1-e^{-\beta\Omega}}(a_0 \rho_S a_0^\dagger - \tfrac{1}{2} a_0^\dagger a_0 \rho_S - \tfrac{1}{2}\rho_S a_0^\dagger a_0). \tag{4.6}$$

This is the quantum M-equation for the density matrix of the system S by itself. It can be used to compute averages. For instance, the average excitation number of S obeys

$$\frac{d}{dt}\langle N_0 \rangle = \operatorname*{Tr}_S a_0^\dagger a_0 \dot\rho_S$$
$$= -2\pi g(\Omega)\{\langle N_0 \rangle - (e^{\beta\Omega}-1)^{-1}\}, \tag{4.7}$$

which is the same as (2.30). In fact, the present treatment involves the same approximations as stipulated in **2**, namely: second order in the coupling; continuous distribution of the bath oscillator frequencies k_n; special initial state with the bath in equilibrium and no correlations between bath and system.

Exercise. Verify that the principal-value term in (4.5) does not contribute.

Exercise. Derive in analogy to (4.7) for the factorial cumulant of the occupation number

$$\frac{d}{dt}[N_0^2] = -4\pi g(\Omega)\left\{[N_0^2] - (e^{\beta\Omega}-1)^{-1}\langle N_0 \rangle\right\}. \tag{4.8}$$

Exercise. It follows from (1.9) that the diagonal elements of a density matrix are a probability distribution. Call the diagonal elements of our present ρ_S in the occupation representation $(N|\rho_S|N) = p_N$. Show that according to (4.6) these p_N obey an M-equation, namely (VI.4.1) including (VI.4.4). (This is an exceptional case: in general the diagonal elements do not obey a closed equation without involving the off-diagonal elements.)

Exercise. The factorial cumulants are suitable for classical particles, but for bosons the generalized cumulants π_m defined in (I.2.20) are better suited. Express (4.8) in terms of π_2.

The first comment briefly describes an alternative (and very popular) derivation of the same result (3.14), using projection operators.[*] Define the

---

[*] S. Nakajima, Prog. Theor. Phys. **20**, 948 (1958); R. Zwanzig, J. Chem. Phys. **33**, 1338 (1960); F. Haake, in: *Quantum Statistics in Optics and Solid State Physics* (Springer Tracts in Modern Physics 60; Springer, Berlin 1973).

4. THE ELIMINATION OF THE BATH – CONTINUED

superprojector $\mathscr{P}$, acting on operators in $\mathbf{H}_T$, by

$$\mathscr{P}\rho_T = \rho_B^e \operatorname*{Tr}_B \rho_T = \rho_B^e \rho_S. \tag{4.9}$$

The evolution equation (3.2) may be decomposed into coupled equations for $\mathscr{P}\rho_T$ and $(1 - \mathscr{P})\rho_T \equiv \mathscr{Q}\rho_T$. Using the properties*⁾

$$\mathscr{P}\mathscr{L}_B = \mathscr{L}_B\mathscr{P} = 0, \qquad \mathscr{P}\mathscr{L}_S = \mathscr{L}_S\mathscr{P}, \qquad \mathscr{P}\mathscr{L}_1\mathscr{P} = 0, \tag{4.10}$$

one has the two equations

$$\partial_t \mathscr{P}\rho_T = \mathscr{L}_S \mathscr{P}\rho_T + \alpha \mathscr{P}\mathscr{L}_1 \mathscr{Q}\rho_T, \tag{4.11a}$$

$$\partial_t \mathscr{Q}\rho_T = \mathscr{Q}\mathscr{L}_T \mathscr{Q}\rho_T + \alpha \mathscr{Q}\mathscr{L}_1 \mathscr{P}\rho_T. \tag{4.11b}$$

The latter may be formally solved to give

$$\mathscr{Q}\rho_T(t) = \alpha \int_0^t d\tau\, e^{\tau \mathscr{Q}\mathscr{L}_T \mathscr{Q}} \mathscr{L}_1 \rho_B^e \rho_S(t - \tau).$$

Substitution in the former yields

$$\dot\rho_S(t) = \mathscr{L}_S \rho_S(t) + \alpha^2 \int_0^t d\tau\, \mathscr{G}(\tau) \rho_S(t - \tau), \tag{4.12a}$$

where $\mathscr{G}$ is a superoperator in $\mathbf{H}_S$,

$$\mathscr{G}(\tau) = \operatorname*{Tr}_B \mathscr{L}_1 e^{\tau \mathscr{Q}\mathscr{L}_T} \mathscr{Q}\mathscr{L}_1 \rho_B^e. \tag{4.12b}$$

This is called the "generalized master equation". It should be clear, however, that the formal resemblance to the master equation used so far, defined in V.1, is misleading. Equation (4.12) is exactly equivalent to the original equation (3.2); the bath variables have been eliminated at the expense of introducing the entire history of ρ_S. On the other hand, the essence of the true M-equation is that it approximates the evolution of ρ_S by a semigroup, which moreover embodies the irreversibility. The "generalized master equation" is merely a rewriting of the original equation of motion in a form convenient for applying the subsequent approximations.

First one assumes again that there exists a τ_c such that $\mathscr{G}(\tau)$ is negligible for $t > \tau_c$, so that the integral in (4.12a) may be extended to infinity as soon as $t > \tau_c$.

Secondly one neglects in the exponent the term with $\alpha \mathscr{L}_1$. That means that powers of α beyond the second are omitted, just as in our (3.10). (This resembles the Born approximation in scattering theory.) One easily sees that the omitted terms are of relative order $\alpha\tau_c$ and of absolute order $\alpha^3 \tau_c^2$. Thus

---

*⁾ The last one states that the bath average of the interaction term vanishes.

one has

$$\dot{\rho}_S(t) = \mathscr{L}_S\rho_S(t) + \alpha^2 \int_0^\infty d\tau \left\{ \underset{B}{\text{Tr}} \mathscr{L}_I e^{\tau \mathscr{L}_0} \mathscr{L}_I \rho_B^e \right\} \rho_S(t-\tau). \tag{4.13}$$

Thirdly one needs a drastic step to turn this integral equation into a differential equation. This is the "Markov approximation", which comes in two varieties. The first variety consists in replacing $\rho_s(t-\tau)$ with $\rho_s(t)$. The error is of relative order τ_c/τ_m (where $1/\tau_m$ is the unperturbed rate of change due to $\mathscr{L}_s$) and of absolute order $\alpha^2 \tau_c^2/\tau_m$. In this approximation one may as well omit the $\mathscr{L}_s$ in the exponent of (4.13) and the result is the same as (3.19). The second variety takes the zeroth order variation of ρ_s into account by setting $\rho_s(t-\tau) = e^{-\tau \mathscr{L}_s}\rho_s(t)$. The result is the same as was obtained in (3.14) by means of the interaction representation, and the only requirement is $\alpha\tau_c \ll 1$.

Other choices for $\mathscr{P}$ are possible[*], but these three assumptions cannot be avoided.

The second comment concerns yet another derivation, based on the idea that the bath variables are much faster than those of the system. One writes for the total Hamiltonian

$$H_T = H_S + \varepsilon^{-1}H_I + \varepsilon^{-2}H_B. \tag{4.14}$$

where ε is small ("singular coupling limit"[**]). The general method for eliminating fast variables can then be applied without difficulty[†] and one is smoothly led to (3.13). The objection to this derivation is that it is *not* true that the bath variables are fast. As shown in **2** the effect of the bath is not due to the fast oscillators but to those in resonance with S. This treatment is therefore only valid when the proper frequencies of S are extremely low, e.g., zero, as in the case of a Brownian particle. In that case, however, the Hamiltonian does not have the form (4.14) and the elimination of the fast bath variables takes a different aspect.[††] The number of systems with Hamiltonian (4.14) seems to be extremely limited.

5. The Schrödinger–Langevin equation and the quantum master equation

What follows is a heuristic derivation of the "quantum master equation" (5.6) suggested by a formal resemblance between the Schrödinger equation and the

---

[*] V. Romero-Rochin, A. Orski, and I. Oppenheim, Physica A **150**, 244 (1989); M. Morilla, R.I. Cukier, and M. Tij, Physica A **179**, 411 (1991); M. Morilla and M. Tij, Physica A **179**, 428 (1991).

[**] R. Dümcke and H. Spohn, Z. Phys. B **34**, 419 (1979).

[†] GARDINER, p. 390.

[††] N.G. van Kampen and I. Oppenheim, Physica A **138**, 231 (1986); I. Oppenheim and V. Romero-Rochin, Physica A **147**, 184 (1987).

5. THE SCHRÖDINGER–LANGEVIN AND THE QUANTUM MASTER EQUATIONS

Langevin approach. It may be skipped by the reader who is satisfied with the subsequent appeal to the mathematical foundation.

In section 2 it was mentioned that a damping term in the Schrödinger equation violates the conservation of probability. It is possible, however, to compensate this loss by adding artificially a suitable fluctuating term. The result will be an equation for ψ resembling the classical Langevin equation.

To find such an equation we try

$$\dot\psi = -iH\psi - U\psi + l(t)V\psi,$$

where U and V are operators to be specified. H is Hermitian and U may also be taken Hermitian, because any anti-Hermitian part could be combined with H. For $l(t)$ we take white noise, not necessarily real,

$$\langle l(t)\rangle = 0, \quad \langle l(t)l^*(t')\rangle = \delta(t-t'). \tag{5.1}$$

One then obtains for small Δt

$$\psi(t+\Delta t) = \left[1 - iH\,\Delta t - U\,\Delta t + \int_t^{t+\Delta t} l(t')\,\mathrm{d}t'\, V\right]\psi(t).$$

The complex conjugate of this equation, written in the adjoint form, is

$$\psi^\dagger(t+\Delta t) = \psi^\dagger(t)\left[1 - iH\,\Delta t - U\,\Delta t + V^\dagger \int_t^{t+\Delta t} l^*(t')\,\mathrm{d}t'\right].$$

Take the scalar product of both equations and average over l,

$$\psi^\dagger(t+\Delta t)\psi(t+\Delta t) = \psi^\dagger(t)\psi(t) - 2\psi^\dagger(t)U\psi(t) + \psi^\dagger(t)V^\dagger V\psi(t).$$

The product implies integration over q as in (1.1); and also summation over the components of ψ if there are more than one. Evidently the average norm is conserved when $2U = V^\dagger V$. Thus we have found a Langevin extension of the Schrödinger equation:

$$\dot\psi = -iH\psi - \tfrac{1}{2}V^\dagger V\psi + l(t)V\psi, \tag{5.2}$$

where H is Hermitian, V may be any operator and $l(t)$ obeys (5.1). There is no need to worry about higher moments of ψ because they never occur in quantum mechanical expressions. Neither is it necessary that $l(t)$ should be Gaussian.

Remark. It is easily seen that the second term of (5.2) by itself causes the norm of ψ to change. In order that this is compensated by the fluctuating term the two terms must be linked, as is done by the relation $U = \tfrac{1}{2}V^\dagger V$. This resembles the classical fluctuation–dissipation theorem, which links both terms by the requirement that the fluctuations compensate the energy loss so as to establish the equilibrium. The difference is that the latter requirement involves the temperature T of the environment; that makes it possible to suppress the fluctuations by taking $T=0$ without losing the damping. This is the reason why in classical theory deterministic equations with damping exist, see XI.5.

Our Schrödinger–Langevin equation (5.2) involves no T or equivalent parameter; it is therefore not possible to obtain a Schrödinger equation with damping.

Exercise. The equation (5.2) may be generalized to

$$\dot{\psi} = -iH\psi - \frac{1}{2}\sum_\alpha V_\alpha^\dagger V_\alpha \psi + \sum_\alpha l_\alpha(t) V_\alpha \psi \tag{5.3}$$

with a set of arbitrary operators V_α and

$$\langle l_\alpha(t)\rangle = 0, \quad \langle l_\alpha(t) l_\beta^*(t')\rangle = \delta_{\alpha\beta}\delta(t-t'). \tag{5.4}$$

This is *the general Schrödinger–Langevin equation*.

Exercise. The Schrödinger–Langevin equation (5.3) implies that the average of any Heisenberg operator (1.24) obeys

$$\partial_t \langle A\rangle = i\langle [H, A]\rangle + \sum_\alpha \{\langle V_\alpha^\dagger A V_\alpha\rangle - \frac{1}{2}\langle V_\alpha^\dagger V_\alpha A\rangle - \frac{1}{2}\langle A V_\alpha^\dagger V_\alpha\rangle\}$$

$$= i[H, \langle A\rangle] + \sum_\alpha \{V_\alpha^\dagger \langle A\rangle V_\alpha - \frac{1}{2}V_\alpha^\dagger V_\alpha \langle A\rangle - \frac{1}{2}\langle A\rangle V_\alpha^\dagger V_\alpha\}. \tag{5.5}$$

The stochastic differential equation (5.2) also generates equations for the second moments $\langle \psi(q)\psi^\dagger(q')\rangle$, see (XVI.2.18). These moments constitute the density matrix. An easier way to find the equation for the density matrix is by writing (5.5) by means of the density matrix,

$$\partial_t \operatorname{Tr} \rho A = i \operatorname{Tr} \rho(HA - AH) + \sum_\alpha \operatorname{Tr} \rho\{V_\alpha^\dagger A V_\alpha - \frac{1}{2}V_\alpha^\dagger V_\alpha A - \frac{1}{2}A V_\alpha^\dagger V_\alpha\}$$

$$= -i \operatorname{Tr}[H, \rho]A + \sum_\alpha \operatorname{Tr}\{V_\alpha \rho V_\alpha^\dagger - \frac{1}{2}\rho V_\alpha^\dagger V_\alpha - \frac{1}{2}V_\alpha^\dagger V_\alpha \rho\}A.$$

Since this is true for all A one may conclude

$$\dot{\rho} = -i[H, \rho] + \sum_\alpha \{V_\alpha \rho V_\alpha^\dagger - \frac{1}{2}\rho V_\alpha^\dagger V_\alpha - \frac{1}{2}V_\alpha^\dagger V_\alpha \rho\}. \tag{5.6}$$

Thus we have found *the general form of the quantum master equation* that corresponds to the Schrödinger–Langevin equation (5.3).

Exercise. Take for ρ a 2×2 matrix and work out (5.6) for the following case involving a single V:

$$H = \begin{pmatrix} \Omega & 0 \\ 0 & 0 \end{pmatrix}, \quad V = \begin{pmatrix} 0 & 0 \\ a & 0 \end{pmatrix}.$$

One finds

$$\rho_{11}(t) = \rho_{11}(0)\exp[-a^2 t], \quad \rho_{12}(t) = \rho_{12}(0)\exp[-i\Omega t - \frac{1}{2}a^2 t],$$

which describes decay from level 1 into level 2.

5. THE SCHRÖDINGER–LANGEVIN AND THE QUANTUM MASTER EQUATIONS

Exercise. Same question for the case

$$H = \begin{pmatrix} \Omega & 0 \\ 0 & 0 \end{pmatrix}, \quad V = \begin{pmatrix} 0 & b \\ a & 0 \end{pmatrix}.$$

This describes a two-level atom interacting with a heat bath such that $b^2/a^2 = \exp[-\beta\Omega]$.

Exercise. An alternative form of (5.6) is

$$\rho = -\mathrm{i}[H, \rho] + \frac{1}{2} \sum_\alpha \{[V_\alpha \rho, V_\alpha^\dagger] + [V_\alpha, \rho V_\alpha^\dagger]\}. \tag{5.7}$$

If among the V_α there is a Hermitian matrix, V_0, its contribution may be written

$$-\frac{1}{2}[V_0, [V_0, \rho]]. \tag{5.8}$$

Exercise. The equation (5.6) keeps the same form under the transformation

$$V_\alpha = \sum_\beta c_{\alpha\beta} \bar{V}_\beta \text{ with unitary matrix } c_{\alpha\beta}. \tag{5.9}$$

It also keeps its form for $V_\alpha = c_\alpha + \bar{V}_\alpha$ with scalar constants c_α. It is therefore permitted to restrict (5.6) to traceless V_α.

Exercise. Show that (3.19) is of the form (5.6). [Hint: Write (3.19) in the form (3.18) and note that $[S, [S, \ldots]+] = [S^2, \ldots]$ and can therefore be incorporated in H.]

We have arrived at the quantum master equation (5.6) heuristically via the Schrödinger–Langevin equation (5.3). It turns out, however, that (5.6) has a firmer foundation: it can be proved mathematically on the basis of the following three general conditions concerning the evolution of ρ.
  (i) The mapping $\rho(0) \to \rho(t)$ is linear.
  (ii) The mapping preserves the essential properties (1.9) of a density matrix: Hermitian, trace = 1, positive definite.
  (iii) The mapping is a semigroup, i.e., it is generated by a differential equation

$$\dot{\rho}(t) = \mathscr{L} \rho(t). \tag{5.10}$$

Under these three conditions[*] it can be rigorously proved that $\mathscr{L}$ must have the form (5.6).[**] The weak element is the condition (iii), since we know that (5.10) is at best true only approximately.

---

[*] The condition that ρ must remain positive definite has to be replaced by the stronger condition that the mapping must be "completely positive", but the physical relevance is unclear.

[**] A. Kossakowski, Bull. Acad. Pol. Sci., Série Math. Astr. Phys. **20**, 1021 (1971); **21**, 649 (1973); V. Gorini, A. Kossakowski, and E.C.G. Sudarshan, J. Mathem. Phys. **17**, 821 (1976); G. Lindblad, Commun. Mathem. Phys. **40**, 147 (1975). A more readable account is given by R. Alicki and K. Lendi, *Quantum Dynamical Semigroups and Applications* (Lecture Notes in Physics **286**; Springer, Berlin 1987).

Exercise. Verify directly from (**5.6**) that the first two properties of (**1.9**) are preserved. To verify the positivity requirement, observe that if $\rho(t)$ at some time t_1 ceases to be positive definite, there must be a Hilbert vector $\phi(q)$ such that $(\phi|\rho(t_1)|\phi) = 0$. It follows that $\rho(t_1)\phi = 0$ and subsequently (5.6) tells us that at $t = t_1$

$$\frac{\mathrm{d}}{\mathrm{d}t}(\phi|\rho(t)|\phi) = \sum_\alpha (V_\alpha^\dagger \phi|\rho(t)|V_\alpha^\dagger \phi) \geq 0.$$

Exercise. In an ill-advised attempt to modify quantum mechanics the following equation for the evolution of ρ (for a Schrödinger particle in one dimension) has been proposed[*]

$$\dot\rho = -\mathrm{i}[H,\rho] - \lambda \left\{ \rho - \sqrt{\frac{\gamma}{\pi}} \int_{-\infty}^\infty \mathrm{e}^{-\frac{1}{2}\gamma(q-s)^2} \rho\, \mathrm{e}^{-\frac{1}{2}\gamma(q-s)^2} \mathrm{d}s \right\}.$$

Here q is the operator for the position of the particle and λ, γ are positive constants of nature. Show that this equation has the required form (5.6). In the q-representation the last term is

$$-\lambda(q|\rho|q')\{1 - \mathrm{e}^{-\frac{1}{4}\gamma(q-q')^2}\},$$

and it is not easy to recognize the form (5.6).

Remark. Normally the name "quantum Langevin equation" is used for equations that are the direct analog of the classical Langevin equation; e.g., in the case of a one-dimensional particle of unit mass in a potential V,

$$\dot q = p, \quad \dot p = -V(q) - \gamma p + \xi(t).$$

Here q and p are Heisenberg operators, γ is the usual damping coefficient, and $\xi(t)$ is a random force, which is also an operator. Not only does one have to characterize the stochastic behavior of $\xi(t)$, but also its commutation relations, in such a way that the canonical commutation relation $[q(t), p(t)] = \mathrm{i}$ is preserved at all times and the fluctuation–dissipation theorem is obeyed.[**] Moreover it appears impossible to maintain the delta correlation in time in view of the fact that quantum theory necessarily cuts off the high frequencies.[†] We conclude that no quantum Langevin equation can be obtained without invoking explicitly the equation of motion of the bath that causes the fluctuations.[††] That is the reason why this type of equation has so much less practical use than its classical counterpart.

---

[*] G.C. Ghirardi, A. Rimini and T. Weber, Phys. Rev. D **34**, 470 (1986).

[**] R. Benguria and M. Kac, Phys. Rev. Letters **46**, 1 (1981); H. Grabert, U. Weiss, and P. Talkner, Z. Phys. B **55**, 87 (1984). A review of the ample literature is given by P. Hänggi and G-L. Ingold. arXiv:quant-ph./04152v1 (2004).

[†] P. Shiktorov. E. Starikov, V. Gružinskis and L. Reggiani, Semicond. Sci. Technol. **19**, 232 (2004).

[††] C.W. Gardiner, *Quantum Noise* (Springer, Berlin 1991) and literature quoted there.

Exercise. Suppose the total system S+B is in equilibrium. Then it follows from (3.14) that to second order in α and for $\tau > 0$,

$$\rho_S(\tau) = \left[e^{\tau \mathscr{L}_S} + \alpha^2 \int_0^\tau e^{(\tau - t')\mathscr{L}_S} \mathscr{K} e^{t' \mathscr{L}_S} dt' \right] \rho_S(0).$$

From this one can conclude that, if F is an operator acting on the system S, the correlation of its values in equilibrium at times t and $t + \tau$ is[*]

$$\langle F(t)F(t+\tau) \rangle_\alpha^e = \langle F(t)F(t+\tau) \rangle_0^e - \alpha^2 \int_0^\tau dt' \int_{-\infty}^0 dt'' \langle F[H_1(t''), [H_1(t'), F(\tau)]] \rangle_0^e \quad (5.11)$$

6. A new approach to noise

A serious difficulty now appears. The quantum master equation (3.14), obtained by eliminating the bath, does not have the required form (5.6) and therefore results in a violation of the positivity of $\rho_S(t)$. Only by the additional approximation $\tau_c \ll \tau_m$ was it possible to arrive at (3.19), which does have that form (see the Exercise). The origin of the difficulty is that (3.14) is based on our assumed initial state (3.4), which expresses that system and bath are initially uncorrelated. This cannot be true at later times because the interaction inevitably builds up correlations between them. Hence it is unjustified to use the same derivation for arriving at a differential equation in time without invoking a repeated randomness assumption, such as embodied in $\tau_c \ll \tau_m$.[**] At any rate it is physically absurd to think that the study of the behavior of a Brownian particle requires the knowledge of an initial state.

Our answer to this difficulty is the following.[†] Consider the equilibrium of the *total* system S+B, compute the fluctuations occurring in S and their time correlations. This provides a more consistent way for obtaining the information for which the Langevin approach was intended. The purpose of this section is to implement these general remarks by explicit calculations. Unfortunately they are rather laborious; I apologize.

The total system is subject to the Hamiltonian (3.1). Its thermal equilibrium has the density matrix

$$\rho_T = Z_\alpha^{-1} e^{-\beta H_T} \qquad \text{with} \quad Z_\alpha = \text{Tr } e^{-\beta H_T}. \quad (6.1)$$

If F is some operator the correlation between its values at time t and $t + \tau$ is

$$\langle F(t)F(t+\tau) \rangle^e = Z_\alpha^{-1} \text{ Tr } e^{-\beta H_T} F \, e^{i\tau H_T} F \, e^{-i\tau H_T}. \quad (6.2)$$

---

[*] N.G. van Kampen, Fluctuation and Noise Lett. **1**, C7 (2001).

[**] A. Suarez, R. Silbey and I. Oppenheim, J. Chem. Phys. **97**, 5101 (1992). Classically this inequality is in N.N. Bogoliubov, *Problems of a Dynamical Theory in Statistical Physics* (Translated by E.K. Gora, Providence, R.I. 1960).

[†] Fluctuation and Noise Lett. **1**, C7 (2001); J. Stat. Phys. **115**, 1057 (2004).

If F is an operator working on the system S alone this autocorrelation function describes the fluctuations in S under influence of the bath. We compute it to second order in the coupling constant α.

Regular perturbation theory yields

$$e^{-i\tau H_T} = e^{-i\tau H_0}\left\{1 - i\alpha\int_0^\tau H_I(t')dt' - \alpha^2 \int_0^\tau H_I(t')dt' \int_0^{t'} H_I(t'')dt'' + O(\alpha^3)\right\}.$$

In particular

$$Z_\alpha(\beta) = Z_0(\beta) - \alpha\int_0^\beta \langle H_I(-i\beta')\rangle_0^e d\beta'$$

$$+ \alpha^2 \int_0^\beta d\beta' \int_0^{\beta'} d\beta'' \langle H_I(-i\beta')H_I(-i\beta'')\rangle_0^e. \qquad (6.3)$$

The subscript $_0$ refers to $\alpha = 0$ and the term with $\langle H_I(-i\beta')\rangle$ may again be omitted, as in XVI.2.2. In the same way the other exponentials in (6.2) can be written explicitly to second order in α and after some laborious algebra one obtains

$$\langle FF(\tau)\rangle_\alpha^e = \langle FF(\tau)\rangle_0^e \left\{1 - \alpha^2 \int_0^\beta d\beta' \int_0^{\beta'} d\beta'' \langle H_I(-i\beta')H_I(-i\beta'')\rangle_0^e\right\}$$

$$- \alpha^2 \int_0^\tau dt' \int_0^{t'} dt'' \langle F[H_I(t''), [H_I(t'), F(\tau)]]\rangle_0^e$$

$$+ \alpha^2 \int_0^\beta d\beta' \int_0^{\beta'} d\beta'' \langle H_I(-i\beta')H_I(-i\beta'')FF(\tau)\rangle_0^e$$

$$- i\alpha^2 \int_0^\beta d\beta' \int_0^\tau dt' \langle H_I(-i\beta')F[H_I(t'), F(\tau)]\rangle_0^e. \qquad (6.4)$$

In this approach the lack of positivity never comes up. Apart from that there is a difference between the present autocorrelation function and the one computed on the basis of (4.12). The latter, now indicated by an upper index [a] (for "ancient"), is

$$\langle FF(\tau)\rangle_\alpha^a = \text{Tr}_S \, \rho_S^e FF(\tau)$$

and is given in (5.11). There are three differences.

(i) The first line takes into account the correction of the partition function Z_0 of the bare system S due to the interaction with B. This correction is small when β is small, that is, when the temperature of the combined system is high compared to the energy of the interaction:

$$kT >> |\alpha H_I|. \qquad (6.5)$$

(ii) The second line is the same as in the ancient expression (4.12), except that it does not contain the additional assumption about short survival time of the correlation, by which the integral was extended to $-\infty$. The present correct expression (6.4) for the correlation function has no need of this assumption.

(iii) The last two lines involve integrals from 0 to β. They also disappear under the condition (**6.5**).

Exercise. Apply this approach by explicit computation to the harmonic oscillator in section **2**.[*]

7. Internal noise

In IX.5 the distinction was made between external and internal noise. *External* noise comes from a source outside the system under consideration; its stochastic properties are given; it can be switched off by means of a coupling parameter; and its effect is described by a stochastic differential equation as in **3**, or by means of the joint equilibrium distribution, as in **6**. *Internal* noise is inherent in the system itself; it is due to the particulate nature of matter and cannot be switched off; and it is described by a master equation which derives from the microscopic equations of motion. In the classical Brownian motion the fluctuations are the internal noise of the total system, but that is not how Langevin proceeded. He considered the Brownian particle by itself as the system to be studied and the surrounding fluid as the external bath. After splitting off that part of the interaction with the bath that causes damping, the remaining part is treated as external noise described by a Langevin force. In quantum mechanics one does not know how to split off a damping term[**] and we had to start from the equations of motion of the total system $S+B$.

In the present section we are concerned with genuine internal noise.[†] We consider a closed, isolated many-body system, whose evolution is given by a Schrödinger equation. Remember that in the classical case in III.2 we gave a macroscopic description in terms of a reduced set of macroscopic variables, which obey an autonomous set of differential equations. These equations are approximate and deviations appear in the form of fluctuations, which are a vestige of the large number of eliminated microscopic variables. Our task is to carry out this program in the framework of quantum mechanics.

Consider a quantum mechanical many-body system, e.g., a gas. There is an enormous number of coordinates q of all molecules, and the state of the system is described by a wave function $\psi(q, t)$ in an enormous Hilbert space **H**. The evolution is governed by a Hamilton operator H whose spectrum is discrete but inordinately dense. The number of eigenvalues of H in a range of the order of

---

[*] See J. Stat. Phys. **115**, 1057 (2004).

[**] Equation (**5.3**) is clearly inappropriate as it does not involve the bath temperature.

[†] N.G. van Kampen, Physica **20**, 603 (1954).

the experimental precision ΔE is unimaginably large.*) This is the microscopic description: the quantum mechanical counterpart of the classical specification of the q and p of all molecules and their Hamilton equations of motion.

Actual experiments and observations are concerned with macroscopic aspects. The level distance δE is so much less than ΔE that it is never possible to prepare a system in a single eigenstate φ_n of H. Rather, the system is always in some superposition of a very large number of them, of the order of $\Delta E/\delta E$. In contrast to simple systems like atoms, *observations of many-body systems deal not with single eigenstates, but with macrostates*, being linear combinations of many φ_n. We outline how these macrostates are constructed.

The φ_n are a complete orthonormal set and define a representation in which H is diagonal, with eigenvalues E_n. We cut up the sequence of E_n in successive segments of length ΔE, labelled ε. A segment ε contains $\Delta E/\delta E$ eigenvalues E_n, which we replace by some intermediate value $\{E\}_\varepsilon$. The eigenfunctions belonging to the eigenvalues in the segment ε span a subspace $\mathbf{H}_\varepsilon$ of $\mathbf{H}$, which we call the *energy shell* ε. We define a *coarse-grained energy operator* $\{H\}$ by stipulating that it is diagonal in the same representation as H, but with the eigenvalues E_n replaced with the $\{E\}_\varepsilon$ and having the energy shells $\mathbf{H}_\varepsilon$ as the corresponding eigenspaces. Thus $\{H\}$, in contrast to H, is highly degenerate. As far as experimental observation of the energy is concerned $\{H\}$ is as good as H, but the connection with the equations of motion is lost.

Next consider a macroscopic observable with its associated operator A. The precise characterization of "macroscopic" is one of the main tasks of statistical mechanics; I merely say that A must have the properties used in the following. Its rate of change must be very small compared to the microscopic motion. That means that

$$\dot{A}_{nm} = \frac{i}{\hbar}(E_n - E_m)A_{nm}$$

must be small, and hence A_{nm} must be small as soon as $|E_n - E_m| > \hbar$. Thus the matrix elements of A in the energy representation form a narrow strip along the diagonal (fig. 50).

In this representation the coarse-grained energy operator $\{H\}$ is diagonal and there are long segments of the diagonal in which its eigenvalues have the same value $\{E\}_\varepsilon$. It follows that there are few matrix elements A_{nm} linking two of these segments, i.e. for which E_n and E_m belong to different energy shells. No serious error is made in replacing these elements by zero; this changes A into a slightly mutilated operator A', which commutes with $\{H\}$.

Subsequently it is possible to perform in each energy shell ε a unitary transformation that diagonalizes A' without affecting $\{H\}$. As a result, in each energy

---

*) A numerical estimate can be found in D. Bohm, *Quantum Theory* (Prentice-Hall, New York 1951) ch. 4.

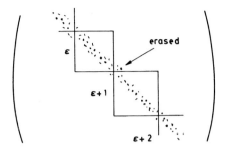

Fig. 50. The operator A in the representation of H.

shell the operator A' appears as a sequence of eigenvalues along the diagonal. Again they lie dense compared to the experimental precision ΔA and one may cut up the diagonal in subsegments in which these eigenvalues differ no more than ΔA. In each subsegment we replace them with an intermediate value $\{A\}_\alpha$ and define in this way the coarse-grained operator $\{A\}$. As far as experimental observation is concerned $\{A\}$ is as good as A, and it commutes with $\{H\}$. This is the reason why an experimentalist can measure both $\{H\}$ and $\{A\}$ although their quantum mechanical operators H and A do not commute.

When there are more macroscopic observables $B, C, \ldots$ the process can be continued. The end result is a collection of coarse-grained observables $\{H\}, \{A\}, \{B\}, \{C\}, \ldots$, which all commute with one another. The Hilbert space $\mathbf{H}$ is decomposed in linear subspaces that are common eigenspaces of these observables. We shall call these subspaces *phase cells* and indicate them with a single label J. They correspond to definite values of the coarse-grained variables, which we shall now denote by $E_J, A_J, B_J, C_J, \ldots$. These phase cells are the macrostates.

Each phase cell J has a dimension g_J and we may choose in it some complete orthogonal set of unit vectors ξ_{J_i}. Thus

$$\{H\}\xi_{J_i} = E_J \xi_{J_i}, \quad \{A\}\xi_{J_i} = A_J \xi_{J_i} \quad (i = 1, 2, \ldots, g_J). \tag{7.1}$$

Any wave function ψ of our system may be expanded in this set,

$$\psi = \sum_{J_i} b_{J_i} \xi_{J_i} \tag{7.2}$$

with coefficients b_{J_i}. The probability for the system to be in the macrostate J is

$$P_J = \sum_{i=1}^{g_J} |b_{J_i}|^2. \tag{7.3}$$

We claim that the set of probabilities P_J constitutes a complete description of the system from a macroscopic point of view.

To justify this claim I show that the expectation value of any macroscopic observable A in a state ψ can be expressed in the P_J alone.

$$(\psi|A|\psi) \approx (\psi|\{A\}|\psi)$$
$$= \sum b_{J_t}^* b_{J't'}(\xi_{J_t}|\{A\}|\xi_{J't'})$$
$$= \sum b_{J_t}^* b_{J't'} \delta_{JJ'} \delta_{tt'} A_{J'}$$
$$= \sum_J A_J \sum_t |b_{J_t}|^2 = \sum_J A_J P_J. \qquad (7.4)$$

The remarkable conclusion is that the microscopic quantum state, specified by the wave function ψ, can be described on a macroscopic level by the probability distribution P_J. *A single pure state corresponds to a macroscopic ensemble.* The interference terms that are typical for quantum mechanics no longer appear. Incidentally, this resolves the paradox of Schrödinger's cat and, in general, the quantum mechanical measurement problem.*⁾

In order to describe the evolution of the system we have to return to the microscopic level. The microscopic coefficients $b_{J_t}(t)$ can be expressed in their initial values $b_{J_t}(0)$ through the matrix of the evolution operator:
$$b_{J_t}(t) = \sum_{J't'} (Jt|e^{-itH}|J't') b_{J't'}(0).$$

The probability distribution at t is, according to (7.3),
$$P_J(t) = \sum_t \sum_{J't'} \sum_{J''t''} (Jt|e^{-itH}|J't')^* (Jt|e^{-itH}|J''t'') b_{J't'}^*(0) b_{J''t''}(0). \qquad (7.5)$$

This does not express $P_J(t)$ in terms of $P_J(0)$; rather the full microscopic specification of the initial state is needed. This shortcoming can only be remedied by some drastic assumptions, akin to the repeated randomness assumption in III.2.

First notice that the sum in (7.5) consists of many complex terms with irregularly varying phases. Our first assumption is that in the summation they average out to zero, or at least to a negligible amount. Only the terms with $J' = J''$, $t' = t''$ are positive by construction and will therefore survive. On the basis of this assumption (7.5) reduces to
$$P_J(t) = \sum_t \sum_{J't'} |(Jt|e^{-itH}|J't')|^2 |b_{J't'}(0)|^2. \qquad (7.6)$$

Next we observe that the summation over t' is a sum over positive terms, each of which is a product of two factors, namely $|b_{J't'}(0)|^2$ and a squared matrix element of e^{-itH}. In the successive terms of the sum these factors appear with irregularly varying magnitudes. Our second assumption is that these magnitudes are not correlated. It is then possible to replace either factor by the mean of the values occurring in the successive terms. That reduces (7.6) to
$$P_J(t) = \sum_t \sum_{J'} \left\{ \frac{1}{g_{J'}} \sum_{t'} |(Jt|e^{-itH}|J't')|^2 \right\} P_{J'}(0). \qquad (7.7)$$

---

*⁾ N.G. van Kampen, Physica A **153**, 97 (1988).

7. INTERNAL NOISE

This expresses the macroscopic distribution at t in terms of the initial one,

$$P_J(t) = \sum_{J'} T_t(J|J') P_{J'}(0), \tag{7.8a}$$

$$T_t(J|J') = \frac{1}{g_{J'}} \sum_{tt'} |(Jt|e^{-itH}|J't')|^2. \tag{7.8b}$$

Exercise. Verify $T_0(J|J') = \delta_{JJ'}$ and $\sum_J T_t(J|J') = 1$.

Exercise. What happened to the conservation of energy, how does it enter into (7.7)?

Exercise. What is the equilibrium distribution? Prove that (7.8) obeys the detailed balance relation (V.**6**.11).

Exercise. For fixed J' the $b_{J't'}(0)$ constitute a complex vector with squared norm $P_{J'}(0)$. Show that if one takes the average of (7.5) over all vectors with this norm the result is (7.7). (This is the quantum analog of the classical averaging over phase cells.)

Exercise. Show that our two assumptions are not independent, because the effect of each of them separately depends on the arbitrary choice of the $\xi_{J't'}$, but their combination does not.

The equation (**7.8**) is not yet a differential equation and the evolution of the P_J is not yet Markovian. As a final step the *repeated randomness assumption* has to be invoked. It assumes that it is possible to partition the time axis into intervals Δt such that

(i) Δt is so small that the P_J vary little during it; more precisely, one may take, as in (V.**1**.1),

$$T_{\Delta t}(J|J') = (1 - a_0 \Delta t) \delta_{JJ'} + \Delta t \, W_{JJ'}.$$

(ii) Δt is so large that at the end of each interval one may apply anew the above averaging postulate; the memory of the precise initial $b_{J't'}$ has been wiped out during Δt.

It is then possible to conclude, as in V.**1**, that for the macroscopic observer the system performs a Markov process with sites J and with an M-equation

$$\dot{P}_J(t) = \sum_{J'} \{W_{JJ'} P_{J'}(t) - W_{J'J} P_J(t)\}.$$

Thus *our closed, isolated, quantum many-body system is described macroscopically by a master equation* of the form discussed extensively in this book.

Normally the possible values of the observables lie so dense that they may be treated as continuous variables. The values $e, a, b, \ldots$ of these variables may be utilized instead of the indices J to identify the phase cells. The connection is given by

$$e < E_J < e + \Delta e, \quad a < A_J < a + \Delta a, \quad b < B_J < b + \Delta b, \ldots$$

The M-equation then takes the form (V.1.5):

$$P(a, t) = \int \{W(a|a')P(a', t) - W(a'|a)P(a, t)\}\, da'.$$

This is the quantum master equation describing internal fluctuations. The energy e enters only as a parameter and *a* stands for the entire set $a, b, \ldots$.

Conclusion. In classical statistical mechanics the evolution of a many-body system is described as a stochastic process. It reduces to a Markov process if one assumes coarse-graining of the phase space (and the repeated randomness assumption). Quantum mechanics gives rise to an additional fine-graining. However, these grains are so much smaller that they do not affect the classical derivation of the stochastic behavior. These statements have not been proved mathematically, but it is better to say something that is true although not proved, than to prove something that is not true.

SUBJECT INDEX

absorbing boundary 153, 154, 209, 292
absorbing state 94, 103, 104, 180
acceptor 186
activation energy 179
additive noise 229, 397
additive stochastic differential equation 397
adjoint equation 127, 286, 298, 323
adsorption 144
Alkemade's diode 164, 235
angular velocity 205
annihilation and creation operators 429
a posteriori probability 19, 21, 130
approach to equilibrium 91, 104, 121
a priori probability 19, 21, 130
Arrhenius factor 179, 276, 328, 334, 347
artificial boundary 148, 153, 180
asymmetric random walk 17, 139, 197
atomic distribution 2
attractor 360
autocatalytic chemical reaction 146, 162, 166, 257
autocorrelation 52, 53, 120, 220
autocorrelation function 54, 127, 269, 282
autocorrelation matrix 277
autocorrelation time = correlation time 53, 241, 397, 405, 416, 418, 439
average 5, 14, 423
  marginal – 419
Avogadro's number 248

backward equation 128
bacterium 100, 340, 421
Bayes' rule 10, 45, 46, 130
Bénard cells 331, 382
Bernoulli distribution 3
Bertrand circle problem 20
binomial distribution 3, 144
birth-and-death process =
  generation–recombination process 100, 134
bistable 326, 335

bivariate Fokker–Planck equation 215, 278, 359
bivariate master equation 152, 177, 263
Boltzmann equation 113, 216, 374
Boltzmann statistics 183
Bose statistics 9, 442
boundary 135, 209, 312
  see absorbing –, artificial – , impure –,
    natural –, pure –, random –,
    reflecting –, totally absorbing –
bound state 160, 179
branching process 69, 104
Bremsstrahlung 145
Brillouin's paradox 235
Brownian motion 1, 55, 74, 200, 219, 451
Brusselator 356

Campbell's process 54, 60
Campbell's theorem 54
canonical distribution 88
canonical form of the transition probability 249, 275
carcinogenesis 153, 161
Casimir effect 66
catalyst 166
Cauchy distribution = Lorentz distribution 5
Cauchy process 86
central limit theorem 26
centrifuge 116, 141
chain reaction 153, 167
chaos 58
Chapman–Kolmogorov equation 78, 79, 97
characteristic function 6, 7, 31, 137, 423
characteristic functional 63, 132, 225
characteristics, method of – 125, 149, 212
Chebyshev polynomials 54
chemical reaction 1, 145, 166
  see autocatalytic –, diffusion-controlled –,
    unimolecular –
chi-square distribution = gamma distribution 8, 47

458 SUBJECT INDEX

chromatography   186, 190, 191
closed, isolated, physical system   108, 142, 283
cluster expansion   42
cluster property   44, 393
collective system   182
colored noise   240
completely random process   86, 238
completely reducible W-matrix   101
completeness relation   159
complex Fokker–Planck equation   359
complex Langevin equation   359, 445
composite Markov process   186, 242
compound distribution   17
compounding moments, method of –   364, 367, 371
compound Poisson process   239
concentration   171, 173, 247
conditional average   384
conditional mean first-passage time   294
conditional probability   10, 62
conservation of atoms   168
conservation of positivity   106, 194, 269, 447
consistency conditions   62
continuous distribution   1, 20
continuous Markov process   194
continuous systems (in space)   363
continuous time random walk   100, 104, 136, 275
convection term   193
convex addition   64, 425
convex function   111, 112
correlation   12, 41, 53
cosmic rays   30, 69, 145
covariance   12, 211, 380
critical fluctuations   345
critical point   329, 345
critical slowing down   345
cross-correlation   53, 199
cross-section   19, 171, 375
cumulant   6, 25
  factorial –   9, 13, 64, 184
cumulant expansion   405, 438
cumulative distribution function   3
curtailed characteristic functional   132

damping   219, 428, 451
Darboux transformation   279
Debye relaxation   121, 271

decay process   93, 98, 156, 236
decomposable   91, 101, 140, 175, 255
degree of advancement   168
density matrix   424, 436, 446
derivate moment   124
detailed balance   83, 109, 114, 118, 142, 197, 284
  extended –   117
deterministic equation   235, 251, 287, 290, 333
diatomic molecule   75, 162, 174
dichotomic Markov process = random telegraph process   79, 83, 91, 99, 386, 419
diffusion   16, 137, 186, 201
  – in bistable potential   332
  – in external field   161, 276
  – in inhomogeneous medium   279
diffusion approximation   274
diffusion-controlled chemical reaction   148, 155, 157
diffusion equation   81, 201, 276
  see generalized –, multivariate –
diffusion noise   365, 367
diffusion term   193
diffusion type   273, 274
diode   164, 236
discrete distribution   1
discrete time random walk   16, 91, 137
dispersion   5
dissipative structures   382
dissociation   75, 174, 178
dissociation and recombination   162, 181
distribution   1
  see atomic –, canonical –, compound –, continuous –, discrete –, grand-canonical – , lattice –, Maxwell –, microcanonical –, multi-dimensional = multivariate – , pair –, Planck –, Rayleigh–Jeans –, ring –, square –
  more under probability distribution
distribution function   3, 61
  cumulative –   3
distribution functions f_n 36, 383
domain of attraction   257, 327, 337
donor   186
Doob's theorem   84
drift term   193

Ehrenfest urn model   91, 165, 185, 247, 302
Einstein relations   201, 202, 221, 260

ensemble 2, 52, 55, 184, 219, 454
  grand – 30, 32, 173
  microcanonical – 11, 108
entropy 105, 111, 185
  non-extensive – 113
epidemic 263, 267
ergodic 22, 56, 93, 108, 191
Erlang k-distribution 47
escape time 328, 333
evolution matrix = evolution operator 213, 407
excluded volume problem 92
exit time 294, 297
expectation value 5, 423
extended detailed balance 117
extensive 248
extent of the reaction 168
external noise 233, 396, 451

factorial correlation function 47
factorial cumulant 9, 13, 184, 365
factorial moment 9, 13
Fermi gas 381
ferromagnetism 331
first passage 241, 292, 319
first-passage time 299, 349
fluctuation–dissipation theorem 89, 213, 221, 260, 445
fluctuation spectrum = spectral density of fluctuations 59, 120, 121
fluctuation term 193
Fokker–Planck equation 85, 193, 197
  see bivariate –, complex – , linear –, multivariate –, nonlinear –, quasilinear –, time-dependent –
free energy 288, 291
Furry process 146

Galton–Watson process 71
gamma distribution 18, 23, 29
Gauss distribution = normal distribution 5, 7, 23, 212, 418
  time-dependent – 214
Gaussian noise 225
Gaussian process 63, 66, 85, 225
gene 421
generalized diffusion equation 193
generalized Langevin equation 57
generalized master equation 443
generating function 6, 8, 137
generating functional 37, 43, 63, 237

generation–recombination process = one-step process 134, 208, 288, 383
geometrical distribution = Pascal distribution 9, 143, 146
Gibbs paradox 185
globally stable 257, 260, 329
Golden Rule 98, 434, 435
Gottlieb polynomials 181
grand canonical distribution 30, 32, 173
gravitation 82, 108, 141, 161, 201, 223

Hammersley's model 147
harmonic oscillator 143, 180, 221
  quantum theory of – 428, 440
  with random frequency 397, 401, 404
Heisenberg representation 128, 427
Hermite polynomials 25
H-function 105, 111
Hohlraum 9
homogeneous Markov process 87, 88
hydrodynamics 228, 331
hypergeometric distribution 3, 162
hysteresis 295

impure boundary 155, 157
independent increments 89, 139, 199, 238
independent (statistically) 10, 24
indicator 31
induction (as in logic) 21
inhomogeneous stochastic differential equation 402
intensive 248
interaction representation 211, 437
internal noise = intrinsic noise 234, 397, 451
intrinsic probability = quantum probability 422
irreversibility 58, 185, 285, 435, 443
isomeric reaction 175
itinerant oscillator 224
Itô equation 233, 236, 396, 415
Itô interpretation 232
Itô–Stratonovich dilemma 232, 243, 262, 275

Janossy density 30
joint probability 10, 62, 215, 242, 423
jump events 383
jump moment 124, 194, 198, 252, 270

kangaroo process 100
Klein's inequality 114
Kolmogorov equations 85, 193

SUBJECT INDEX

Kramers' equation   215, 226, 278, 283, 347, 415
Kramers' escape problem   347
Kramers–Moyal expansion   199, 209, 229, 253
Kubo–Anderson process   100
Kubo number   395, 399, 416
Kubo relation (linear response)   57, 88, 401
Kuramoto length   370
kurtosis   13

Lamb shift   435
Landau equation   194
Landau–Ginzburg potential   337, 359
Langevin approach   219, 227
Langevin equation   220, 396, 428
  see complex –, generalized – , multivariate –, nonlinear –, quasilinear –, quantum – , Schrödinger –
  see also Itô equation
Langevin force = Langevin term   220, 228, 375
laser   95, 358
lattice distribution   8
law of mass action   165, 173
least squares method   21
life time   11
limbo   154, 210, 316
limit cycle   355, 358
linear Fokker–Planck equation   194, 246, 258, 287, 288
linear noise approximation   246, 258, 387
linear one-step process   135, 143, 184
line broadening   203, 361, 397, 419
line narrowing   420
Liouville equation   352, 411, 412
  stochastic –   418
locally stable   327
Lorentz distribution   5, 22, 29, 85, 203, 418, 420, 434
Lorentz gas   381
Lyapunov function   105, 257

MacDonald's theorem   61
macroscopic   56, 77, 245, 452
macroscopic equation = macroscopic law   122, 124, 196, 254
macrostate   326, 330, 452, 453
Malthus' law   145, 200

Malthus–Verhulst problem   163, 338, 415
marginal average   419
marginal distribution   10, 177
Markov approximation   444
Markov chain   89
Markov process   73, 96
  see composite–, continuous – , dichotomic = two-valued –, homogeneous –, stationary –
Markov property   73, 77, 92
maser   144
master equation = M-equation   97, 244, 456
  see bivariate –, multivariate – , nonlinear –, quantum –
master equation of diffusion type   273
matching problem   4
Maxwell distribution   2, 11, 29, 171, 206
mean   5
memory kernel   226, 409, 443
M-equation   98
mesoscopic   57, 98, 185
mesostate   326, 330
metastable   178, 330, 339
microcanonical distribution   11, 108
microscopic   56, 185, 454
mixed state   425
modified Bessel function I_n 6, 17, 138, 247, 316
molecular chaos   59
moment   5, 367
  see derivate –, factorial – , jump –
moment generating function   6
moment generating functional   63
monitoring (stochastic)   129
monomer   170
multi-dimensional distribution = multivariate distribution   1, 10
multinomial distribution = multivariate binomial distribution   13, 183
multiplicative noise   229
multiplicative stochastic differential equation   397, 399
multivariate diffusion equation   282, 296, 305
multivariate Fokker–Planck equation   210, 265, 290, 377
multivariate Langevin equation   222, 232
multivariate master equation   172, 263, 364

natural boundary   147, 312
natural attractive boundary   313, 319

SUBJECT INDEX 461

natural repulsive boundary 313, 318
negative binomial distribution = Pólya distribution 9, 144
network theory 105
Neumann equation 427
neurons 293
neutrons 69, 73, 135, 189, 374
noise term = noise source 222, 359
nonlinear Fokker–Planck equation 195, 274, 290, 346
nonlinear Langevin equation 230, 235, 415
nonlinear master equation 135, 161, 245, 388
nonlinear one-step process 135, 161
nonlinear stochastic differential equation 398, 410
normal distribution 23
 see Gauss distribution

one-step process 134, 207, 250, 338
 see linear –, nonlinear –, birth-and-death process, generation–recombination process
Onsager's reciprocity relations 260, 283, 290, 291
open systems 176, 245
orbitally stable 356
Oregonator 356
Ornstein–Uhlenbeck process 83, 194, 205, 226, 361
overdamped 321

pair distribution 47
Palm function 47
parabolic approximation 334, 337, 349, 354
paramagnetism 88, 397
partition function 144, 173, 177
Pascal distribution 9, 143, 146
Perron–Frobenius theorem 90, 104
phase cell 109, 121, 184, 371
phase shift 158
phase slip 358
phase space 1, 56, 115, 215, 371
phenomenological equation 122, 235
photoconductor 162, 268, 383, 388
photon statistics 388, 392
Planck distribution 67, 144, 434, 442
plasma 126, 194, 413
Poincaré period 435
point process = random dots 30

Poisson distribution 7, 28, 33, 173
Poisson process 39, 80, 136
Pólya distribution = negative binomial distribution 9, 144
polymer 92, 170
population 11, 69, 145, 163, 414
principle of insufficient reason 20
principle of maximum entropy 22
probability 1
 see a posteriori –, a priori –, conditional –, joint –, splitting , transition –
probability density 3
probability distribution 1
 see Bernoulli –, binomial – , Cauchy = Lorentz –, chi-square = gamma –, Erlang k –, Gauss = normal –, geometrical = Pascal – , hypergeometric –, marginal –, negative binomial = Pólya – , Poisson –
 more under distribution
probability distribution function 3
probability flow = probability flux 140, 193
probability generating function 8, 137
probability measure 4
process = stochastic process = random function 52
 see branching –, decay –, one-step –, Markov –, point –
product property 44
progress variable 168
projection 15, 443
propagator 213
pure birth process 173
pure boundary 153
pure death process 163
pure state 425, 454

quantum Langevin equation = Schrödinger–Langevin equation 444, 446, 447
quantum master equation 437, 442, 444, 446, 449, 456
 modified – 448, 451
quantum state 422, 425
quasilinear diffusion equation 278
quasilinear Fokker–Planck equation 215, 289, 333, 337
quasilinear Langevin equation 229
queuing problems 44, 155

radiation field 66, 143, 435
radioactive decay, *see* decay process
random = stochastic 1
random boundary 398
random dots = random events 30, 238, 388
random field 67, 414
random function, *see* process
random matrix 36, 399, 407
random phase 58, 454
random telegraph process = dichotomic Markov process 79
random walk 17
  see asymmetric –, continuous time –, discrete time –
random walk with persistence 29, 91
range 1, 135, 152
rate coefficients 171
rate equations for chemical reaction 171, 181
Rayleigh–Jeans distribution 144
Rayleigh particle = Rayleigh piston 204, 257, 283
Rayleigh's equation 204
reaction coordinate 347
reaction parameter 168
realization 2, 52, 187
Redfield equation 437
reducible W-matrix 101, 102, 140
reflecting boundary 141, 153, 157, 202, 209, 318
reflection principle 156, 160
regular boundary 312, 313, 318
relative stability 332
relaxation 121
rencontre problem 4
renewal equation 307, 310
renormalized coefficients 393, 401
repeated randomness assumption 57, 449, 455
ring distribution 11
rotating wave approximation 436

sample 1, 2, 52
Schlögl reaction 257, 327, 347
Schottky's theorem 61
Schrödinger equation 218, 279, 336, 426, 427
Schrödinger–Langevin equation 444, 446
Schrödinger's cat 454
second quantization 182
self-averaging 56

semiconductor 1, 161, 165, 208
semigroup 437, 447
shot noise 34, 39, 45, 47, 136, 176
singular perturbation theory 216, 254
site 326
skewness 13
S-matrix 159
Smoluchowski equation 78, 193, 295, 306, 310
  boundaries 312
spectral density = fluctuation spectrum 59, 120, 223
splitting master equation 103, 255
splitting probability 294, 298, 321, 337
splitting process 103
spurious drift = spurious flow 232, 280
square distribution 5
stable 256, 260
  see globally –, locally –, orbitally –, bistable, metastable
stable distribution = Lévy distribution 29
standard deviation 5
state 1, 167, 304, 425
  see absorbing –, mixed –, pure –, quantum –, transient –, macrostate, mesostate
stationary 34, 39, 53, 63, 81, 90
stationary Markov process 81
step operator 139, 245
stimulated emission 41, 71, 144
stochastic differential equation 220, 396
  see additive –, multiplicative –, inhomogeneous –, nonlinear –
stochastic eigenvalue problem 398
stochastic Liouville equation 418
stochastic matrix 90
  see also 'random matrix'
stochastic partial differential equation 398
stochastic process, *see* process
stochastic resonance 227
stochastic variable = random variable 1
stoichiometric coefficients 166, 174
Stosszahlansatz 58, 146, 375, 381
strength function 432
subensemble 10, 46, 62, 86, 132
sum function 36
superoperator 428, 436
superradiance 163
supersymmetry 279

sure 15
survival 94, 187, 339, 421
Suzuki limit 337
symmetry breaking 331

Taylor instability 331
telephone exchange 156
tensor product 427
thermodynamic limit 111, 175
time-dependent Fokker–Planck equation
 213, 246, 258
time ordering 390, 405
 anti-chronological – 391
time reversal 115, 285
totally absorbing boundary 155
transient 126, 273, 408
transient state 94, 103, 104
transition probability 73, 120
 canonical form 249, 275
 – per unit time 96, 134
transport term 193
tridiagonal matrix 143
tunnel diode 327

ultraviolet catastrophe 66
uncorrelated 12, 24, 220
unimolecular chemical reaction 178
unstable 256, 273, 326

Van der Waals force 66
variance 5, 125
variational equation 257, 265, 345
Vlasov equation 126

waiting times 44
white noise 225, 234, 445
 non-Gaussian – 237
Wiener–Khinchin theorem 59, 60
Wiener process = Wiener–Lévy process
 80, 139, 201, 225
Wigner–Weisskopf formula 434
W-matrix 100
 completely reducible = decomposable –
 101, 140
 reducible – 102, 140
 splitting – 103

zero-point energy 66, 428
Zhabotinskii reaction 356, 382